FUNCTIONAL COMPOUNDS AND FOODS OF PLANT ORIGIN

FUNCTIONAL COMPOUNDS AND FOODS OF PLANT ORIGIN

Krishan Dutt Sharma, PhD
Satish Kumar, PhD
Rakesh Sharma, PhD
Raj K. Keservani, PhD, MPharm
Editors

AAP APPLE ACADEMIC PRESS

First edition published 2025

Apple Academic Press Inc.
1265 Goldenrod Circle, NE,
Palm Bay, FL 32905 USA

760 Laurentian Drive, Unit 19,
Burlington, ON L7N 0A4, CANADA

CRC Press
2385 NW Executive Center Drive,
Suite 320, Boca Raton FL 33431

4 Park Square, Milton Park,
Abingdon, Oxon, OX14 4RN UK

Library and Archives Canada Cataloguing in Publication

CIP data on file with Canada Library and Archives

Library of Congress Cataloging-in-Publication Data

CIP data on file with US Library of Congress

ISBN: 978-1-77964-362-9 (hbk)
ISBN: 978-1-77952-012-8 (pbk)
ISBN: 978-1-00364-004-2 (ebk)

About the Editors

Krishan Dutt Sharma, PhD, JSPS Post-Doc
Retired Professor and Head, Faculty of Food Science and Technology, Dr. Y. S. Parmar University of Horticulture and Forestry, Nauni, Solan, Himachal Pradesh, India

Krishan Dutt Sharma, PhD, is a retired Professor and Head from the Faculty of Food Science and Technology at Dr. Y. S. Parmar University of Horticulture and Forestry, Nauni, Solan, Himachal Pradesh, India. He obtained his master's degree in Food Science and Nutrition and a doctorate in Horticulture, specializing in Fruit and Vegetable Preservation, in India, and subsequently completed postdoctoral research and training in Japan, New Zealand, and Israel. Dr. Sharma has 35 years of professional experience in various roles in India and abroad. He has supervised the research of 19 postgraduate students, including four PhD candidates, as Chairman of the Advisory Committee, and led approximately a dozen externally funded research projects as Principal Investigator and Co-Principal Investigator. Notable contributions include establishing the Functional Foods Research and Development (R&D) and Dissemination Center at Solan and the Institute of Food Science and Technology in Dehradun. Dr. Sharma has an extensive record of publications and presentations, including 70 peer-reviewed research papers, four patents, 36 conference presentations, two books, eight book chapters, 20 booklets, 36 guest lectures, 12 technical articles, and 180 technology transfer lectures and demonstrations. He is a member of several professional organizations, including the Japan Society for the Promotion of Science and the India-Israel Shalom Club, and has received numerous national and international fellowships in recognition of his contributions to the advancement of science.

Satish Kumar, PhD
Assistant Professor, Faculty of Food Science and Technology, Dr. Y. S. Parmar University of Horticulture and Forestry, Nauni, Solan, Himachal Pradesh, India

Satish Kumar, PhD, is an Assistant Professor of Food Science and Technology at Dr. Y. S. Parmar University of Horticulture and Forestry, Solan, India. He earned his doctorate in Food Technology from Dr. Y. S. Parmar University of Horticulture and Forestry, Solan (Himachal Pradesh), one of India's prestigious State Agricultural Universities (SAUs). With over six years of experience in teaching and research, Dr. Kumar has supervised 19 MSc scholars and published over 72 papers in reputed national and international journals. His contributions include two edited books, 18 book chapters, and six abstracts, alongside delivering several brainstorming sessions on current topics in food technology and nutrition at various national and international events. Dr. Kumar is a Master Trainer in fruit and vegetable processing for the PM Formalization of Micro Food Processing Enterprises (PMFME) Scheme under the Ministry of Food Processing Industries (MoFPI), Government of India. His cumulative NAAS score is 390, and his impact factor is 220. He is a lifetime member of various national and international scientific organizations and serves as an editorial board member and reviewer for esteemed journals. His research contributions focus on innovative technologies for food waste valorization, recovery protocols for functional compounds, nanostructures to improve the delivery of bioactive compounds, and herbal nanoemulsions to extend the shelf life of fresh horticultural produce.

Rakesh Sharma, PhD
Associate Professor, Faculty of Food Science and Technology, Dr. Y. S. Parmar University of Horticulture and Forestry, Nauni, Solan, Himachal Pradesh, India

Rakesh Sharma, PhD, is an Associate Professor and Head of Food Science and Technology at Dr. Y. S. Parmar University of Horticulture and Forestry, Nauni, Solan, Himachal Pradesh, India. He earned his degrees in Horticulture with a specialization in postharvest technology, along with a one-year diploma in Agricultural Marketing and a six-month certificate in Food and Nutrition. With 15 years of experience in research and development (R&D) and teaching undergraduate and graduate students, Dr. Sharma has guided over 20 MSc and PhD students. He has led six externally funded projects and contributed new knowledge in areas such as low-calorie functional food products, herbal processing, alcoholic beverages, and vinegar production. His publications include 67 research papers, 20 review articles, 15 technical articles, two books, 20 book chapters, eight technology brochures, three teaching manuals, and 22 conference presentations. Dr. Sharma has received numerous awards, including the N. N. Mohan Memorial Award, J. S. Pruthi Memorial Award, AFSTI Award, Eminent Scientist Award, and the

Institutional Award for Excellence in Research Publications. He has filed two Indian patents and successfully transferred apple cider vinegar technology to two entrepreneurs. Additionally, Dr. Sharma serves as a Coordinator and Master Trainer in fruit and vegetable processing under the PM Formalization of Micro Food Processing Enterprises (PMFME) scheme of the Government of India.

Raj K. Keservani, PhD, MPharm
Professor, Faculty of B. Pharmacy, CSM Group of Institutions, Prayagraj, India

Raj K. Keservani, PhD, MPharm, working as Professor, Faculty of B. Pharmacy, CSM Group of Institutions, Prayagraj, India. He has more than 16 years of academic (teaching) and research experience from various institutes of India in pharmaceutical education. He has supervised more than 25 UG students for their research work, Dr. Keservani published more than 50 review and research papers in the field of pharmaceutical sciences in national and international journals of repute, 10 patent, 90 book chapters, three co-authored books, and 54 edited books. He is also active as a reviewer for several international scientific journals. Dr. Keservani has been recognized among the world's top 2% of scientists published by Stanford University, USA. Dr. Keservani graduated with a pharmacy degree from the Department of Pharmacy, Kumaun University, Nainital (Uttarakhand), India. He received his Master of Pharmacy (MPharm) (specialization in pharmaceutics) from the School of Pharmaceutical Sciences, Rajiv Gandhi Proudyogiki Vishwavidyalaya, Bhopal, India. He received his Ph.D. degree from IKG Punjab Technical University, Jalandhar, Punjab, India. Dr. Keservani is a life member of Society of Pharmaceutical Education and Research (SPER) and Pharmaceutical Educational Society Originated by Teachers & Students (pesots). His research interests include nutraceutical and functional foods, novel drug delivery systems (NDDS), transdermal drug delivery/drug delivery, health science, cancer biology, and neurobiology.

Contents

Contributors

Ghanshyam Abrol
Rani Lakshmi Bai Central Agricultural University, Jhansi, Uttar Pradesh, India

Rakesh Kumar Ameta
Sri M. M. Patel Institute of Sciences and Research, Kadi Sarva Vishwavidyalaya, Gandhinagar, Gujarat, India

S. Anandakumar
Department of Food Packaging and System Development, NIFTEM, Thanjavur, Tamil Nadu, India

B. L. Attri
Indian Council of Agriculture Research (ICAR), Directorate of Mushroom Research (DMR), Solan, Himachal Pradesh, India

Soumitra Banerjee
Center for Incubation, Innovation, Research, and Consultancy, Jyothy Institute of Technology, Thathaguni, Karnataka, India

Vishesh Bhadariya
Chemical Engineering, Lovely Professional University, Phagwara, Punjab, India

Sunakshi Gautam
School of Bioengineering and Food Technology, Shoolini University, Solan, Himachal Pradesh, India

Antima Gupta
Food Science and Technology, Punjab Agricultural University, Ludhiana, Punjab, India

Rachna Gupta
Department of Biotechnology, Visva-Bharti Shantiniketan, Bolpur, West Bengal, India

Hamid
Food Technology and Nutrition, Lovely Professional University, Phagwara, Punjab, India

Abida Jabeen
Division of Food Science and Technology, Sher-E-Kashmir University of Agricultural Science and Technology, Kashmir, Jammu and Kashmir, India

T. Jyothirmayi
CSIR–Central Food Technological Research Institute, Resource Center, Hyderabad, Telangana, India

Ayush Kashyap
M. S. Swaminathan School of Agriculture, Shoolini University, Solan, Himachal Pradesh, India

Deepika Kathuria
Department of Food Science and Technology, Graphic Era University, Dehradun, Uttarakhand, India

Jaspreet Kaur
Food Technology and Nutrition, Lovely Professional University, Phagwara, Punjab, India

Pratap G. Kenchappa
Department of Studies in Biochemistry, Davangere University, Shivagangothri, Davangere, Karnataka, India

Mumtahin Ul Kousar
Division of Food Science and Technology, Sher-E-Kashmir University of Agricultural Science and Technology, Kashmir, Jammu and Kashmir, India

Anil Kumar
Department of Microbiology, DAV University, Jalandhar, Punjab, India; School of Chemical and Metallurgical Engineering, University of the Witwatersrand, Johannesburg, South Africa

Deepak Kumar
Department of Microbiology, DAV University, Jalandhar, Punjab, India

Harshini V. Kumar
Jagadguru Jayadeva Murugharajendra Medical College, Davangere, Karnataka, India

Sanjeev Kumar
Department of Vegetable Science, Aspee College of Horticulture and Forestry, Navsari Agricultural University, Navsari, Gujarat, India

Satish Kumar
Department of Food Science and Technology, Dr. Y. S. Parmar University of Horticulture and Forestry, Nauni, Solan, Himachal Pradesh, India

Vikas Kumar
Food Science and Technology, Punjab Agricultural University, Ludhiana, Punjab, India

D. Madhusudhan
CSIR–Central Food Technological Research Institute, Resource Center, Hyderabad, Telangana, India

B. V. C. Mahajan
Punjab Horticultural Postharvest Technology Center, Punjab Agricultural University, Ludhiana, Punjab, India

Parth Malik
School of Chemical Sciences, Central University of Gujarat, Gandhinagar, Gujarat, India

Bushra Manzoor
Division of Food Science and Technology, Sher-E-Kashmir University of Agricultural Science and Technology, Kashmir, Jammu and Kashmir, India

Mariya Nayeem
Division of Food Science and Technology, Sher-E-Kashmir University of Agricultural Science and Technology, Kashmir, Jammu and Kashmir, India

Avnish K. Pandey
Department of Fruit Science, Navsari Agricultural University, Navsari, Gujarat, India

Pooja
Punjab Horticultural Postharvest Technology Center, Punjab Agricultural University, Ludhiana, Punjab, India

Sridhar Rachakonda
CSIR–Central Food Technological Research Institute, Resource Center, Hyderabad, Telangana, India

Aasima Rafiq
Division of Food Science and Technology, Sher-E-Kashmir University of Agricultural Science and Technology, Kashmir, Jammu and Kashmir, India

Dev Raj
Department of Post-Harvest Technology, Aspee College of Horticulture and Forestry, Navsari Agricultural University, Navsari, Gujarat, India

Pooja Rani
Department of Microbiology, DAV University, Jalandhar, Punjab, India
A. Harshavardhan Reddy
Faculty of Agricultural Sciences, DAV University, Jalandhar, Punjab, India

Shubham Samkariya
Department of Food Science and Technology, Dr. Y. S. Parmar University of Horticulture and Forestry, Solan, Himachal Pradesh, India

Shailja Sankhyan
Department of Biotechnology, Chandigarh University, Mohali, Punjab, India

Ashok K. Senapati
Department of Post-Harvest Technology, Aspee College of Horticulture and Forestry, Navsari Agricultural University, Navsari, Gujarat, India

Neha Sharda
Department of Microbiology, DAV University, Jalandhar, Punjab, India

Ajay Sharma
Department of Chemistry, Career Point University, Hamirpur, Himachal Pradesh, India

Ashutosh Sharma
Faculty of Agricultural Sciences, DAV University, Jalandhar, Punjab, India

Deepika Sharma
M. S. Swaminathan School of Agriculture, Shoolini University, Solan, Himachal Pradesh, India

Indu Sharma
Department of Botany, Sant Baba Bhag Singh University, Jalandhar, Punjab, India

Kartik Sharma
CSIR–Institute of Himalayan Bioresource Technology, Palampur, Himachal Pradesh, India

Krishan Dutt Sharma
Department of Food Science and Technology, Dr. Y. S. Parmar University of Horticulture and Forestry, Solan, Himachal Pradesh, India

Paras Sharma
Food Chemistry Division, ICMR–National Institute of Nutrition, Hyderabad, Telangana, India

Rakesh Sharma
Department of Food Science and Technology, Dr. Y. S. Parmar University of Horticulture and Forestry, Solan, Himachal Pradesh, India

Sakshi Sharma
Department of Food Science and Technology, Dr. Y. S. Parmar University of Horticulture and Forestry, Solan, Himachal Pradesh, India

Shweta Sharma
M. S. Swaminathan School of Agriculture, Shoolini University, Solan, Himachal Pradesh, India

Suresh Sharma
CSIR–Institute of Himalayan Bioresource Technology, Palampur, Himachal Pradesh, India

Bharti Shree
CSK Himachal Pradesh Krishi Vishvavidyalaya, Palampur, Himachal Pradesh, India

Shweta Suri
College of Community Science, G. B. Pant University of Agriculture and Technology, Pantnagar, Uttarakhand, India

Priyanka Suthar
Department of Food Science and Technology, Dr. Y. S. Parmar University of Horticulture and Forestry, Solan, Himachal Pradesh, India

Prathap H. Manjunatha Swamy
Department of Studies in Biochemistry, Davangere University, Shivagangothri, Davangere, Karnataka, India

Abhimanyu Thakur
Department of Food Science and Technology, Dr. Y. S. Parmar University of Horticulture and Forestry, Solan, Himachal Pradesh, India

Abinaya Venkatesh
Department of Food Packaging and System Development, NIFTEM, Thanjavur, Tamil Nadu, India

Ritika Verma
ICAR–Indian Agricultural Research Institute, Regional Station, Shimla, Himachal Pradesh, India

S. V. N. Vijayendra
CSIR–Central Food Technological Research Institute, Resource Center, Hyderabad, Telangana, India

Poornima D. Vijendra
Department of Studies in Food Technology and Biochemistry, Davangere University, Shivagangothri, Davangere, Karnataka, India

Abbreviations

ACE	Angiotensin-Converting Enzyme
ACE-II	Angiotensin-Converting Enzyme-II Activity
AG	Agmatine
AGEs	Advanced Glycation End Products
AMB	Aerobic Mesophilic Bacteria
ARE	Antioxidant Response Element
ASG	Acylated Steryl Glycosides
ASLT	Accelerated Shelf-Life Test
BDMC	Bisdemethoxycurcumin
BF	Binding Forces
BMI	Body Mass Index
Ca	Calcium
CCS	Capsanthin-Capsorubin Synthesis
CGA	Chlorogenic Acid
CHF	Congestive Heart Failure
CHI	Chalcone Isomerase
CHO	Carbohydrates
CLA	Conjugated Linoleic Acid
CRP	C-Reactive Protein
CTAB	Cetyl Trimethylammonium Bromide
CVD	Cardiovascular Diseases
DFP	Dragon Fruit Peel Powder
DG	Degree of Gelatinization
DM	Diabetes Mellitus
DMC	Demethoxycurcumin
DPP-4	Dipeptidyl Peptidase-4
DPPH	2,2-Diphenyl-1-Picrylhydrazyl
EGCG	Epigallocatechin Gallate
EGFR	Epidermal Growth Factor Receptor
ERK1/2	Extracellular Signal-Regulated Kinase 1/2
F3H	Flavanone 3-Hydroxylase
FCR	Food Conversion Ratio
FIM	Foundation for Innovation in Medicine
FOS	Fructo-Oligosaccharides
FS	Free Alcohol
GABA	γ-Aminobutyric Acid
GAE	Gallic Acid Equivalent

GI	Glycemic Index
GLP-1	Glucagon-Like Peptide-1
GOS	Galacto-Oligosaccharides
GRAS	Generally Regarded as Safe
GSH	Glutathione
HDLs	High-Density Lipoproteins
HE	Hydroxyecdysone
HH	High Heating
HIV-1	Human Immunodeficiency Virus Type 1
HP	Homogenized-Pasteurized
HPH	High-Pressure Homogenization
HPLC	High-Pressure Liquid Chromatography
HPP	High Pressure Processing
HSE	Hydroxycinnamic-Acid Esters
HTST	High-Temperature Short Time
IBD	Inflammatory Bowel Disease
IBS	Irritable Bowel Syndrome
IDF	International Diabetes Federation
IL	Interleukin
IR	Insulin Resistance
ISO	International Organization for Standardization
LAB	Lactic Acid Bacteria
LC	Liquid Chromatography
LDL	Low-Density Lipoprotein
LF	Lachrymatory Factor
LFS	Lachrymatory Factor Synthase
LH	Low Heating
MDA	Malondialdehyde
MP	Minimally Processed
MUFA	Monounsaturated Fatty Acid
MVA	Mevalonic Acid
Na_2SeO3	Sodium Selenite
NaHSeO3	Sodium Hydrogen Selenite
NCDs	Non-Communicable Diseases
NDCs	Nondigestible Carbohydrates
NDRI	National Dairy Research Institute
NH4+	Ammonium Cations
NK	Natural Killer
NMDA	N-Methyl-D-Aspartate
NO3–	Nitrate
NPs	Nanoparticles
NPU	Net Protein Utilization
NRF2	Nuclear Factor Erythroid 2-Related Factor 2

NSPs	Non-Starch Polysaccharides
OTR	Oxygen Transmission Rate
PAD	Peripheral Arterial Disease
PC3	Prostate Cancer Cell Line
PEP	Phosphoenolpyruvate
PI3K	Phosphatidyl Inositol 3-Kinase
PMA	Phorbol 12-Myristate 13-Acetate
PME	Pectin Methyl Esterase
POD	Peroxidase
PPARγ	Peroxisome Proliferator-Activated Receptor-γ
PPO	Polyphenol Oxidase
PR	Protein Efficiency Ratio
PS	Polysaccharides
PUFA	Polyunsaturated Fatty Acid
PWV	Pulse Wave Velocity
RDI	Recommended Dietary Intake
RO	Reverse Osmosis
ROS	Reactive Oxygen Species
RS	Resistant Starch
RTE	Ready-to-Eat
SCFAs	Short-Chain Fatty Acids
SE	Fatty-Acid Esters
SeO2	Selenium Dioxide
SG	Steryl Glycosides
SPD	High Spermidine
STAT	Signal Transducer and Activator of Transcription
STZ	Streptozotocin
TDM	Type-II Diabetes Mellitus
TFC	Total Flavonoid Content
TIU	Trypsin Inhibiting Unit
TMR	Transparency Market Research
TNF	Tumor Necrosis Factor
TOS	Trans-Galacto-Oligosaccharides
TPC	Total Phenolic Content
TRAMP	Transgenic Adenocarcinoma of the Mouse Prostate
UHPH	Ultra High-Pressure Homogenization
UNGA	United Nations General Assembly
US	United States
WHO	World Health Organization
WS	Water-Solubility

Preface

With rapid urbanization and continuously changing food habits, food scientists are developing safe, healthy, and nutritious foods that offer higher convenience, enhanced functionality, favorable sensory attributes, and extended shelf-life to people worldwide. During the 20th century, most nations achieved food security either by increasing domestic production or by importing food from surplus-producing countries. Now, the research focus has shifted from food security to nutritional security. Globally, scientists and crop engineers are working tirelessly to achieve nutritional security, aiming to improve public health and reduce pressure on healthcare services. Numerous innovations in functional and nutraceutical food products have been introduced over the past two decades to maximize the retention of bioactive compounds in food and to ensure their effective delivery at the highest possible quality.

The field of innovative research in functional food processing encompasses a wide range of disciplines, including material science, technology, engineering, chemistry, microbiology, and biotechnology. This field caters to professionals involved in product and process development, engineering design, toxicology, quality assurance and safety, marketing, and food regulatory legislation. Food is no longer viewed solely as a source of energy; instead, consumers today are increasingly interested in the functional benefits of what they eat. Functional diet ingredients may include natural foods or food components, as well as additives, removals, or modifications made through technological means to address specific health concerns. A functional food is similar in appearance to conventional food but offers physiological benefits and/or reduces the risk of chronic disease beyond its basic nutritional function. The rise in industrialization, per capita income, corporate culture, limited consumer knowledge about healthy food choices, and easy access to fast food have led to the adoption of unhealthy dietary habits among youth, making them more susceptible to chronic diseases at an early age.

The role of food in maintaining health and preventing or curing diseases has been well-documented in both Hippocratic and Vedic texts. Many traditional healthcare systems, such as Ayurveda, Siddha, Unani, and Homeopathy, are rooted in the concept of the functionality of herbal medicines.

Numerous plants, their parts, or derived products have been used to treat various diseases and to support overall health. These traditional concepts continue to inspire the pharmaceutical industry to diligently search for, characterize, and isolate compounds with significant potential for combating chronic health conditions. Similarly, the food processing sector is actively working to develop food products that retain the highest quality and bioavailability of these beneficial compounds. This field is continuously evolving, with a wide range of innovations and interventions emerging to enhance the quality, safety, and functionality of food.

In India, the shift from a traditional to a modern lifestyle, along with a diet high in fat and calories and elevated levels of mental stress, has led to an increase in lifestyle-related diseases, such as coronary heart disease, obesity, diabetes, cancer, and hypertension. Addressing these widespread chronic health concerns requires substantial investment in medical infrastructure by the central and/or state governments, which presents significant challenges. A more accessible approach to tackling these issues is through educating people about healthy food choices. Recently, special emphasis has been placed on developing functional foods and evaluating their efficacy and safety. Establishing an effective database to support systematic research and development (R&D) and the dissemination of functional food technologies to stakeholders and consumers is essential. There is an urgent need to explore and investigate current innovations, existing shortcomings, and future challenges in the field of functional foods.

This book serves as a comprehensive collection of emerging and innovative techniques for the development of functional foods, providing valuable insights into various aspects and categories of both fresh and processed functional foods, as well as their transformation from traditional to modern products. It offers an in-depth exploration of plant- and animal-based functional foods, detailing their characterization, validation, and evolution into novel products. The book is divided into three parts, each covering distinct categories of functional foods, safety considerations, regulatory issues, and more.

Part I focuses on the importance of fresh produce as a potential ingredient in functional diets. Part II examines processed food products and their functional significance, while Part III highlights specialty functional foods tailored for specific health conditions. Each chapter includes tables, figures, and illustrations to enhance clarity and understanding.

This resource aims to provide academia, researchers, and industry professionals with the latest research and advancements in functional foods. In summary, the book explores key concepts essential for understanding current trends and developments in the functional foods field to ensure quality, safety, and effective delivery of functional foods and bioactive compounds to the public. It also provides a comprehensive overview of the progress in the functional food industry, striving to achieve the mission of nutritional and health security for the betterment of humankind.

This book has been conceptualized to meet the diverse needs of the food community in developing novel food products with enhanced functionality. It is intended to serve as an informative reference for the food community, including scientists, students, and professionals in processing industries. Additionally, this compendium is expected to be a valuable resource for scholars in food science, food technology, food biotechnology, food microbiology, and food nanoscience, supporting academic and research advancements in curriculum and methodology.

The book represents the collective expertise and dedication of distinguished professionals in research and academia, whose passion for science has shaped its content. The names of individual contributors are listed in the affiliation section of each chapter.

We extend our heartfelt thanks to all contributors who devoted their valuable time and shared their vision with the utmost dedication and enthusiasm to complete this draft. We also sincerely thank the editorial office of Apple Academic Press for their invaluable support and guidance throughout this project, as well as the production staff for their work in typesetting the manuscript to create this quality publication. We invite readers to share their constructive feedback, which will help guide improvements in future editions.

We would like to thank our families for their continuous support, motivation, and blessings, which helped us overcome every obstacle encountered during this project. We also express our gratitude to the entire research community for setting high standards for aspiring scientists and for providing continuous encouragement and invaluable scientific support as needed. Lastly, we are grateful to the administration of Dr. YS Parmar University of Horticulture and Forestry, Solan, for providing essential infrastructural support during the compilation of this book.

PART I

Nutraceutical Potential of Plant-Based Functional Compounds and Foods

CHAPTER 1

Phytochemicals: The Functional Food Ingredients and Their Health Benefits

RAKESH SHARMA, KRISHAN DUTT SHARMA, SATISH KUMAR, and ABHIMANYU THAKUR*

Department of Food Science and Technology, Dr. Y. S. Parmar University of Horticulture and Forestry, Solan, Himachal Pradesh, India

*Corresponding author.

ABSTRACT

Phytochemicals present in plants are non-nutrient secondary metabolites that have potential therapeutic properties, and historically, they have played a role in the treatment of various diseases. The polyphenols, carotenoids, tannins, flavones, terpenoids, glucosinolates polyacetylenes, nondigestible carbohydrates (CHO), phytoestrogen, steroids, saponins, and alkaloids are the major phytochemicals with distinct bioactivities providing numerous benefits in maintaining good human health and which are referred to as functional ingredients or nutraceutical compounds. In the last 3 decades, various epidemiological studies have suggested that appropriate consumption of fruits and vegetables in the daily diet reduces the risk of various cancers, various CVDs, and other aging-related macular degeneration diseases. Functional foods of plant origin, like fruits, vegetables, oilseeds, nuts, and so on, are rich in various phytochemicals and have nutraceutical importance, which provides medical health benefits along with the prevention and/or treatment of many diseases and physiological disorders. Throughout the world, the non-communicable chronic diseases that are associated with changing lifestyles and diets are increasing day by day. The pharmacological properties like anti-inflammatory, anti-allergic, antibacterial, anti-fungal, anti-spasmodic, chemo-preventive, hepatoprotective, hypolipidemic, neuroprotective, hypotensive, anti-aging nature with potential antioxidant activity and free

Functional Compounds and Foods of Plant Origin:Krishan Dutt Sharma, Satish Kumar, Rakesh Sharma, and Raj K. Keservani (Eds.)

radical scavenging abilities are associated with the phytochemicals and helps in lowering down the incidence of these chronic diseases. In addition, the potential of phytochemicals-based medicines not only contributes to the maintenance of good health and wellbeing of humans but has also encouraged the scientific community in the expansion of novel drug development.

1.1 INTRODUCTION

The 21st century is the age of transformation and modernization and in recent years, lifestyle is being considered an important factor in health and wellbeing. Changes in lifestyle and patterns of food consumption, along with increasing levels of mental stress, have increased various human health problems like CVDs, diabetes, obesity, cancer, and hypertension. The World Health Organization (WHO) has identified four significant non-communicable diseases (NCDs), that is, cardiovascular disease (CVD), cancer, diabetes, and chronic lung disease/chronic obstructive pulmonary disease, which are mainly due to common lifestyle-related behavioral risk factors (Pappachan, 2011; Sharma et al., 2021). Accordingly, there is an increasing interest in a specific food or its component that can improve and maintain health with its consumption for a specific period. Hence, there is a huge focus on the components which can improve health and quality of life. Earth is dominated by plants, and it is estimated that out of all plant species, approximately 10% of these are used for food and medicinal purposes. The phytochemicals or bioactive compounds obtained from plants are non-nutritive compounds and have significant effects on the health and nutrition of humans as they protect against various chronic degenerative diseases (Almodaifer et al., 2017; Hai Liu, 2004). The application or utilization of phytochemicals in the form of nutraceutical or functional foods is extending day by day (Das et al., 2011). The great possibilities of extracting bioactive compounds from plants and their by-products with their health-improving properties are raising interest in the development of new and innovative products with encouraging pharmacological effects (Hamid et al., 2020). Nutraceutical term consists of two words: *nutrition* and *pharmaceutical* Stephen De Felice, who was the founder and chairman of the Foundation for Innovation in Medicine (FIM) (Keservani et al., 2010a, b, 2017, 2020), Cranford in 1989 has described it as any substance that is a food or a part of food which provides medical or health benefits including the prevention and treatment of disease. A nutraceutical may also be defined as a product that is isolated or purified from foods that are usually sold commercially in medicinal forms not typically linked with foods.

There are various functional ingredients present in food that have been consumed with intent or unintentionally over recent times in some form of food. The functional food components are mainly known for their immunomodulatory and antiviral properties, which play a significant role in maintaining a healthy immune system (Arya et al., 2022; Hamid et al., 2021; Keservani et al., 2015; Kumar et al., 2022; Rane et al., 2022). To date, about 10,000 phytochemicals have been successfully identified, which include polyphenols (phenolic acid, stilbenes, flavonoids, etc.), carotenoids, tannins, flavones, terpenoids, glucosinolates polyacetylenes, nondigestible carbohydrates (CHO), phytoestrogen, steroids, saponins, alkaloids, and dietary fiber as important compounds and still, a large number is yet to be explored. The various phytochemicals found in plant foods often possess strong antioxidant, anti-inflammatory, and free radical scavenging abilities, which provide health benefits to human beings, such as anti-aging, anti-cancer, and protective action against a wide range of other diseases like cardiovascular, obesity, diabetes mellitus (DM) and neurodegenerative diseases (Thakur et al., 2019; Zhang et al., 2015). Various researchers have identified the functional ingredients/bioactive or nutraceutical compounds, and some of them are mentioned in Table 1.1. With respect to the perspective of these compounds in human health, this chapter gives a brief introduction to various health-promoting effects of phytochemicals with their potential utilization in the form of foods or food supplements.

TABLE 1.1 List of Important Healthful Bioactive and Nutraceutical Compounds Present in Different Foods

Class	Source(s)	Potential Health Benefit
	1. Fatty Acids	
Conjugated linoleic acid (CLA)	Dairy products (cheese, milk) and various meat products.	• Improved body composition. • Reduction in different types of cancers.
Fatty acids (EPA, DHA)	Oilseeds (linseed, mustard, rapeseed) and tree nuts.	• Lowers down the risk of CVDs. • Improvement in visual and mental health.
	2. Polyphenolic Compounds	
Catechins	Mustard cake, tea, and rape seed.	• Act as an antioxidant. • Anticarcinogenic activity.
Flavones and flavanone	Citrus fruits like oranges, limes, lemons, and soybeans.	• Act as an antioxidant. • Anticarcinogenic activity.

TABLE 1.1 *(Continued)*

Class	Source(s)	Potential Health Benefit
Phenolic compounds	Coffee, wine, artichokes, basil, kale, mentha, rose, rosemary, rice, and strawberries.	• Analgesic, anti-inflammatory, prevention of arrhythmia, cancer. • Helps in reducing weight. • Antioxidant properties and prevention of cancer. • Lowers down LDL (low-density lipoprotein), that is, unhealthy cholesterol.
Stilbenes	Grapes, peanuts, and many berries.	• Effective in the prevention of chronic diseases like cardiovascular, cancer, and neurodegenerative pathologies.
	3. Flavonoids	
i. Anthocyanidin	Apple, peach, black olive, and berries (blueberry, cherry, etc.).	• Free radical neutralization. • Anticarcinogenic.
ii. Flavan-3-ols	Apples (un-peeled), tea, chocolate, wine, and red Shiraz.	• Beneficial for metabolic and cardiovascular health.
iii. Flavonols	Blueberries, spinach, broccoli, chili peppers, kale, cowpea.	• Act as an anti-inflammatory, antimicrobial, anticancerous, cardioprotective, neuroprotective, and antidiabetic agent.
iv. Flavanones	Citrus fruits (oranges, grapefruits, lemons, etc.).	• Strong antioxidant activity with very high free radical scavenging potential. • Helps in up-regulation of the expressions of SOD, catalase, and glutathione (GSH) peroxidize.
v. Flavones	Celery, peppermint, parsley, and thyme.	• Act as a neuroprotective agent with anti-inflammatory properties. • prevention of neuroinflammation. • Improvement of cognitive and amnestic functions.
vi. Isoflavones	Legumes like soybeans and various soy-based fermented foods.	• Helps in lowering the LDLs. • Reduce the incidence of endometrial and breast cancer. • Lower the hot flush symptoms during the menopausal phase.
	4. Terpenoid Compounds	
	Salvia divinorum, citral, menthol, camphor, ginkgo biloba, cannabis, turmeric, and mustard seeds.	• Plays a significant role as an anti-feedant mechanism in plants. • Antimalarial, anti-ulcer, hepaticidal, antimicrobial, and anti-diuretic activities.

TABLE 1.1 *(Continued)*

Class	Source(s)	Potential Health Benefit
	5. Glucosinolates	
	Leafy and root vegetables like broccoli, cauliflower, cabbage, bok choy, turnip, kohlrabi, rapeseed, and radish.	• Act as an antibacterial and anti-fungal compound due to the presence of isothiocyanates. • Helps in the detoxification of undesirable toxic compounds from the human body. • Enhance the antioxidant defense system.
	6. Polyacetylene	
	Spices like asafoetida, coriander, bishop's weed, celery, and ajwain.	• Boost up the immune system (anti-platelet-aggregatory, anti-inflammatory, antibacterial neurotoxicity activity).
	7. Phytosterols and Phystotanols	
	Mostly in oilseed crops (rapeseed, soybean, sunflower), nuts, seeds, and cereals.	• Component of food supplements.
	8. Nondigestible Carbohydrates	
	Starchy fruits and vegetables (e.g., bananas, potatoes, etc.), legumes, and other food products having modified starches as ingredients (e.g., bread, and nutrition bars).	• Prebiotic action and increased stool bulk. • Lowers down LDL cholesterol levels and hypoglycemic effect. • Reduction of CHD risk factors and some cancers.
	9. Saponins	
	Soybeans, alfalfa, chickpea, quinoa, and haricot bean.	• Cholesterol lowering property and aids in cholesterol degradation. • Anticancerous and acts as an antibiotic agent with fungicidal properties.
	10. Phytoestrogen	
Isoflavones – daidzein and genistein	Leguminous crops like soybean, flax seeds, lentils, and maize.	• Influences various menopause symptoms. • Helps in improving bone strength and brain health. • Boosts the immunity.
Lignans	Vegetables, rye, and flax seeds.	• Anticancerous • Helps in the prevention of the occurrence of heart diseases.
	11. Carotenoids	
β-carotene	Orange-yellow colors vegetables and fruits like carrots, mango, papaya, and so forth.	• Binds with free radicals for its neutralization.

TABLE 1.1 *(Continued)*

Class	Source(s)	Potential Health Benefit
Lutein, zeaxanthin	Plant sources like vegetables, marigolds, citrus, and animal sources like eggs.	• Major role in maintaining healthy vision.
Lycopene	Tomato and its products.	• Lowers down the occurrence of prostate cancer.
	12. Dietary Fiber	
Insoluble dietary fiber	Cereals and grains like wheat bran and rice bran are also present in fruits (raw).	• Reduce the risk of breast cancer. • Helps in maintaining a healthy digestive system.
Whole grains (β-glucan)	Cereal grains and oats.	• Lowers down the risk of CVD.

1.2 CLASSIFICATION AND POTENTIAL HEALTH BENEFITS OF PHYTOCHEMICALS

The word *Phytochemical* originated from the Greek word "*phyto,*" which means plant, and it may be defined as a naturally occurring biologically active chemical compound produced in plants during primary or secondary metabolic processes and plays a major role in plant growth or defense mechanism with respect to various pathogens or predators attacking them (Molyneux et al., 2007). These compounds possess some pharmacological properties, and they are usually not considered an essential component of the human diet for sustaining normal life (Dillard and German, 2000). Various researchers concluded from their epidemiological studies that phytochemicals have the potential to prevent chronic diseases (Vinson et al., 2001) and protect cellular systems by acting as an antioxidant, antimicrobial, anti-inflammatory, anticancerous, detoxifying agent, immunity-potentiating agent, and neuro-pharmacological agent (Koche et al., 2016).

Phytochemicals have diverse forms and structures, and about 4000 phytochemicals have been identified to date, which are mostly classified based on their various characteristics like biological activity, physical characteristics, and chemical characteristics (Saxena et al., 2013). Based on their role in various metabolic activities of the plants, the phytochemicals are classified into two categories: primary and secondary metabolites. The primary metabolites includes compounds like common sugars, purines, pyrimidines of nucleic acids, chlorophylls, amino acids, proteins, and so on, whereas, the remaining other plant chemicals like alkaloids, flavonoids, terpenes, plant steroids, lignans, saponins, curcumins (Kesharwani and Misra, 2011; Kesharwani et al., 2015, 2018; Mishra et al., 2019; Singh et al., 2013a, b, 2015, 2016; Upadhyaya et al., 2009; Vyas et al., 2010), phenolics, and glucosides have been considered as secondary metabolites (Koche et al., 2016; Ramawat et al., 2009). During the last decade, the total number of identified phytochemicals has increased tremendously, especially concerning secondary plant metabolites. Based on particular distinctions and characters, the classification of phytochemicals is shown in Figure 1.1 and has been discussed further.

1.2.1 PHENOLICS

Phenolic compounds refer to simple phenolic molecules, or they may be as highly polymerized compounds with benzene rings containing one or more than one hydroxyl substituents (Velderrain-Rodríguez et al., 2014). These are produced in the shikimic acid and pentose phosphate pathways through phenylpropanoid metabolization and are considered secondary metabolites (Lin et al., 2016; Randhir et al., 2004). Phenolic compounds are the leading category of phytochemicals that are most extensively reported in plants (Walton et al., 2003). The most important phenolics are flavonoids, phenolic acids, and stilbenes (also called tannins), and in various foods, they are present in a wide range from not as much as 1 mg/kg to up to 3000 mg/kg (King and Young, 1999). Plant phenolic compounds exhibit several beneficial effects and various properties like antioxidants, structural polymers like lignin, attractants (carotenoids and flavonoids), defense response chemicals (tannins and phytoalexins), UV screens (flavonoids), and signal compounds (salicylic acid and flavonoids). These compounds are vital in defense responses in the human body as they have anti-aging, anti-inflammatory, antioxidant, and antiproliferative activities (Lin et al., 2016). The phenolics can be classified according to their biosynthetic origin or structure. Flavonoids and non-flavonoids are the two major classes concerning their structure, which constitutes a large number of compounds in each group (Kumar et al., 2022).

1.2.1.1 PHENOLIC ACIDS

Phenolic acids include a sub-class of a larger category of metabolites that describes phenols that possess carboxylic acid functionality with two distinctive carbon frameworks, including hydroxybenzoic and hydroxycinnamic acids. The prominent naturally-occurring phenolic acids like caffeic, p-coumaric,

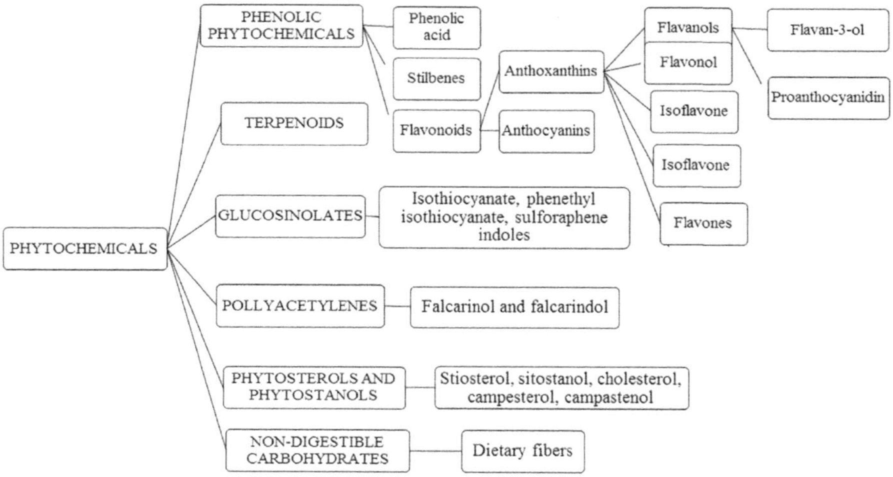

FIGURE 1.1 Classification of common dietary phytochemicals.

vanillic, ferulic, and protocatechuic are mostly found in most plants, whereas other acids like cinnamic, o-coumaric, m-coumaric, sinapic, benzoic, salicylic, p-hydroxybenzoic, syringic, gentisic, gallic, veratric, syringaldehyde, and vanillin are present in selected foods or plants (Robbins, 2003). The sources, along with the concentration of a few phenolic acids, have been mentioned in Table 1.2. One-third of these phenolic acid compounds include dietary phenols, and there is an increasing scientific research interest in the potential of dietary antioxidants and the health benefits associated with their intake (Robbins, 2003). The literature investigations revealed a correlation between a diet high in phenolic acids and the potential for oxidative damage, which may cause various degenerative diseases. The various phenolic compounds effectively inhibited mouse pre-adipocytes and also resulted in enhanced apoptosis (Hsu and Yen, 2006). Other researchers reported the role of these dietary phenolic compounds in controlling weight gain and prevention of fatty acid biosynthesis (Baile et al., 2011; Mulvihill and Huff, 2010). Dietary polyphenols have recently gained major attention due to their supplementation, as they are associated with improved cognitive status and play an important role in the prevention of cognitive decline (Caruso et al., 2022).

The phenolic compound ferulic acid, found predominantly in rice bran oil and also in wheat, oats, coffee beans, apples, artichoke, peanuts, oranges, and pineapple, has been used extensively for its high antioxidant activity (Nicholson et al., 2008; Srinivasan et al., 2007). Besides, it also possesses

TABLE 1.2 Some Commonly Found Phenolic Acids in Nature

Phenolic Acid	Source and Quantity	References
Caffeic acid	• Tea stems from *Moringa oleifera* L. • *Yerba mate* (South American herb): 1.5 g/kg. • Herbal leaf tea. • Thyme: 1.7 mg/kg.	Berté et al. (2011); Žugic et al. (2014)
Carnosic acid	• Rosemary and common salvia species. • Dried sage leaves 1.5 to 2.5%.	Loussouarn et al. (2017); Raes et al. (2015)
Ferulic acid	• Black beans: 0.8 g/kg. • Flaxseed: 4.1 ± 0.2 g/kg. • *Angelica sinensis*.	Flanagan et al. (2013); Kiokias and Oreopoulou (2021); Mojica et al. (2015)
Gallic acid	• Tea. • *Cynomorium coccineum*, *Myriophyllum spicatum* and *Microcystis aeruginosa*. • Medicinal halophyte (*Thespesia populnea*).	Kiokias and Oreopoulou (2021)
p-Coumaric acid	• Garlic, basil, and amaranth (leaves and stem): 28–44 mg/kg. • *Bambusae caulis*. • Husked oat (*Avena sativa* L.).	Kavita and Gandhi (2017)
Rosmarinic acid	• *Origanum* spp., *Rosmarinus officinalis*, *Perilla* spp., and *Salvia officinalis*: 0.05 to 26 g/kg dry weight. • Pink savory leaves (*Satureja thymbra* L.): 19.5 g/kg.	Tsimogiannis et al. (2016); Yashin et al. (2017)
Vanillic acid	• Fruits, olives, wine, and cereal grains.	Chengke et al. (2018); Kiokias and Oreopoulou (2021)

hypolipidemic properties, which help in lowering the obesity linked to a diet with high-fat content, reduce serum cholesterol, protect the liver from various injuries, and also act as a tumor inhibitor (Son et al., 2010; Srinivasan et al., 2007). Caffeic acid and its derivatives, like caffeoylquinic and caffeoyl tartaric acids, are inhibitors of human immunodeficiency virus type 1 (HIV-1) integrase (King et al., 1999). In recent times, clinical research has focused on exploring phenolic acids by conducting *in vitro* studies, that is, in human cells and *in vivo* (animal) studies, for their health-protective effects against many severe human diseases. As the phenolic acids are present ubiquitously in plant-based foods, they are part of the daily diet of human beings, and depending on a diet including fruit, vegetables, grain, tea, coffee, spices, and so on, the average consumption is 25 mg to 1 g per day (Clifford, 1999).

1.2.1.2 STILBENES

Stilbenes are naturally found as plant monomers and oligomers, are low molecular weight (approximately 200–300 g/mol) phenolics that are widely distributed as secondary metabolites in plants. These are included in natural defense polyphenols that mostly occur as heartwood constituents in heterogeneous plant assemblies. During environmental stress conditions like exposure to excessive UV light and infection process, they are produced in plants, and more than 400 stilbene derivatives have been identified (Roupe et al., 2013). The stilbene structure consists of C6–C2–C6 backbone, defined by two aromatic rings linked by an ethylene chain, and carries various substituents occurring at different positions like methyl, glycosyl, hydroxyl, or isopropyl groups (Khawand et al., 2018). The most prevalent stilbene is resveratrol, which consists of a para-hydroxylated compound (estrogen-like activity), while piceatannol, isorhapontigenin, pinosylvin, rhapontigenin, and pterostilbene are other commonly found stilbenes in many plant species (Kumar et al., 2022; Reinisalo et al., 2015). These are widely distributed in Fabaceae, Pinaceae, Myrtaceae, and Vitaceae families and predominantly found in peanut, grape wine, sorghum, and many other tree species like *Pinus and Picea* (Parage et al., 2012). The commercial sources of stilbenes include plants like *Rheum undulatum, Polygonum cuspidatum, Melaleuca leucadendron, Rhodomyrtus tomentosa*, and *Euphorbia lagascae* which are cultivated in the countries of the Asian continent as folk medicines. Pterostilbene, which is mainly found predominantly in blueberries (*Vaccinium* species), bilberries (*Vaccinium myrtillus*), and grapes, acts as an important source of stilbene for use in nutraceutical applications (Reinisalo et al., 2015).

Stilbenes are commercially used in aroma therapy products and the nutraceutical industry; they are widely used as dietary supplements. Normally, they are referred to as tannins and further classified into two classes: (i) hydrolyzable and (ii) condensed tannins. Dietary stilbenes exhibit potential health benefits, but due to their rapid metabolism and as they are excreted in partially digested forms, their bioavailability is very limited, which restricts their use as potential nutraceuticals (Wen and Walle, 2006). There are many reports of the therapeutic potential of stilbenes against various disorders like oxidative stress, age-related diseases like obesity, Type-2 diabetes, diabetic vascular diseases, aging, neurodegenerative diseases, eye, and brain-like age-related macular degeneration, and Alzheimer's disease (Reinisalo et al., 2015). Resveratrol has anticancerous and antioxidant activities and is one of the most important stilbenes found in the skin of grapes (Petrovski et al., 2011; Rahman et al., 2006). It has been widely reported for its chemo-preventive activity and increased the rate of survival in mice with subcutaneous neuroblastomas (0 to 70%) with a daily administered dose of resveratrol at the rate of 40 mg/kg body weight (Chen et al., 2004). Stilbenes oligomers (viniferin and pallidol) present in botrytized grapes or wines have an anticancerous property that protects lipoproteins from oxidative damage. During grape processing, there is an accumulation of high resveratrol content in wine, and its regular appropriate consumption is related to the "French Paradox" in French people in which there is a low risk of CVDs despite high saturated fat intake in their diet (Sharma et al., 2012, 2021). Regular consumption of alcoholic beverages helps in maintaining cardiovascular health, and red wine with a higher amount of resveratrol can promote vasorelaxation (Lekakis et al., 2005), reduce lipid peroxidation, and decrease platelet aggregation (Miura et al., 2000), suppress atherosclerosis (Wang et al., 2005) and improve triglyceride concentrations and serum cholesterol (Zern et al., 2003). In recent animal trials, it has been reported that various stilbene compounds, particularly resveratrol, modulate insulin levels and help in reducing blood glucose levels in various test animals like mice, rats, and rodents with hyperglycemia (Reinisalo et al., 2015). Other structurally similar stilbene compounds (piceatannol, rhapontigenin, pinosylvin, and pterostilbene occurring in different plant species also have great pharmaceutical potential (Roupe et al., 2013).

1.2.1.3 FLAVONOIDS

Flavonoids constitute a diverse group of polyphenolic phytonutrients, and these compounds have wide spectrum occurrence and low molecular weight (Dai and Mumper, 2010). These are mainly present in the genus citrus and impart vibrant colors in various horticultural crops with a recommended daily allowance of 25 to 130 mg per day in various therapeutic applications (Chun et al., 2007; Gupta and Guliani, 2022). Their biological and pharmacological properties, such as antioxidant, anti-inflammatory, anti-cancer, antimicrobial, anti-allergic, and anti-diarrheal activities, have been studied extensively; however, recently, their potential in the form of inhibitory substances for various stages of tumor development is of concern. On basis of their structure, the various classes of flavonoids include flavones, flavonols, flavanones, flavanols, isoflavones, anthocyanidins, chalcones, catechins, flavans, dihydrochalcones, and bioflavonoids (Keservani and Sharma, 2014; Keservani et al., 2016a, b; Panche et al., 2016). Concerning broad classification, flavonoids are classified as anthocyanins and anthoxanthins.

1.2.1.3.1 *Anthocyanins*

Anthocyanins are water-soluble plant pigments, with flavylium (2-phenylch romenylium) ions being their basic core structure and present as a vacuolar pigment exhibiting red, black, purple, or blue color (Mattoo et al., 2022). They have been studied for their significant anti-inflammatory and anti-obese properties, while various epidemiological studies have reported a correlation between anthocyanin intake and improved cardiovascular risk, type-2 diabetes, and myocardial infarction (Ockermann et al., 2021).

1.2.1.3.2 *Anthoxanthins*

They are grouped into five sub-classes, which include flavonols, flavanols, flavanones, flavones, and isoflavones, which are generally colorless or white to yellow molecules. In various fruits and vegetables, the flavonols and flavones are most commonly distributed as anthoxanthins. The most important flavonols in various plant foods are kaempferol, quercetin, and myricetin. Quercetin, kaempferol, luteolin, and total flavonoid content (TFC) reported in some important fruits have been shown in Figure 1.2.

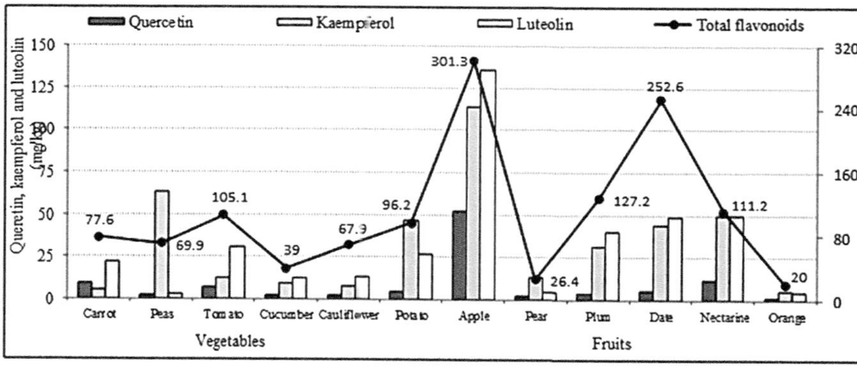

FIGURE 1.2 Quercetin, kaempferol, luteolin, and total flavonoid content of some important vegetables and fruits.

Source: Adapted from Cao et al., 2010.

Flavonols and flavones vary from each other by the existence of a hydroxyl group at C_3 in flavonols. Dehydrogenation of flavanones and dihydroflavonols give rise to flavones, 2-aryl-4H-chromen-4-ones, and flavonols, 2-aryl-3-hydroxy-4H-chromen-4-ones, respectively. Flavones are the most widely reported flavonoids, and out of them, 3-hydroxyflavones are the most important (Dias et al., 2021). Based on their concentration present in different fruits and vegetables, they are classified into three subclasses, that is, low (<10 mg per kg or L), medium (10–50 mg per kg or L), and high (>50 mg per kg or L). Fruits and vegetables like strawberries, cabbage, carrots, spinach, peas, and mushrooms have been reported to have low levels of flavonol and flavone content whereas white wine, tomato, orange, brewed coffee, red pepper, apples, broad beans, grapes, cherries have medium levels. The plant food species with high levels of flavonol and flavone content are celery, tomato juice, red wine, tea, kale, and French beans (Hertog et al., 1992, 1993). These compounds are sensitive to heat, and depending upon the type of cooking method, there are different levels of losses during food preparation, which are up to 20% with average cooking methods and can be as high as 50% during mechanical food processing (Hertog et al., 1992). In immature fruits, the concentration of flavonols is highest, which starts to decrease with advancing maturity, and in fully ripe fruits, about half of their content remains present in immature fruits (King and Young, 1999).

The risk of occurrence of coronary heart disease can be reduced with the intake of flavonol and flavones in the diet. Flavonoids and quercetin inhibit adipogenesis and induce apoptosis (Fang et al., 2008; Hsu and Yen, 2006), whereas other researchers have reported their anticarcinogenic, anti-obesity, and antiproliferative effects. The various flavonoids like quercetin, kaempferol, resveratrol, silibinin, galangin, baicalein, and luteolin have been reported for preclinical and clinical evidence for chemoprevention in hepatocellular carcinoma (Kumar et al., 2021). The flavones include various compounds like luteolin, apigenin, nobiletin, sinensetin, galangin, isosinensetin, tangeretin, chrysin, and so on, and these flavones exhibit anti-adipogenic activity (Hsu and Yen, 2007; Yun, 2010). The two major isoflavones, genistein and daidzein, are widely used as they have potential functional properties due to their numerous pharmacological benefits. Ivanov et al. (2021) carried out studies on some flavonols (quercetin), flavones (apigenin, luteolin), and their glycosylated derivatives like quercitrin, isoquercitrin, rutin, apigetrin, and so on, and reported promising anti-fungal activity. The biofilm-forming potential of *Candida albicans* was lowered by apigetrin and isoquercitrin, and it has anti-virulence potential because of its anti-hyphal properties, elucidating the prospective of these natural products to be used in anti-virulence therapeutics.

Only limited and scattered information is available with respect to the recommended dietary intake (RDI) of polyphenols like phenolic acids, proanthocyanidins, flavonoids, and ellagitannins which help in lowering the various chronic diseases along with other health benefits. Kuhnau (1976) reported a daily intake of flavonoids between 1 and 1.1 g/day, whereas Hertog et al. (1993) reported that an intake of 23 mg/day of flavonol and flavones inflicted the nutraceutical action effectively in the Dutch diet. In Denmark, the total combined intake of polyphenols like flavones, flavonols, and flavanones has been reported as 28 mg/day (Justesen et al., 2000). Dietary intake of one gram of polyphenols per day has been estimated in the studies carried out by Scalbert and Williamson (2000). This intake is comparatively very high compared to the other recognized dietary antioxidants, that is, 10 and 100 times higher than ascorbic acid (vitamin C) and tocopherol (vitamin E) intake. The different health effects associated with the intake of polyphenols from different plant sources reported by different researchers have been mentioned in Table 1.3.

1.2.2 TERPENOIDS

Terpenoids are one of the structurally diverse and major groups of naturally occurring plant metabolites that are essential for the various processes like normal growth, development, and metabolism mechanisms of plants. Terpenes consist of isoprene units only, while terpenoids derived from mevalonic acid (MVA) have ketone or heterocyclic and hydroxyl rings as an additional functional group, and nearly 50,000 terpenoids exist in nature which are mostly of plant origin (Sun et al., 2017; Yang et al., 2020; Yeshi et al., 2022). In nature, they are mostly prevalent in the form of volatile oils (in higher medicinal plants) and mainly exist in the plant species of Ranunculaceae, Compositae, Araliaceae, Oleaceae, Lauraceae, Magnoliaceae, Aristolochiaceae, Rutaceae, Pinaceae, Labiatae, Umbelliferae, Celastraceae, Taxaceae, Acanthaceae family (Yang et al., 2020). Commercially the terpenoids are used as pesticides, insect attractants or repellents, antifeedants, industrial raw materials, toxins, or antibiotic agents, and so on, while limonoids and pyrethrin are often used as insecticides (Gershenzon and Dudareva, 2007; Yang et al., 2020). The volatile terpenoids like menthol and perillyl alcohol are used as functional foods, colorants, pharmaceuticals, cosmetics, disinfectants, fragrances, agrichemicals, raw material for spices, and flavoring agents in ice creams (68 ppm), non-alcoholic beverages (31 ppm), baked foods (120 ppm), candy (49 ppm), gelatines and puddings (48–400 ppm), and 2300 ppm in chewing gum (Bowen, 2006; Maggi et al., 2009; Martin et al., 2003;

TABLE 1.3 Major Health Benefits of Polyphenols

Health Effect	Mechanism	References
Anti-cancer effect	• Inhibit the growth of tumors. • Cytotoxic action on cancer cells. • Apoptosis induction on cancer cells. • Enzymatic activity stimulation.	Kumar et al. (2022, 2023); Nirmala et al. (2018)
Anti-atherosclerotic effect	• Protects cardiovascular system. • Prevents oxidation of LDL, reduces LDL and triglycerides, and increases HDL.	Bahramsoltani et al. (2019)
Anti-metabolic syndrome	• Caused due to the clustering of various cardiometabolic factors like high blood pressure, dyslipidemia, blood sugar regulation disorder, and abdominal obesity. • intake of polyphenols is linked with the reduction of the above factors.	Anon (2005)
Anti-obesity effect	• Reduction of adipocyte proliferation inhibits the pre-adipocyte differentiation and stimulates adipocyte apoptosis. • Suppress the lipogenesis process.	Güneş Bayır et al. (2019)
Anti-hypertensive effect	• Improvement of endothelial function. • Reduce the LDL sensitivity towards oxidation. • Increased endothelium-dependent vasorelaxation.	Grasi et al. (2008); Williamson (2017)
Antimicrobial effect	• Antimicrobial action against a broad range of microorganisms. • Flavonoids have higher antimicrobial action with respect to other polyphenols. • Synergetic effect with antibiotics and suppress microbial virulence.	Daglia (2012)
Antidiabetic effect	• Positive regulation mechanism on insulin. • Increase insulin sensitivity in peripheral tissues.	Avila et al. (2017)
Anti-inflammatory effect	• Effects on cells of the immune system and other inflammatory cells. • Affects cytokine and its receptors. • Influences the secretory process.	Ferguson (2001)

NTP, 1990). Based on structural construction, terpenoids are classified as aliphatic (geraniol) and cyclic (limonene) terpenoids. The general formula for terpenoids is $(C_5H_8)_n$, and they are categorized on the basis of the value of n (the number of carbon atoms in the structure). The terpenoids consist of many isoprene units; they are further divided into different groups which are described in Table 1.4 (Yeshi et al., 2022).

TABLE 1.4 Group Classification of Terpenoids

Group	Isoprene Units	Carbon Atom	Example
Monoterpenoids	2	C-10	Linalool
Sesquiterpenoids	3	C-15	Caryophyllene
Diterpenoids	4	C-20	Abietic acid
Sesterterpenoids	5	C-25	Ophiobolin A
Triterpenoids	6	C-30	Ganoderic acid
Tetraterpenoids	8	C-40	Carotene
Polyterpenoids	>8	>C-40	Trans-1,4-polyisoprene

Terpenoids have their use as pharmacological agents and play a prominent increasing role in the medical field to benefit human health. These compounds exhibit various biological activities such as anti-inflammatory, antitumor, antiviral, antibacterial, and antimalarial, which further promote transdermal absorption and thus help in the prevention and treatment of CVDs and blood sugar. In other studies, terpenoids have been reported to have immunomodulatory, insecticidal, antioxidant, anti-aging, and neuroprotective effects, and the terpenoid paclitaxel is commonly used in various clinical practices (Yang et al., 2020). Vitamin E has been studied for its antioxidant potential and is further reported as an efficient apoptotic inducer for breast cancer cells (Yu et al., 1999), whereas artemisinin terpenoid is used as an antimalarial drug (Van et al., 1999). Terpenoids have also been reported to exhibit an anti-cancer effect by different mechanisms, such as triggering cancer progression stages with suppression of the tumorigenesis (early stage) via induction of cell cycle arrest, inhibition of cancer cell differentiation, and activating apoptosis (Kamran et al., 2022).

Carotenoids, present in various fruits and vegetables are one of the important classes of terpenoids that show different colors (yellow, orange, and red) and are synthesized by most of the photosynthetic organisms (plants, algae, and cyanobacteria) with a very rich antioxidant profile (Sun et al., 2022). α-carotene and β-carotene act as the precursors of vitamin A, which act as essential components for the healthy eyes and immune system (Sun et al., 2022). Vitamin A deficiency has been reported to affect about one-third of preschool children worldwide, causing serious consequences including blindness and death (https://news.un.org/en/story/2018/05/1008782). The different steps involved in carotenoid biosyntheses, like condensation followed by desaturation/isomerization, hydroxylation, oxidation, and epoxidation, further generate carotenes and xanthophylls compounds (Sun et al., 2022). Carotenes are carotenoids containing hydroaromatic rings at both ends of the four isoprene molecules, so in this way, each carotene molecule contains two hydroaromatic rings (Kabir et al., 2022). Carotenes (lycopene, γ-carotene, and lutein) have been reported in the prevention of various diseases like uterine, breast, prostate, colorectal, and lung cancers. Xanthophylls are carotenoids which are yellow pigments that occur naturally and possess oxygen atoms (Milani et al., 2017). The mode of action of xanthophylls includes symbiotic association with other classes of antioxidants, which protects them from the oxidation process and, at the same time, may exhibit tissue-specific protection. The various xanthophylls, like zeaxanthin, cryptoxanthin, and astaxanthin, play an important role as functional food ingredients in various preparations. Dietary carotenoids have high antioxidant activity, which lowers the risk of various diseases such as CVD and cancer (Eggersdorfer and Wyss, 2018), whereas zeaxanthin and lutein (macular pigments) help in delaying the onset of age-related vision diseases (Sauer et al., 2019).

Limonoids, mostly bitter, are a prominent group of secondary metabolites in citrus fruit which are highly oxygenated triterpenes, classed as tetranorterpenoids, and are responsible for the scent of fresh peels of citrus fruits (Manners et al., 2007; Roy and Saraf, 2006). They are effective chemotherapeutic detoxification agents and present in two forms, that is, neutral (noncarboxylated/aglycon) which is insoluble and bitter, and the other one is acidic form (carboxylated/glucoside) which is soluble and tasteless. It has a wide range of biological activities as it acts as an antibacterial, antifungal, antimalarial, anti-cancer, and antiviral agent and has several other pharmacological health effects on humans (Roy and Saraf, 2006). The most common monocyclic monoterpene is D-Limonene, which is present in high concentrations in orange peel oil and prevents gastric carcinogenesis induced by *N*-methyl-*N*-nitro-*N*-nitrosoguanidine in waster rats and pancreatic carcinogenesis induced by *N*-nitrosobis (2-oxopropyl) amine in the hamster (Uedo et al., 1999). The limonoids present in citrus essentials have antioxidant, antimicrobial, and flavoring properties, which play an important role in protecting lung tissues against many ailments (Saini et al., 2022). Citrus limonoids like limonexic acid, nomilin, and limonin have been reported to inhibit the proliferation (IC_{50} values < 50 μM after 72 h) of pancreatic cancer cells Panc-28 and induce apoptosis by the up-regulation of the expression of B-cell lymphoma 2 (Bcl-2)-associated X protein (Bax)/Bcl-2 proteins and enhanced cleavage of caspase-3, decreased mitochondrial membrane potential (Murthy et al., 2021).

1.2.3 GLUCOSINOLATES

Glucosinolates (water-soluble anions) consist of two parts: a common -D-thio-glucose moiety and other is a variable aglycone side chain derived from amino acids with or without an aliphatic, aromatic, or indolyl side chain (Ishida et al., 2014). Sulfur-containing glucosides are referred to as glucosinolates (Gls), which are found in cruciferous vegetables of *Brassica* spp. *like* cabbage and broccoli. It is also found in some oilseed crops and condiments such as rapeseed and mustard seed. Glucosinolates are most commonly found in cruciferous vegetables, and it includes various compounds like sinigrin (allyl isothio-cyanate), glucoraphanin (sulforaphane), glucoraphasatin, glucobrassicin, and glucoiberin compounds (Favela-González et al., 2020). To date, more than 120 different glucosinolates have been identified in nature, and among different plant species, agronomic practices, and climatic conditions, there is wide diversity in its occurrence and composition. Aliphatic, indole, and aromatic glucosinolates are the three most commonly classified glucosinolates.

Connolly et al. (2021) have reported that glucosinolates act *via* several mechanisms that further exhibit antioxidant, anti-inflammatory, and chemo-protective effects, which in turn have beneficial outcomes on cardiometabolic disorders (hyperglycemia, diabetes, hypertension, and dyslipidemia), neurological and psychiatric disorders, musculoskeletal disorders and different types of prostates, breast, liver, and colon cancers. Isothiocyanates obtained from different sources have their use in preventing different cancers, and in the recent era, more consideration has been made to sulforaphane (isothiocyanate present in broccoli), which is the principal inducer of a Phase II enzyme (quinone reductase) with strong anticancerous activity. During the chopping, cutting, mixing, or chewing process of plant tissue, thioglucosidase, namely, myrosinase, is released, which improves the bioavailability and bio-accessibility of glucosinolates present in cruciferous crops via the formation of anti-cancer metabolites like indole-3-carbinol and isothiocyanates (Matusheski et al., 2004). Glucosinolates have anti-cancer properties and exert their preventive effects against gastrointestinal cancers, and the various breakdown products formed from them (isothiocyanates and indoles) are currently under intensive investigation (Melim et al., 2022).

1.2.3 POLYACETYLENES

Polyacetylenes, biosynthesized from fatty acid and polyketide precursors, are derived from different compounds like crepenynic acid, tartaric acid, or stea-rolic acid, also referred to as acetylenic natural products with alkyne functional groups (Christensen and Brandt, 2006; Minto and Blacklock, 2008). They are chemically reactive compounds and 1400 different polyacetylenes have been isolated from different flora. The most commonly reported polyacetylenes are polyacetylenic alcohols and include other compounds like panaxydol, falcarinol (panaxynol), and panaxytriol. They have been reported in more than 21 other families with high concentrations in A*raliaceae* (ginseng, *hedera* spp.), A*piaceae* (carrot, celery, and fennel), and A*steraceae* (lettuce, chicory, and artichoke) families. Concerning the natural defense mechanism in carrots, polyacetylenes (falcarinol, falcarindiol-3-acetate, and falcarindiol) are released as natural pesticides which have strong functional properties and are gaining scientific attention (Acworth et al., 2011). Falcarinols have biological activities such as anti-inflammatory, and anti-platelet-aggregatory, which stimulate the immune system (Bernart et al., 1996) and have cytotoxic properties (Zidorn et al., 2005). At higher non-toxic concentrations, the beneficial effects of these compounds are exhibited (Zidorn et al., 2005), and they represent pharmacologically useful properties like anti-fungal, antibacterial, antitumor, cytotoxic, antiviral, neurotoxic, immunosuppressant, and piscicidal activities (Christensen and Brandt, 2006).

1.2.4 PHYTOSTEROLS AND PHYTOSTEROLS

"Phytosterols" are a large variety of sterols present in plants, and the term phytosterols is derived from two Greek words. The first word is "phyton," which means plant, and the second word is "stereos," meaning solid. A large number of phytosterols (>250) have been identified to date, and further, they have been successfully isolated from various plant species (Brufaua et al., 2008). These represent a portion of un-saponifiable matter in plant lipids, and they are referred to as plant-derived fatty compounds (steroids), such as cholesterol, in animals (Piironen et al., 2003). Phytosterols are formed of a steroid skeleton with the saturated bond between C-5 and C-6 of the sterol moiety and with a hydroxyl group attached to the C-3 atom and an aliphatic side chain attached to the C-17 atom. Plant sterols and cholesterol are formerly found as esterified and non-esterified forms of cinnamic acid or fatty acids or as glycosides (Salehi et al., 2021).

In nature, phytosterols and phytosterols are present in most plants, but their content varies widely in different plant parts. The good sources of phytosterols include vegetable oils, beans, nuts, cereals, fruits, and seeds of plants (Figures 1.3 and 1.4), whereas phytosterols are abundantly present in corn, wheat, rye, and rice. A very high amount of phytosterol content, 235.9

mg/100 g, has been reported in olive oil, whereas in the potato, it has been reported as only 4.3 ± 0.2 mg/100 g (Jiménez-Escrig et al., 2006). Sitosterol, campesterol, and stigmasterol are the most commonly found phytosterols in higher plants, and their content may go beyond 80% collectively (Li et al., 2022). The phytosterols in plants occur in five different forms *viz.*, (i) fatty-acid esters (SE); (ii) free alcohol (FS); (iii) steryl glycosides (SG); (iv) phytosteryl hydroxycinnamic-acid esters (HSE); and (v) acylated steryl glycosides (ASG) which have similar functioning as that of cholesterol in human body (Pascal and Segal, 2006).

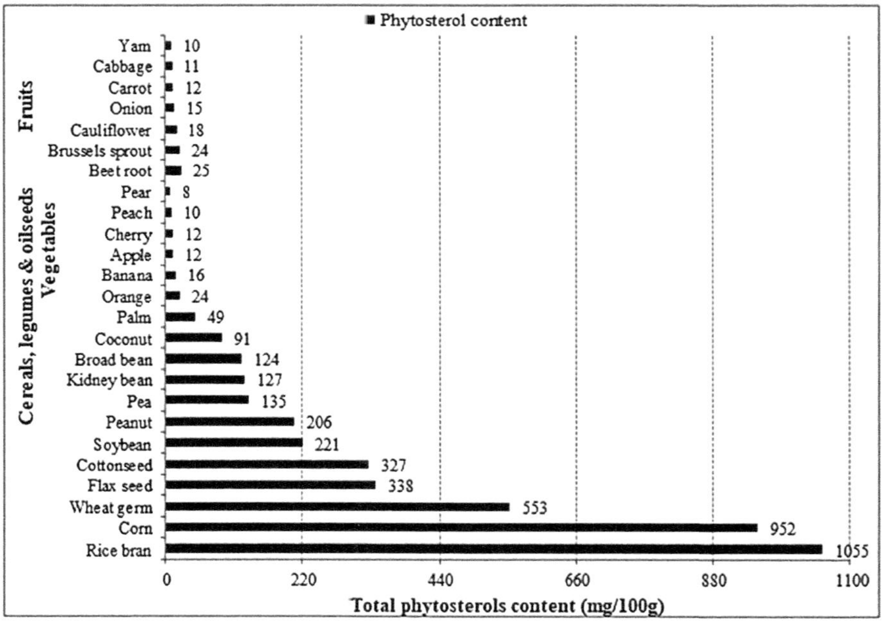

FIGURE 1.3 Total phytosterol content of some important plant foods.

Source: Adapted from Kritchevsky, 2004; Kuna and Poshadri, 2012; Laakso, 2005.

Salehi et al. (2021) have reported the remarkable pharmacological effects of phytosterols, including their action as chemo-preventive, anti-inflammatory, antioxidant, antidiabetic, and anti-atherosclerotic agents. Phytosterols play an important role in lowering the cholesterol and LDL cholesterol plasma levels and help in the prevention of CVD and non-alcoholic fatty liver diseases (Li et al., 2022). Alvarez-Sala et al. (2018) conducted an *in vitro* study by using a selected phytosterol mixture (β-sitosterol, stigmasterol, campesterol, or β-cryptoxanthin) and stated that the β-cryptoxanthin enhanced the volume and number of eryptotic cell, hemolysis, and glutathione depletion (GSH). In a recent study, it was reported that phytosterols-enriched milk-based fruit beverages with/without galacto-oligosaccharides (GOS) reduced the levels of interleukin IL-6 and (IL)-8 followed by IL-1β-induction and blocked cholesterol oxidation products oxidation-induced oxidative stress (López-García et al., 2020). An intake of 1.5 to 3.0 g/day of phytosterols has been recommended to reduce the incidence of coronary heart disease in several countries (Jones et al., 2018).

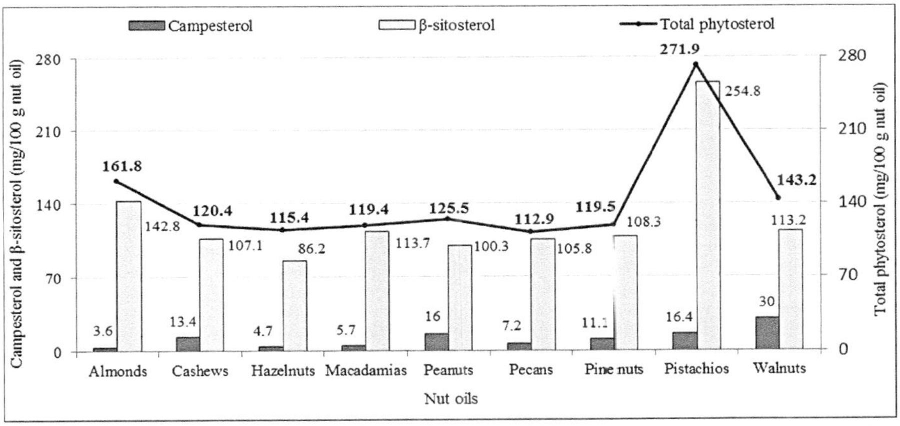

FIGURE 1.4 Campesterol, β-sitosterol, and total phytosterol (mg/100 g) content of nut oils.
Source: Adapted from Kornsteiner-Krenn et al., 2013

1.2.5 NONDIGESTIBLE CARBOHYDRATES

Nondigestible carbohydrates (NDCs) are present in the form of plant fiber (soluble and insoluble), and resistant starch (RS) are complex heterogeneous dietary substances that cannot be digested or broken down in the stomach or small intestine. Further, in the large intestine (colon), gut microbiota acts on them where they are fermented into short-chain fatty acids (SCFAs), and across the colonic epithelial cells, they are absorbed by specific transporters (Flint et al., 2007; Johnstone et al., 2020). NDCs are further classified into three types *viz*., (i) resistant starch (RS), (ii) non-starch polysaccharides (NSPs), and (iii) nondigestible oligosaccharides (NDOs). NDCs affect gut function through microbial-produced propionate and butyrate to improve satiety (Morrison and Preston, 2016). Nondigestible carbohydrates (NDC) have a significant role in human health, and insufficient intake in the diet may lead to gastrointestinal disorders like irritable bowel syndrome (IBS), constipation, diverticular disease, and colorectal cancer (Mussatto et al., 2007). RS, which is a potential source of NDCs, can improve insulin sensitivity (Johnstone et al., 2020). Dietary fibers are analogous carbohydrates, and on the basis of their solubility, they are classified into two categories, that is, (a) water-insoluble fiber which includes cellulose, lignin, hemicelluloses, and arabinoxylan; and (b) water-soluble fiber consisting of non-starchy polysaccharides like β-glucan.

Due to almost similar nomenclature and structural complexity, the classification of NSPs is difficult. However, on their distinct mode of action, they are classified as (i) cellulose, (ii) non-cellulosic polymers, and (iii) pectic polysaccharides. The main source of dietary fibers is cereal grains, and most of them are attached to the bran layer, and the concentration of these compounds starts decreasing towards the inner endosperm (Table 1.5). The content of total dietary fiber in some commonly consumed grains has been presented in Figure 1.5.

TABLE 1.5 Different Sources and Types of Non-Starch Polysaccharides (NSP)

Source	Group and Components	Structural Compounds	Allocation in Food Matrix
Plant cell wall (structural materials)	Cellulose	Galacturonans	Cell walls of fruits and vegetables
	Pectic substances	Long-chain β-glucans	
	Polysaccharides (non-cellulosic)	Glucuronoxylans and glucurono-arabinoxylans	Cereals, fruits, and vegetables
	Hemicelluloses	Arabinoxylans and arabinogalactans	
Polysaccharides (non-structural)	Pectic polysaccharides, gums	Heteropolysaccharides	Fruits and seeds

β-glucan is nutritionally one of the most important dietary fiber present in bacteria, algae, plants (cereal seeds), micro-fungi, and macro-fungi (mushrooms), which include a group of polysaccharides (PS) of natural origin and are characterized by diverse structures and therapeutic properties with protection against heart diseases and reduction of glycemic response and cholesterol (Charalampopoulos et al., 2002). Beta-glucan is a generally used term for PS that consists of β-1, 3 and β-1, 6 bonds in combined form to make long linear glucans (Suzuki et al., 2021). Their metabolic and immune-modulatory properties include anticancerous, antiviral, and antibacterial activities, and recently, it has been suggested that beta-glucans have a potential for the prevention and treatment of COVID-19 (Miro'nczuk-Chodakowska et al., 2021). In the human body, β-glucan plays an important role in improving immunity as it activates the immune system (acquired and innate), and the glucan receptors have been reported on macrophages, natural killer (NK) cells, neutrophils, and so forth (Herre et al., 2004). Further, glucan attaches with these cells on its receptors, and it activates macrophages, and all other activities involved in immunity mechanisms like phagocytosis, the release of specific cytokines like GM-CSF, interleukin-1, interleukin-6, and interferons (Khanjani et al., 2021).

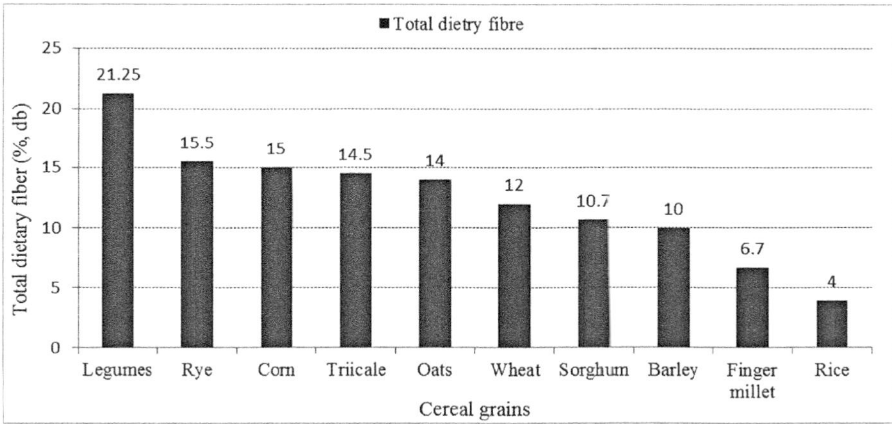

FIGURE 1.5 Total dietary fibers (%) of some commonly consumed cereals.

Source: Adapted from Nelson, 2001

The other class of nondigestible carbohydrates is RS, which does not undergo digestion or absorption in the stomach and small intestine as it can't be hydrolyzed to D-glucose after consumption. Further, in the large intestine, it is fermented, and there is production of SCFAs. RS is present in most starchy foods, and it forms a linear molecule of α-1, 4-D-glucan, having a low molecular weight of 1.2×10^5 Da, which is derived from its retrograded amylose fraction (Tharanathan 2002). Further, RS are categorized into four different classes given in Figure 1.6, and the extent of resistance towards digestion is generally affected by various processing conditions like milling, chewing, and other factors such as processing temperature and time, pH, cooling cycles, freezing, and drying. The most common sources of RSs are milled grains (whole or partial), legumes, potatoes, green bananas, and so forth. Based on RS content, food can be categorized as low, medium, and high RS-containing foods with different amounts of starch content (Figure 1.6).

The modified RSs are incorporated into different foods to improve their functional properties like crispness and expansion of products, mouth feel, color, and flavor (Fuentes-Zaragoza et al., 2010). RS provides various health benefits to the human body, such as control of plasma triglyceride during fasting, glycemic control, and absorption of various minerals. Jaiturong et al. (2020) studied the biochemical and prebiotic properties of RS obtained from *Musa sapientum* Linn. and further reported that unpeeled raw banana powder and banana starch have prebiotic properties like commercial inulin so they can be used for the development of health-promoting products.

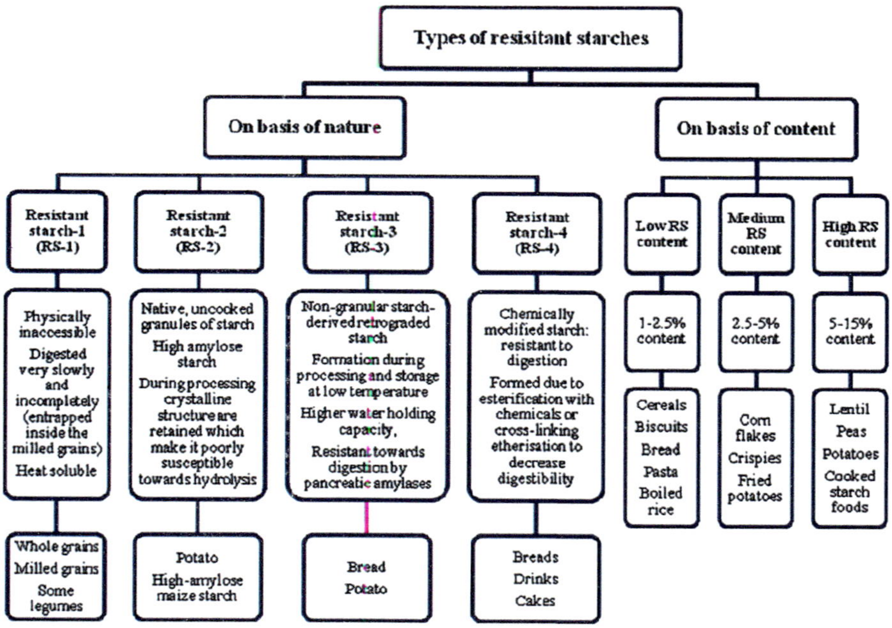

FIGURE 1.6 Classification and types of resistant starch (RS), along with their sources.

Source: Adapted from Lunn and Buttriss, 2007; Nugent, 2005; Sajilata et al., 2006; Sharma et al., 2008.

The other important class of nondigestible oligosaccharides (NDOs) includes fructans, which are fructose units linked to sucrose. They exist as heterogeneous mixtures with wide variation in their molecular structures and mass, different degrees of polymerization, and types of linkages between fructosyl residues and in predominantly found in plants like artichoke, onion, garlic, chicory, banana, rye, and barley (Márquez-López et al., 2022). Due to their selective stimulation nature towards the micro-organisms and indigestibility in the GI tract, these carbohydrates can be regarded as prebiotic dietary fiber sources with various positive health-promoting effects for human beings (Zimmermann et al., 2021). Simple oligosaccharides are also having they're used as pre-biotics in the food industry with their wide use in some infant products with the aim of providing health benefits that are similar to that of oligosaccharides present in the breast milk of human beings (Rivero-Urgell and Santamaria-Orleans, 2001). In the recent era, NDOs have the most commercial food applications in confectionary products like candy, cookies, and biscuits; dessert preparations like jellies, puddings, and sherbets; non-vegetarian-based food products like fish, and meat paste; some break-fast cereal produces and sweets, jams, marmalades, tofu, and so forth.

1.3 CONCLUSION

In nature, there is a huge diversity of phytochemicals that have interesting medicinal properties and biological activities. Many of them have been successfully isolated from different flora and there are still huge research opportunities for studying their complex behavior and relationship or interactions among different food components. These phytochemicals can combat specific diseases and have a huge potential to maintain or improve health. In the present era, the advancement of sophisticated isolation, synthetic methodology, and analytical techniques represents a huge gap in the knowledge available with respect to the existence and use of these dietary substances and phytochemicals. These compounds may be expected to be identified and explored in the near future and could be used commercially in the development of formulated functional food products. In the modern era, awareness about the health benefits of phytochemicals-rich foods is also important to include them as healthy food choices in the daily diet. Most of the research studies conducted with respect to the health benefits of various phytochemicals or their role in disease prevention have been generated from *in vitro* studies, which further need to be verified through *in vivo* studies. Hence, extensive research efforts are necessary for the use of phytochemicals as functional food/nutraceuticals. In this way, the specific functional health benefits associated with the intake of phytochemicals can be achieved successfully.

KEYWORDS

- **anti-inflammatory agents**
- **antioxidant agents**
- **cardiovascular diseases**
- **cholesterol oxidation products**
- **eryptotic cells**
- **functional food**
- **galacto-oligosaccharides**
- **glutathione depletion**
- **hemolysis**
- **non-alcoholic fatty liver diseases**
- **oxidative stress**
- **pharmacological effects**
- **phytosterols**

REFERENCES

Acworth, I., Plante, M., Bailey, B., Crafts, C., & Waraska, J. (2011). Simple and direct analysis of falcarinol and other polyacetylenic oxylipins in carrots by reversed-phase HPLC and charged aerosol detection. *Planta Medica, 77*(12), 1263–1270.

Almodaifer, S., Alsibaie, N., Alhoumendan, G., Alammari, G., & Kavita, M. S. (2017). Role of phytochemicals in health and nutrition. *BAO J Nutr., 3*, 28–34.

Alvarez-Sala, A., López-García, G., Attanzio, A., Tesoriere, L., Cilla, A., Barberá, R., & Alegría, A. (2018). Effects of plant sterols or β-cryptoxanthin at physiological serum concentrations on suicidal erythrocyte death. *Journal of Agricultural and Food Chemistry, 66*(5), 1157–1166.

Anonymous. (2005). *International Diabetes Federation: The IDF Consensus Worldwide Definition of the Metabolic Syndrome*. International Diabetes Federation. http://www.idf.org/webdata/docs/Metabolic_syndrome_definition.pdf.

Arya, R. K. K., Sati, D., Bisht, D., & Keservani, R. K. (2022). Nanotechnology-based bacterial immunotherapy. In R. K. Kesharwani, R. K. Keservani, & A. K. Sharma (Eds) *Nutraceuticals and Functional Foods in Immunomodulators* (pp. 3–19). Springer. https://doi.org/10.1007/978-981-19-2507-8_1.

Bahramsoltani, R., Ebrahimi, F., Farzaei, M. H., Baratpourmoghaddam, A., Ahmadi, P., Rostamiasrabadi, P., Rasouli Amirabadi, A. H., & Rahimi, R. (2019). Dietary polyphenols for atherosclerosis: A comprehensive review and future perspectives. *Critical Reviews in Food Science and Nutrition, 59*(1), 114–132.

Baile, C. A., Yang, J. Y., Rayalam, S., Hartzell, D. L., Lai, C. Y., Andersen, C., & Della-Fera, M. A. (2011). Effect of resveratrol on fat mobilization. *Annals of the New York Academy of Sciences*, *1215*(1), 40–47.

Bernart, M. W., Cardellina, J. H., Balaschak, M. S., Alexander, M. R., Shoemaker, R. H., & Boyd, M. R. (1996). Cytotoxic falcarinol oxylipins from Dendropanax arboreus. *Journal of Natural Products*, *59*(8), 748–753.

Berté, K. A., Beux, M. R., Spada, P. K., Salvador, M., & Hoffmann-Ribani, R. (2011). Chemical composition and antioxidant activity of yerba-mate (Ilex paraguariensis A. St.-Hil., Aquifoliaceae) extract as obtained by spray drying. *Journal of Agricultural and Food Chemistry*, *59*(10), 5523–5527.

Bowen, H. R. (2010). *U.S. Patent No. 7,727,401*. U.S. Patent and Trademark Office.

Brufau, G., Canela, M. A., & Rafecas, M. (2008). Phytosterols: Physiologic and metabolic aspects related to cholesterol-lowering properties. *Nutrition Research*, *28*(4), 217–225.

Cao, J., Chen, W., Zhang, Y., Zhang, Y., & Zhao, X. Content of selected flavonoids in 100 edible vegetables and fruits. *Food Science and Technology Research*, *16*(5), 395–402.

Caruso, G., Godos, J., Privitera, A., Lanza, G., Castellano, S., Chillemi, A., Bruni, O., Ferri, R., Caraci, F., & Grosso, G. (2022). Phenolic acids and prevention of cognitive decline: Polyphenols with a neuroprotective role in cognitive disorders and Alzheimer's disease. *Nutrients*, *14*(4), 819.

Charalampopoulos, D., Wang, R., Pandiella, S. S., & Webb, C. (2002). Application of cereals and cereal components in functional foods: A review. *International Journal of Food Microbiology*, *79*(1, 2), 131–141.

Chen, Y., Tseng, S. H., Lai, H. S., & Chen, W. J. (2004). Resveratrol-induced cellular apoptosis and cell cycle arrest in neuroblastoma cells and antitumor effects on neuroblastoma in mice. *Surgery*, *136*(1), 57–66.

Christensen, L. P., & Brandt, K. (2006). Bioactive polyacetylenes in food plants of the Apiaceae family: Occurrence, bioactivity and analysis. *Journal of Pharmaceutical and Biomedical Analysis*, *41*(3), 683–693.

Chun, O. K., Chung, S. J., & Song, W. O. (2007). Estimated dietary flavonoid intake and major food sources of US adults. *The Journal of Nutrition*, *137*(5), 1244–1252.

Clifford, M. N. (1999). Chlorogenic acids and other cinnamates–nature, occurrence, and dietary burden. *Journal of the Science of Food and Agriculture*, *79*(3), 362–372.

Connolly, E. L., Sim, M., Travica, N., Marx, W., Beasy, G., Lynch, G. S., Bondonno, C. P., Lewis, J. R., Hodgson, J. M., & Blekkenhorst, L. C. (2021). Glucosinolates from cruciferous vegetables and their potential role in chronic disease: Investigating the preclinical and clinical evidence. *Frontiers in Pharmacology*, *12*, 767975.

Daglia, M. (2012). Polyphenols as antimicrobial agents. *Current Opinion in Biotechnology*, *23*(2), 174–181.

Dai, J., & Mumper, R. J. (2010). Plant phenolics: Extraction, analysis and their antioxidant and anti-cancer properties. *Molecules*, *15*(10), 7313–7352.

Das, L., Bhaumik, E., Raychaudhuri, U., & Chakraborty, R. (2012). Role of nutraceuticals in human health. *Journal of Food Science and Technology*, *49*, 173–183.

Dias, M. C., Pinto, D. C., & Silva, A. M. (2021). Plant flavonoids: Chemical characteristics and biological activity. *Molecules*, *26*(17), 5377.

Dillard, C. J., & German, J. B. (2000). Phytochemicals: Nutraceuticals and human health. *Journal of the Science of Food and Agriculture*, *80*(12), 1744–1756.

Eggersdorfer, M., & Wyss, A. (2008). Carotenoids in human nutrition and health. *Archives of Biochemistry and Biophysics, 652*, 18–26.

El Khawand, T., Courtois, A., Valls, J., Richard, T., & Krisa, S. (2018). A review of dietary stilbenes: Sources and bioavailability. *Phytochemistry Reviews, 17*(5), 1007–1029.

Fang, T., Worm, V., & Tung, R. L. Changing success and failure factors in business negotiations with the PRC. *International Business Review, 17*(2), 159–169.

Favela-González, K. M., Hernández-Almanza, A. Y., & De la Fuente-Salcido, N. M. (2020). The value of bioactive compounds of cruciferous vegetables (Brassica) as antimicrobials and antioxidants: A review. *Journal of Food Biochemistry, 44*(10), e13414.

Ferguson, L. R. (2001). Role of plant polyphenols in genomic stability. *Mutation Research/ Fundamental and Molecular Mechanisms of Mutagenesis, 475*(1, 2), 89–111.

Flanagan, J., Bily, A., Rolland, Y., & Roller, M. (2004). Lipolytic activity of Svetol®, a decaffeinated green coffee bean extract. *Phytotherapy Research, 28*(6), 946–948.

Flint, H. J., Duncan, S. H., Scott, K. P., & Louis, P. (2007). Interactions and competition within the microbial community of the human colon: Links between diet and health. *Environmental Microbiology, 9*(5), 1101–1111.

Fuentes-Zaragoza, E., Riquelme-Navarrete, M. J., Sánchez-Zapata, E., & Pérez-Álvarez, J. A. (2010). Resistant starch as a functional ingredient: A review. *Food Research International, 43*(4), 931–942.

Gershenzon, J., & Dudareva, N. (2010). The function of terpene natural products in the natural world. *Nature Chemical Biology, 3*(7), 408–414.

Grace Nirmala, J., Evangeline Celsia, S., Swaminathan, A., Narendhirakannan, R. T., & Chatterjee, S. (2018). Cytotoxicity and apoptotic cell death induced by Vitis vinifera peel and seed extracts in A431 skin cancer cells. *Cytotechnology, 70*, 537–554.

Grassi, D., Desideri, G., Necozione, S., Lippi, C., Casale, R., Properzi, G., Blumberg, J. B., & Ferri, C. (2008). Blood pressure is reduced and insulin sensitivity is increased in glucose-intolerant, hypertensive subjects after 15 days of consuming high-polyphenol dark chocolate. *The Journal of Nutrition, 138*(9), 1671–1676.

Gupta, D., & Guliani, E. (2022). Flavonoids: Molecular mechanism behind the natural chemoprotective behavior-a mini-review. *Biointerface Res. Appl. Chem., 12*, 5983–5995.

Hamid, H., Thakur, A., & Thakur, N. S. (2019). Role of functional food components in COVID-19 pandemic: A review. *Ann. Phytomed. Int. J., 10*(1), 5240–5250.

Hamid, T., NS, T., Sharma, C., Bhatt, K., & Khan, A. A. (2020). Pomegranate and its wild genotypes: Nutraceutical opportunities and challenges. *Annals of Phytomedicine, 9*(1), 32–43.

Herre, J., Gordon, S., & Brown, G. D. (2004). Dectin-1 and its role in the recognition of β-glucans by macrophages. *Molecular Immunology, 40*(12), 869–876.

Hertog, M. G., Feskens, E. J., Kromhout, D., Hollman, P. C. H., & Katan, M. B. (1993). Dietary antioxidant flavonoids and risk of coronary heart disease: The Zutphen Elderly Study. *The Lancet, 342*(8878), 1007–1011.

Hertog, M. G., Hollman, P. C., & Katan, M. B. (1992). Content of potentially anticarcinogenic flavonoids of 28 vegetables and 9 fruits commonly consumed in the Netherlands. *Journal of Agricultural and Food Chemistry, 40*(12), 2379–2383.

Hsu, C. L., & Yen, G. C. (2006). Induction of cell apoptosis in 3T3-L1 pre-adipocytes by flavonoids is associated with their antioxidant activity. *Molecular Nutrition & Food Research, 50*(11), 1072–1079.

Hsu, C. L., & Yen, G. C. (2007). Effects of flavonoids and phenolic acids on the inhibition of adipogenesis in 3T3-L1 adipocytes. *Journal of Agricultural and Food Chemistry*, *55*(21), 8404–8410.

Ishida, M., Hara, M., Fukino, N., Kakizaki, T., & Morimitsu, Y. (2014). Glucosinolate metabolism, functionality, and breeding for the improvement of Brassicaceae vegetables. *Breeding Science*, *64*(1), 48–59.

Ivanov, M., Kannan, A., Stojković, D. S., Glamočlija, J., Calhelha, R. C., Ferreira, I. C., Sanglard, D., & Soković, M. (2020). Flavones, flavonols, and glycosylated derivatives—Impact on Candida albicans growth and virulence, expression of cdr1 and erg11, cytotoxicity. *Pharmaceuticals*, *14*(1), 27.

Jaiturong, P., Laosirisathian, N., Sirithunyalug, B., Eitssayeam, S., Sirilun, S., Chaiyana, W., & Sirithunyalug, J. (2020). Physicochemical and prebiotic properties of resistant starch from Musa sapientum Linn., ABB group, cv. Kluai Namwa Luang. *Heliyon*, *6*(12), 05789.

Jiménez-Escrig, A., Santos-Hidalgo, A. B., & Saura-Calixto, F. (2006). Common sources and estimated intake of plant sterols in the Spanish diet. *Journal of Agricultural and Food Chemistry*, *54*(9), 3462–3471.

Johnstone, A. M., Kelly, J., Ryan, S., Romero-Gonzalez, R., McKinnon, H., Fyfe, C., Naslund, E., Lopez-Nicolas, R., Bosscher, D., Bonnema, A., & Frontela-Saseta, C. (2020). Nondigestible carbohydrates affect metabolic health and gut microbiota in overweight adults after weight loss. *The Journal of Nutrition*, *150*(7), 1859–1870.

Jones, P. J., Shamloo, M., MacKay, D. S., Rideout, T. C., Myrie, S. B., Plat, J., Roullet, J. B., Baer, D. J., Calkins, K. L., Davis, H. R., & Barton Duell, P. (2018). Progress and perspectives in plant sterol and plant stanol research. *Nutrition Reviews*, *76*(10), 725–746.

Justesen, U., Knuthsen, P., Lyhne Andersen, N., & Leth, T. (2000). Estimation of daily intake distribution of flavonols and flavanones in Denmark. *Näringsforskning*, *44*(1), 158–160.

Kabir, M. T., Rahman, M. H., Shah, M., Jamiruddin, M. R., Basak, D., Al-Harrasi, A., Bhatia, S., Ashraf, G. M., Najda, A., El-Kott, A. F., & Mohamed, H. R. (2022). Therapeutic promise of carotenoids as antioxidants and anti-inflammatory agents in neurodegenerative disorders. *Biomedicine & Pharmacotherapy*, *146*, 112610.

Kamran, S., Sinniah, A., Abdulghani, M. A., & Alshawsh, M. A. (2022). Therapeutic potential of certain terpenoids as anti-cancer agents: A scoping review. *Cancers*, *14*(5), 1100.

Keservani, R. K., & Sharma, A. K. (2014). Flavonoids: Emerging trends and potential health benefits. *Journal of Chinese Pharmaceutical Sciences*, *23*(12), 815.

Keservani, R. K., Kesharwani, R. K., Sharma, A. K., & Jarouliya, U. (2015d). Dietary supplements, nutraceuticals, and functional foods in immune response (immunomodulators). In D. Bagchi, H. G. Preuss, & A. Swaroop (Eds.), *Nutraceutical and Functional Foods in Human Life and Disease Prevention* (Vol. 20, pp. 343–358). CRC Press, Taylor & Francis.

Keservani, R. K., Kesharwani, R. K., Sharma, A. K., Vyas, N., & Chadoker, A. (2010b). Nutritional Supplements: An Overview. *International Journal of Current Pharmaceutical Review and Research*, *1*(1), 59–75.

Keservani, R. K., Kesharwani, R. K., Vyas, N., Jain, S., Raghuvanshi, R., & Sharma, A. K. (2010a). Nutraceutical and functional food as future food: A review. *Der Pharmacia Lettre*, *2*(1), 106–116.

Keservani, R. K., Sharma, A. K., & Kesharwani, R. K. (2016a). Nutraceutical and functional foods for cardiovascular health. In *Food Process Engineering* (pp. 257–278). Apple Academic Press, CRC Press.

Keservani, R. K., Sharma, A. K., & Kesharwani, R. K. (2016b). Medicinal effect of nutraceutical fruits for cognition and brain health. *Scientifica, 3109254*, 10. doi: 10.1155/2016/3109254.

Keservani, R. K., Sharma, A. K., & Kesharwani, R. K. (2017). An overview and therapeutic applications of nutraceutical and functional foods. *Recent Advances in Drug Delivery Technology*, 160–201. Doi: 10.4018/978-1-5225-0754-3.ch006.

Keservani, R. K., Sharma, A. K., & Kesharwani, R. K. (2020). *Nutraceuticals and Dietary Supplements: Applications in Health Improvement and Disease Management* (pp. 1–342). CRC Press.

Kesharwani, R. K., & Misra, K. (2011). Prediction of the binding site for curcuminoids at human topoisomerase II α protein; an *in silico* approach. *Current Science*, 1060–1065.

Kesharwani, R. K., Singh, D. B., Singh, D. V., & Misra, K. (2018). Computational study of curcumin analogs by targeting DNA topoisomerase II: A structure-based drug designing approach. *Network Modeling Analysis in Health Informatics and Bioinformatics*, *7*(1), 1–7.

Kesharwani, R. K., Srivastava, V., Singh, P., Rizvi, S. I., Adeppa, K., & Misra, K. (2015). A novel approach for overcoming drug resistance in breast cancer chemotherapy by targeting new synthetic curcumin analogs against aldehyde dehydrogenase 1 (ALDH1A1) and glycogen synthase kinase-3 β (GSK-3β). *Applied Biochemistry and Biotechnology*, *176*(7), 1996–2017.

Khanjani, M. H., Sharifinia, M., & Ghaedi, G. (2022). β-glucan as a promising food additive and immunostimulant in the aquaculture industry. *Annals of Animal Science*, *22*(3), 817–827.

King, A. M. Y., & Young, G. (1999). Characteristics and occurrence of phenolic phytochemicals. *Journal of the American Dietetic Association*, *99*(2), 213–218.

King, P. J., Ma, G., Miao, W., Jia, Q., McDougall, B. R., Reinecke, M. G., Cornell, C., Kuan, J., Kim, T. R., & Robinson, W. E. (1999). Structure–activity relationships: Analogues of the dicaffeoylquinic and dicaffeoyltartaric acids as potent inhibitors of human immunodeficiency virus type 1 integrase and replication. *Journal of Medicinal Chemistry*, *42*(3), 497–509.

Kiokias, S., & Oreopoulou, V. (2021). A review of the health protective effects of phenolic acids against a range of severe pathologic conditions (including coronavirus-based infections). *Molecules*, *26*(17), 5405.

Koche, D., Shirsat, R., & Kawale, M. (2016). An overview of major classes of phytochemicals: Their types and role in disease prevention. *Hislopia Journal*, *9*(1/2), 976–2124.

Köksel, H., Edney, M. J., & Özkaya, B. (1999). Barley bulgur: Effect of processing and cooking on chemical composition. *Journal of Cereal Science*, *29*(2), 185–190.

Kornsteiner-Krenn, M., Wagner, K. H., & Elmadfa. (2013). Phytosterol content and fatty acid pattern of ten different nut types. *Int. J. Vitam. Nutr. Res.*, *83*(5), 263–270.

Kritchevsky, D. (2004). Safety of phytosterols and phytosterol esters as functional food components. In *Phytosterols as Functional Foods Components and Nutraceuticals* (pp. 347–363). Marcel Dekker, Inc.

Kuhnau, J. (1976). Flavonoids. A class of semi-essential food components: Their role in human nutrition. *World Review of Nutrition and Dietetics, 24*, 117–191.

Kumar, P., Dixit, J., Saini, R., Verma, P., Mishra, A. K., & Nath Tiwari, K. (2021). Potential of flavonoids as anti-cancer drugs. In *Phytopharmaceuticals: Potential Therapeutic Applications* (pp. 135–159).

Kumar, P., Rai, S., Verma, S. K., Prakash, P. S., & Chitara, D. (2022). Classification, mode of action, and uses of various Immunomodulators. In R. K. Kesharwani, R. K. Keservani, &

A. K. Sharma (Eds.), *Immunomodulators and Human Health* (pp. 3–38). Springer. https://doi.org/10.1007/978-981-16-6379-6_1.

Kumar, S., Saini, R., Suthar, P., Kumar, V., & Sharma, R. (2022). Plant secondary metabolites: Their food and therapeutic importance. In *Plant Secondary Metabolites: Physico-Chemical Properties and Therapeutic Applications* (pp. 371–413). Springer Nature Singapore.

Kumar, S., Sharma, S., Kumar, V., Sharma, A., Kaur, R., & Saini, R. (2023). Jamun (*Syzygium cumini* (L.) Skeels): The conventional underutilized multifunctional plant–An exotic gleam into its food and functional significance. *Industrial Crops and Products*, *191*, 115873.

Kumar, S., Sharma, S., Kumar, V., Sharma, R., Minhas, A., & Boddu, R. (2022). Cruciferous vegetables: A mine of phytonutrients for functional and nutraceutical enrichment. In *Current Advances for Development of Functional Foods Modulating Inflammation and Oxidative Stress* (pp. 401–426). Academic Press.

Kuna, A., & Poshadri, A. (2012). Phytosterol–A functional ingredient for the Indian food industry. *Indian Food Industry*, *30*(3), 21–28.

Laakso, P. (2005). Analysis of sterols from various food matrices. *European Journal of Lipid Science and Technology*, *107*(6), 402–410.

Lekakis, J., Rallidis, L. S., Andreadou, I., Vamvakou, G., Kazantzoglou, G., Magiatis, P., Skaltsounis, A. L., & Kremastinos, D. T. (2005). Polyphenols compounds from red grapes acutely improve endothelial function in patients with coronary heart disease. *European Journal of Preventive Cardiology*, *12*(6), 596–600.

Li, X., Xin, Y., Mo, Y., Marozik, P., He, T., & Guo, H. (2022). The bioavailability and biological activities of phytosterols as modulators of cholesterol metabolism. *Molecules*, *27*(2), 523.

Liebler, D. C. (1993). Antioxidant reactions of carotenoids A. *Annals of the New York Academy of Sciences*, *691*(1), 20–31.

Lin, D., Xiao, M., Zhao, J., Li, Z., Xing, B., Li, X., Kong, M., Li, L., Zhang, Q., Liu, Y., & Chen, H. (2016). An overview of plant phenolic compounds and their importance in human nutrition and management of type 2 diabetes. *Molecules*, *21*(10), 374.

López-García, G., Cilla, A., Barberá, R., & Alegría, A. (2019). Anti-inflammatory and cytoprotective effect of plant sterol and galactooligosaccharides-enriched beverages in caco-2 cells. *Journal of Agricultural and Food Chemistry*, *68*(7), 1862–1870.

Loussouarn, M., Krieger-Liszkay, A., Svilar, L., Bily, A., Birtić, S., & Havaux, M. (2017). Carnosic acid and carnosol, two major antioxidants of rosemary, act through different mechanisms. *Plant Physiology*, *175*(3), 1381–1394.

Lunn, J., & Buttriss, J. L. (2007). Carbohydrates and dietary fiber. *Nutrition Bulletin*, *32*(1), 21–64.

Maggi, L., Carmona, M., Del Campo, C. P., Kanakis, C. D., Anastasaki, E., Tarantilis, P. A., Polissiou, M. G., & Alonso, G. L. (2009). Worldwide market screening of saffron volatile composition. *Journal of the Science of Food and Agriculture*, *89*(11), 1950–1954.

Manners, G. D. (2007). Citrus limonoids: Analysis, bioactivity, and biomedical prospects. *Journal of Agricultural and Food Chemistry*, *55*(21), 8285–8294.

Márquez-López, R. E., Loyola-Vargas, V. M., & Santiago-García, P. A. (2022). Interaction between fructan metabolism and plant growth regulators. *Planta*, *255*(2), 49.

Martin, V. J., Pitera, D. J., Withers, S. T., Newman, J. D., & Keasling, J. D. (2003). Engineering a mevalonate pathway in *Escherichia coli* for the production of terpenoids. *Nature Biotechnology*, *21*(7), 796–802.

Mattoo, A. K., Dwivedi, S. L., Dutt, S., Singh, B., Garg, M., & Ortiz, R. (2022). Anthocyanin-rich vegetables for human consumption-focus on potatoes, sweet potatoes, and tomatoes. *International Journal of Molecular Sciences*, *23*(5), 2634.

Matusheski, N. V., Juvik, J. A., & Jeffery, E. H. (2004). Heating decreases epithiospecifier protein activity and increases sulforaphane formation in broccoli. *Phytochemistry*, *65*(9), 1273–1281.

Melim, C., Lauro, M. R., Pires, I. M., Oliveira, P. J., & Cabral, C. (2022). The role of glucosinolates from cruciferous vegetables (Brassicaceae) in gastrointestinal cancers: From prevention to therapeutics. *Pharmaceutics*, *14*(1), 190.

Milani, A., Basirnejad, M., Shahbazi, S., & Bolhassani, A. (2017). Carotenoids: Biochemistry, pharmacology and treatment. *British Journal of Pharmacology*, *174*(11), 1290–1324.

Minto, R. E., & Blacklock, B. J. (2008). Biosynthesis and function of polyacetylenes and allied natural products. *Progress in Lipid Research*, *47*(4), 233–306.

Mirończuk-Chodakowska, I., Kujawowicz, K., & Witkowska, A. M. (2021). Beta-glucans from fungi: Biological and health-promoting potential in the COVID-19 pandemic era. *Nutrients*, *13*(11), 3960.

Mishra, H., Kesharwani, R. K., Singh, D. B., Tripathi, S., Dubey, S. K., & Misra, K. (2019). Computational simulation of inhibitory effects of curcumin, retinoic acid, and their conjugates on GSK-3 beta. *Network Modeling Analysis in Health Informatics and Bioinformatics*, *8*(1), 3.

Miura, T., Muraoka, S., Ikeda, N., Watanabe, M., & Fujimoto, Y. (2000). Antioxidative and prooxidative action of stilbene derivatives. *Pharmacology and Toxicology*, *86*(5), 203–208.

Mojica, L., Meyer, A., Berhow, M. A., & de Mejía, E. G. (2015). Bean cultivars (*Phaseolus vulgaris* L.) have similar high antioxidant capacity, *in vitro* inhibition of α-amylase and α-glucosidase while diverse phenolic composition and concentration. *Food Research International*, *69*, 38–48.

Molyneux, R. J., Lee, S. T., Gardner, D. R., Panter, K. E., & James, L. F. (2007). Phytochemicals: The good, the bad and the ugly? *Phytochemistry*, *68*(22–24), 2973–2985.

Morrison, D. J., & Preston, T. (2016). Formation of short-chain fatty acids by the gut microbiota and their impact on human metabolism. *Gut Microbes*, *7*(3), 189–200.

Mulvihill, E. E., & Huff, M. W. (2010). Antiatherogenic properties of flavonoids: Implications for cardiovascular health. *Canadian Journal of Cardiology*, *26*, 17A–21A.

Murthy, K. C., Jayaprakasha, G. K., Safe, S., & Patil, B. S. (2021). Citrus limonoids induce apoptosis and inhibit the proliferation of pancreatic cancer cells. *Food and Function*, *12*(3), 1111–1120.

Mussatto, S. I., & Mancilha, I. M. (2007). Nondigestible oligosaccharides: A review. *Carbohydrate Polymers*, *68*(3), 587–597.

National Toxicology Program. (1990). *National Toxicology Program Technical Report on the Toxicology and Carcinogenesis Studies of D-Limonene (CAS No. 5989-27-5) in F344/N Rats and B6C3F1 Mice (Gavage Studies).* (NTP TR 347, NIH Publication No. 90-2802).

Nelson, A. L. (2001). High-fiber properties and analyses. *High-Fiber Ingredients*.

Nicholson, S. K., Tucker, G. A., & Brameld, J. M. (2008). Effects of dietary polyphenols on gene expression in human vascular endothelial cells. *Proceedings of the Nutrition Society*, *67*(1), 42–47.

Nugent, A. P. (2005). Health properties of resistant starch. *Nutrition Bulletin*, *30*(1), 27–54.

Ockermann, P., Headley, L., Lizio, R., & Hansmann, J. (2021). A review of the properties of anthocyanins and their influence on factors affecting cardiometabolic and cognitive health. *Nutrients*, *13*(8), 2831.

Panche, A. N., Diwan, A. D., & Chandra, S. R. (2016). Flavonoids: An overview. *Journal of Nutritional Science*, *5*, 47.

Pappachan, M. J. (2011). Increasing prevalence of lifestyle diseases: High time for action. *The Indian Journal of Medical Research*, *134*(2), 143.

Parage, C., Tavares, R., Réty, S., Baltenweck-Guyot, R., Poutaraud, A., Renault, L., Heintz, D., Lugan, R., Marais, G. A., Aubourg, S., & Hugueney, P. (2012). Structural, functional, and evolutionary analysis of the unusually large stilbene synthase gene family in grapevine. *Plant Physiology*, *160*(3), 1407–1419.

Pascal, S. L., & Segal, R. (2006). Phytosterols–biologically active compounds in food. *Journal of Agroalimentary Processes and Technologies*, *12*(1), 149–158.

Petrovski, G., Gurusamy, N., & Das, D. K. (2011). Resveratrol in cardiovascular health and disease. *Annals of the New York Academy of Sciences*, *1215*(1), 22–33.

Piironen, V., Toivo, J., Puupponen-Pimiä, R., & Lampi, A. M. (2003). Plant sterols in vegetables, fruits, and berries. *Journal of the Science of Food and Agriculture*, *83*(4), 330–337.

Raes, K., Doolaege, E. H., Deman, S., Vossen, E., & De Smet, S. (2015). Effect of carnosic acid, quercetin and α-tocopherol on lipid and protein oxidation in an *in vitro* simulated gastric digestion model. *International Journal of Food Sciences and Nutrition*, *66*(2), 216–221.

Rahman, I., Biswas, S. K., & Kirkham, P. A. (2006). Regulation of inflammation and redox signaling by dietary polyphenols. *Biochemical Pharmacology*, *72*(11), 1439–1452.

Ramawat, K. G., Dass, S., & Mathur, M. (2009). The chemical diversity of bioactive molecules and therapeutic potential of medicinal plants. *Herbal Drugs: Ethnomedicine to Modern Medicine*, 7–32.

Randhir, R., Lin, Y. T., & Shetty. Stimulation of phenolics, antioxidant, and antimicrobial activities in dark germinated mung bean sprouts in response to peptide and phytochemical elicitors. *Process Biochemistry*, *39*(5), 637–646.

Rane, B. R., Patil, R. R., Jain, A. S., Keservani, R. K., & Kesharwani, R. K. (2022). Nutraceuticals as disease preventive food and immunity boosters. In R. K. Kesharwani, R. K. Keservani, & A. K. Sharma (Eds.), *Nutraceuticals and Functional Foods in Immunomodulators*. Springer. https://doi.org/10.1007/978-981-19-2507-8_7.

Reinisalo, M., Kårlund, A., Koskela, A., Kaarniranta, K., & Karjalainen, R. O. (2015). Polyphenol stilbenes: Molecular mechanisms of defense against oxidative stress and aging-related diseases. *Oxidative Medicine and Cellular Longevity*.

Rivero-Urgell, M., & Santamaria-Orleans, A. (2001). Oligosaccharides: Application in infant food. *Early Human Development*, *65*, S43–S52.

Robbins, R. J. (2003). Phenolic acids in foods: An overview of analytical methodology. *Journal of Agricultural and Food Chemistry*, *51*(10), 2866–2887.

Roupe, K. A., Remsberg, C. M., Yáñez, J. A., & Davies, N. M. (2006). Pharmacometrics of stilbenes: Seguing towards the clinic. *Current Clinical Pharmacology*, *1*(1), 81–101.

Roy, A., & Saraf, S. (2006). Limonoids: Overview of significant bioactive triterpenes distributed in plants kingdom. *Biological and Pharmaceutical Bulletin*, *29*(2), 191–201.

Saini, R. K., Ranjit, A., Sharma, K., Prasad, P., Shang, X., Gowda, K. G. M., & Keum, Y. S. (2022). Bioactive compounds of citrus fruits: A review of composition and health benefits of carotenoids, flavonoids, limonoids, and terpenes. *Antioxidants, 11*(2), 239.

Sajilata, M. G., Singhal, R. S., & Kulkarni, P. R. (2006). Resistant starch–A review. *Comprehensive Reviews in Food Science and Food Safety, 5*(1), 1–17.

Salehi, B., Quispe, C., Sharifi-Rad, J., Cruz-Martins, N., Nigam, M., Mishra, A. P., Konovalov, D. A., Orobinskaya, V., Abu-Reidah, I. M., Zam, W., & Sharopov, F. (2021). Phytosterols: From preclinical evidence to potential clinical applications. *Frontiers in Pharmacology, 11*, 599959.

Sauer, L., Li, B., & Bernstein, P. S. (2019). Ocular carotenoid status in health and disease. *Annual Review of Nutrition, 39*, 95–120.

Saxena, M., Saxena, J., Nema, R., Singh, D., & Gupta, A. (2013). Phytochemistry of medicinal plants. *Journal of Pharmacognosy and Phytochemistry, 1*(6), 168–182.

Scalbert, A., & Williamson, G. (2008). Dietary intake and bioavailability of polyphenols. *The Journal of Nutrition, 130*(8), 2073S–2085S.

Sharma, A., Yadav, B. S., & Ritika. (2008). Resistant starch: Physiological roles and food applications. *Food Reviews International, 24*(2), 193–234.

Sharma, P., Jha, A. B., Dubey, R. S., & Pessarakli, M. (2012). Reactive oxygen species, oxidative damage, and antioxidative defense mechanism in plants under stressful conditions. *Journal of Botany, 2012*, 26.

Sharma, S., Kumar, S., Kumar, V., & Sharma, R. (2021). Pesticides and vegetables: Ecological and metabolic fate with their field and food significance. *International Journal of Environmental Science and Technology, 20*(3), 1–26.

Singh, D. B., Gupta, M. K., Kesharwani, R. K., & Misra, K. (2013a). Comparative docking and ADMET study of some curcumin derivatives and herbal congeners targeting β-amyloid. *Network Modeling Analysis in Health Informatics and Bioinformatics, 2*(1), 13–27.

Singh, D. V., Agarwal, S., Kesharwani, R. K., & Misra, K. (2013b). 3D QSAR and pharmacophore study of curcuminoids and curcumin analogs: Interaction with thioredoxin reductase. *Interdisciplinary Sciences: Computational Life Sciences, 5*(4), 286–295.

Singh, P., Kesharwani, R. K., Misra, K., & Rizvi, S. I. (2015). The modulation of erythrocyte Na+/K+-ATPase activity by curcumin. *Journal of Advanced Research, 6*(6), 1023–1030.

Singh, P., Kesharwani, R. K., Misra, K., & Rizvi, S. I. (2016). Modulation of erythrocyte plasma membrane redox system activity by curcumin. *Biochemistry research international*, 702039.

Son, M. J., Rico, C. W., Nam, S. H., & Kang, M. Y. (2010). Influence of oryzanol and ferulic acid on the lipid metabolism and antioxidative status in high fat-fed mice. *Journal of Clinical Biochemistry and Nutrition, 46*(2), 150–156.

Southgate, D. A. (1995). The diet is a source of dietary fiber. *European Journal of Clinical Nutrition, 49*, S22–S26.

Srinivasan, M., Sudheer, A. R., & Menon, V. P. (2007). Ferulic acid: Therapeutic potential through its antioxidant property. *Journal of Clinical Biochemistry and Nutrition, 40*(2), 92–100.

Sun, L. C., Li, S. Y., Wang, F. Z., & Xin, F. J. (2007). Research progresses in the synthetic biology of terpenoids. *Biotechnology Bulletin, 33*(1), 64.

Sun, T., Rao, S., Zhou, X., & Li, L. (2022). Plant carotenoids: Recent advances and future perspectives. *Molecular Horticulture, 2*(1), 3.

Suzuki, T., Kusano, K., Kondo, N., Nishikawa, K., Kuge, T., & Ohno, N. (2021). Biological activity of high-purity β-1, 3-1, 6-glucan derived from the black yeast *Aureobasidium pullulans*: A literature review. *Nutrients, 13*(1), 242.

Thakur, A., & Sharma, R. (2018). Health-promoting phytochemicals in vegetables: A mini-review. *International Journal of Food and Fermentation Technology*, *8*(2), 107–117.

Thakur, A., Joshi, V. K., & Thakur, N. S. (2019). Immunology and its relation with food components: An overview. *International Journal of Food and Fermentation Technology*, *9*(1), 1–16.

Tharanathan, R. N. (2022). Food-derived carbohydrates—Structural complexity and functional diversity. *Critical Reviews in Biotechnology*, *22*(1), 65–84.

Tsimogiannis, D., Choulitoudi, E., Bimpilas, A., Mitropoulou, G., Kourkoutas, Y., & Oreopoulou, V. (2017). Exploitation of the biological potential of *Satureja thymbra* essential oil and distillation by-products. *Journal of Applied Research on Medicinal and Aromatic Plants, 4*, 12–20.

Uedo, N., Tatsuta, M., Iishi, H., Baba, M., Sakai, N., Yano, H., & Otani, T. (1999). Inhibition by d-limonene of gastric carcinogenesis induced by N-methyl-N′-nitro-N-nitrosoguanidine in Wistar rats. *Cancer Letters*, *137*(2), 131–136.

Upadhyaya, J., Kesharwani, R. K., & Misra, K. (2009). Metabolism, pharmacokinetics, and bioavailability of ascorbic acid; synergistic effect with tocopherols and curcumin. *Journal of Computational Intelligence and Bioinformatics*, *2*(1), 77–84.

Van Agtmael, M. A., Eggelte, T. A., & Van Boxtel, C. J. (1999). Artemisinin drugs in the treatment of malaria: From medicinal herb to registered medication. *Trends in Pharmacological Sciences*, *20*(5), 199–205.

Velderrain-Rodríguez, G. R., Palafox-Carlos, H., Wall-Medrano, A., Ayala-Zavala, J. F., Chen, C. O., Robles-Sánchez, M., Astiazaran-García, H., Alvarez-Parrilla, E., & González-Aguilar, G. A. (2014). Phenolic compounds: Their journey after intake. *Food and Function*, *5*(2), 189–197.

Vinson, J. A., Su, X., Zubik, L., & Bose, P. (2001). Phenol antioxidant quantity and quality in foods: Fruits. *Journal of Agricultural and Food Chemistry*, *49*(11), 5315–5321.

Vyas, N., Keservani, R. K., Nayak, A., Jain, S., & Singhal, M. (2010). Effect of *Tamarindus indica* and *Curcuma longa* on stress-induced Alopecia. *Pharmacologyonline*, *1*, 377–384.

Walton, N. J., Mayer, M. J., & Narbad, A. (2003). Vanillin. *Phytochemistry*, *63*(5), 505–515.

Wang, Z., Zou, J., Cao, K., Hsieh, T. C., Huang, Y., & Wu, J. M. (2005). Dealcoholized red wine containing known amounts of resveratrol suppresses atherosclerosis in hypercholesterolemic rabbits without affecting plasma lipid levels. *International Journal of Molecular Medicine*, *16*(4), 533–540.

Wen, X., & Walle, T. (2006). Methylated flavonoids have greatly improved intestinal absorption and metabolic stability. *Drug Metabolism and Disposition*, *34*(10), 1786–1792.

Williamson, G. (2017). The role of polyphenols in modern nutrition. *Nutrition Bulletin*, *42*(3), 226–235.

Yang, W. Z., Beauchemin, K. A., & Rode, L. M. (2001). Barley processing, forage: Concentrate, and forage length effects on chewing and digesta passage in lactating cows. *Journal of Dairy Science, 84*(12), 2709–2720.

Yashin, A., Yashin, Y., Xia, X., & Nemzer, B. (2017). Antioxidant activity of spices and their impact on human health: A review. *Antioxidants*, *6*(3), 70.

Yeshi, K., Crayn, D., Ritmejerytė, E., & Wangchuk, P. (2022). Plant secondary metabolites produced in response to abiotic stresses have potential applications in pharmaceutical product development. *Molecules*, *27*(1), 313.

Yu, W., Simmons-Menchaca, M., Gapor, A., Sanders, B. G., & Kline, K. (1999). Induction of apoptosis in human breast cancer cells by tocopherols and tocotrienols. *Nutrition and Cancer, 33*(1), 26–32.

Yun, J. W. (2010). Possible anti-obesity therapeutics from nature–A review. *Phytochemistry, 71*(14, 15), 625–1641.

Zern, T. L., West, K. L., & Fernandez, M. L. (2003). Grape polyphenols decrease plasma triglycerides and cholesterol accumulation in the aorta of ovariectomized guinea pigs. *The Journal of Nutrition, 133*(7), 2268–2272.

Zhang, Y. J., Gan, R. Y., Li, S., Zhou, Y., Li, A. N., Xu, D. P., & Li, H. B. Antioxidant phytochemicals for the prevention and treatment of chronic diseases. *Molecules, 20*(12), 21138–21156.

Zhao, C., Jia, Y., & Lu, F. (2018). Angelica stems A potential low-cost source of bioactive phthalides and phytosterols. *Molecules, 23*(12), 3065.

Zidorn, C., Jöhrer, K., Ganzera, M., Schubert, B., Sigmund, E. M., Mader, J., Greil, R., Ellmerer, E. P., & Stuppner, H. (2005). Polyacetylenes from the Apiaceae vegetables carrot, celery, fennel, parsley, and parsnip and their cytotoxic activities. *Journal of Agricultural and Food Chemistry, 53*(7), 2518–2523.

Zimmermann, A., Visscher, C., & Kaltschmitt, M. (2021). Plant-based fructans for increased animal welfare: Provision processes and remaining challenges. *Biomass Conversion and Biorefinery, 13*(Suppl 2), 1–19.

Žugić, A., Đorđević, S., Arsić, I., Marković, G., Živković, J., Jovanović, S., & Tadić, V. (2014). Antioxidant activity and phenolic compounds in 10 selected herbs from Vrujci Spa, Serbia. *Industrial Crops and Products, 52,* 519–527.

CHAPTER 2

Millets as Functional Food

ASHUTOSH SHARMA,[1] AJAY SHARMA,[2] PARAS SHARMA,[3] A. HARSHAVARDHAN REDDY,[1] and INDU SHARMA[4*]

[1]*Faculty of Agricultural Sciences, DAV University, Jalandhar, Punjab, India*

[2]*Department of Chemistry, Career Point University, Hamirpur, Himachal Pradesh, India*

[3]*Food Chemistry Division, ICMR–National Institute of Nutrition, Hyderabad, Telangana, India*

[4]*Department of Botany, Sant Baba Bhag Singh University, Jalandhar, Punjab, India*

*Corresponding author.

ABSTRACT

Millets are the plants of the family Poaceae, bearing small-seeded grains. Due to their low water requirements, they are grown in less fertile and marginal soils. Compared to traditional cereals like rice, wheat, and maize, they relatively have more phytochemical diversity owing to their bio-active properties. They are gluten-free, fiber-rich, antioxidant-rich foods that can be grown well even under rain-fed conditions and have a short maturity time. The relatively high content of dietary fiber can be used as a potential source of prebiotics. The dietary fiber present in them is considered important for reducing the glycemic index of the food. Many types of millet have different hermetically active properties like antioxidant, antimicrobial, antidiabetic, anti-inflammatory, cyto-protective, anti-lithiatic, cholesterol-lowering, and anti-cancer properties. Nowadays, millet is also used for weight management. The present chapter, therefore, deals with the classification of millets, their bio-active properties, their chemical profile, and their role in fighting

Functional Compounds and Foods of Plant Origin:Krishan Dutt Sharma, Satish Kumar, Rakesh Sharma, and Raj K. Keservani (Eds.)

malnutrition, especially under rain-fed conditions. The present chapter presents the millets as an excellent candidate for functional foods, due to their rich plant secondary metabolic profile.

2.1 INTRODUCTION

Millets are the annual, warm-weather, small-grained, coarse cereals belonging to the family Poaceae, which can flourish well under rainfed adverse-climatic conditions and are more commonly cultivated by tribal or small-scale farmers (Banerjee and Maitra, 2020; Bora et al., 2019). Being, C-4 plants, millets combat harsh climatic conditions, and they can easily adapt to various climatic changes. Therefore, they can fulfill the goal of ecologically sustainable nutrition and food security even under harsh climatic conditions (Maitra, 2020). Millets have also been documented to have grown in East-Asian regions for the last 10,000 years. Since ancient civilizations, millets have been the staple food throughout Asia, Africa, and Europe (Banerjee and Maitra, 2020). Proso, pearl, Kodo, finger, foxtail, and little millet are examples of commercially cultivated millets (Singh et al., 2019). In some developed countries, proso millet is used as bird seed, while in certain regions in Asia, it is used as food cereal. The foxtail millet is extensively grown in Europe and China. In the high-altitude cold regions of Asia and Africa, finger millet is regularly cultivated, which is also a crucial constituent of traditional beer.

In the Indian and African regions, pearl millet is a widely cultivated crop, whereas, in the USA, proso millet is the most cultivated millet that is used as a bird seed (Mall and Tripathi, 2016). Although millets are indigenously cultivated in various regions throughout the world, they are considered to have originated in the tropical West African region. These are also cultivated as a source of staple food, mainly on the two continents of Asia and Africa. Millets possess relatively low amounts of phytic acids, but a higher amount of dietary fibers, vitamins, calcium (Ca), and iron. Proso millet with a short growth period (10–11 weeks), the crop can be efficiently grown in barren or saline soils by effectively utilizing water under high temperatures and drought conditions (Wang et al., 2021). It is the 6th most vital cultivated food crop around the globe. In India, millet is an important ancient crop that was cultivated. Compared with traditional cereals, millet is rich in some of the dietary constituents. Therefore, they have recently emerged as "Nutri-cereals," owing to their richness in dietary value (Maitra, 2020). They contain high contents of minerals, proteins, antioxidants, polyphenols, phenolic acids, polyphenols, flavonoids, and vitamins as well as they have been reported to be effective in the management of various non-communicable diseases (NCDs) (Singh et al., 2019). Owing to the presence of micro-nutrients, the millets may be mixed in various baked food products, such as biscuits, bread, cakes, cookies, and pastries, or may be employed for the biofortification of staple crops to overcome the micro-nutrient deficiency.

A healthy alternative to traditional, regular food is functional food, which is gaining more popularity in modern society for a disease-free and healthier life. Recently, millets have also emerged as a healthy alternative to the existing commonly consumed cereals, with better biochemistry, thereby securing a strong potential of a well-accepted functional food candidate for the future. However, the product type, its stability, compatibilities, regional acceptance, and consumer behavior also determine the potential of a broad acceptance and recognition of functional food. Many groups are trying to blend the nutraceutical properties of functional food with other food products to further increase their acceptability (Keservani et al., 2010a, b, 2017, 2020). The cookies prepared with flour from foxtail millet mixed with flying fish have also been used as food with higher functional food properties (Patimah et al., 2019).

According to Doyon and Labrecque (2008), functional food mimics conventional food and is often consumed as part of a regular diet with proven health benefits. It can interfere with the incidences of chronic diseases or can be beneficial for certain conditions beyond basic nutritional functions. Therefore, functional food may be a missing link between conventional food and the traditional medicine system, which should be consumed adequately for a healthy lifestyle, keeping diseases at bay. The functional foods have been categorized as (i) natural functional foods that can be consumed immediately without any processing, such as vegetables and fruits, (ii) traditional functional foods that can be consumed after processing in a traditional way such as tempeh, tofu, and herbs (Rane et al., 2020a, b, c, 2023; Sharma et al., 2020); and (iii) modern functional foods that are prepared by using novel ingredients in more innovative ways (Patimah et al., 2019). In this chapter, the different types of millets, their nutritional values, major bioactive constituents, pharmaceutical properties, and their potential as nutria-cereals have been discussed.

2.2 MILLET TYPES AND THEIR NUTRITIONAL VALUE

Millets are considered potential functional foods because they contain valuable nutritional and hypoglycemic ingredients, owing to their rich nutrient profiles and improved starch digestibility (Bora et al., 2019). Millets are recognized as a rich source of lipids (11.5–31.7%), insoluble dietary fibers (9.3–56.7 mg/g), minerals (2.1–8.0%), unsaturated fatty acids (such as linoleic and oleic acids), and phenolic acids (both free and bound), though these constituents may vary by millet type. Millets comprise extremely variable, small-seeded grasses broadly classified into two categories: minor (or small) and major millets (Singh et al., 2019).

The major millet types include *Sorghum bicolor* L. (common name: Sorghum) and *Pennisetum typhoides* L. (common name: pearl millet). Minor or small millet types consist of *Eleusine coracana* L. Gaertn (common names: finger millet or ragi), *Echinochloa frumentacea* L. (common name: barnyard millet), *Setaria italica* L. (common name: foxtail or Italian millet), *Paspalum scrobiculatum* L. (common name: Kodo millet), *Panicum sumatrense* L. (common name: little millet), *Panicum miliaceum* L. (common name: proso millet), and *Brachiaria ramosa* L. Stapf (synonym: *Panicum ramosum* L., common name: brown-top millet). Various millet varieties are generally grouped into these two categories (Mall and Tripathi, 2016), as shown in Figure 2.1.

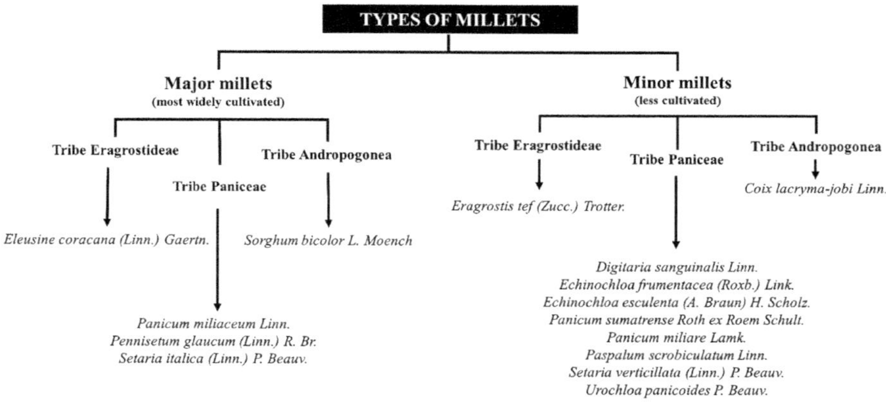

FIGURE 2.1 Different types of millets are classified as major and minor millets.

Although millets are considered a rich source of dietary fibers, however, its quantity and quality may vary among different millets. The dietary fiber content of finger millets is higher than in barnyard millet, foxtail millet, little millet, Kodo millet, polished rice, and brown rice (Lansakara et al., 2016; Shobana et al., 2013).

In certain aspects, millets are considered better than the commonly consumed cereals. The minor millets have been documented to be nutritionally superior to wheat and rice (Maitra 2020). In an investigation, when the nutritional composition, α-amylase inhibitory activity, fermentation ability, and antioxidant activity of two varieties of finger millets (Rawana and Oshada) were compared with two rice varieties (Basmati and BG-300), many differences were recorded. It was revealed that the contents of soluble and insoluble dietary fibers, as well as the total flavonoids (Keservani and Sharma, 2014; Keservani et al., 2016a, b), were significantly higher in the finger millet varieties as compared to the rice varieties. However, a variation in terms of dietary fiber content was recorded within varieties of finger millet (Lansakara et al., 2016). It has been documented that maximum levels of free and bound phenolic acids are present in Kodo millet (Bora et al., 2019). Both linoleic and oleic acids were present in almost all the millet types. With the presence of a low glycemic index (GI) (42.7–58.3), the millets may be used by diabetic patients, and therefore, the millets are promising functional food ingredients. Further, millet proteins are known to have a higher content of essential amino acids, including sulfur-containing amino acids (Geervani and Eggum, 1989).

In the foxtail millet, husk and bran layers contribute to about 13.5% and 1.5–2% of the total content, respectively (Liang and Liang, 2019). The millet has a good volume of carbohydrates (60–70%), proteins (7–11%), fat (1.5–5%), vitamins, and minerals. Among millets, foxtail millet has the highest lysine (an essential amino acid) content, whereas proso millet has more isoleucine, leucine, and methionine content. Starch is a polysaccharide carbohydrate that is made up of amylose and amylopectin. Proso millet has the maximum amylose content, followed by Sorghum (24.0%), Kodo millet (24.0%), pearl millet (21.0%), foxtail millet (17.5%), and the least amylose content is present in finger millets (16%) (Banerjee and Maitra, 2020). The proteins present in the proso millet w139 variety have better amounts of amino acids and superior functional properties, namely, foaming capacity, emulsifying properties, and solubility (Wang et al., 2021). Further, treatment with nitrogen fertilizer enhanced the levels of hydrophobic amino acids, increased water/oil absorption capacity, high β-sheet proportion in the protein, and greater thermal stability.

The phenolic extracts of different millets, that is, Kodo millet, finger millet, foxtail millet, proso millet, pearl millet, and little millets, have been reported to contain good amounts of ferulic and p-coumaric acids

(Chandrasekara and Shahidi, 2010). The insoluble bound phenolic fractions reported higher volumes of ferulic and p-coumaric acids. Among these millets, Kodo millets possess maximum total phenolic content (TPC), and white proso millet contains the least TPC. In an experiment, it was found that the finger millet possesses the Xylan, which is comparable to rice bran (Palaniappan et al., 2017). A comparative study on the phytochemical and antioxidant profile of both aqueous and methanolic extracts of flours of two millet varieties (MB-87 and AF-POP) demonstrated that flavonoids, saponins, terpenoids, and tannins were present in both of them, but the methanolic extract of millet variety MB-87 possessed a better antioxidant profile (Nosheen et al., 2020).

2.3 FUNCTIONAL PROPERTIES OF MILLETS

Millets are the natural sources of antioxidants, anti-inflammatory, cytoprotective, and anti-lithiatic agents (Bahuguna et al., 2009; Jayawardana et al., 2021). The regular consumption of millet may be useful in the dietary management and prevention of oxidative stress-associated diseases in human beings (Jayawardana et al., 2021). Burgeoning studies have indicated that due to the presence of various bioactive constituents, millets possess different therapeutic properties, thereby making them a futuristic nutraceutical. The major bio-active properties of some important millets have been listed in Table 2.1.

In the experimental studies carried out on male albino rats, the aqueous and alcohol extracts of *Eleusine coracana* Linn. were found to be effective against calcium oxalate nephrolithiasis (Bahuguna et al., 2009). When supplemented with ethylene glycol, it resulted in hyperoxaluria and enhanced renal excretion of calcium and phosphate in rats. However, further supplementation with *E. coracona* extracts resulted in a decline in the increased urinary oxalate, thereby suggesting its role in regulating endogenous oxalate synthesis. Both prophylactic and therapeutic supplementations of *E. coracana* extracts revealed their inhibitory impact on crystal growth with improved kidney function, cyto-protective, and anti-lithiatic potential. The phenolic extracts of different millets are rich in phenolic compounds and antioxidants (Chandrasekara and Shahidi, 2010). A recent study highlighted the anti-inflammatory and antioxidant properties of extracts of three varieties (Ravi, Rawana, and Oshadha) of finger millet (Jayawardana et al., 2021). The finger millet varieties have

TABLE 2.1 Important Bioactive Properties of Millets to Act as Functional Foods

SL. No.	Millet	Scientific Name	Vernacular Name (Local Name)	Bioactive Properties	References
1.	Jowar	*Sorghum bicolor* (L.) Moench subsp. *bicolor*	Great millet, Egyptian millet, Guinea corn, jowar, juwar, dari, jola, jonna, feterita, durra, cholam, dura, jwari, kafir corn, solam, mtama, kaoliang, shallu, milo, and gaoliang	Antimicrobial	Radhajeyalakshmi et al. (2003)
				Antioxidant activity	Hegde et al. (2005)
2.	Finger millet	*Eleusine coracana* (L.) Gaertn.	Ragi, Hunsa, Marua, wimbi, Mandua, Nagli, Nachni, African bird's foot, Kapai, rapoko, telebun, koracan, kurakkan, and bulo.	Antioxidant activity, alpha (α) amylase inhibitory activity.	Hegde et al. (2005); Lansakara et al. (2016)
				Antimicrobial	Siwela et al. (2010); Viswanath et al. (2009)
				Antilithiatic effect	Bahuguna et al. (2009)
				Management of hyperglycemia	Kumari and Sumathi (2002)
				Antidiabetic, hypocholesterolemic, nephroprotective, and anticataractogenic properties.	Mathangi and Sudha (2012); Shobana et al. (2010)
				Cataractogenesis Inhibition	Chethan et al. (2008)
				Wound healing	Rajasekaran et al. (2004)
				Anti-inflammatory	Jayawardana et al. (2021)
				Anti-tumerogenic, atherosclerogenic	Mathangi and Sudha (2012)
				Protection against aging	Hegde et al. (2002)
				Prevention of cancer initiation	Chandrasekara and Shahidi (2011)
				Control of lipid peroxidation in human LDL cholesterol	Chandrasekara and Shahidi (2012)
				Prevention of cardiovascular disease (CVD)	Lee et al. (2010)

TABLE 2.1 *(Continued)*

SL. No.	Millet	Scientific Name	Vernacular Name (Local Name)	Bioactive Properties	References
3.	Pearl millet	*Pennisetum glaucum* (L.) R. Br. [synonym = *P. americanum* (L.)]	Bajra, bulrush, cattail, munga, candlestick, seno, and sanyo.	Antioxidant activity	Chandrasekara and Shahidi (2010)
				Antimicrobial	Radhajeyalakshmi et al. (2003)
				Prevention of cancer initiation	Chandrasekara and Shahidi (2011)
				Control of lipid peroxidation in human LDL cholesterol.	Chandrasekara and Shahidi (2012)
4.	Foxtail millet	*Setaria italica* (L.) P. Beauv. ssp. *Italica*	Kauni, korra, kangni, tenai, Italian, rala, Hungarian, German, and sibe.	Antioxidant activity	Amadou et al. (2011); Chandrasekara and Shahidi (2010); Hegde et al. (2005); Mohamed et al. (2012); Suma and Urooj (2011)
				Antidiabetic activity	Senevirathne et al. (2021)
				Antimicrobial	Radhajeyalakshmi et al. (2003); Xu et al. (2011)
				Prevention of cancer initiation	Chandrasekara and Shahidi (2011)
				Control of lipid peroxidation in human LDL cholesterol	Chandrasekara and Shahidi (2012)
				Antihyperglycemic activity	Sireesha et al. (2011)
5.	Little millet	*Panicum sumatrense* Roth ex Roem. & Schult. subsp. *sumatrense* (syn. *P. miliare* auct. pl.)	Kutki, same, samalu, heenmeneri, samai, and blue panic.	Antioxidant activity	Chandrasekara and Shahidi (2010); Hegde et al. (2005)
				Antimicrobial	Radhajeyalakshmi et al. (2003)
				Prevention of cancer initiation	Chandrasekara and Shahidi (2011)
				Control of lipid peroxidation in human LDL cholesterol	Chandrasekara and Shahidi (2012)

TABLE 2.1 *(Continued)*

SL. No.	Millet	Scientific Name	Vernacular Name (Local Name)	Bioactive Properties	References
6.	Proso millet	*Panicum miliaceum* L. subsp. *Miliaceum*	Panivaragu, cheena, variga, hog, baragu, common broom, samai, Russian, panic, and mahameneri.	Hepatoprotective and reduced risk of elevated serum cholesterol	Nishizawa and Fudamo (1995)
				Antioxidant activity	Chandrasekara and Shahidi (2010)
				Antidiabetic activity	Park et al. (2008); Senevirathne et al. (2021)
				Antimicrobial	Radhajeyalakshmi et al. (2003)
				Prevention of cancer initiation	Chandrasekara and Shahidi (2011)
				Control of lipid peroxidation in human LDL cholesterol	Chandrasekara and Shahidi (2012)
				Prevention of CVD	Lee et al. (2010)
7.	Kodo millet	*Paspalum scrobiculatum* L.	Varagu, Indian paspalum, bastard, water couch, ditch, creep, and naraka.	Antioxidant activity	Chandrasekara and Shahidi (2010); Hegde et al. (2005)
				Antidiabetic activity	Senevirathne et al. (2021)
				Protection against aging	Hegde et al. (2002)
				Prevention of cancer initiation	Chandrasckara and Shahidi (2011)
				Control of lipid peroxidation in human LDL cholesterol	Chandrasekara and Shahidi (2012)
8.	Barnyard millet	*Echinochloa crus-galli* (L.) P. Beauv. [synonyms: *E. esculenta* (A. Braun), *E. frumentacea*]	Japanese, oodalu, jhingora, kudraivali, Korean, kweichou, sanwa, and sawan.	Antimicrobial	Radhajeyalakshmi et al. (2003)
				Lowering of blood glucose, serum cholesterol and triglycerides	Kumari and Thayumanavan (1997)
				Beneficial in type II diabetics	Ugare et al. (2011)
				Antioxidant activity	Hegde et al. (2005)

been reported to have significantly higher DPPH and ABTS scavenging activity and alpha (α) amylase inhibitory activity than rice varieties (Lansakara et al., 2016). Furthermore, *in vitro,* microbial fermentation studies suggested the relatively better health-promoting effects of finger millet than rice.

In another study, the methanolic extracts of five millet types (foxtail, finger, proso, Kodo, and white finger millets), and sweet Sorghum and Sorghum ICSV 112 were tested for their antidiabetic potential and their role in the management of diabetes-associated complications. A dose-dependent anti-amylase activity, early-stage anti-glycation activity, middle-stage anti-glycation potential, early-stage glycation reversing property, and middle-stage glycation reversing potential of the millet and sorghum varieties were confirmed in the study (Senevirathne et al., 2021). Besides diabetes, gluten intolerance is another problem, which can worsen by increasing the consumption of common cereal grains. However, it can be managed better by increasing the consumption of millet in the food. Gluten intolerance in some people may lead to celiac disease, an auto-immune disorder that is based on genetic predisposition. The only treatment for such patients is to follow a strict gluten-free diet. In commonly consumed cereals such as wheat, barley, rye, and some oat varieties, a complex mixture of proteins known as gluten is present. Therefore, some cereals (rice, Sorghum, and corn), minor cereals (millet, fonio, job's tears, and teff), and pseudo-cereals (amaranth, quinoa, and buckwheat) have emerged as potential alternative gluten-free grains for the patients with gluten-related health problems (Moreno et al., 2014).

Millets can be a better and more economical source of nutrition and environmentally more sustainable due to their hardy nature. They can be cultivated even in rainfed environments. Pearl millet grain contains about 70% starch, and therefore, it can be employed as a cheaper source of starch in comparison to other cereals (Punia et al., 2021). Moreover, its promising techno-functional characteristics (pasting properties, gelatinization, solubility, digestibility, and swelling power) and their modifications by physical, enzymatic, or mechanical methods improve the perspective of pearl millet starch as a functional food.

The functional food properties may also be improved further by fermentation. In an experiment, the solid-state fermentation method by employing a fungus, *Aspergillus sojae* was used to ferment pearl millet cultivar PUSA-415, and its effect was analyzed on the contents of phenolics and condensed tannins, protective properties on DNA damage, and antioxidant potential. The bio-transformed pearl millet hence obtained showed increased TPC, enhanced antioxidant potential, and greater DNA damage protection potential (Salar et al., 2017). Significant levels of α-amylase, xylanase, β-glucosidase, ascorbic acid, p-coumaric acid, and gallic acid were also observed in the bio-transformed *A. sojae* fermented pearl millet, emphasizing the role of fermentation in enhancing the nutraceutical properties of this millet.

2.4 BIOACTIVE CONSTITUENTS

Millets have been known to possess many bio-active properties. Efforts have been made to identify the bioactive constituents from the millets, which can be broadly categorized into primary (peptides/proteins, dietary fibers, unsaturated fatty acids, etc.) and secondary metabolites (mainly phenolic acids and flavonoids). Besides, they are also rich in some important minerals. The bioactive properties of millet-derived peptides/proteins have been revised elsewhere (Majid and Priyadarshini, 2020). Millets are also rich in free as well as bound phenolics, mainly phenolic acids, and flavonoids. Out of various millets, Kodo millet has the highest phenolic content (Bora et al., 2019). Diverse phenolic compounds contribute to the antioxidant properties of millets, and these phenolics, therefore, serve as an important functional food ingredient. In an experiment, the phenolic profile of proso millets was studied, and 12 phenolic derivatives in the free extract and 15 phenolic acid derivatives in the bound fraction were found. Further p-coumaric and ferulic acid were the important phenolics present (Yuan et al., 2022). The distribution of phenolics varies within the millet grains within different milling fractions. The bran layers are mostly the storage houses of the phenolics, and the bran of foxtail millet reported the highest free phenolic content, while the husk reported the highest bound phenolics. Further characterization revealed the presence of 18 phenolic compounds in the free phenolic extracts, while 13 were reported in the bound phenolic extracts in different parts of foxtail millet (Zhang et al., 2021a).

Structures of the major bio-active secondary metabolites present in millets have been presented in Figure 2.2 with their chemical classification. The distribution of major bioactive compounds present in different millets is shown in Figure 2.3.

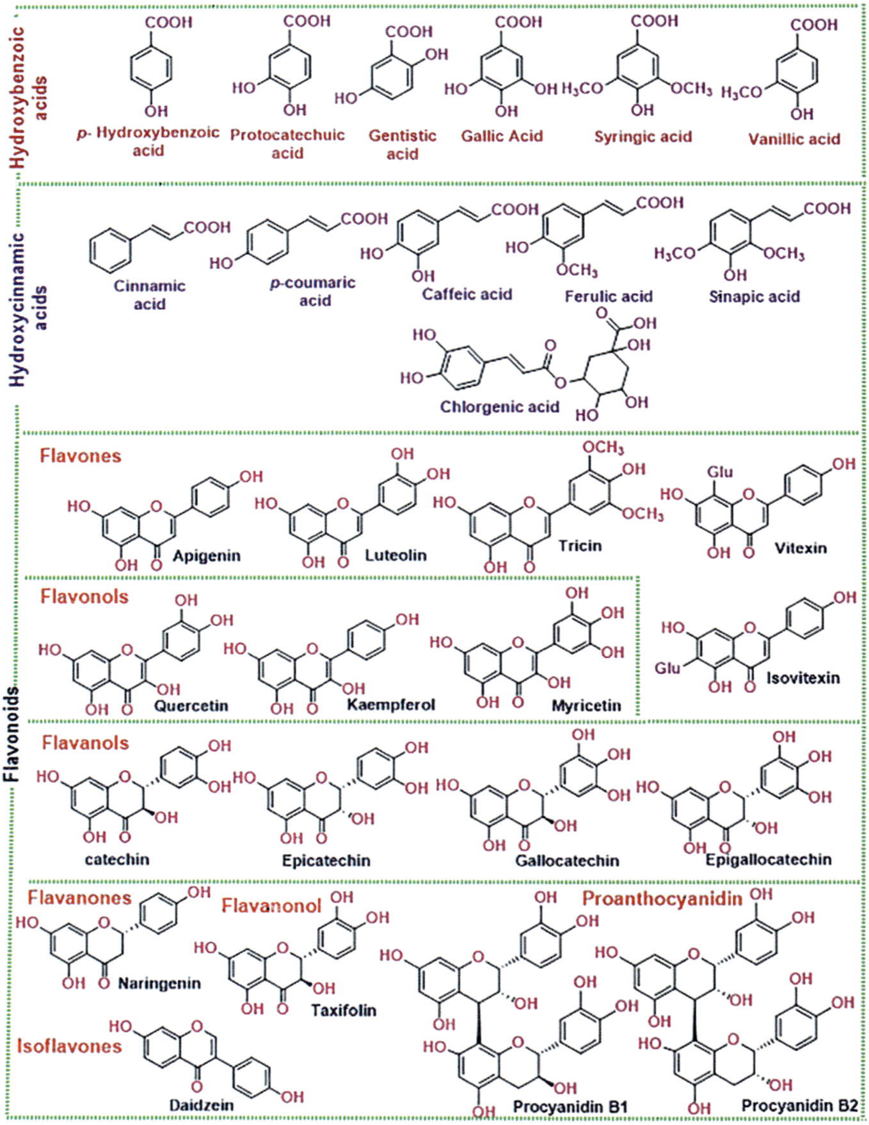

FIGURE 2.2 Classification of major bioactive constituents in millets that have nutraceutical properties.

Besides, these secondary metabolites, some millets have been studied as an important source of minerals. Calcium mal-nutrition is a major issue in elderly populations marginalized and poor people, young children, and

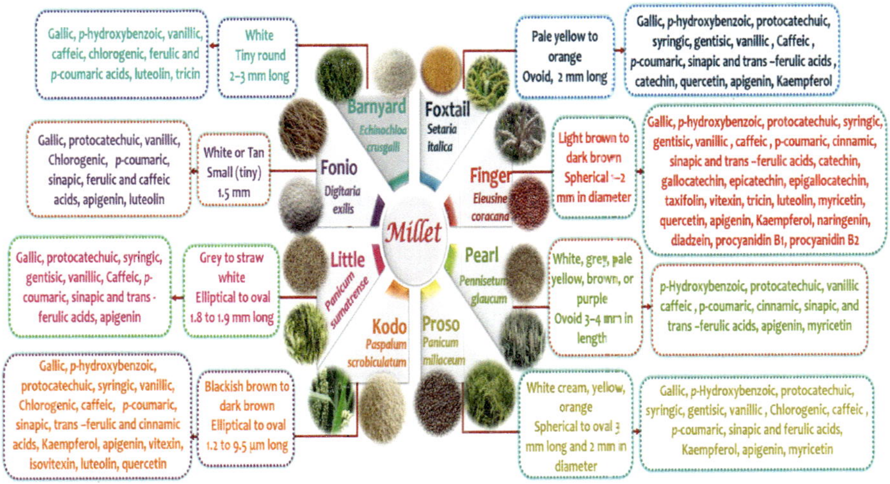

FIGURE 2.3 Type of millets with the major bioactive constituents.

pregnant and nursing women, who are relatively more risk-prone to calcium malnutrition. Severe deficiency may lead to conditions like osteoporosis. The dietary intake of calcium can be increased by consuming finger millet grains that are rich in calcium (Puranik et al., 2017; Singh et al., 2019).

2.5 MECHANISM OF ACTION OF SOME BIOACTIVE FROM MILLETS

Millets are rich in polyphenols. These polyphenols act as good antioxidants (Scalbert et al., 2005). It was found that millet shell polyphenols are helpful in the regulation of the gut barrier integrity and affect the gut microbiota in Apo E mice, inhibiting the development of atherosclerotic plaques, suggesting their potential cardiovascular protective role (Liu et al., 2021). It was found that the bound polyphenols from millet bran (p-coumaric and ferulic acid) can increase the bioactivity of oxaliplatin (a commonly used chemotherapy drug against colorectal cancer) as it effectively blocks the ganglioside GM3 catabolism. Therefore, millets may increase the effectiveness of such chemotherapy drugs too (Zhang et al., 2021b). Further, the immunomodulatory roles of phenolic acid-bound arabinoxylans extracted from little and Kodo millets have also been recorded (Arya et al., 2022; Keservani et al., 2015; Kumar et al., 2022; Rane et al., 2022; Srinivasan et al., 2021). Further, millet has been reported to release sugar slowly into the blood and lessen the absorption of glucose (Mall and Tripathi, 2016). Foods with a low GI (containing more slowly digestible or RS) are recommended for managing type-2 diabetes due to the slower release of glucose (Yadav et al., 2021). Some millet-based foods have a low GI, thereby making them safer for people suffering from type-2 diabetes (Arya and Shakya, 2021). Proso millet proteins have been documented to prevent liver damage by affecting cholesterol metabolism (Wang et al., 2021).

2.6 CONCLUSION AND FUTURE PROSPECTS

Among various cereal grains, the millets have been considered to have tremendous potential to sustain under poor resources and adverse ecological conditions. Due to their rich phytochemical diversity than commonly used cereals like rice, wheat, and maize, they have more bioactive properties. Due to their ability to grow in varied environments, even under rainfed conditions and other harsh environments, their cultivation is environmentally more sustainable. However, at present, the area under their cultivation is low, and they give lesser yield. Their varietal improvement has a lot of scope. Even then, they are quite helpful in ensuring nutritional and economic security because of their lower cost of production and better cultivation under various biotic or abiotic stress conditions. Since millet is gluten-free, it may serve as a better food option for people suffering from gluten intolerance. Also, because of their numerous medicinal properties owing to their rich phytochemical diversity, millets are a potential candidate for cereal-based medicines. Further exhaustive research needs to be carried out to connect the missing information to strategically devise the potential of the millet crops as Nutri-cereals and their implications in the food-based pharmaceutical industry.

KEYWORDS

- **antioxidant activity**
- **blood glucose**
- **calcium oxalate nephrolithiasis**
- **endogenous oxalate synthesis**
- **millets**
- **nutritional security**
- **phenolic compounds**
- **renal excretion**
- **therapeutic supplementation**
- **triglycerides**

REFERENCES

Amadou, I., Amza, T., Shi, Y. H., & Le, G. W. (2011). Chemical analysis and antioxidant properties of foxtail millet bran extracts. *Songklanakarin Journal of Science and Technology, 33*(5), 509–515.

Arya, R. K. K., Sati, D., Bisht, D., & Keservani, R. K. (2022). Nanotechnology-based bacterial immunotherapy. In R. K. Kesharwani, R. K. Keservani, & A. K. Sharma (Eds.), *Nutraceuticals and Functional Foods in Immunomodulators* (pp. 3–19). Springer. https://doi.org/10.1007/978-981-19-2507-8_1.

Bahuguna, Y. M., Rawat, M. S. M., Juyal, V., & Gnanarajan, G. (2009). Antilithiatic effect of grains of *Eleusine coracana. Saudi Pharmaceutical Journal, 17*(2), 182–188.

Banerjee, P., Maitra, S., & Banerjee, P. (2020). The role of small millet as a functional food to combat malnutrition in developing countries. *Indian Journal of Natural Sciences, 10*(60), 20412–20417.

Bora, P., Ragaee, S., & Marcone, M. (2019). Characterization of several types of millet as functional food ingredients. *International Journal of Food Sciences and Nutrition, 70*(6), 714–724.

Chandrasekara, A., & Shahidi, F. (2010). Content of insoluble bound phenolics in millets and their contribution to antioxidant capacity. *Journal of Agricultural and Food Chemistry, 58*(11), 6706–6714.

Chandrasekara, A., & Shahidi, F. (2011). Antiproliferative potential and DNA scission inhibitory activity of phenolics from whole millet grains. *Journal of Functional Foods, 3*(3), 159–170.

Chandrasekara, A., & Shahidi, F. (2012). Antioxidant phenolics of millet control lipid peroxidation in human LDL cholesterol and food systems. *Journal of the American Oil Chemists' Society, 89*, 275–285.

Chethan, S., Dharmesh, S. M., & Malleshi, N. G. (2008). Inhibition of aldose reductase from cataracted eye lenses by finger millet (*Eleusine coracana*) polyphenols. *Bioorganic and Medicinal Chemistry, 16*(23), 10085–10090.

Doyon, M., & Labrecque, J. (2008). Functional foods: A conceptual definition. *British Food Journal, 110*(11), 1133–1149.

Geervani, P., & Eggum, B. O. (1989). Nutrient composition and protein quality of minor millets. *Plant Foods for Human Nutrition, 39*, 201–208.

Hegde, P. S., Chandrakasan, G., & Chandra, T. S. (2002). Inhibition of collagen glycation and crosslinking *in vitro* by methanolic extracts of Finger millet (*Eleusine coracana*) and Kodo millet (*Paspalum scrobiculatum*). *The Journal of Nutritional Biochemistry, 13*(9), 517–521.

Hegde, P. S., Rajasekaran, N. S., & Chandra, T. S. (2005). Effects of the antioxidant properties of millet species on oxidative stress and glycemic status in alloxan-induced rats. *Nutrition Research, 25*(12), 1109–1120.

Jayawardana, S. A. S., Samarasekera, J. K. R. R., Hettiarachchi, G. H. C. M., Gooneratne, J., Choudhary, M. I., & Jabeen, A. (2021). Anti-inflammatory and antioxidant properties of finger Millet (*Eleusinecoracana* (L.) Gaertn.) varieties cultivated in Sri Lanka. *BioMed Research International, 2021*, 7744961.

Keservani, R. K., & Sharma, A. K. (2014). Flavonoids: Emerging trends and potential health benefits. *Journal of Chinese Pharmaceutical Sciences, 23*(12), 815.

Keservani, R. K., Kesharwani, R. K., Sharma, A. K., & Jarouliya, U. (2015d). Dietary supplements, nutraceuticals, and functional foods in immune response (immunomodulators). In D. Bagchi, H. G. Preuss, & A. Swaroop (Eds.), *Nutraceutical and Functional Foods in Human Life and Disease Prevention* (Chapter 20, pp. 343–358). CRC Press, Taylor & Francis.

Keservani, R. K., Kesharwani, R. K., Sharma, A. K., Vyas, N., & Chadoker, A. (2010b). Nutritional Supplements: An Overview. *International Journal of Current Pharmaceutical Review and Research, 1*(1), 59–75.

Keservani, R. K., Kesharwani, R. K., Vyas, N., Jain, S., Raghuvanshi, R., & Sharma, A. K. (2010a). Nutraceutical and functional food as future food: A review. *Der Pharmacia Lettre, 2*(1), 106–116.

Keservani, R. K., Sharma, A. K., & Kesharwani, R. K. (2016a). Nutraceutical and functional foods for cardiovascular health. In *Food Process Engineering* (pp. 291–312). Apple Academic Press, CRC Press.

Keservani, R. K., Sharma, A. K., & Kesharwani, R. K. (2016b). Medicinal effect of nutraceutical fruits for cognition and brain health. *Scientifica, 3109254*, 10. doi: 10.1155/2016/3109254.

Keservani, R. K., Sharma, A. K., & Kesharwani, R. K. (2017). An overview and therapeutic applications of nutraceutical and functional foods. *Recent Advances in Drug Delivery Technology*, 160–201.

Keservani, R. K., Sharma, A. K., & Kesharwani, R. K. (2020). *Nutraceuticals and Dietary Supplements: Applications in Health Improvement and Disease Management* (pp. 1–342). CRC Press.

Kumar, P., Rai, S., Verma, S. K., Prakash, P. S., & Chitara, D. (2022). Classification, mode of action, and uses of various immunomodulators. In R. K. Kesharwani, R. K. Keservani, A. K. Sharma, (eds) *Immunomodulators and Human Health.* Springer. https://doi.org/10.1007/978-981-16-6379-6_1.

Kumari, S. K., & Thayumanavan, B. (1997). Comparative study of resistant starch from minor millets on intestinal responses, blood glucose, serum cholesterol, and triglycerides in rats. *Journal of the Science of Food and Agriculture*, *75*(3), 296–302.

Lakshmi Kumari, P., & Sumathi, S. (2002). Effect of consumption of finger millet on hyperglycemia in non-insulin dependent diabetes mellitus (NIDDM) subjects. *Plant Foods for Human Nutrition*, *57*, 205–213.

Lansakara, L. H. M. P. R., Liyanage, R., Perera, K. A., Wijewardana, I., Jayawardena, B. C., & Vidanarachchi, J. K. (2016). Nutritional composition and health related functional properties of *Eleusine coracana* (Finger Millet). *Procedia Food Science*, *6*, 344–347.

Lee, S. H., Chung, I. M., Cha, Y. S., & Park, Y. (2010). Millet consumption decreased serum concentration of triglyceride and C-reactive protein but not oxidative status in hyperlipidemic rats. *Nutrition Research*, *30*(4), 290–296.

Liang, S., & Liang, K. (2019). Millet grain as a candidate antioxidant food resource: A review. *International Journal of Food Properties*, *22*(1), 1652–1661.

Liu, F., Shan, S., Li, H., Shi, J., Hao, R., Yang, R., & Li, Z. (2021). Millet shell polyphenols prevent atherosclerosis by protecting the gut barrier and remodeling the gut microbiota in ApoE$^{-/-}$ mice. *Food and Function*, *12*(16), 7298–7309.

Maitra, S. (2020). Potential horizon of brown-top millet cultivation in drylands: A review. *Crop Research*, *55*(1 and 2), 57–63.

Mall, T. P., & Tripathi, S. C. (2016). Millets the nutrimental potent ethno-medicinal grasses: A review. *World Journal of Pharmaceutical Research*, *5*(2), 495–520.

Mohamed, T. K., Issoufou, A., & Zhou, H. (2012). Antioxidant activity of fractionated foxtail millet protein hydrolysate. *International Food Research Journal*, *19*(1), 59–66.

Moreno Amador, M. D. L., Comino Montilla, I. M., & Sousa Martín, C. (2014). Alternative grains as potential raw material for gluten-free food development in the diet of celiac and gluten–sensitive patients. *Austin Journal of Nutrition and Food Sciences*, *2*(3), 1016.

Naz, N., Naz, A., Rukha, A., Ahmad, S., Waheed, U., Razzaq, K., Farooq, U., & Raza, N. (2020). Phytochemical screening and antioxidant evaluation of millet varieties of Pakistan. *Croatian Journal of Food Science and Technology*, *12*(1), 118–122.

Nishizawa, N., & Fudamoto, Y. (1995). The elevation of plasma concentration of high-density lipoprotein cholesterol in mice fed with protein from proso millet. *Bioscience, Biotechnology, and Biochemistry*, *59*(2), 333–335.

Palaniappan, A., Yuvaraj, S. S., Sonaimuthu, S., & Antony, U. (2017). Characterization of Xylan from rice bran and finger millet seed coat for functional food applications. *Journal of Cereal Science*, *75*, 296–305.

Park, K. O., Ito, Y., Nagasawa, T., Choi, M. R., & Nishizawa, N. (2008). Effects of dietary Korean proso-millet protein on plasma adiponectin, HDL cholesterol, insulin levels, and gene expression in obese type 2 diabetic mice. *Bioscience, Biotechnology, and Biochemistry*, *72*(11), 2918–2925.

Patimah, S., Arundhana, A. I., Mursaha, A., & Syam, A. (2019). Development of foxtail millet and flying fish flour-based cookies as a functional food. *Current Research in Nutrition and Food Science Journal*, *7*(2), 504–516.

Punia, S., Kumar, M., Siroha, A. K., Kennedy, J. F., Dhull, S. B., & Whiteside, W. S. (2021). Pearl millet grain as an emerging source of starch: A review on its structure, physicochemical properties, functionalization, and industrial applications. *Carbohydrate Polymers*, *260*, 117776.

Puranik, S., Kam, J., Sahu, P. P., Yadav, R., Srivastava, R. K., Ojulong, H., & Yadav, R. (2017). Harnessing finger millet to combat calcium deficiency in humans: Challenges and prospects. *Frontiers in Plant Science*, *8*, 1311.

Radhajeyalakshmi, R., Yamunarani, K., Seetharaman, K., & Velazhahan, R. (2003). Existence of thaumatin-like proteins (TLPs) in seeds of cereals. *Acta Phytopathologica et Entomologica Hungarica*, *38*(3,4), 251–257.

Rajasekaran, N. S., Nithya, M., Rose, C., & Chandra, T. S. (2004). The effect of finger millet feeding on the early responses during the process of wound healing in diabetic rats. *Biochimica et Biophysica Acta (BBA)–Molecular Basis of Disease*, *1689*(3), 190–201.

Rane, B. R., Ahirrao, R. A., Gujarathi, N. A., Jain, A. S., & Keservani, R. K. (2023). Pathophysiology and application of herbs in the metabolic syndrome. In R. K. Keservani, D. Yadav, R. K. Keservani, S. Singh, & K. Sandeep (Eds.), *The Metabolic Syndrome: Dietary Supplements and Food Ingredients* (Chapter 10, pp. 199–230). Apple Academic Press, CRC Press, Taylor & Francis Group.

Rane, B. R., Bharath, M. S., Patil, R. R., Keservani, R. K., & Jain, A. S. (2020b). Novel approaches in nutraceuticals. In R. K. Kesharwani, R. K. Keservani, & A. K. Sharma (Eds.), *Enhancing the Therapeutic Efficacy of Herbal Formulations Through Novel Drug Delivery Systems* (Chapter 11, pp. 241–266). IGI Global International Publisher.

Rane, B. R., Patil, A. S., Keservani, R. K., & Jain, A. S. (2020a). Novel approaches in herbal formulation. In R. K. Kesharwani, R. K. Keservani, & A. K. Sharma (Eds.), *Enhancing the Therapeutic Efficacy of Herbal Formulations Through Novel Drug Delivery Systems* (Chapter 2, pp. 43–68). IGI Global International Publisher.

Rane, B. R., Patil, R. R., Jain, A. S., Keservani, R. K., & Kesharwani, R. K. (2022). Nutraceuticals as Disease Preventive Food and Immunity Boosters. In R. K. Kesharwani, R. K. Keservani, & A. K. Sharma (Eds.), *Nutraceuticals and Functional Foods in Immunomodulators*. Springer. https://doi.org/10.1007/978-981-19-2507-8_7.

Rane, B. R., Tadavi, S. A., & Keservani, R. K. (2020c). Naturopathy. In A. K. Sharma, R. K. Keservani, & S. P. Gautam (Eds.), *Herbal Product Development* (Chapter 12, pp. 321–347). Apple Academic Press, CRC Press, Taylor & Francis Group.

Salar, R. K., Purewal, S. S., & Sandhu, K. S. (2017). Fermented pearl millet (*Pennisetum glaucum*) with *in vitro* DNA damage protection activity, bioactive compounds, and antioxidant potential. *Food Research International*, *100*, 204–210.

Scalbert, A., Johnson, I. T., & Saltmarsh, M. (2005). Polyphenols: Antioxidants and beyond. *The American Journal of Clinical Nutrition*, *81*(1), 215S–217S.

Senevirathne, I. G. N. H., Abeysekera, W. K. S. M., Abeysekera, W. P. K. M., Jayanath, N. Y., Galbada Arachchige, S. P., & Wijewardana, D. C. M. S. I. (2021). Antiamylase, antiglucosidase, and antiglycation properties of millets and Sorghum from Sri Lanka. *Evidence-Based Complementary and Alternative Medicine*, 5834915.

Sharma, A. K., Keservani, R. K., & Gautam, S. P. (2020). *Herbal Product Development*. Apple Academic Press, CRC Press, Taylor & Francis Group.

Shobana, S., Harsha, M. R., Platel, K., Srinivasan, K., & Malleshi, N. G. (2010). Amelioration of hyperglycemia and its associated complications by finger millet (*Eleusine coracana* L.) seed coat matter in streptozotocin-induced diabetic rats. *British Journal of Nutrition*, *104*(12), 1787–1795.

Singh, R. B., Khan, S., Chauhan, A. K., Singh, M., Jaglan, P., Yadav, P., Takahashi, T., & Juneja, L. R. (2019). Millets as functional food, a gift from Asia to the Western World. In *The Role of Functional Food Security in Global Health* (pp. 457–468). Academic Press.

Sireesha, Y., Kasetti, R. B., Nabi, S. A., Swapna, S., & Apparao, C. (2011). Antihyperglycemic and hypolipidemic activities of *Setaria italica* seeds in STZ diabetic rats. *Pathophysiology*, *18*(2), 159–164.

Siwela, M., Taylor, J. R., de Milliano, W. A., & Duodu, K. G. (2010). Influence of phenolics in finger millet on grain and malt fungal load, and malt quality. *Food Chemistry*, *121*(2), 443–449.

SK, M., & Sudha, K. (2012). Functional and phytochemical properties of finger millet (*Eleusine coracana* L.) for health. *International Journal of Pharmaceutical, Chemical and Biology Sciences*, *2*(4), 431–438.

Srinivasan, A., Aruldhas, J., Perumal, S. S., & Ekambaram, S. P. (2021). Phenolic acid-bound arabinoxylans extracted from Little and Kodo millets modulate immune system mediators and pathways in RAW 264.7 cells. *Journal of Food Biochemistry*, *45*(1), 13563.

Suma, P. F., & Urooj, A. (2012). Antioxidant activity of extracts from foxtail millet (*Setaria italica*). *Journal of Food Science and Technology*, *49*, 500–504.

Ugare, R., Chimmad, B., Naik, R., Bharati, P., & Itagi, S. (2014). Glycemic index and significance of barnyard millet (*Echinochloa frumentacae*) in type II diabetics. *Journal of Food Science and Technology*, *51*, 392–395.

Viswanath, V., Urooj, A., & Malleshi, N. G. (2009). Evaluation of antioxidant and antimicrobial properties of finger millet polyphenols (*Eleusine coracana*). *Food Chemistry*, *114*(1), 340–346.

Wang, H., Li, D., Wan, C., Luo, Y., Yang, Q., Gao, X., & Feng, B. (2021). Improving the functionality of proso millet protein and its potential as a functional food ingredient by applying nitrogen fertilizer. *Foods, 10*(6), 1332.

Xu, W., Wei, L., Qu, W., Liang, Z., Wang, J., Peng, X., Zhang, Y., & Huang, K. (2011). A novel antifungal peptide from foxtail millet seeds. *Journal of the Science of Food and Agriculture, 91*(9), 1630–1637.

Yadav, C. B., Srivastava, R. K., Beynon, S., Englyst, K., Gangashetty, P. I., & Yadav, R. S. (2022). Genetic variability and genome-wide marker association studies for starch traits contributing to low glycaemic index in pearl millet. *Food and Energy Security, 11*(1), 341.

Yuan, Y., Xiang, J., Zheng, B., Sun, J., Luo, D., Li, P., & Fan, J. (2022). The diversity of phenolics including hydroxycinnamic acid amide derivatives, and phenolic acids contribute to the antioxidant properties of proso millet. *LWT, 154*, 112611.

Zhang, M., Xu, Y., Xiang, J., Zheng, B., Yuan, Y., Luo, D., & Fan, J. (2021). Comparative evaluation on phenolic profiles, antioxidant properties, and α-glucosidase inhibitory effects of different milling fractions of foxtail millet. *Journal of Cereal Science, 99*, 103217.

Zhang, X., Shan, S., Shi, J., Li, H., & Li, Z. (2021). Polyphenol from millet bran increases the sensitivity of colorectal cancer cells to oxaliplatin by blocking the ganglioside GM3 catabolism. *Food & Function, 12*(1), 291–301.

CHAPTER 3

Spices, Herbs, and Aromatic Plants: Excellent Functional Food Incorporations

ASHOK K. SENAPATI, DEV RAJ,* and SANJEEV KUMAR

Aspee College of Horticulture and Forestry, Navsari Agricultural University, Navsari, Gujarat, India

*Corresponding author.

ABSTRACT

India possesses a diverse array of medicinal plants, which have played a significant role in the advancement of traditional Indian medicine throughout history. The majority of spices and herbs exhibit significant antioxidant activity, primarily due to phenolic compounds, particularly flavonoids, which impact nutrition through various mechanisms, including regulating nutrient absorption. Over the past two decades, the pharmaceutical sector has allocated substantial resources to global pharmacological, chemical, and clinical research. Spices are seeds, roots, bark, fruits, or other plant parts primarily used as flavoring and coloring agents in food. Spices differ from herbs, which include leaves, stems, and flowers used for flavoring and garnishing. Spices are also used in medicine, religious rituals, cosmetics, and perfume production. Their therapeutic properties arise from complex chemical compounds found in one or more plant parts as secondary metabolites. Based on their composition, these metabolites are classified as glycosides, alkaloids, essential oils, corticosteroids, and so forth.

The largest group of alkaloids includes compounds such as quinine (cinchona), codeine (poppy), cocaine (coca plant), strychnine and brucine (nux vomica), ergotamine (ergot), emetine (ipecac), caffeine (tea dust), lobeline (lobelia), vasacine (vasaka), hyoscyamine (belladonna),

Functional Compounds and Foods of Plant Origin: Krishan Dutt Sharma, Satish Kumar, Rakesh Sharma, and Raj K. Keservani (Eds.)

reserpine (rauwolfia), santonin (artemisia), scopolamine (datura), and numerous others. Certain essential oils, such as peppermint and valerian kutch, exhibit medicinal properties and can thus be used in pharmaceutical industries. It must be acknowledged that our understanding of the genetic and physiological characteristics of most medicinal plants is limited, and knowledge of the biosynthetic pathways responsible for producing the active constituents in these crops is even more deficient. Agronomic techniques for cultivating medicinal plants such as belladonna, cinchona, ergot, isabgol, senna, ipecac, and others have been developed, leading to localized commercial cultivation of these medicinal herbs. Over the past five decades, industries involving plant-based pharmaceuticals, medicines, cosmetics, perfumery, and food flavors have experienced remarkable growth. Consequently, this sector stands out as one of the high-growth industries within the agri-business sector, achieving significant annual expansion rates.

3.1 INTRODUCTION

India is home to a multitude of spices, herbs, and aromatic plants with therapeutic value due to their functional or nutraceutical properties (Keservani et al., 2010a, b, 2017, 2020; Kumar and Thakur, 2020; Peter, 2007; Thakur et al., 2018). These plants have significantly contributed to traditional medicines due to their functional qualities. Many spices and herbs have strong antioxidant properties, mainly due to phenolic compounds, particularly flavonoids. These compounds affect nutrition through a variety of mechanisms, including the regulation of nutrient absorption. Over the last two decades, the pharmaceutical sector has made significant investments in pharmacological, clinical, and chemical research worldwide.

Spices typically derived from seeds, roots, fruits, and bark are commonly used to add flavor and color to food products, whereas herbs, which include leaves, flowers, or stems, are used as garnishes and flavoring agents. Spices, herbs (Rane et al., 2020a, b, c, 2023; Sharma et al., 2020), and aromatic plants can also be used in cosmetics, perfume, and religious ceremonies (Ballabh et al., 2008; Davidson, 1999; Emamghoreishi et al., 2005; Panwar et al., 2023). The medicinal properties of these plants stem from chemical compounds of varied composition. These secondary metabolites are typically found in one or more parts of the plants (Peter, 2007) and are classified as glycosides, alkaloids, essential oils, corticosteroids, and so forth.

Several essential oils, like valerian kutch and peppermint, have medicinal properties and are used in the pharmaceutical industry. Over the past five decades, industries involved in plant-based pharmaceuticals, medicines, cosmetics, perfumery, and food flavors have experienced remarkable growth. As a result, this sector stands out as one of the high-growth industries within the agri-business sector, achieving significant annual growth rates (Elgayyar et al., 2001; Karnick, 1994; Peter, 2007).

This chapter discusses the functional and pharmaceutical properties of spices, herbs, and aromatic plants in detail.

3.2 SPICES

Spices and condiments have long held a crucial place among horticultural commodities. Since ancient times, they have been regarded as essential in the culinary realm for enhancing the flavor of dishes. Additionally, they find applications in pharmaceuticals, perfumery, cosmetics, and various other industries. Certain spices also offer attributes such as coloration, preservation, antioxidation, antiseptic, and antibiotic properties. Furthermore, they wield considerable influence in the economies of nations involved in spice production, export, and import worldwide. As per the International Organization for Standardization (ISO) report, approximately 109 spice species are cultivated in various regions globally. India, in particular, cultivates over 60 different types of spices. The word 'Condiments and Spices' refers to organic substances or blends employed either in ground form or whole as culinary additives, primarily for enhancing the taste, scent, and spiciness of dishes. These substances are also utilized to season foods and enhance the flavor of beverages. Spices encompass dried seeds, roots, barks, fruits, or flowers derived from plants or herbs. They are employed sparingly to elevate flavor, add color, and offer preservative qualities to various dishes. Many of these chemical compounds are integral to traditional medicinal treatments. Moreover, the process of globalization has expanded the availability and use of these spices across the globe. Throughout history, spices have held significant importance in the cultures of specific regions, serving diverse functions as colorants, flavor enhancers, preservatives, food supplements, and medicinal ingredients (Kumar et al., 2023; Leung and Foster, 1996). The active phytochemicals extracted from these spices underlie their various effects. Spices, natural plant products, are employed to enhance the taste, scent, and appearance of food items. They also have applications in beverages, alcoholic beverages, pharmaceuticals, cosmetics, and perfumes. India, often referred to as the 'land of spices,' boasts an unparalleled diversity of spice crops not found in any other country worldwide. Indian spices are distinguished for their exceptional aroma, flavor, and spiciness, surpassing those of other nations. Even in small quantities, spices offer a genuine sensory delight, elevating the palatability, flavor, and digestibility of food. While their usage is commonplace in households in various forms, there has been a notable surge in demand for spices within the global food processing industry. Consequently, the demand for spices has been steadily on the rise in recent years (Ballabh et al., 2008; Elagayyar et al., 2001).

India is leading in the production and exportation of spices around the world and contributes around 25% of the world's spice trade. India is blessed with varied agro-climatic conditions, tropical, subtropical, and temperate, having great scope for the production of the major spices. Indian spices have been in high demand for export due to their exceptional quality, boasting maximum essential oil, oleoresin, and active components. Recent shifts in people's dietary preferences and habits have led to a greater inclination toward spicy foods, driven by their enticing flavors and aromas, as well as a fondness for fast food.

Spices and condiments hold a distinct position in the realm of fast-food preparation, particularly in bakery products and confectionery items. Some spices and condiments, such as pepper, cardamom, saffron, and vanilla, are considered costly, less-volume crops that yield substantial profit. These crops possess the unique characteristic of adaptability to various cropping systems (Jafri et al., 2001). Spices can be cultivated in various ways, including as standalone crops, intercrops, mixed crops, and in multi-storied cropping systems. Within the category of spices, the production of tree spices such as nutmeg, cinnamon, allspice, and clove is currently limited to around 5000 metric tons, which falls short of the domestic demand, estimated to be around 7000–8000 metric tons. This indicates significant potential for expanding the cultivation of tree spices. Also, the demand for seed spices, both domestically and internationally, is increasing. As a result, there is a promising opportunity to increase the cultivation of these seed spices to meet the growing utilization and export demands (Ballabh et al., 2008; Bensky et al., 2004; Elagayyar et al., 2001).

3.2.1 FUNCTIONAL PROPERTIES OF SPICES

Spices, herbs, and their derivatives are employed both in their whole and powdered forms. Essential oils and oleoresins, in particular, are extensively utilized to season foods and infuse them with aroma, flavor, and taste. These products have also gained prominence in the cosmetic and pharmaceutical sectors. Moreover, spices are utilized not only for their culinary attributes but also as colorants or dyes in various industries, including cotton textiles, tobacco, bakery products, condiments, and the processing of meat and fish products. The demand for value-added products within the food industry is substantial. India boasts a rich array of important spices, including pepper, cardamom, chili, ginger, turmeric, coriander, cumin, fennel, fenugreek, celery, saffron, tamarind, and garlic. These spices play a significant role in enhancing the flavor and appeal of a wide range of culinary creations (Devrim and Durak, 2007; Emamghoreishi et al., 2005; Gray and Flatt, 1999; Grzanna et al., 2005; Liu et al., 2007; Schleicher and Saleh, 1998). There are several other spices that are cultivated and exported in relatively smaller quantities. These include ajwain, dill seeds, tejpata, curry leaves, cinnamon, anise plant, kokam, poppy seeds, and a few other culinary herbs. Additionally, tree spices such as mace, allspice, clove, nutmeg, and star anise, as well as some herbal spices like basil, sage, rosemary, chive, thyme, marjoram, oregano, chive, parsley, savory, tarragon, and are also cultivated in limited amounts, primarily for domestic consumption. The medicinal properties of spices are illustrated in Table 3.1, and the chemical structure and active principles of spices are presented in Table 3.2 (Balch and Balch, 2000; Bensky et al., 2004; Elagayyar et al., 2001; Jafri et al., 2001; Karnick, 1994; Kassahun et al., 2013; Lans et al., 2008; Wichtl and Bisset, 1994). Further, Table 3.3 illustrates the compounds that are most prevalent in spices. Each spice has a precise flavor that matches the complex mix of organic compounds.

3.3 HERBS

The plants used in various herbal remedies are termed medicinal plants (Karnick 1994; Lans et al., 2008). Any part of the plant, like seed, flower, stigma, fruit, leaf, bark, stem, or root, as well as non-woody plants, is referred to as an herb. Initially, the term 'herb' was solely applied to softwood plants, including those that are derived from trees and shrubs. These therapeutic plants are also utilized for food, flavonoids (Keservani and Sharma, 2014; Keservani et al., 2016a, b), pharmaceuticals, and fragrances, as well as for spiritual purposes. For nearly 4000 years, Indian vaids, Unani hakims, and European and Mediterranean cultures have used these herbs as medicine.

TABLE 3.1 Medicinal Properties of Some Important Spices

Spice	Medicinal Use	References
Fenugreek (*Trigonella foenumgraecum*)	Diuretic, demulcent, useful in heart diseases, and promotes menstrual discharge.	Eigner and Scholz (1999); Phillips et al. (1993)
Ginger (*Zingiber officinale*)	Ptyalagogue is helpful in cardiovascular disease (CVD).	Elagayyar et al. (2001); Kim et al. (2002)
Garlic (*Allium sativum*)	Major contributors to flavor are dipropyl disulfide. It also has antibiotic and anti-cancerous properties, prevents dyspepsia, prevents flatulence, can cure ear infections and peptic ulcers, and acts as a rubefacient in skin diseases.	Bensky et al. (2004); Park et al. (2004)
Turmeric (*Curcuma longa*)	Antibacterial, anti-inflammatory, diuretic, laxative, good for infections of the liver, jaundice, diseases of blood, and anti-cancer.	Aggarwal et al. (2007); Grzanna et al. (2005); Tolan et al. (2004)
Red pepper (*Capsicum annuum*)	Anti-inflammatory, for pain relief (rheumatism and neuralgia), useful in indigestion and rubefacient.	Devrim and Durak (2007); Liu et al. (2007)
Onion (*Allium cepa*)	Antibacterial, diuretic, expectorant, used for the treatment of piles and sulfur-containing organic compounds (dipropyl disulfide) that give flavor. Also promotes menstrual discharge.	Ahui et al. (2008); Ali et al. (2008)
Asafoetida (*Ferula asafoetida*)	It prevents muscle spasms, diuretics, and expectorants, is antibacterial and purgative, and is useful in coronary artery disease (angina) and bronchial asthma.	Ballabh et al. (2008)
Pepper (*Piper nigrum*)	Antipyretic and rubefacient.	Al-Qattan (2009)

TABLE 3.2 Chemical Structure and Active Principles of Some Important Spices

Spice	Chemical Structure	Active Principle
Onion	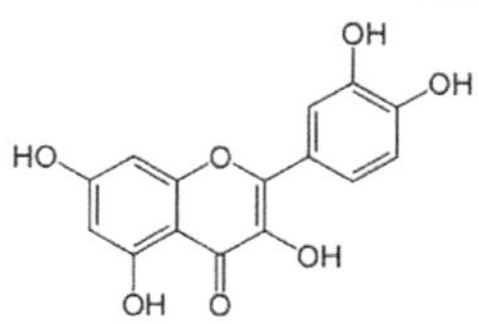	Quercetin and sulfides
Garlic	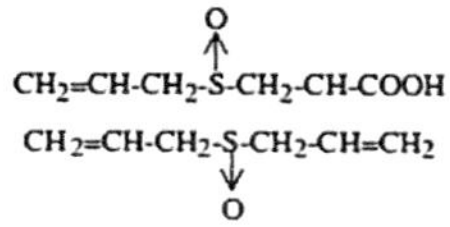	Allicin

TABLE 3.2 *(Continued)*

Spice	Chemical Structure	Active Principle
Asafoetida	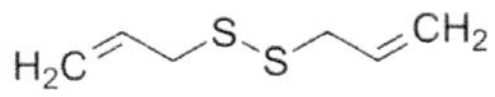	2-butyl-propenyl-disulfide, diallyl sulfide, diallyl disulfide
Ginger	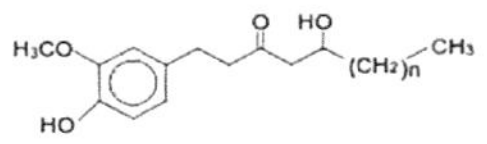	Gingerol
Red pepper	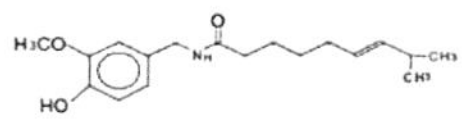	Capsaicin
Fenugreek	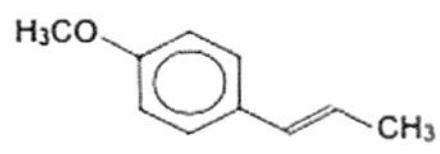	Anethole
Turmeric	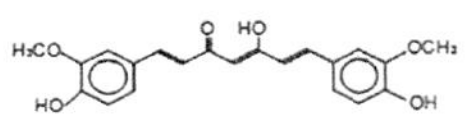	Curcumin
Black pepper	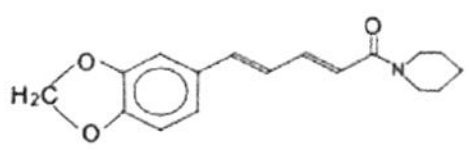	Piperine

Source: Ali et al. (2008); Al-Qattan (2009); Devrim and Durak (2007); Duke et al. (2003); Eigner and Scholz (1999); Elagayyar et al. (2001); El-Sabbon and Abouazra (2008); Grzanna et al. (2005); Gülçin (2005); McNamara et al. (2005); Thangapazham et al. (2008); Xu et al. (2005).

TABLE 3.3 Principal Phytochemicals in Some Spices

Spice	Active Chemical Compound(s)	References
Cardamom	The primary constituent found in it is 1,8-cineole. The distinctive cardamom aroma is the result of the interaction of 1,8-cineole with monoterpenoid, α-terpinyl acetate.	Suekawa et al. (1984)
Chili pepper	Capsaicin is the predominant constituent, providing characteristic spiciness.	Al-Zuhair et al. (1996)
Clove	Eugenol is a primary compound found in cloves. Other minor constituents, such as methyl amyl ketone and methyl salicylate, also contribute to the characteristic odor. Eugenol and its derivative compounds are commonly employed in perfumery and as a flavoring agent.	Schleicher and Saleh (1998)
Cinnamon	Cinnamaldehyde is responsible for imparting the flavor and aroma of cinnamon. Approximately 90% of oil extracted from the bark of cinnamon is composed of cinnamaldehyde. Cinnamon is widely utilized as a flavoring agent in different products such as beverages, chewing gum, and confectionery products.	Davidson (1999)
Coriander seed	Coriander seeds primarily contain linalool as their main compound. Linalool has two isomers (mirror-image), one of which is indeed found in coriander and contributes to its characteristic aroma and flavor. The other mirror-image isomer is commonly found in plants like lavender and sweet basil, providing their distinct scents.	Gray and Flatt (1999)

TABLE 3.3 *(Continued)*

Spice	Active Chemical Compound(s)	References
Dill	Carvone is one of the key compounds present in dill. Spearmint, an isomer of carvone, is utilized for the manufacturing of chewing gum, which imparts the classic spearmint flavor by being soaked in carvone.	Elagayyar et al. (2001)
Cumin	The primary compound responsible for the warm aroma of cumin is cuminaldehyde. Cumin also contains a variety of other aldehyde compounds, which collectively contribute to its overall flavor and fragrance.	Jafri et al. (2001)
Ginger	The major organic compound found in ginger is known as Zingiberene. Further, the pungent sensation associated with ginger is attributed to a group of compounds known as gingerols.	Duke et al. (2003)
Nutmeg	Sabinene is one of the primary components found in the essential oil of nutmeg. Another compound present in nutmeg is myristicin, which, in large quantities, can lead to hallucination.	Bensky et al. (2004)
Oregano	A warm and strong odor of oregano is due to Carvacrol. Also, it is a constituent of thyme oil and is present in tequila.	Emamghoreishi et al. (2005)
Paprika	The reddish-orange color of paprika is due to two compounds, capsanthin, and capsorubin. These compounds become more concentrated as peppers ripen further, intensifying the reddish color of the spice.	Grzanna et al. (2005)
Parsley	The major constituent found in parsley leaves is 1,3,8-p-mentha-triene. Also, contains other constituents such as myristicin, α-pinene, limonene, and so on.	Gülçin (2005)
Pepper	The primary compound that can be isolated from black pepper is peperine, and it is the primary constituent responsible for the spice's pungent taste. Also, an isomer of peperine, chavicine, is responsible for the pungency of black pepper.	McNamara et al. (2005)
Saffron	The color of saffron is due to Crocin. The color of saffron is deep red, and when dissolved in water, it gives an orange color. It possesses antioxidant properties.	Xu et al. (2005, 2008)
Star anise	Anethol constitutes 80–90% of star anise essential oil extract. It can be used as a flavoring agent in alcoholic beverages such as sambuca, as well as licorice sweets.	Ali et al. (2008)
Turmeric	The yellow color of turmeric is due to compounds known as curcuminoids. They serve as a source of natural antioxidants. The E number for curcumin as a food additive is E100.	Ballabh et al. (2008)

Herbs have played a vital role in therapeutic practices across ancient civilizations such as Egypt, America, Iran, Rome, and Africa. Additionally, other advanced medicine systems such as Ayurveda, Unani, and Chinese medicine systematically utilized herbal remedies as integral components of their healing practices.

Among ancient civilizations, India has been known to be a rich repository of medicinal plants. Indian forests are the primary reservoir of an enormous variety of therapeutic and aromatic plants, primarily harvested as raw material for the production of pharmaceuticals and perfumery items. In India, approximately 8000 herbal remedies have been documented within the AYUSH systems (Karnick, 1994). These encompass Ayurveda, Unani, Siddha, and Folk (tribal) medicines, constituting the principal indigenous medical traditions. Of these, Ayurveda and Unani Medicine are the most advanced and commonly followed healthcare systems in the country. As per the reports of the World Health Organization (WHO), nearly 80% of the worldwide population depends on herbal medicines for various aspects of their primary healthcare needs. Around 21,000 plant species are utilized as medicinal plants, and due to their minimal or negligible side effects, they are widely regarded as safe for various treatments. One significant aspect is that the utilization of herbal remedies is not limited by age or gender. This explains the increasing global popularity of herbal treatments. Medicinal herbs, known for their therapeutic properties, offer a logical approach to addressing numerous internal ailments that may otherwise be regarded as challenging to treat. Plants possessing medicinal properties, such as Tulsi, neem, aloe, ginger, and turmeric, provide relief from a wide array of diseases (Mowrey and Clayson, 1982). Numerous parts of the country consider these plants as home remedies. It is well known that many people use Tulsi (Basil) leaves in their daily lives to make medicines, black tea, puja (religious rituals), and other products (Grzanna et al., 2005; Lans et al., 2008). Medicinal plants are recognized as valuable reservoirs of compounds that can be utilized for drug development, whether for pharmacopoeial drugs, non-pharmacopoeial substances, or synthetic medications. Certain plants and their derivatives serve as significant sources of active components employed in addressing common health issues such as diarrhea, constipation, hypertension, low sperm count, dysentery, erectile dysfunction, hemorrhoids, oral hygiene, menstrual irregularities, bronchial asthma, leucorrhea, fevers, and even in the formulation of toothpaste, among other applications. Medicinal uses of some important herbal medicinal plants, parts to be used, and active ingredients are presented in Table 3.4 (Lans et al., 2008).

TABLE 3.4 Medicinal Uses of Some Important Herbal Plants

Crop	Part Used	Medicinal Use	References
Satavri (*Asparagus recemosus*)	Tuberous roots	• A rejuvenating herb that helps reduce the adverse impacts of stress. • Promotes the well-being of the immune system and contributes to healthy sexual functions. • Aids in alleviating PMS, menstrual, and menopausal irregularities. • The primary use of Shatavari roots is as a lactogogue, which stimulates the production of breast milk. • It is beneficial for enhancing body weight loss and is additionally recognized as an aphrodisiac. • The root is effective in treating dysentery, diabetes, and tuberculosis. Overall, it contributes to maintaining health by bolstering immunity against diseases.	Gagnier et al. (2006)
Chirata (*Swertia Chirayita*)	Whole plant	• It serves as one of the best remedies for a fragile or feeble stomach. • It also strengthens the liver. • This plant is extensively employed as a revitalizing tonic for the digestive system. • It also possesses properties to manage malaria and tuberculosis, as well as prevent carbon tetrachloride poisoning in the liver. • It is an exceptionally potent digestive medicine with strong astringent properties, highly effective in managing various types of fevers.	Gagnier et al. (2006)
Ashwagandha (*Withania somnifera*)	Root, leaf, and seed	• It serves various purposes, functioning as an aphrodisiac, diuretic, sedative, revitalizing agent, sleep-inducer, and rejuvenating tonic. • Beneficial for alleviating rheumatism and cough-related debility in elderly individuals, as well as addressing conditions like dropsy and overall weakness.	Pieroni and Torry (2007)
Giloi (*Tinaspora cordololia*)	Stem	• Stem is employed for its laxative properties, as well as its ability to reduce fever and inflammation.	Mengesha et al. (2009)
Stevia (*Stevia rebaudiana*)	Leaves	• It does not affect blood sugar levels, so it is safe for diabetic people. • Does not cause any neurological or renal disorder like some artificial sweeteners. • A mild tea made from stevia leaves offers remarkable relief for the stomach. • In addition to its diverse applications, it possesses antifungal and antibacterial properties as well. • It is also incorporated into everyday products such as mouthwash and various types of toothpaste.	Mengesha et al. (2009)

TABLE 3.4 *(Continued)*

Crop	Part Used	Medicinal Use	References
Reetha (*Sapindus mukorossi*)	Soapnuts	• It can be beneficial for removing stains from hands and may offer relief for skin conditions like eczema, psoriasis, itchy skin, and sensitive skin. • It can be used as an antibacterial mild foaming agent, exfolaint, and cleanser. • It can be used as a hair tonic.	Beemnet et al. (2014)
Bahera (*Terminalia bellirica*)	–	• It is effective in addressing a range of conditions, including ascites, asthma, bleeding, ulceration of the gums, blood pressure, burns, providing a cooling wash for the eyes, cough, dysentery, diarrhea, and worms.	Raj et al. (2016)
Harad (*Terminalia chebula*)	–	• It possesses astringent, stomachic, and laxative properties, making it beneficial against diseases such as asthma, piles, and bronchitis. • Used to treat injuries and scalds. • It can be used to cure stomatitis.	Pieroni and Torry (2007)
Amla (*Emblica officinalis*)	Fruits	• It is an abundant source of Vitamin C. • The fruits are employed in treating various ailments, including hemorrhages, diarrhea, dysentery, anemia, jaundice, dyspepsia, and cough. • Enhances the production of red blood cells. Strengthens teeth, hair, and nails. • It helps regulate blood sugar levels and prevents lipid peroxidation in cell membranes. • It also enhances immunity and provides protection against heart and nervous disorders. • It assists in regulating blood sugar levels and boasts potent anti-inflammatory properties. • Used mainly in addressing conditions like hemorrhoids, gastritis, and colitis. • It promotes an increase in red blood cell count and prevents premature greying of hair, making it stronger and free from dandruff.	Pieroni and Torry (2007); Raj et al. (2016)

3.4 PRESERVATIVES PROPERTIES OF SPICES AND HERBS

Numerous spices possess antimicrobial properties that aid in food preservation. Among these, garlic, onion, allspice, and oregano are notably effective at eliminating bacteria, serving as potent antibacterial and antifungal agents. Additionally, thyme, cinnamon, tarragon, and cumin also exhibit significant antibacterial and antifungal properties, contributing to their role in food preservation (El-Sabon and Abouazra, 2008; Goreja, 2003). Capsicums, which include chilies and other hot peppers, exhibit intermediate antimicrobial activity, capable of eliminating up to 90% of bacteria. In contrast, white or black pepper, as well as ginger, anise seed, celery seed, lemon juice, and lime juice, inhibit around 25% of bacteria. The inhibitory effects of spices on microorganisms can either be bacteriostatic or fungistatic, and numerous spices indeed possess antimicrobial and/or antifungal properties (Kassahun et al., 2013a; Kim et al., 2002; Mowrey and Clayson, 1982; Park et al., 2004; Wiesenauser, 1998). Numerous bioactive compounds with antibacterial and therapeutic properties have been identified from herbs and spices (Devrim and Durak, 2007). Antimicrobial agents that have the potential to be used as bio-preservatives can be divided into several useful categories, including phenolics, polyphenols, quinines, flavones, flavonoids, tannins, coumarins, and alkaloids. The main part of the spices used in food processing is presented in Table 3.5. The role of major spices in food preservation is discussed ahead (Chadwick, 1976; Elagayyar et al., 2001; Grzanna et al., 2005).

TABLE 3.5 Parts of Major Spices Used in Food Preservation

Spice	Scientific Name	Part Used	References
Allspice	*Pimenta dioica*	Fruit	Mowrey and Clayson (1982)
Bay leaf	*Laurus nobilis*	Leaves	Phillips et al. (1993)
Caper	*Capparis spinosa*	Flower bud	Al-Zuhair et al. (1996)
Cardamom	*Elettaria cardamomum*	Pod	Elagayyar et al. (2001)
Cinnamon	*Cinnamomum verum*	Bark	Jafri et al. (2001)
Cloves	*Syzygium aromaticum*	Flower bud	Bensky et al. (2004)
Ginger	*Zingiber officinale*	Rhizome	Al-Zuhair et al. (1996)
Mace	*Mysistica fragrans*	Aril	Grzanna et al. (2005)
Nutmeg	*Mysistica fragrans*	Endosperm	Al-Zuhair et al. (1996)
Pepper	*Piper nigrum*	Berry/drupes	Elagayyar et al. (2001)
Red/hot pepper	*Capsicun annuum, C. frutescens*	Berry	Xu et al. (2005)
Saffron	*Crocus sativus*	Stigmas	Elagayyar et al. (2001)
Sesame	*Sesamum indicum*	Seeds and seed oil	Spiller et al. (2008)
Star anise	*Illicium verum*	Immature fruit	Grzanna et al. (2005)
Turmeric	*Curcuma longa*	Root	Grzanna et al. (2005)
Vanilla	*Vanilla plantifolia*	Berry	Grzanna et al. (2005)

Spices like coriander, anise, carvone, peppermint, caraway, cardamom, cumin, fennel, celery, dill, mustard, and pimento have also been known to possess antimicrobial properties (Goreja, 2003; Gray and Flatt, 1999). However, normal inhibitory concentrations used in these cases are very high, which indicates that these spices have little antimicrobial activity. Spices like rosemary, turmeric, ginger, and nutmeg also possess antimicrobial activities (Beemnet et al., 2013; Davidson, 1999; Emamghoreishi et al., 2005; Grzanna et al., 2005; Muhammad et al., 2008; Xu et al., 2005). The inhibitory effect of rosemary is attributed to its terpene fraction, which is comprised of borneol, cineole, pipene, camphene, camphor, and so forth. Gram-negative bacteria are more sensitive to rosemary extract, and 1% of rosemary extract has moderate antifungal activity and no effect on yeasts. The rosemary extract is a useful preservative in refrigerated foods. The ginger extract inhibits *Escherichia coli, Streptococcus faecalis, Salmonella typhimureum, Staphylococcus aureus, Pseudomonas aeruginosa, Bacillus cereus, Clostridium perfringens* in meat and meat products. Black pepper has bacteriostatic properties against *E. coli, Salmonella virchow,* and *Staphylococcus aureus* in the meat preparation (Gülçin, 2005). Mustard, along with its pungent component allyl-isothiocyanate, exhibits antimicrobial properties that hinder the growth of molds, reduce microbial counts, and deter insect infestation in various food items like bread, pasta, lettuce, grains, dried mushrooms, and fruits. Additionally, red pepper, containing capsaicin, is recognized for its antimicrobial attributes (Duke et al., 2003). Menthol and anethol from anise and fennel have antimicrobial activity although slow, against *Salmonella typhimureum, Staphylococcus aureus,* and *Vibrio parachaemolyticus.* Spices such as onion, sage, garlic, cinnamon, cloves, thyme, and so on can inhibit the growth of food-borne bacteria (gram-positive and gram-negative), molds, and yeast (Bensky et al., 2004; Devrim and Durak, 2007; Elagayyar et al., 2001). The presence of these spices and herbs can be noticed in foods such as bread, rice, pickles, and meat products. The antimicrobial effectiveness of various spices and the inhibitory effect of various spices on different microorganisms has been shown in Tables 3.6 and 3.7.

TABLE 3.6 Relative Antimicrobial Effectiveness of Some Spices and Herbs

Spice and Herb	Inhibitory Effect	References
Mustard, cinnamon, and cloves	Strong	Al-Zuhair et al. (1996); Bown (1995); Gray and Flatt (1999)
Cumin, thyme, allspice, coriander, bay leaf, caraway, oregano, and sage	Medium	Bensky et al. (2004); Duke et al. (2003); Elagayyar et al. (2001); Emamghoreishi et al. (2005); Goreja (2003)
Ginger, black pepper, and red pepper	Weak	Beemnet et al. (2013); Gagnier et al. (2006); Grzanna et al. (2005); Saini et al. (2022)

TABLE 3.7 Inhibitory Effect of Spices and Herbs On Microorganisms

Spice and Herb	Microorganisms Inhibited	References
Garlic	*Salmonella typhymurium*, *Escherichia coli*, *Staphylococcus aureus*, *Bacillus cereus*, *Bacillus subtilis*, mycotoxigenic *Aspergillus flavus*, and *Candida albicans*	Bown (1995)
Onion	*Aspergillus flavus*, *Aspergillus parasiticus*	Elagayyar et al. (2001)
Cinnamon	Mycotoxigenic *Aspergillus flavus*, *Aspergillus parasiticus*	–
Clove	Mycotoxigenic *Aspergillus*	Duke et al. (2003)
Mustard	Mycotoxigenic *Aspergillus*	Bensky et al. (2004)
Allspice	Mycotoxigenic *Aspergillus*	Gagnier et al. (2006)
Oregano	Mycotoxigenic *Aspergillus*, *Salmonella* spp., *Vibrio parahaemolyticus*	Devrim and Durak (2007)
Rosemary	*Bacillus cereus*, *Staphylococcus aureus*, *Vibrio parahaemoliticus*	El-Sabbon and Abouazra (2008)
Bay leaf	*Clostridium botulinum*	–
Sage	*Bacillus cereus*, *Staphylococcus aureus*, *Vibrio parahaemoliticus*	Beemnet et al. (2013b)
Thyme	*Vibrio parahaemoliticus*	Gagnier et al. (2006)

The antimicrobial properties of spices residing in their essential oils are presented in Table 3.8. These essential oils consist not only of a single chemical compound but also of several compounds. The inhibitory action of spices is due to the major components of the essential oils. The various chemical constituents responsible for the antimicrobial properties of essential oils can be categorized under the following classes:

1. **Sulfur Compounds:** Sulfides and thiols serve as the primary preservative compounds within the Alliaceae family, which includes onions.
2. **Terpenes and Terpene Derivatives:** These are major components of essential oils in spices. Compounds with 10, 15, and 20 carbon (C) atoms are called monoterpenes, sesquiterpenes, and diterpenes, respectively. Monoterpenes are significant constituents of the mint and parsley families, while sesquiterpenes play a vital role in cinnamon and the ginger family. Terpene derivatives, such as carotenoids, include compounds like thymol—a preservative compound found in thyme (Davidson, 1999; Reuter et al., 2008).
3. **Phenol:** Phenolic compounds occur prominently in several products. These are the most important and abundant anti-microbials present in spices.

4. **Glycosides:** These compounds are composed of a sugar molecule and a non-sugar component referred to as an aglycone. An example of this is mustard oil glycosides.
5. **Aldehydes:** It includes citral, cinnamaldehyde, benzaldehyde, and citronellal.
6. **Esters:** These compounds are commonly present in various essential oils and consist of substances like eugenol acetate, bornyl acetate, and linalyl acetate.

TABLE 3.8 Active Component in the Essential Oil of Some Spices

Spice	Active Component	References
Allspice	Eugenol, thymol, phellandrene, caryophyllene, cineol, and methyl eugenol.	Gray and Flatt (1999)
Anise	Anethole, methyl chavicol, anise aldehyde, and limonene.	Elagayyar et al. (2001)
Basil	Methyl chavicol, linalool, cineol, and anethole.	Duke et al. (2003)
Caraway	D-carvone, limonene, carveol, cinnamaldehyde, eugenol, and pinene.	Uchibayashi (2010)
Clove	Eugenol, caryophylene, and acetyleugenol.	Bensky et al. (2004)
Coriander	Linalool, pinene, and cymene.	Devrim and Durak (2007)
Cumin	Cuminaldehyde and limonene.	El-Sabon and Abouazra (2008)
Fennel	Anethole, limonene, fenchone, and camphene.	Reuter et al. (2008)
Garlic	Diallyl disulfide, allyl propyl disulfide, and allicin.	Uchibayashi (2010)
Ginger	Gingeberone, borneol, linalool, shogaol, and gingeroene.	Bensky et al. (2004)

3.5 AROMATIC PLANTS

Aromatic plants possess odorous volatile substance, which occurs as an essential oils, gum exudates, balsam, and oleoresin in one or more parts. Important aromatic plants are basil, bay leaves, caraway, coriander leaves, lemon grass, mace, marjoram, mint, rosemary, sage, tarragon, thyme, and vanilla (Kassahun et al., 2013a; Uchibayashi, 2010). The art and science of perfumery and flavoring in India have flourished since ancient times with the advent of civilization and perfected under the patronage of Moghul emperors. The aromatic plants and therapeutic herbs are commercially vital since they produce the basic raw material for cosmetic, soap, pharmaceutical, flavor, and perfumery industries globally. The sequence of these industries starts right from naturally aromatic plants to finished products such as food and cosmetics. The sequence is as follows: Primary processing of natural aromatic plants to produce essential oil, absolute, oleoresin, and resinoids. Although aromatic plants are known for their flavor and aroma, even then these plants have great potential to cure different diseases owing to the functional properties of the aroma compounds. The Chemical Active principles of aromatic plants and their functional properties have been presented in Table 3.9. Maximum final moisture content (% wet basis) for various aromatic plants prescribed in the European Pharmacopoeia (2005) to maintain the functional properties are presented in Table 3.10. Moreover, aromatic plants contain active compounds with antimicrobial properties, and the European Pharmacopoeia specifies the allowable threshold for microbial counts in aromatic plant materials (Davidson, 1999) has been presented in Table 3.11.

TABLE 3.9 Chemical Active Principles of Aromatic Plants and Their Functional Properties

Aromatic Plant	Active Compound	Functional Properties	References
Basil (*Ocimum basilicum*)	Basil oil is rich in estragole, and it also contains a significant amount of the compound linalool.	Used in the treatment of snakebites, common colds, and alleviating inflammation in nasal passages. Additionally, basil contributes essential macronutrients like calcium and vitamin K to the diet and offers a diverse array of antioxidants.	Umberto (1947)
Bay leaves (*Cinnamomum tamala*)	The primary constituent found in cigarettes is 1,8-cineole and is also present in numerous mouthwashes.	People use bay leaf for diabetes, cancer, stomach problems, and pain.	Kanjanapothi et al. (1981)
Caraway (*Carum carvi*)	Caraway seeds primarily contain the compound carvone, which exists in two mirror-image isomers. One of these isomers emits the aroma of caraway, while the other has a scent reminiscent of spearmint. In caraway seeds, only the S-carvone isomer, which carries the caraway fragrance, is present.	Used to treat everything from indigestion to bloating to gas and also helps in the reduction of chronic inflammation.	Ammon (1991)
Coriander leaves (*Coriandrum sativum*)	Coriander leaf oil extract consists mainly of 2-decenoic acid, along with a variety of aldehydes, with a particular emphasis on those containing 9 to 10 carbon atoms in their structure.	Coriander leaves are good for antibiotics to overcome digestive, colic pain, and urinary tract problems.	Gray and Flatt (1999)

TABLE 3.9 *(Continued)*

Aromatic Plant	Active Compound	Functional Properties	References
Lemon grass (*Cymbopogon citrates*)	A combination of two distinct isomeric aldehydes, that is, neral and geranial, is known as citral, due to its citrus-like fragrance used in perfumery.	Stomachaches, diarrhea, gas, bowel spasms, vomiting, fever, the flu, headaches, and other types of pain, including rheumatism.	Akdogan et al. (2003)
Mace (*Myristica fragrans*)	Terpinene-4-ol is the predominant compound found in the essential oil of mace. It is also present in the essential oil of the tea tree.	Used for diarrhea, nausea, stomach spasms and pain, and intestinal gas.	Bhattacherjee and De (2005)
Marjoram (*Origanum majorana*)	Sabinene hydrate serves as the primary component in the oil of *Origanum majorana* and imparts the herb's distinctive taste in conjunction with other compounds of the terpene family.	Used as analgesic, anaphrodisiac, and antibacterial. It is also used as a steam inhalant to relieve laryngitis.	Guney et al. (2006)
Mint (*Mentha* spp.)	Menthol is the predominant compound found in mint leaves and is widely utilized as a flavoring agent in chewing gum and toothpaste. Also, it is a key ingredient in menthol cigarettes.	Used for the treatment of fevers, headaches, and digestive disorders.	Eguale et al. (2007)
Rosemary (*Rosmarinus officinalis*)	Alpha-pinene is present in numerous coniferous tree species, notably in pine trees, and it also constitutes one of the primary compounds in rosemary. Together with other members of the monoterpene compound family, it is emitted in significant quantities by vegetation.	Used to improve the health of hair and skin. Rosemary has also stopped dandruff.	Pieroni and Torry (2007)
Sage (*Salvia officinalis*)	Sage contains several key chemical compounds, with manool being one of the prominent constituents. Additionally, sage includes compounds such as eucalyptol and thujone, the latter known for its psychoactive properties.	Primarily used for problems related to digestion, including stomach pain, bloating, loss of appetite, flatulence, diarrhea, and pyrosis.	Ammon (2008)
Tarragon (*Artemisia dracunculus*)	Estragole, an isomeric form of anethole, is present in both *Artemisia* and *Ocimum basilicum*. While there are suspicions of its carcinogenic and genotoxic properties, these effects are only observed at levels significantly higher (up to 1000 times) than the estimated public exposure to the constituent.	Eugenol is an essential oil present in the herb used in dentistry as an anesthetic and antiseptic. Also used for the stimulation of appetite and as a remedy for dyspepsia, flatulence, and singultus.	Uchibayashi (2010)

TABLE 3.9 *(Continued)*

Aromatic Plant	Active Compound	Functional Properties	References
Thyme (*Thymus vulgaris*)	Thymol is the chemical constituent that is responsible for the certain flavor of thyme. Also, it is utilized as an additive in the manufacturing of cigarettes.	Used as a germ-killer in mouthwashes to prevent tooth decay. Used for treating coughs and colds.	Kassahun et al. (2013a)
Vanilla (*Vanilla planifolia*)	The predominant aroma in vanilla is attributed to vanillin, constituting a significant portion, ranging from 74–96%, of the flavor and aroma compounds. Additionally, more than 100 other volatile compounds have been identified, including acids, phenolic compounds, alcohols, and aldehydes.	Possess antimicrobial, antioxidant, anti-mutagenic, and anti-sickle cell effects. Vanillin has shown an effect in reducing cholesterol levels and thus preventing cardiovascular diseases (CVDs). Helps to prevent atherosclerosis, inflammation in the arteries, and blood clots.	Beemnet et al. (2013)

TABLE 3.10 Moisture Content for Various Aromatic Plant Parts Prescribed in the European Pharmacopoeia (2005)

Plant Species	Plant Part Used as a Functional Food	Maximum Moisture Content (% Wet Basis)	References
Althaea officinalis L.	Root	10	Duke et al. (2003); Eguale et al. (2007); Gray and Flatt (1999); Guney et al. (2006); Kanjanapothi et al. (1981); Umberto (1947)
Arnica montana L.	Flower	10	
Calendula officinalis L.	Flower	12	
Chamomilla recutita [L.] Rauschert	Flower	12	
Coriandrum sativum L.	Seed	10	
Foeniculum vulgare Mill.	Seed	8	
Hypericum perforatum L.	Herb	10	
Levisticum officinale Koch	Leaves	12	
Malva silvestris L.	Leaves and flower	12	
Melissa officinalis L.	Leaves	10	
Mentha x *piperita* L.	Leaves	11	
Plantago lanceolata L.	Herb	10	
Valeriana officinalis L.	Root	12	
Verbascum phlomoides L.	Herb	12	

TABLE 3.11 Thresholds of Microbial Count of Aromatic Plant Material According to the European Pharmacopoeia (2005)

Category	Application	Aerobic Bacteria	Molds	Entero Bacteria	E. coli	References
4A	Herbal medicinal products to which boiling water is added before use.	10^7	10^5	–	10^2	Duke et al. (2003); Lans et al. (2008)
4B	Herbal medicinal products to which boiling water is not added before use.	10^5	10^4	10^3	Nil in 1 g	
3B	Preparations for oral administration containing raw materials of natural origin.	10^5	10^2	10^2	Nil in 1 g	
3A	Preparations for oral and rectal administration.	10^3	10^2	–	Nil in 1 g	

3.6 BIOSAFETY

Throughout history, spices, herbal remedies, and aromatic substances have served as medicinal agents for treating a wide spectrum of illnesses. According to the WHO, due to factors like poverty, limited access to modern medical care, and a strong belief in traditional remedies, approximately 65–80% of the global population residing in both developing and developed nations primarily rely on plants for their primary healthcare needs. This preference is attributed to the ease of access, affordability, compatibility with the human body, and fewer associated side effects, among other factors (Elagayyar et al., 2001; Vasudevan et al., 2000).

Despite the common notion that natural products are safe, it's important to recognize that herbal medicines can lead to notable toxic effects, spanning from allergic responses to disorders affecting the cardiovascular, hepatic, renal, neurological, and dermatologic systems. Therefore, there is a pressing need for the standardization of herbal medicines or traditional remedies. The current imperative for herbal drug research and industry in the nation lies in prioritizing rigorous standardization and scientific validation of promising medicinal plants to truly realize their health benefits. The misidentification of medicinal plants, the presence of steroids in herbal products, and the potential for adverse reactions and toxicity can pose challenges when utilizing these remedies. Frequently, adverse effects associated with herbal products stem from either subpar quality or improper usage. Consequently, it becomes challenging to differentiate between genuine adverse reactions to herbal medicines and product-related issues until the root cause is identified. Therefore, ensuring the safety of herbal medicine necessitates several crucial elements: accurate identification and understanding of adverse effects, effective management of risks and problems, the establishment of organizations to prevent adverse events, and clear communication of issues to promote the safe and beneficial use of herbal medicines.

3.7 FUTURE PERSPECTIVES

With its abundant biodiversity and substantial potential for commercial utilization, India has the opportunity to emerge as a global frontrunner in providing raw materials for the phytopharmaceutical industry. Through the formulation of a comprehensive strategy in collaboration with forest departments for the cultivation and preservation of medicinal plants, numerous challenges mentioned earlier can be transformed into opportunities for profitable commercial utilization, all while safeguarding the interests of the communities engaged in the collection of these plants (Lans et al., 2008). Incorporating medicinal plants into crop rotation, particularly in arid regions watershed areas, has the potential to significantly boost the cultivation of spices, aromatic plants, and herbs.

KEYWORDS

- **active components**
- **allicin**
- **anise aldehyde**
- **caryophyllene**
- **cinnamaldehyde**
- **essential oils**
- **esters**
- **functional properties**
- **phellandrene**
- **phytochemicals**

REFERENCES

Aggarwal, B. B., Surh, Y. J., & Shishodia, S. (2007). *The Molecular Targets and Therapeutic Uses of Curcumin in Health and Disease* (pp. 595). Springer Science & Business Media.

Ahui, M. L. B., Champy, P., Ramadan, A., Van, L. P., Araujo, L., Andre, K. B., Diem, S., Damotte, D., Kati-Coulibaly, S., Offoumou, M. A., & Dy, M. (2008). Ginger prevents Th2-mediated immune responses in a mouse model of airway inflammation. *International Immunopharmacology, 8*(12), 1626–1632.

Akdogan, M., KWlWnc, I., Oncu, M., Karaoz, E., & Delibas, N. (2003). Investigation of biochemical and histopathological effects of *Mentha piperita* L. and Mentha spicata L. on kidney tissue in rats. *Human and Experimental Toxicology, 22*(4), 213–219.

Ali, B. H., Blunden, G., Tanira, M. O., & Nemmar, A. (2008). Some phytochemical, pharmacological and toxicological properties of ginger (*Zingiber officinale* Roscoe): A review of recent research. *Food and Chemical Toxicology, 46*(2), 409–420.

Al-Qattan, M. M. (2009). Garlic burns: Case reports with an emphasis on associated and underlying pathology. *Burns, 35*(2), 300–302.

Al-Zuhair, H., El-Sayeh, B., Ameen, H. A., & Al-Shoora, H. (1996). Pharmacological studies of cardamom oil in animals. *Pharmacological Research, 34*(1, 2), 79–82.

Ammon, H. P. (2008). Cinnamon in type 2 diabetics. *Medizinische Monatsschrift fur Pharmazeuten, 31*(5), 179–183.

Ammon, H. P., & Wahl, M. A. (1991). Pharmacology of *Curcuma longa. Planta Medica, 57*(01), 1–7.

Balch, P. A. (2006). *Prescription for Nutritional Healing*. Penguin.

Ballabh, B., Chaurasia, O. P., Ahmed, Z., & Singh, S. B. (2008). Traditional medicinal plants of cold desert Ladakh—are used against kidney and urinary disorders. *Journal of Ethnopharmacology, 118*(2), 331–339.

Beemnet, M. K., Basazenew, D., & Desta, F. (2013). Variability in Ethiopian rosemary (*Rosmarinus officinalis* L.) collections for agronomic and chemical traits. In *Proceedings of the Fourth Biennial Conference of the Ethiopian Horticultural Science Society on Sustainable Training, Research and Development towards Achieving the Growth and Transformation Plan (GTP)*.

Bensky, D., Clavey, S., & Stõger, E. (2004). Materia medica. *Chinese Herbal Medicine*, 3–6.

Bhattacharjee, S. K., & De, L. C. (2005). *Post-Harvest Technology of Flowers and Ornamental Plants*. Distributed by Aavishkar Publishers Distributors.

Bown, D. (1995). *The Royal Horticultural Society Encyclopedia of Herbs & Their Uses*. Dorling Kindersley Limited.

Chadwick, J. (1976). *The Mycenaean World*. Cambridge University Press.

Davidson, A. (2014). *The Oxford Companion to Food*. OUP Oxford.

Devrim, E., & Durak, I. (2007). Is garlic a promising food for benign prostatic hyperplasia and prostate cancer? *Molecular Nutrition and Food Research, 51*(11), 1319–1323.

Duke, J. A., Bogenschutz-Godwin, M. J., deCellier, J., & Duke, P. (2003). *Piper nigrum* L. (Piperaceae) black pepper. *CRC Handbook of Medicinal Spices*, 253–263.

Duke, J. A., Bogenschutz-Godwin, M. J., deCellier, J., & Duke, P. K. (2003). *Trigonella foenum-graecum* L.(Fabaceae) fenugreek, Greek clover, Greek hay. *CRC Handbook of Medicinal Spices*, 296.

Duke, J. A., Bogenschutz-Godwin, M. J., deCellier, J., & Duke, P. K. (2003). *Nigella sativa* L. (Ranumculaceae) Black caraway, Black cumin, Fennel flower, Nutmeg flower, and Roman coriander. *CRC Handbook of Medicinal Spices,* 227–228.

Eguale, T., Tilahun, G., Debella, A., Feleke, A., & Makonnen, E. (2007). *In vitro* and *in vivo* anthelmintic activity of crude extracts of *Coriandrum sativum* against *Haemonchus contortus*. *Journal of Ethnopharmacology*, *110*(3), 428–433.

Eigner, D., & Scholz, D. (1999). Ferula asa-foetida and *Curcuma longa* in traditional medical treatment and diet in Nepal. *Journal of Ethnopharmacology*, *67*(1), 1–6.

El Sabban, F., & Abouazra, H. (2008). Effect of garlic on atherosclerosis and its factors. *EMHJ-Eastern Mediterranean Health Journal, 14*(1), 195–205.

Elgayyar, M., Draughon, F. A., Golden, D. A., & Mount, J. R. (2001). Antimicrobial activity of essential oils from plants against selected pathogenic and saprophytic microorganisms. *Journal of Food Protection*, *64*(7), 1019–1024.

Emamghoreishi, M., Khasaki, M., & Aazam, M. F. (2005). *Coriandrum sativum*: Evaluation of its anxiolytic effect in the elevated plus-maze. *Journal of Ethnopharmacology*, *96*(3), 365–370.

Gagnier, J. J., van Tulder, M. W., Berman, B. M., & Bombardier, C. (2006). Herbal medicine for low back pain. *Cochrane Database of Systematic Reviews*, *2*.

Ghayur, M. N., Gilani, A. H., & Janssen, L. J. (2008). Ginger attenuates acetylcholine-induced contraction and Ca^{2+} signaling in murine airway smooth muscle cells. *Canadian Journal of Physiology and Pharmacology*, *86*(5), 264–271.

Goreja, W. G. (2003). *Black Seed: Nature's Miracle Remedy*. Karger Publishers.

Gray, A. M., & Flatt, P. R. (1999). Insulin-releasing and insulin-like activity of the traditional anti-diabetic plant *Coriandrum sativum* (coriander). *British Journal of Nutrition*, *81*(3), 203–209.

Grzanna, R., Lindmark, L., & Frondoza, C. G. (2005). Ginger: An herbal medicinal product with broad anti-inflammatory actions. *Journal of Medicinal Food*, *8*(2), 125–132.

Gülçin, İ. (2005). The antioxidant and radical scavenging activities of black pepper (*Piper nigrum*) seeds. *International Journal of Food Sciences and Nutrition*, *56*(7), 491–499.

Güney, M., Oral, B., Karahanli, N., Mungan, T., & Akdogan, M. (2006). The effect of *Mentha spicata Labiatae* on uterine tissue in rats. *Toxicology and Industrial Health, 22*(8), 343–348.

Gupta, P., Dhawan, S. S., & Lal, R. K. (2015). Adaptability and stability based differentiation and selection in aromatic grasses (*Cymbopogon* species) germplasm. *Industrial Crops and Products*, *78*, 1–8.

Jafri, M. A., Javed, K., & Singh, S. (2001). Evaluation of the gastric antiulcerogenic effect of large cardamom (fruits of *Amomum subulatum* Roxb). *Journal of Ethnopharmacology*, *75*(2, 3), 89–94.

Kanjanapothi, D., Smitasiri, Y., Panthong, A., Taesotikul, T., & Rattanapanone, V. (1981). Postcoital antifertility effect of *Mentha arvensis*. *Contraception*, *24*(5), 559–567.

Karnick, C. R. (1994). *Pharmacopoeial Standards of Herbal Plants. Vol. 2* (No. Ed. 1). Sri Satguru Publications.

Kassahun, B. M., Egata, D. F., Lulseged, T., Yosef, W. B., & Tadesse, S. (2014). Variability in agronomic and chemical characteristics of Spearmint (*Mentha spicata* L.) genotypes in Ethiopia. *International Journal of Advanced Biological and Biomedical Research*, *2*(10), 2704–2711.

Kassahun, B. M., Yosef, W. B., & Mekonnen, S. A. (2013). Performance of lemon Verbena (*Aloysia triphylla* L.) for morphological, economic, and chemical traits in Ethiopia.

American-Eurasian Journal of Agricultural and Environmental Sciences, 13(11), 1576–1581.

Keservani, R. K., & Sharma, A. K. (2014). Flavonoids: Emerging trends and potential health benefits. *Journal of Chinese Pharmaceutical Sciences, 23*(12), 815.

Keservani, R. K., Kesharwani, R. K., Sharma, A. K., Vyas, N., & Chadoker, A. (2010b). Nutritional Supplements: An Overview. *International Journal of current pharmaceutical review and research, 1*(1), 59–75.

Keservani, R. K., Kesharwani, R. K., Vyas, N., Jain, S., Raghuvanshi, R., & Sharma, A. K. (2010a). Nutraceutical and functional food as future food: A review. *Der Pharmacia Lettre, 2*(1), 106–116.

Keservani, R. K., Sharma, A. K., & Kesharwani, R. K. (2016a). Nutraceutical and functional foods for cardiovascular health. In *Food Process Engineering* (pp. 257–278). Apple Academic Press, CRC Press.

Keservani, R. K., Sharma, A. K., & Kesharwani, R. K. (2016b). Medicinal effect of nutraceutical fruits for cognition and brain health. *Scientifica, 3109254*, 10. doi: 10.1155/2016/3109254.

Keservani, R. K., Sharma, A. K., & Kesharwani, R. K. (2017). An overview and therapeutic applications of nutraceutical and functional foods. *Recent Advances in Drug Delivery Technology*, 160–201.

Keservani, R. K., Sharma, A. K., & Kesharwani, R. K. (2020). *Nutraceuticals and Dietary Supplements: Applications in Health Improvement and Disease Management* (pp. 1–342). CRC Press.

Kim, K. S., Koo, M. S., Jeon, J. W., Park, H. S., & Seung, I. S. (2002). Capsicum plaster at the Korean hand acupuncture point reduces postoperative nausea and vomiting after abdominal hysterectomy. *Anesthesia and Analgesia, 95*(4), 1103–1107.

Kumar, S., & Thakur, K. S. 92020). Effect of 1-methylcyclopropene (1-MCP) application and periodic cold storage on ripening of 'Bartlett' pear during ambient shelf life periods. *Journal of Food Processing and Preservation, 44*(6), e14467.

Kumar, S., Sharma, S., Kumar, V., Sharma, A., Kaur, R., & Saini, R. (2023). Jamun (*Syzygium cumini* (L.) Skeels): The conventional underutilized multifunctional plant: An exotic gleam into its food and functional significance. *Industrial Crops and Products, 191*, 115873.

Lans, C., Turner, N., & Khan, T. (2008). Medicinal plant treatments for fleas and ear problems of cats and dogs in British Columbia, Canada. *Parasitology Research, 103*, 889–898.

Leung, A. Y. (1980). *Encyclopedia of Common Natural Ingredients Used in Food, Drugs, and Cosmetics*. Wiley.

Liu, C. T., Sheen, L. Y., & Lii, C. K. (2007). Does garlic have a role as an antidiabetic agent? *Molecular Nutrition and Food Research, 51*(11), 1353–1364.

McNamara, F. N., Randall, A., & Gunthorpe, M. J. (2005). Effects of piperine, the pungent component of black pepper, at the human vanilloid receptor (TRPV1). *British Journal of Pharmacology, 144*(6), 781.

Mengesha, B., Mohammed, O., Tessema, T., & Abate, S. (2009). *Production, Processing and Utilization Technologies of Aromatic Plants*. Ethiopian Institute of Agricultural Research (EIAR).

Mowrey, D., & Clayson, D. (1982). Motion sickness, ginger, and psychophysics. *The Lancet, 319*(8273), 655–657.

Nadkarni, K., & Nadkarni, A. K. (1976). *Indian Materia Medica* (Vol. 1, p. 799). Popular Prakashan Pvt. Ltd.

Pandir, D. R., Rakesh, S., & Patel, N. L. (2016). *Handbook of Food Science and Technology: Vol. 2 Processing techniques.*

Panwar, A., Kumar, V., Dhiman, A., Thakur, P., Sharma, A., & Kumar, S. (2023). Nanoemulsion-based edible coatings for quality retention of fruits and vegetables-decoding the basics and advancements in the last decade. *Environmental Research*, 117450.

Park, H. S., Kim, K. S., Min, H. K., & Kim, D. W. (2004). Prevention of postoperative sore throat using capsicum plaster applied at the Korean hand acupuncture point. *Anaesthesia*, *59*(7), 647–651.

Phillips, S., Hutchinson, S., & Ruggier, R. (1993). *Zingiber officinale* does not affect gastric emptying rate: A randomized, placebo-controlled, crossover trial. *Anaesthesia*, *48*(5), 393–395.

Pieroni, A., Houlihan, L., Ansari, N., Hussain, B., & Aslam, S. (2007). Medicinal perceptions of vegetables traditionally consumed by South-Asian migrants living in Bradford, Northern England. *Journal of Ethnopharmacology*, *113*(1), 100–110.

Rane, B. R., Ahirrao, R. A., Gujarathi, N. A., Jain, A. S., & Keservani, R. K. (2023). Pathophysiology and application of herbs in the metabolic syndrome. In R. K. Keservani, D. Yadav, R. K. Keservani, S. Singh, & K. Sandeep (Eds.), *The Metabolic Syndrome: Dietary Supplements and Food Ingredients* (Chapter 10, pp. 199–230). Apple Academic Press, CRC Press, Taylor & Francis Group.

Rane, B. R., Bharath, M. S., Patil, R. R., Keservani, R. K., & Jain, A. S. (2020b). Novel approaches in nutraceuticals. In R. K. Kesharwani, R. K. Keservani, & A. K. Sharma (Eds.), *Enhancing the Therapeutic Efficacy of Herbal Formulations Through Novel Drug Delivery Systems* (Chapter 11, pp. 241–266). IGI Global International Publisher.

Rane, B. R., Patil, A. S., Keservani, R. K., & Jain, A. S. (2020a). Novel approaches in herbal formulation. In R. K. Kesharwani, R. K. Keservani, & A. K. Sharma (Eds.), *Enhancing the Therapeutic Efficacy of Herbal Formulations Through Novel Drug Delivery Systems* (Chapter 2, pp. 43–68). IGI Global International Publisher.

Rane, B. R., Tadavi, S. A., & Keservani, R. K. (2020c). Naturopathy. In A. K. Sharma, R. K. Keservani, & S. P. Gautam (Eds.), *Herbal Product Development* (Chapter 12, pp. 321–347). Apple Academic Press, CRC Press, Taylor & Francis Group.

Reuter, J., Huyke, C., Casetti, F., Theek, C., Frank, U., Augustin, M., & Schempp, C. (2008). Anti-inflammatory potential of a lipolotion containing coriander oil in the ultraviolet erythema test. *JDDG: Journal der Deutschen Dermatologischen Gesellschaft*, *6*(10), 847–851.

Saini, R., Kumar, S., Sharma, A., Kumar, V., Sharma, R., Janghu, S., & Suthar, P. (2022). Deep eutectic solvents: The new generation sustainable and safe extraction systems for bioactive compounds in the agri-food sector: An update. *Journal of Food Processing and Preservation, 46*(10), e16250.

Schleicher, P., & Saleh, M. (2000). *Black Cumin: The Magical Egyptian Herb for Allergies, Asthma, and Immune Disorders*. Inner Traditions/Bear & Co.

Sharma, A. K., Keservani, R. K., & Gautam, S. P. (2020). *Herbal Product Development*. Apple Academic Press, CRC Press, Taylor & Francis Group.

Siddeeg, A., Salih, Z. A., Mukhtar, R. M., & Ali, A. O. (2018). Extraction and characterization of peppermint (*Mentha piperita*) essential oil and its assessment as antioxidant and antibacterial. 1–14.

Spiller, F., Alves, M. K., Vieira, S. M., Carvalho, T. A., Leite, C. E., Lunardelli, A., Poloni, J. A., Cunha, F. Q., & de Oliveira, J. R. (2008). Anti-inflammatory effects of red pepper

(*Capsicum baccatum*) on carrageenan-and antigen-induced inflammation. *Journal of Pharmacy and Pharmacology*, *60*(4), 473–478.

Srivastava, K. C., & Mustafa, T. (1992). Ginger (*Zingiber officinale*) in rheumatism and musculoskeletal disorders. *Medical Hypotheses*, *39*(4), 342–348.

Suekawa, M., Ishige, A., Yuasa, K., Sudo, K., Aburada, M., & Hosoya, E. (1984). Pharmacological studies on ginger. I. Pharmacological actions of pungent constituents,(6)-gingerol and (6)-shogaol. *Journal of pharmacobio-dynamics*, *7*(11), 836–848.

Szallasi, A. (2005). Piperine: Researchers discover new flavor in an ancient spice. *Trends in Pharmacological Sciences*, *26*(9), 437–439.

Thakur, K. S., Kumar, S., & Kumar, N. (2018). Effect of active packaging and refrigerated storage on quality attributes of kiwifruits (*Actinidia deliciosa* Chev). *Journal of Pharmacognosy and Phytochemistry*, *7*(2), 1372–1377.

Thangapazham, R. L., Sharma, A., & Maheshwari, R. K. (2007). Beneficial role of curcumin in skin diseases. *The Molecular Targets and Therapeutic Uses of Curcumin in Health and Disease*, 343–357.

Tolan, I., Ragoobirsingh, D., & Morrison, E. Y. S. A. (2004). Isolation and purification of the hypoglycaemic principle present in *Capsicum frutescens*. *Phytotherapy Research: An International Journal Devoted to Pharmacological and Toxicological Evaluation of Natural Product Derivatives*, *18*(1), 95–96.

Uchibayashi, M. (2001). The coriander story. *Yakushigaku Zasshi*, *36*(1), 56–57.

Vasudevan, K., Vembar, S., Veeraraghavan, K., & Haranath, P. S. (2000). Influence of intragastric perfusion of aqueous spice extracts on acid secretion in anesthetized albino rats. *Indian Journal of Gastroenterology: Official Journal of the Indian Society of Gastroenterology*, *19*(2), 53–56.

Wichtl, M. (2004). *Herbal Drugs and Phytopharmaceuticals: A Handbook for Practice on a Scientific Basis* (No. Ed. 3). Medpharm GmbH Scientific Publishers.

Wiesenauer, M. (1998). Comparison of solid and liquid forms of homeopathic remedies for tonsillitis. *Advances in Therapy*, *15*(6), 362–371.

Xu, Y., Ku, B. S., Yao, H. Y., Lin, Y. H., Ma, X., Zhang, Y. H., & Li, X. J. (2005). The effects of curcumin on depressive-like behaviors in mice. *European Journal of Pharmacology*, *518*(1), 40–46.

CHAPTER 4

Cereal and Millet-Based Traditional and Novel Foods

AASIMA RAFIQ,* MARIYA NAYEEM, MUMTAHIN UL KOUSAR, BUSHRA MANZOOR, and ABIDA JABEEN

Division of Food Science and Technology, Sher-E-Kashmir University of Agricultural Science and Technology, Kashmir, Jammu and Kashmir, India

*Corresponding author.

ABSTRACT

Cereals, millets, and their ingredients have been recognized as functional foods and nutraceuticals in recent years due to their ability to provide dietary fiber, protein, energy, minerals, vitamins, and antioxidants that are essential for human health. Foods containing cereal and millet components also provide opportunities to incorporate probiotics, prebiotics, and fiber into the human diet. Various growth studies have been conducted on cereal and millet-based substrates with probiotic lactic acid bacteria (LAB), as well as the use of whole-grain or components as high-fiber foods in the development of novel food products. The development of a wide range of foods using local food crops and other biological resources is very common on the Indian subcontinent. Science and traditional knowledge must collaborate to achieve mutually beneficial outcomes.

4.1 INTRODUCTION

The demand for functional foods is growing, fueled in part by the commercial potential for foods that might improve customers' health and well-being. Functional foods are foods or dietary ingredients that influence the host's health and/or lessen the risk of chronic disease in addition to their basic nutritional functions. In recent years, cereals have also been studied concerning

Functional Compounds and Foods of Plant Origin: Krishan Dutt Sharma, Satish Kumar, Rakesh Sharma, and Raj K. Keservani (Eds.)

their possible application in the development of functional foods. Cereals are cultivated on 73% of the world's harvested land and account for more than 60% of global food output (Charalampopoulos et al., 2002). Cereals are grains or edible seeds from the Gramineae family of grasses. Cereals are farmed for their nutrient-dense edible seeds, which are commonly referred to as grains. Since the dawn of civilization, several grains have staple foods, both directly for human consumption and indirectly through livestock feed. Cereals are the world's most important food sources (FAO, 2002), and cereal-based foods provide a significant amount of energy, protein, B-complex vitamins, and minerals to the global population. Wheat and rice are the most significant crops on the planet, accounting for more than half of all cereal production. The embryo (or germ), which holds the genetic material for a new plant, and the endosperm, which is packed with starch grains, are structurally similar in all cereals (McKevith, 2004). Wheat, rye, rice, barley, millet, and oats are the most common grains. Wheat and rye play a unique function because they are the only grains that can be used to make bread. Rice and corn have been grown in South America, tropical Southeast Asia, and Central America for over 5000 years. Millet cereals have played a role in Asia and Africa's sub-tropical and tropical regions since ancient times (Belitz et al., 2009).

Millet, along with maize and sorghum, is a generic word for small grains that form a heterogeneous group and are referred to as coarse cereals. Millets are a minor item in Western diets, but they are a mainstay in African and Asian diets. Their agricultural relevance stems from their hardiness, ability to withstand extreme weather, and ability to be cultivated with few inputs in low-rainfall areas. Millets are regarded as a food-security crop due to their capacity to thrive in a variety of agro-climatic situations. Although millets are not a common part of the daily diet of Americans and Europeans, millets are increasingly widely used as a component in multigrain and gluten-free cereals in these nations. However, millet is the staple diet of the people in millet-producing areas in many Asian and African countries, and it is used to make a variety of traditional dishes and beverages such as papad, idli, chakli, dosa, bread, porridges, snacks, and baby foods (Jaybhaye et al., 2014). Maize is a major staple food crop farmed in a variety of agroecological zones and farming techniques in Sub-Saharan Africa, and it is consumed by people of various food preferences and socio-economic backgrounds. The key importance of maize as a staple grain in Sub-Saharan Africa is comparable to that of rice or wheat in Asia, with eastern and southern Africa having the greatest consumption rates. Maize provides over half of the calories and protein consumed in Eastern and Southern Africa and one-fifth of the calories and protein consumed in West Africa (Macauley, 2015). Millet grains are gluten-free, non-acidifying, and have a low glycemic index (GI), making them easy to digest. Its low GI food feature is said to make it a suitable choice for those with celiac disease and diabetes since it helps to regulate blood glucose levels (Abah et al., 2020). Wheat isn't usually divided into varieties; instead, classifications are used to differentiate between wheat farmed at different times of the year and the milling and baking quality of the flour produced. Within each class, there are several different wheat cultivars with similar traits. Because of its distinctive qualities, wheat is used to make a wide range of ingredients and foods, including wheat germ, spelt (a coarse form of wheat), couscous, cracked wheat or bulgur, and wheat starch.

To avoid grain germination, pest infestation, and mold deterioration, proper grain storage is essential after harvest. If dry grains are stored for only a few months, only minor nutritional changes occur; however, if the grains are stored with more moisture, grain quality can degrade due to starch breakdown by grain and microbial amylases (enzymes). Milling is the most common method for processing cereals, but a variety of methods are also employed to create a variety of products. The various grains are milled in slightly different ways, but the process can be summarized as grinding, sifting, separation, and regrinding. After milling, the final nutrient value of cereal is determined by how much of the outer bran and aleuronic layers are removed, as here is where the fiber, vitamins, and minerals are concentrated (McKevith, 2004).

4.2 RICE AND RICE-BASED PRODUCTS

Rice is mostly used as cooked table rice, especially in Asiatic countries, but of late, it has also been introduced to consumers as a breakfast and snack meal. There are many varieties of rice snacks and breakfast items available. However, there are some fascinating variances, such as pancakes, crackers, puddings, fries, cakes, and other products like noodles and so forth. In South Asia, which consumes rice products in abundance, flaked, popped, and puffed rice products dominate the markets. Rice-based breakfast and snack meals have recently made their way into wheat-utilizing parts of the globe. This made the diet of people more diverse and healthier as wheat is associated with various allergies, and it added economic benefits because most of these products can work just fine with broken. Some of these products are prepared from rice flour, others from broken and granular milled rice, and so on, as the starting material. Based on that, here's a brief overview of products that are prepared from rice and their substitutes. The cooked rice and substitutes include retort rice, frozen rice, and canned rice. At the same time, another popular form of rice is instant or quick-cooking rice. The rice flour is sometimes converted into batter, which is used to manufacture products like noodles, puddings, fries, and fermented cakes. At the same time, dry flour is mostly utilized for the formation of baby foods and baked products, *viz.*, cookies, bread, crackers, and cakes. Another important category of Rice-based products includes its utilization as breakfast cereal and snacks in the form of shredded rice, rice flakes, Hurum, piping, and emping.

4.2.1 COOKED RICE AND SUBSTITUTES (RETORT, FROZEN, AND CANNED RICE)

One of the leading forms of rice distribution is through the retort pouches. The principle of retort rice is simple: rice is first prepared and then placed in pouches, or rice and water are placed in pouches. The bag is then sealed and retorted under the required variables. Retort rice has been popular in recent years since it is highly easy for single people or even small families to utilize for dinner. All that is required is to soak the pouch in hot water for a couple of minutes or to pierce and microwave the pouch for roughly a minute or two, and the rice is ready to eat. This method is extremely adaptable. Not only can plain rice be made in this manner, but it can also be used to make a whole meal by combining it with suitable spices, vegetables, meats, sauces, or a blend of them.

Frozen rice is another option for distributing ready-to-eat (RTE) rice. In this, rice is first cooked and then frozen. It is, therefore, maintained and delivered in a frozen state. The frozen rice is cooked in microwave ovens and then presented to consumers either at homes or eating centers and diners, and so forth. Rice can be distributed either in individual pouches or is centrally produced in cooking systems and then transported. This method may be used to prepare plain rice or rice with a variety of veggies, meat, and sauces. Rice in soups, rice dishes, salads, and infant meals is possibly used for canned rice. The principle of canned rice is the same as retort pouches. Due to some delimitations of the technique, retort pouches were introduced, and therefore, limited research is based on the canning process of rice.

4.2.2 *INSTANT COOKING RICE*

Instant rice reduces rather than eliminates home cooking. Depending on the size and form of the variety, milled rice might take anywhere from 15 to 30 minutes to cook in water. Uncle Ben's company invented a new concept of milled parboiled rice that produces quick-cooking rice without the need for additional cook-processing (Lin and Jacops, 2002). A major advantage of this process over all other processes above is that there is no alteration in textural properties, and the processed rice resembles the typical milled parboiled rice. Yet, these products are precooked and rather porous and are more likely to get rancid at a faster rate.

4.2.3 *WET BATTER-BASED PRODUCTS FROM RICE (NOODLES, PUDDINGS, FRIES, AND FERMENTED PANCAKES)*

Noodles are structured food formed from flour that has been pulverized from whole grains. Pasta, macaroni, and noodles are quite popular structured wheat flour culinary items in the West. Rice is crushed and made into noodles almost the same way; thus, it is a widely consumed dish in Southeast Asia. The fundamental method for making noodles is wet milling rice, then shaping dough and steaming or partially cooking it in water. After that, the dough is kneaded and extruded in a basic press, either mechanically or automatically in industry. The extruded noodles are then cooked in boiling water and finally dried. Dry milling is not feasible as it results in partial surface gelatinization, making the extremely fine particles mushy and sticky. Aside from that, the particle sizes are not as homogeneous as they are after wet milling. As a result, several of the particles are coarser, resulting in rough materials.

Pudding is a cuisine that can be served either as a dessert delicacy or as a savory side dish during a meal. Puddings made of rice are usually blended with milk or water, and raisins, vanilla, cinnamon, and so on are added. In the case of desserts, sugar as a sweetener is added, and the pudding is then boiled or baked. Recently, rice fries that are equivalent to conventional potato fries have been produced. The method is basic; first, rice flour blends are extruded, shaped in the desired forms, and fried. After that, the fries are frozen and fried again. Rice fries are a healthier alternative to their contemporary potato fries as the fat content of the finished rice fries was reported to be around two-thirds that of potato fries (Kadan et al., 2001).

The most well-known fermented rice cake is idli from South India, which is now popular around the world. It is typically composed of about two parts rice and one part dehulled black gram (*Phaseolus mungo*). The rice is normally parboiled and coarsely ground, while the black gram is finely crushed. The batter is mixed and left to ferment overnight before being steamed. Another product, dosa, is prepared similarly but is cooked in oil, giving it a pancake-like appearance. Other fermented cakes available in the market include puto, a product from the Philippines, and Lao-Chao, a fermented sweet rice cake from China made from steamed, milled waxy rice fermented with fungus and yeast.

4.2.4 RICE FLOUR-BASED BABY FOOD

As the first solid diet for newborns, different combinations of basic cereals and other foods blended properly have been utilized for a long. However, the infant food industry has emerged to offer infants a safe, healthy, and consistent diet. This approach offers a systematic way to include numerous micronutrients, mainly iron and other minerals and vitamins, in the infant's diet, in addition to providing a proper texture. Rice is the most common ingredient in these precooked newborn cereals due to its well-known hypoallergenic properties and ease of digestion. To lower the viscosity and boost calories in a given volume, diastatic enzymes are often mixed into the blend as well. The main premise of the production of precooked baby cereals is to generate slurry using powdered components, such as rice flour, mildly precook it, and then dry the paste in a drum drier. After that, the dry sheet is softly crushed, and the resultant flakes are packed. The biggest plus of the technique is that it allows the blending of diverse materials. Extrusion processing has recently made it feasible to develop infant products, but the fundamental concepts remain the very same.

4.2.5 RICE FLOUR-BASED BAKED PRODUCTS (COOKIES, BREAD, CRACKERS, AND CAKES)

Rice flour, rather than wheat flour, is used in the preparation of certain cookies. One such example is *nan-e berenji*, a sweet rice cookie enjoyed in Iran. A blend of cardamom and sugar syrup mixed with rose water is used to make these flaky rice-flour biscuits. They are among the most cherished cookies in Persian culture and are traditionally eaten at significant events like weddings and the Persian New Year. A meat mallet is gently pressed into the dough to create a grid pattern. Gluten-free rice cookies are also available on the market.

Wheat gluten protein is special in the preparation of bread as it allows the wheat dough to leaven during baking. Rice lacks a comparable protein, and unlike gluten, rice proteins do not form an interlinked framework when mixed; therefore, allowing fermentation gases to escape. As a result, rice cannot be used to replace wheat in yeast-leavened bread. For this objective, several gums and surfactants have been tested. Only hydroxypropyl methylcellulose, a gum, allowed gas to be retained in rice flour (Nishita et al., 1976). Rice flour prepared from low-GT rice (70°C) has proved to be reliable in replacing 60% of wheat flour in the bread manufacturing industry in Japan (Yeh, 2004). Gujral et al. (2003) investigated the use of cyclodextrin glycosyl transferase in rice bread. The enzyme had a positive impact on the product's quality and crumb texture. Likewise, Gujral and Rosell (2004) demonstrated that glucose oxidase altered rice protein to increase the quality of the product. Rice bread has gained popularity as it is a healthier alternative for people suffering from celiac disease. Moreover, unleavened rice bread is also quite popular in many South Asian countries like India and Pakistan.

Many Asian countries, particularly Japan, produce rice crackers in large quantities. Rice crackers such as *senbei* and *arare* are especially popular in Japan. These items are both cooked and baked rice-based products. The former is made using regular japonica rice, while the latter is made with waxy rice. Both are boiled, crushed, kneaded, and baked, and they come in a variety of sizes, shapes, and forms. Numerous other varieties are found throughout Southeast and East Asia, including China.

Different varieties of cooked and semi-cooked rice cakes have been produced for centuries, such as *mochi* in Japan, *puto* and *suman* in the Philippines, and *idli* in India. These items are now being produced on a larger scale in catering facilities and small businesses. Most of these foods are made from fully or partially cooked rice flour, or from cooked rice mashed into dough. The dough is then kneaded and either steamed or boiled. The prepared dough may then be wrapped in various leaves, boiled, or sliced into pieces, roasted, seasoned, fried, or fermented. A wheat-free layer cake recipe using 100% rice flour has also been developed for specialized diets.

4.2.6 RICE-BASED BREAKFAST CEREALS AND SNACKS (SHREDDED RICE, RICE FLAKES, HURUM, PINPIG, AND EMPING)

Shredded rice is quite popular in current markets and is made by pressing washed milled rice with sugar, malt syrup, and salt in a rotary cooker till tender. A preparatory drying of the cooked rice grits in a louver oven produces a firm grain surface that helps the grits to flow uniformly throughout the operation. The modified rice is pushed between two shredding rolls, one smooth and the other with a pattern of shallow square or rectangular corrugations around its edge. Individual grains become trapped between the rollers and dragged into the corrugations, forming a lattice-work ribbon. The toasting of dough pieces then proceeds in an oven.

Rice flakes are also known as flattened rice. Crushed raw, roasted, or parboiled rice grains are pounded into flat flakes to form rice flakes. They can be eaten plain, roasted, fried, or added to other meals as ingredients or garnishes. They might have a crispy, crunchy, chewy, or soft texture with a mild nutty flavor depending on the need. Poha, pauwa, chira, and aval are all examples of flattened rice, which is a staple in South Asia. The current Western rice flakes are a rather different product. It's prepared by heating milled rice with malt and other flavorings, then drying, flaking, and toasting it like many other cereal flakes.

Hurum is a flaked and puffed rice product. It's formed by dry-heating waxy rice types and soaking them overnight before sand roasting them. To make intermediate flaked rice, the roasted paddy is first crushed. The flakes are then coated with lard and roasted to enlarge them. Pinpig is immensely popular in the Philippines. It's prepared from glutinous rice that has been pressed flat before being roasted. It is widely used as a topping for a variety of sweets in Filipino cuisine, but it may also be eaten straight, baked into cakes, or blended with beverages and other meals. In Malaysia, a comparable product is on the market known as emping. Emping snacks come in a variety of flavors, including plain (original), salty, spicy, and sweet, based upon whether salt or caramelized sugar is added.

4.2.7 OTHER PRODUCTS FROM RICE (GERMINATED BROWN RICE, BREWING, AND PET FOOD)

In recent years, germinated brown rice has become a very sought-after product in countries like Japan and Korea. Due to the presence of GABA, germinated brown rice has various health benefits like reduced blood sugar and insomniac conditions, and aids in menopausal and presenile conditions (Bhattacharya, 2011). Another advantage of germinated brown rice is that it can be cooked in a regular rice cooker, unlike untreated brown rice. Rice-based alcoholic beverages are produced in households and cottage businesses across Asia's rice-producing countries, especially in Japan. In current brewing enterprises in the West, finely broken pieces of rice (brewer's rice) are also utilized as an auxiliary. The rice GT is the most significant feature of this application. A low GT is advantageous because it eliminates the need for high-temperature cooking, which can inactivate enzymes. Rice is now commonly utilized in pet diets, which is a relatively new commercial product. In the West, this is most likely the product that provides the greatest industrial outlet for rice besides table rice. Rice has a unique benefit as a pet food component in that it is easily digestible and has a minimal allergic response. Because of its inexpensive cost, brewer's rice is most ordinarily used but flour, bran, or whole-grain rice can also be used. The most prevalent type of processing in the pet food industry is extrusion.

4.3 WHEAT AND WHEAT-BASED PRODUCTS

Wheat, an important crop farmed all over the world is probably the first in production in cereals. For thousands of years, wheat products have been the staple food for the world population, and they are a key source of nutrients as well as energy, particularly in developing nations. Several potential advantages, including improvement in weight and lower risks of cardiovascular disease (CVD) and type-2 diabetes, and so on, have been linked to the intake of wheat-based products, particularly those consisting of whole grains (Reynolds et al., 2020). Common wheat (bread) accounts for over 95% of all wheat grown worldwide, with durum (pasta) wheat accounting for the majority of the remaining 5% (Wieser et al., 2020). By 2050, an expected rise in the world population by 2 billion (Asseng et al., 2015) calls for immediate action to improve food production and develop quality food products to avoid the ever-looming risk of starvation for countries listed high in the global hunger index. Therefore, a growing need to utilize the available food crops to their maximum and create a variety for consumers in terms of novelty and nutrition persists.

Wheat is primarily used as flour milled from the grain, with a minor amount turned into breakfast cereals like flakes and puffed wheat. Wheat proteins are glutamic acid-rich and tryptophan-deficient. Thiamin, riboflavin, niacin, folic acid, calcium, phosphorus, zinc, copper, and iron are all found in whole wheat. It is also a high-fiber food. Gluten is a protein found primarily in wheat that is required for the fundamental structure of dough used in bread, buns, and other baked products. Different parts and types of the wheat grain are used in the development of a variety of wheat-based products such as whole wheat flour which contains the finely pulverized bran, germ, and endosperm used to make chapatis, puris, and whole wheat bread, among other things. Wheat bran is a rich form of insoluble fiber that is good for health. Wheat germ is high in fiber and an excellent source of vegetable protein. Vitamin E, magnesium, thiamine, folate, potassium, and phosphorus are also abundant in it. Broken wheat is used to make upma, bisi bela bath, pongal, and other Indian dishes. Wheat flakes are used in breakfast cereals. They're high in dietary fiber and are fortified with a variety of vitamins and minerals. Maida is also referred to as refined flour. When white flour or maida is made, the bran and germ are separated. It produces a more homogeneous loaf with a larger volume, a blander flavor, and is easier to digest. Semolina is an endosperm that has been roughly mashed and has a chemical makeup comparable to white flour. Macaroni products are also known as pasta and examples are macaroni, spaghetti, vermicelli, and noodles.

Different types of wheat include hard red winter (used for bread and all-purpose flour), hard red spring (used for bagels, and croissants), soft red winter (used for cookies, crackers, pretzels, and pastries, soft white (used for cakes, pastries, and Middle-East flatbreads), hard white (used for Asian noodles, tortillas, and flatbreads) and durum wheat (used for pasta preparation). Some of the products prepared from wheat are – flatbread, bread, bagels, pretzels, pizza crusts, tortillas, cake, and other products, *viz*., layer cake, pound cake, angel food cake, muffin, and quick bread, waffle and pancake, doughnut, crackers, cookie, biscuit, and pasta.

4.3.1 FLAT BREAD

Flatbread is prepared with a basic recipe that generally consists of just four components. Whole wheat flour or extremely high extraction white flour that have high absorption capacity may be used in these varieties of bread. Although yeast is used in flatbread recipes, the unique gap in the middle is created by steam leavening caused by the extremely high processing temperatures. Examples are pita bread, barbari (Iran), gözleme (Turkey), and naan (Afghanistan) among others. Wheat flour, yeast, salt, and water are the four fundamental components needed in bread making. Wheat flour, notably wheat flour protein, is distinctive in its potential to make a dough that holds gas. This is a necessary quality for the manufacture of all leavened dough products. Yeast utilizes sugar and forms carbon dioxide and ethanol in the process. As a result, yeast produces the gas required to leaven the dough, expand air cells, and eventually generate the desired crumb structure found in bread. Salt is typically used as a flavor enhancer in bread. It also affects gluten as it enhances gluten's gas holding capacity.

4.3.2 BAGEL, PRETZEL, PIZZA CRUSTS, AND TORTILLA

Bagel was first introduced in Poland and grew quite popular among the Jews. Wheat dough is kneaded by hand into the shape of a ring, briefly boiled in water, and baked. Bagel doesn't raise as much as other bread because the crust is set from the boiling process. Sometimes, fruits or other ingredients are added to improve the taste. Examples are blueberry bagel, cinnamon bagel, and onion bagel, among others. Pretzel originated in Germany and had a knotted shape. Maillard's reaction plays an important part in the formation of Pretzel. Salt is usually used as a seasoning complemented sometimes by sugar, mustard, cheese, sweet glaze, and so forth. Soft pretzels are eaten right away, hard pretzels have a considerable shelf-life. Bagels, soft pretzels, and pizza crusts are high in gluten. Hard red spring with high protein content is usually preferred for the making of these products. This results in a final product that has the necessary chewy bite. It's also possible to use some dough-enhancing chemicals. Shortening is used sparingly, if at all, in these recipes.

Tortilla is a light, round unleavened flatbread prepared traditionally from maize but presently from wheat flour as well. It originated initially in the Mesoamerican region of North and Central America and has only gained popularity ever since. In a wheat tortilla, high-protein wheat flour, salt, chemical leavening agents, and shortening are commonly used. Tortilla, like bread, is sensitive to mold growth; hence a preservative may well be added. A whole wheat tortilla, as opposed to a white flour tortilla, is thought to be a healthier option. It is prepared using whole grain wheat flour, which is less refined than white flour. Whole wheat tortilla is not only low in calories and fats but high in fiber as well.

4.3.3 CAKE AND OTHER PRODUCTS

A cake is a soft, sweet food made from a mixture of flour, fat, eggs, sugar, and other ingredients, which is baked and sometimes iced or decorated. The mixture of ingredients is called batter which forms a base for making cake, pancake, cake doughnuts, quick bread, muffin, and waffles. A layer cake also known as a sandwich cake is a multi-layered stack of cakes that are held together by a frosting, cream, or jam. The sugar content of a layer cake recipe is high. It is a high ratio cake, that is, it contains more sugar than flour. Baking powder, egg white, emulsified shortening, and water or milk are also essential in a layer cake. Popular examples are black forest, German chocolate cake, and red velvet, among others. Pound cake traditionally spans centuries but presently, different novelties are available and relished particularly in France (quatre-quarts), Germany (Nusskuchen), and Mexico (panqué con nueces). The pound cake is made using about equal parts flour, sugar, shortening, and eggs. The only leavening in this product comes from the air cells that were introduced during the mixing process.

A kind of sponge cake, angel food cake typically includes five times the amount of egg whites and sucrose as flour, with just the addition of tartar used to keep the pH in check. There is no addition of butter, and it has a characteristic fluffy and soft texture. Muffin and quick bread both are prepared with cake formulas that have considerably lower sugar content. Minor components in the two couldn't be any more different; from fruit fragments, to bran, to other taste agents. Examples include bran muffin, poppy-seed muffin, flatbread muffin, and so forth. Most popular in Belgium, the waffle is a leavened batter or dough product fried between two plates that have been designed to give it a certain surface impression. A pancake is a flat cake baked on a hot surface with butter or oil. Both waffle and pancake batters are often produced using basic ingredients such as flour, water, baking powder, and modest amounts of sugar, oil or shortening, and flavor enhancing additives.

4.3.4 DOUGHNUT AND CRUST

The doughnut is fried dough that has been leavened. Because all doughnuts are fried rather than baked, oil is required and not conditional. The cake doughnut recipe is extremely similar to the layer cake recipe. But it comprises less water and has a higher viscosity than cake batter. On the other hand, a yeast doughnut is produced using sweet dough that has been fermented. It typically contains yeast nutrients and various dough conditioners. Lard or shortening is the principal component in a pie crust, aside from flour. When it comes to the flavor of the finished crust, salt is also crucial. The dough is quite dry, and its thickness is mostly determined by the consistency of the shortening or lard. Many pie crust producers use lard because it gives a distinct flavor and texture.

4.3.5 CRACKERS, COOKIES, AND BISCUITS

A flat and dry baked product, crackers are marketed as healthy, tasty, and easy-to-eat cereals. Crackers come in a variety of shapes and sizes, but they are always formed from the dough. In cracker production, water interactions are crucial, and soft wheat flour with a low and consistent absorption rate is desirable. To lessen the elasticity of the dough and create a distinctive flavor, certain crackers, such as saltine crackers, incorporate yeast in addition to sodium bicarbonate in the recipe. Snack crackers normally do not include yeast, and they contain more shortening and flavoring components than the former.

A cookie is a flat, sweet, baked snack or dessert. Because soft wheat flour binds less water, it is favored for cookie making. The water content in cookie dough has a significant impact on quality parameters such as cookie texture and spread. They usually have a lot of flour, shortening, sugar, and very little water. Normally, doughs are used to make cookies, except for wafer cookies, such as ice cream cones, which are prepared using batters. Characteristics of the US: The biscuit is bread that is small in size and prepared from chemically leavened dough. Soft wheat flour is favored for a short texture in biscuits, while hard wheat flour is sometimes replaced when high-speed processing is necessary. The majority of biscuits are formed from dough that is comparable to bread dough. In these biscuits, flakiness is generally a desirable feature. Shortening can be added to achieve this flakiness. Some biscuits, including drop biscuits, are produced with wetter dough, more like a batter in composition.

4.3.6 PASTA

Prominent in Italy, pasta is produced from an unleavened dough of wheat flour combined with eggs or water, shaped into sheets or various forms, and finally baked or boiled. Semolina from durum wheat is the primary raw material used in the creation of high-quality pasta. Semolina made from durum wheat uses less water to make the dough and yields transparent pasta with good cooking and eating qualities. However, a diverse array of wheat milled to varying granulations may be used to make a bunch of pasta products.

4.4 MAIZE PRODUCTS

Maize, often known as corn, is the most widely produced crop in the world, and it is used for food, livestock feed, and fuel. To obtain precooked refined maize flour, dehydrated nixtamalized flour, fermented maize flours, and other maize products, several food technologies are now employed to process industrially produced maize flours and corn meals in various areas of the world. These goods have varying intrinsic vitamin and mineral levels, and their processing follows diverse pathways from raw grain to finished consumer product, resulting in nutrient composition alterations. Maize processing methods and products are as varied as the maize crop itself. Many of the outputs of industrial dry maize–milling techniques can also be made on a small scale or at home. Wet items with high moisture content from small-scale manufacturers in the area have a short shelf-life and must be consumed quickly. However, users are still interested in the products because they help the homemaker save time in the kitchen. When made at home, the process is usually completed to the point where the result is ready to eat (Gwirtz and Garcia-Casal, 2014).

Some of the products prepared from maize are:

- Soups, stews, and porridge (chowder, ginataang mais, and mush);
- Bread and cakes (arepa and cookie);
- Tortilla dishes (taco and tostada);
- Fried dishes (fritters, dog, and nut);
- Corn-based beverages (tea and tejuino);
- Other products from corn (flakes, sauce, hominy, and gofio).

4.4.1 SOUPS, STEWS, AND PORRIDGE (CHOWDER, GINATAANG MAIS, AND MUSH)

In the United States (US), corn chowder is a mass-produced canned food item. Corn chowder is a soup with a thick cream base, typically made from corn, chopped onion and celery, milk or cream, butter, flour, salt, and pepper. It frequently includes potatoes and can also contain other vegetables (Hooker, 1978).

Ginataang mais is a sweet corn and rice gruel from the Philippines. It is prepared by cooking sticky rice (*malagkit*) until nearly done, then adding sweet corn, coconut milk (*gata*), and sugar. The heat is reduced just before the rice finishes cooking, though in some recipes, coconut milk is added after cooking (Abagail, 2018).

Mush is a cornmeal pudding that is boiled with water or milk. It is often allowed to set into a semi-solid state before being cut into flat squares or rectangles and pan-fried. This dish is popular in the Eastern and Southeastern US, traditionally eaten with maple syrup or molasses in the Midwest. In Eastern Europe, milk is poured over the mush once it has been served and chilled, rather than being boiled in milk. Cornmeal mush is also widely consumed in Latin America and Africa (Fussel, 2004).

4.4.2 BREAD AND CAKES (AREPA AND COOKIE)

Arepa is a flat, round, and unleavened patty made from soaked, ground maize kernels, or more commonly nowadays maize meal or maize flour, which can be baked, fried, grilled, boiled, or steamed. Color, flavor, size, and the food with which it can be stuffed differ by area. Simple arepa with butter or cheese filling is baked. Depending on the meal, more filling variants can be added by combining foods such beans, beef, avocados, eggs, tomatoes, salad, shrimp, or fish (Janer, 2008).

A corn cookie is a type of cookie made from corn flour. It is a sort of sugar cookie popular in the US and Indonesia. The color and flavor of the corn cookie come from corn products such as cornmeal, rather than wheat flour, which is frequently used in cookie making. Corn cookies, like their traditional cousins, are frequently flavored with herbs, spices, and fruits such as lemon verbena, apricot, and rosemary. Corn cookies can also be made by using cornbread batter and frying it on a hot griddle, in addition to baking (Suszko, 2006).

4.4.3 TORTILLA DISHES (TACO AND TOSTADA)

A little hand-sized maize or wheat tortilla is topped with a filling in this traditional Mexican delicacy. After that, the tortilla is folded around the filling and eaten with one hand. Tacos come in a variety of non-traditional types, such as hard-shell tacos, soft-shell tacos, breakfast tacos, Indian tacos, and so on. Tostada refers to a deep-fried or toasted flat or bowl-shaped tortilla, but it can also refer to any dish that uses a tostada as a base. It can be eaten on its own or as a base for other dishes. Tostadas are frequently made with corn tortillas. A stale tortilla can be repurposed as a tostada by frying it in hot oil until it becomes brown, stiff, and crunchy, just way stale bread can be made palatable as toast (Bayless et al., 2000).

4.4.4 FRIED DISHES FROM CORN (FRITTERS, DOG, AND NUT)

Corn fritters are cakes that are fried consisting of or containing a significant amount of maize in the dough or batter. In the southern US, as well as Indonesia, they are a traditional sweet and salty snack made from fresh corn, eggs, shallots, garlic, scallion, celery, all-purpose flour, rice flour, salt, ground white pepper, and ice-cold water (Anita, 2013). The corn dog is a sausage on a stick that has been deep-fried after being coated in a thick layer of cornmeal batter. Corn dogs are a popular street snack and fast-food item. Some vendors or restaurants dip and fry their corn dog right before serving for the greatest and freshest results. Some corn dog vendors sell frozen corn dogs that have been thawed and then fried or browned in the oven. Corn dogs are also available frozen in practically every supermarket in North America and can be heated and served. The cornbread coating on pre-made frozen corn dogs can also be heated in a microwave oven, but the texture will be lost (Neal, 2002). Corn nuts are also called toasted corn, quico, or cracker. Corn nuts are made by soaking whole corn kernels in water for three days and then deep-frying them until hard and brittle. Because the kernels shrink throughout the harvesting and cleaning procedure, soaking them restores their original size (Jason, 2000).

4.4.5 CORN-BASED BEVERAGES (TEA AND TEJUINO)

Corn tea is a corn-based Korean drink. Corn kernels, corn silk, or a combination of the two can be used to make it. In the winter, caffeine-free infusion is a popular hot beverage. Corn tea is made by drying and roasting corn kernels. The tea is then cooked with roasted corn kernels until it turns yellow. After that, the tea is strained, and the boiling corn is tossed out. Although the drink is inherently sweet, sugar is occasionally added to give it a sweeter flavor (Jung, 2017).

Tejuno is a cool beverage prepared from fermented corn that is popular in some regions of Mexico. Tejuino is typically produced with the same corn dough that is used to make tortillas and tamales. The dough is combined with water and piloncillo (unrefined cane sugar in the shape of a cone) and cooked until the liquid is quite thick. After that, the liquid is allowed to ferment very slowly. The finished drink is usually served cold, with lime juice, a pinch of salt, and a scoop of shaved ice or lime sorbet on top.

4.4.6 OTHER PRODUCTS FROM CORN (FLAKES, SAUCE, HOMINY, AND GOFIO)

Corn flakes are a commercial cereal made of small, toasted corn flakes that are often served cold with milk and sugar. The simple flakes have been flavored with salt, sugar, and malt since their inception, and various subsequent versions with extra ingredients, such as sugar-frosted flakes and honey and nut corn flakes, have been produced. Crushed flakes can be used in place of breadcrumbs in recipes and added to a variety of prepared foods. Corn sauce known as fermented maize sauce; corn sauce is made with corn starch as the principal substrate. It comes in paste and powder forms and is used as a food condiment and ingredient. Corn sauce has a savory flavor, similar to soy sauce. It is used to flavor soups, broths, and gravies, among other things (Charve et al., 2018).

Hominy is a food made from dried maize kernels that have been nixtamalized, a process that involves treating them with an alkali. Hominy is eaten whole, ground into sand-sized grits, or ground into flour (Fussell and Harper, 2004). Gofio is roasted maize flour from the Canary Islands. It can be used in a variety of ways, including kneading, cooking, and baking. It can be used as a thickening agent as well. Vitamins, protein, fiber, and minerals are abundant in this food. Argentina, Chile, Panama, Venezuela, Uruguay, and several other nations drink gofio dry with sugar or with milk.

4.5 BARLEY PRODUCTS

Barley, the world's oldest and most diversified food crop, was largely utilized as animal feed. Dietary fiber, niacin, magnesium, copper, and iron are all found in abundance in this cereal. Barley fiber, particularly β-glucan in whole-grain barley, has been associated with many health advantages (Makeri et al., 2013). Phytochemicals such as phenolic acids, flavonoids (Keservani and Sharma, 2014; Keservani et al., 2016a, b), lignans, tools, phytosterols, and folate are also present. These phytochemicals have substantial antioxidant, antiproliferative, and cholesterol-lowering properties, which could help to reduce the risk of certain diseases (Idehen et al., 2017; Zhu, 2017). Furthermore, barley proteins can improve the rheological qualities of food items due to their functional properties, such as emulsifying capacity and stability, foaming, elasticity, cohesiveness, and water-holding capacity. Barley can be an excellent substrate for the growth of fastidious lactic acid bacteria (LAB) due to its nutritional properties. LAB has been shown to ferment grain substrates in several investigations. Several researchers have reported on the fermentation of whole-grain barley flour, barley, and barley malt medium with various species of lactobacilli (Hole et al., 2012; Salari et al., 2014). The nutritional and functional properties of barley, as well as its potential as a substrate for LAB growth, have prompted researchers to investigate barley as a functional ingredient for the development of functional fermented meals.

Barley products have traditionally been consumed in countries and regions of the world where other crop production is difficult. The barley grains were pearled, roasted, rolled, flaked, or ground to grits and flour for use as food and incorporated into a variety of dishes. The processed varieties of barley are shown in Table 4.1.

TABLE 4.1 Processed Varieties of Barley

Processed Barley	Description
Hulled	It is whole-grain barley that's merely had the outermost hull removed. This barley type is high in fiber and protein and is very healthy.
Hulless	The outer hull of this variety of barley is loosely linked to the kernel.
Grits	This type of barley is made out of toasted kernels that have been shattered into little bits.
Flakes	This variety of barley resembles rolled oats in appearance. The barley kernels were cut, cooked, and rolled into flakes before being dried.
Flour	Dehulled, hulled, and pearled barley grain is ground to flour using pin or hammer mills to make whole and pearled barley flours, or roller milled to produce white barley flour and fiber-rich fractions.
Pearl	Pearled grain is made from dehulled or hull-less barley grains that have been abraded to remove the rough, fibrous bran layers and germ, resulting in smooth, white grain (endosperm).

Compared to hulled barley and oats, bare barley has a larger potential for food product development. This is because the food is of higher quality and requires less processing. Unlike hulled cultivars, which may lose several nutritional layers during pearling, hull-less barley cultivars require minimum processing and leave the bran layer intact. Barley, unlike wheat, lacks the protein gluten, hence it lacks some of the qualities of bread made with wheat flour. Despite future revolutionary product development, particularly those from selectively selected cultivars, barley is used in a variety of food items, including flatbread, muffins, pasta, noodles, pearled barley, and rice extenders (Sullivan et al., 2013).

4.5.1 POTENTIAL BARLEY-BASED HEALTH PRODUCTS

1. **Multigrain Atta:** Hulled barley can be used to boost the amount of soluble and insoluble fiber in wheat atta (flour). The glycemic index (GI) of chapattis, which is a staple cuisine in northern India, may be reduced as a result of this.
2. **Barley Infusion:** This is a coffee alternative, but it does not contain caffeine because it is made from toasting and grinding barley grains. The name barley coffee is quite common in Europe and Italy.

3. **Barley Tea:** The lightly roasted barley grains are steeped in boiling water to make this. It goes by the name mugicha in Asia.
4. **Barley Water:** This beverage is made by boiling barley grains in water and flavoring it with various fruits of your choice at the time of boiling or after boiling.
5. **Multigrain Biscuits and Bread:** The addition of hulless barley to both bread and biscuits may improve their health benefits. The practicality of barley-based biscuits has been demonstrated in preliminary trials at the ICAR-IIWBR. In the same way, barley malt can be used to flavor biscuits.
6. **Barley Flakes:** Hulless barley can be used to make breakfast cereals such as flakes, but it must be flavored to make it more appealing.
7. **Ready-to-Drink Sattu:** Sattu is a traditional drink made from barley and/or gram flour, but if a ready-to-drink barley-based sattu is developed, it might compete with health drinks. Traditional barley meals made with different types of processed barley grain ingredients is shown in Table 4.2.

TABLE 4.2 Food Products from Various Types of Processed Barley Grain Ingredients

Processed Barley Grain Ingredient	Food Products
Flour	• Pinni;
	• Bread;
	• Injera (pancake-like product);
	• Kitta (unleavened, thin, dehydrated bread) Shorba (soup);
	• Tihilo (flour dough balls).
Roasted hulled grain	• Boricha and mugicha (barley tea).
Whole grain	• Jhang chang;
	• Nakia (beverage);
	• Sanchak tukba (porridge).
Malt	• Borde, bequre (an unfermented or slightly fermented beverage).
Roasted grain	• Kolo (grain mixed with roasted legumes and oil seeds);
	• Yue (popcorn-like products).
Roasted grain flour	• Tsangpa-ba (dough balls with butter and tea);
	• Chiko, besso (dough balls with water, butter, or oil).

4.6 OAT-BASED FOOD PRODUCTS (MUESLI, MILK, CAKE, OATMEAL, PARKIN, PORRIDGE, INFANT FOODS, AND ROLLED OATS)

The high quantity of dietary fiber, phytochemicals, and nutritional value of oats (*Avena sativa* L.) has attracted a lot of interest. Oats are thought to have a variety of health benefits, including hypocholesterolemic and anti-cancer characteristics. Oat has also just been approved for celiac disease sufferers' diets. Oat-based food products such as bread, biscuits, cookies, probiotic drinks, morning cereals, flakes, and infant food are gaining popularity due to their high nutritional content (Rasane et al., 2015). Cosmetics, nutraceuticals, and pharmaceuticals are just a few of the various uses for oats (Webster, 2016). Oat grains are used in a variety of industrial and nonfood sectors like their application in the cosmetics industry, oats are used as fodder and unanimously processed into thousands of value-added products including oat milk, oatcake, oatmeal, parkin, porridge, infant foods, rolled oat.

Muesli is a chilled oatmeal dish made with rolled oats, grains, nuts, seeds, and fresh or dried fruits. Muesli was historically made with milk or cream, a splash of citrus juice, and a sweetener like honey, which was steeped overnight. Yogurt or other milk products are now routinely used in muesli recipes, both purchased and handmade (Kurmann et al., 1992). Oat milk is a plant milk made by extracting the plant material from entire oat grains with water. Oat milk comes in a variety of tastes, including sweetened, unsweetened, vanilla, and chocolate, and has a creamy texture and oatmeal-like flavor. Oat milk can be used to replace dairy milk in vegan diets or in circumstances where dairy is incompatible with medical issues, such as lactose intolerance or a cow milk allergy. Oat milk has a low environmental effect when compared to dairy milk and other vegan milk products since it requires less land and water to produce (Sethi et al., 2016). Oatcake is a flatbread that resembles a cracker or a biscuit, and in some cases, resembles a pancake. Oatmeal is used as the main ingredient, with plain or whole meal flour added occasionally. On a griddle or in the oven, oatcakes are prepared. Nairn's, Stockan's, Paterson's, and Walkers are just a few examples of commercially available oatcake brands. Aside from the bigger commercial oatcake manufacturers, numerous small-town bakers provide variants on the fundamental recipe (Cauvian and Young, 2009). Oatmeal is a de-husked, steamed, and flattened oat product, or coarse flour formed from hulled oat grains (groats) that have either been milled (ground) or steel-cut. White oats are another name for ground oats. It can be eaten raw, as in muesli, or cooked, as in porridge, with water or milk.

Oatmeal is also employed as a thickening ingredient in thick, savory Arabic or Egyptian meat-and-vegetable soups, as well as a low-cost source of fiber and nutritional content in meatloaf (Chakraverty et al., 2003). Parkin is a typical gingerbread dessert made with oatmeal and black treacle. Parkin is baked into a hard cake that becomes moist and sticky after resting. It has a drier, more biscuit-like texture in Hull and East Yorkshire than it does elsewhere. Porridge is a cereal meal produced by heating or boiling ground, crushed, or chopped starchy plants, usually grain, in milk. It's commonly prepared or served with added flavors like sugar, honey, (dried) fruit, or syrup to produce a sweet cereal or it's blended with spices, meat, or vegetables to make a savory dish. Depending on its consistency, it's typically served hot in a bowl. One of the most popular forms of porridge is oat porridge (also known as oatmeal). Gruel, Groats, and Stirabout are some types of oat porridge. Drum-dried oatmeal is commonly used as an infant's first solid food introduction in many regions of the Western world. Many marketed infant meals contain oat flour as a thickening. Other ingredients are frequently combined with oats to accomplish a specific nutritional goal in newborn meals. Rolled oats are used for the preparation of granola cereals and bars. They are also used as a thickener in soups, gravies, and sauces, as well as a healthy meat extender in meatloaf and meat patties, and as a breading for fried meats (Burnette, 1992).

4.7 MILLET AND MILLET-BASED PRODUCTS

Millets are a category of small-seeded crops that are thought to be the world's first cultivated food plants. Their ability to flourish in the world's hardest climatic conditions has earned them the moniker famine crops throughout Africa and Asia's semi-arid tropics and drylands. Millets are nutrient-dense and play a significant role in people's diets. In adverse agro-climatic circumstances, it is considered a crop (Kumari et al., 2004). Millets are employed as nutraceuticals because they contain far more antioxidants than the other cereal crops. Millets have a unique role in human nutrition due to their many characteristics and high mineral content. Millet is a high-nutrient, non-glutenous, non-acid-forming food that is easy to digest and suitable for diabetics. Millets are high in fiber, B vitamins (particularly niacin and B_6), calcium, iron, potassium, magnesium, and zinc. Minor millets are not only nutritious, but they also have a wide range of medical applications. They're used to treat ailments including cancer, leprosy, and pneumonia, among others (Bachar et al., 2013). Despite their high nutritional value, millets are difficult to process and cook due to natural characteristics such as stiff seed coats, antinutritional factors, poor digestibility, and low micronutrient bioavailability. Value-added RTE millet products are not commonly available in the market due to the lack of primary, secondary, and tertiary processing technology (Malleshi, 2014). As people become more aware of millets' nutraceutical (Keservani et al., 2010a, b, 2017, 2020) benefits, they can be marketed and branded in the form of well-processed value-added goods (Sehgal and Kwatra, 2003). Sorghum (*Sorghum bicolor*), pearl millet (*Pennisetum glaucum*), finger millet (*Eleusine coracana*), foxtail millet (*Setaria italica*), barnyard millet (*Echinochloa frumentacea*), kodo millet (*Paspalum scrobiculatum*), proso millet (*Panicum miliaceum*), and little millet (*Panicum miliare*) are the important millets cultivated in African and Asian countries (Jaybhaye et al., 2014) and are used for the manufacture of the following food products:

- Puffed/popped millets: Puffs from sorghum, foxtail millet, pearl millet;
- Extruded products: Extruded snacks and flakes;
- Millets semolina (rawa/suji);
- Millet-based baked products;
- Millet-based fermented products.

4.4.1 PUFFED/POPPED MILLETS (PUFFS FROM SORGHUM, FOXTAIL MILLET, AND PEARL MILLET)

In both industrialized and developing countries, staple grains are consumed in a variety of puffed and popped products. Millet and cereal grains are primarily composed of starch. Various processing procedures, primarily heat treatments, cause physicochemical and structural changes in the starch–protein matrix, resulting in grain expansion and puffed products. Expanded flakes and other popped products are commonly made using the high-temperature short-time (HTST) method (Jaybhaye et al., 2014).

Some examples of puffed millets are as follows:

1. **Sorghum Puffs:** These are a byproduct of explosive or cannon puffing, in which the sorghum grain is expanded to its maximum potential while retaining its original shape. It is a RTE snack created using a puff gun machine. This machine loads dehulled sorghum grain into a rotating barrel, then roasts and rapidly ejects the grain,

producing puffed sorghum (Sharma et al., 2015). These products offer unique advantages, such as:

i. Similar to puffed rice, sorghum puffs are white in color and crisp in texture.
ii. When packed in airtight metalized polyester (MET) pouches at room temperature, they have a shelf life of up to 4 months.
iii. They are a good source of protein and fiber, available in varieties such as masala-coated and fried or flour-coated and fried (including bajra and channadhal).
iv. They make an excellent in-flight or general evening snack.

2. Foxtail Puffs: These are a byproduct of explosive or gun puffing, in which foxtail millet grains are expanded to their maximum extent while maintaining their original shape. This RTE snack is produced using a puff gun machine. The machine loads dehulled foxtail millet into a rotating barrel, which is then roasted and rapidly ejected to produce the puffs. Advantages and unique qualities of foxtail puffs include:

   i. Similar to puffed rice, foxtail puffs are white in color and have a crisp texture.
   ii. When packaged in airtight metalized polyester (MET) pouches at ambient temperatures, they have a shelf life of up to 2 months.
   iii. They are a good source of protein and fiber, with masala-coated and fried varieties available.
   iv. They make an excellent in-flight or general evening snack.

3. **Pearl Millet (Bajra) Puffs:** These puffs are also a byproduct of explosive or gun puffing, in which bajra grains are expanded to their fullest while maintaining their shape. Produced with a puff gun machine, bajra grains are fed into a rotating barrel, roasted, and then rapidly ejected to form the puffs. Unique features of bajra puffs include:

   i. Bajra puffs have a greenish-creamy color and a crisp texture.
   ii. When packed in sufficiently tight MET pouches at ambient temperatures, they have a shelf life of up to 3 weeks.
   iii. They are available in masala-coated and roasted varieties.
   iv. They are suitable as in-flight snacks or general evening snacks.

4.4.2 EXTRUDED PRODUCTS (SNACKS AND FLAKES)

Extrusion cooking is a post-harvest processing method that involves heating starchy and protein-rich materials at high temperatures for a short duration. This process improves protein digestibility, quality, and the variety of processed foods (Dahlin and Lorenz, 1992). Heat is applied either directly via steam injection or indirectly through a jacket, with mechanical energy dissipated through shearing within the mixture.

Extruded snacks are RTE snacks made using a twin-screw hot extruder, which combines heating and extrusion to create a cooked, shaped product through round or minus-shaped dies. Although corn is the primary ingredient in most commercial extruded snacks, sorghum grits, rice, ragi, wheat, and corn flour are also used in these snacks. To produce expanded snacks, the mixture is blended and passed through a twin-screw extruder. Flavor and taste modifications are made by coating the snack with preferred spices. Unique qualities of these extruded snacks include:

- Crisp texture with colors ranging from white to cream;
- Suitable as in-flight snacks or general evening snacks;
- High in protein, fiber, iron, zinc, and magnesium;
- A shelf life of up to 6 months;
- Available in masala-coated varieties.

Extruded flakes are also RTE products made with a twin-screw hot extruder. This method combines heating and extrusion to produce a circular product that is subsequently flattened using a roller flaker machine. Sorghum grits, wheat, and maize flour are commonly used to make extruded flakes. Flavor and taste can be modified by coating the flakes with preferred spices. Key features of extruded flakes include:

- They can be used as breakfast cereals in place of cornflakes;
- Available in masala-coated, chocolate-coated, and essence-coated varieties.

4.4.3 MILLETS SEMOLINA (RAWA/SUJI)

Semolina is a ready-to-cook grain. Dry milling is used to process millet grains (Pearl millet, finger millet, and foxtail millet). Cleaning the grains is the first step in the dry milling process. To make semolina, the cleaned grain is ground in hammer mills to separate the endosperm, germ, and bran from each other. Semolina made from 3 variants of millet grain is pulverized to make semolina, which comes in a variety of particle sizes. The mesh size of the mill is modified according to the variant that has to be processed. The products and byproducts are:

i. Coarse semolina (kichidi rawa) yield is 68–72%; byproduct yield is 32–28% (contains medium/fine semolina, flour, and bran).
ii. Medium semolina (upma rawa) yield is 71–76%; byproduct yield is 29–24% (contains coarse/fine semolina, flour, and bran).
iii. Fine semolina (idli rawa) yield is 74–80%; byproduct, yield is 26–20% (contains coarse/medium semolina, flour, and bran).

The advantages and uniqueness of product are (i) upma, khichidi, rawa laddu, idli, dosa, kesari, and other dishes can be made using utility; (ii) protein, fiber, iron, and zinc are all abundant in them; and (iii) for pearl millet and finger millet, all three rawa can be preserved for four months.

4.4.4 MILLET-BASED BAKED PRODUCTS

Baked goods are popular among people of all ages around the world due to their longer shelf-life, ease of marketing, and attractive packaging (Patel and Rao, 1996). The baking sector, on the other hand, is dominated by wheat's various value-added products. These products are not preferred by celiac disease sufferers, especially in industrialized countries, due to their high gluten content. Millet-based bread items will not only be more nutritious, but they will also command a higher market price (Verma and Patel, 2013). Biscuits and muffins are made with foxtail millet and finger millet flour, while cakes are made with barnyard millet flour. In recent years, demand for finger millet-based baked items has surged in metropolitan areas.

4.4.5 MILLET-BASED FERMENTED PRODUCTS

Millets are a decent source of protein in general, however, the quality of the protein in terms of the necessary amino acid profile is low (Jaybhaye et al., 2014). Probiotic fermentation and germination of millets have been shown to improve protein digestibility, as well as the level of lysine, thiamine, and niacin, sugars, protein fractions, soluble fiber, and *in vitro* micronutrient availability (Arora et al., 2011). In many parts of India, fermented food products such as idli and dosa are the most popular and typical breakfast (Antony et al., 1996; Mbithi-Mwikya et al., 2000).

4.8 CONCLUSION

Cereals and millets have a wide range of dietary applications and also have good functional properties that may aid to improve the health of humans. New food development necessitates greater marketing campaigns, and consumers may require time to adjust to the new product. Nutritional food components enable producers to meet and exceed the demands of today's health-conscious consumer by either inventing new and inventive products or simply reshaping existing products. Although sodium levels in these and other processed cereal meals should continue to be reduced to assist people to minimize their overall sodium consumption, fortified cereal commodities such as breakfast cereals and white bread are significant sources of nutrition for both children and adults. In addition, novel food items will continue to be a prominent study focus in the innovative product development category of food science and technology to achieve customer acceptability through technological interventions. For industrial production, research should focus on enhancing their shelf-life.

KEYWORDS

- **anti-cancer characteristics**
- **celiac disease**
- **dietary fiber**
- **health benefits**
- **morning cereals**
- **novel food**
- **nutraceuticals**
- **nutritional value**
- **pharmaceuticals**
- **phytochemicals**
- **value-added products**

REFERENCES

Abah, C. R., Ishiwu, C. N., Obiegbuna, J. E., & Oladejo, A. A. (2020). Nutritional composition, functional properties and food applications of millet grains. *Asian Food Science Journal, 14*(2), 9–19.

Antony, U., Sripriya, G., & Chandra, T. S. (1996). The effect of fermentation on the primary nutrients in foxtail millet (*Setaria italica*). *Food Chemistry, 56*(4), 381–384.

Arora, S., Jood, S., & Khetarpaul, N. (2011). Effect of germination and probiotic fermentation on nutrient profile of pearl millet-based food blends. *British Food Journal, 113*(4), 470–481.

Asseng, S., Ewert, F., Martre, P., Rötter, R. P., Lobell, D. B., Cammarano, D., Kimball, B. A., Ottman, M. J., Wall, G. W., White, J. W., & Reynolds, M. P. (2015). Rising temperatures reduce global wheat production. *Nature Climate Change, 5*(2), 143–147.

Bayless, R., Brownson, J., & Bayless, D. G. (2000). *Mexico One Plate at a Time*. Simon and Schuster.

Belitz, H. D., Grosch, W., & Schieberle, P. (2009). Cereals and cereal products. *Food Chemistry*, 670–745.

Bhattacharya, K. R. (2011). Product-making quality of rice. *Rice Quality: A Guide to Rice Properties and Analysis*, 298–336.

Burnette, D., Lenz, M., Sisson, P. F., Sutherland, S., & Weaver, S. H. (1992). Marketing, processing, and uses of oat for food. *Oat Science and Technology, 33*, 247–263.

Cauvain, S. P., & Young, L. S. (2009). *More Baking Problems Solved*. Elsevier.

Chakraverty, A., Mujumdar, A. S., & Ramaswamy, H. S. (2003). *Handbook of Postharvest Technology: Cereals, Fruits, Vegetables, Tea, and Spices* (p. 93). CRC press.

Charalampopoulos, D., Wang, R., Pandiella, S. S., & Webb, C. (2002). Application of cereals and cereal components in functional foods: A review. *International Journal of Food Microbiology, 79*(1, 2), 31–141.

Charve, J., Manganiello, S., & Glabasnia, A. (2018). Analysis of umami taste compounds in a fermented corn sauce by means of sensory-guided fractionation. *Journal of Agricultural and Food Chemistry, 66*(8), 1863–1871.

Dahlin, K., & Lorenz, K. (1993). Protein digestibility of extruded cereal grains. *Food Chemistry. 48*(1), 13–18.

Gujral, H. S., & Rosell, C. M. (2004). Improvement of the breadmaking quality of rice flour by glucose oxidase. *Food Research International, 37*(1), 75–81.

Gujral, H. S., Guardiola, I., Carbonell, J. V., & Rosell, C. M. (2003). Effect of cyclodextrinase on dough rheology and bread quality from rice flour. *Journal of Agricultural and Food Chemistry, 51*(13), 3814–3818.

Gwirtz, J. A., & Garcia-Casal, M. N. (2014). Processing maize flour and corn meal food products. *Annals of the New York Academy of Science, 1312*(1), 66–75.

Hole, A. S., Rud, I., Grimmer, S., Sigl, S., Narvhus, J., & Sahlstrøm, S. (2014). Improved bioavailability of dietary phenolic acids in whole grain barley and oat groat following fermentation with probiotic *Lactobacillus acidophilus*, *Lactobacillus johnsonii*, and *Lactobacillus reuteri*. *Journal of Agricultural and Food Chemistry, 60*(25), 6369–6375.

Hooker, R. J. (1978). *The Book of Chowder*. Harvard Common Press.

Janer, Z. (2008). *Latino Food Culture*. ABC-CLIO.

Jaybhaye, R. V., Pardeshi, I. L., Vengaiah, P. C., & Srivastav, P. P. (2014). Processing and technology for millet based food products: A review. *Journal of Ready to Eat Food, 1*(2), 32–48.

Kadan, R. S., Bryant, R. J., & Boykin, D. L. (2001). Effects of processing conditions on qualities of rice fries. *Journal of Food Science*, *66*(4), 610–613.

Keservani, R. K., & Sharma, A. K. (2014). Flavonoids: Emerging trends and potential health benefits. *Journal of Chinese Pharmaceutical Sciences*, *23*(12), 815.

Keservani, R. K., Kesharwani, R. K., Sharma, A. K., Vyas, N., & Chadoker, A. (2010b). Nutritional supplements: An overview. *International Journal of Current Pharmaceutical Review and Research*, *1*(1), 59–75.

Keservani, R. K., Kesharwani, R. K., Vyas, N., Jain, S., Raghuvanshi, R., & Sharma, A. K. (2010a). Nutraceutical and functional food as future food: A review. *Der Pharmacia Lettre*, *2*(1), 106–116.

Keservani, R. K., Sharma, A. K., & Kesharwani, R. K. (2016a). Nutraceutical and functional foods for cardiovascular health. In *Food Process Engineering* (pp. 257–278). Apple Academic Press, CRC Press.

Keservani, R. K., Sharma, A. K., & Kesharwani, R. K. (2016b). Medicinal effect of nutraceutical fruits for the cognition and brain health. *Scientifica*, *3109254*, 10. doi:10.1155/2016/3109254.

Keservani, R. K., Sharma, A. K., & Kesharwani, R. K. (2017). An overview and therapeutic applications of nutraceutical and functional foods. *Recent Advances in Drug Delivery Technology*, 160–201.

Keservani, R. K., Sharma, A. K., & Kesharwani, R. K. (2020). *Nutraceuticals and Dietary Supplements: Applications in Health Improvement and Disease Management*(pp. 1–342). CRC Press.

Kurmann, J. A., Rasic, J. L., & Kroger, M. (1992). *Encyclopedia of Fermented Fresh Milk Products: An International Inventory of Fermented Milk, Cream, Buttermilk, Whey, and Related Products*. Springer Science & Business Media.

Lin, Y. H. E., & Jacops, L. (2002). *Method of Making Quick Cooking and Instant Rice* (US Patent 6,416,802). Uncle Bens Inc.

Macauley, H., & Ramadjita, T. (2015). Cereal crops: Rice, maize, millet, sorghum, wheat.

Makeri, M. U., Nkama, I., & Badau, M. H. (2013). Physico-chemical, malting and biochemical properties of some improved Nigerian barley cultivars and their malts. *International Food Research Journal, 20*(4).

Malleshi, N. G. (2014). *Post-Harvest Processing of Millets for Value Addition.* Harvard University. Retrieved from: https://isites.harvard.edu/fs/docs/icb.topic867074.files/millet (accessed on 5 November 2024).

Mbithi-Mwikya, S., Ooghe, W., Van Camp, J., Ngundi, D., & Huyghebaert, A. (2000). Amino acid profiles after sprouting, autoclaving, and lactic acid fermentation of finger millet (*Eleusine coracan*) and kidney beans (*Phaseolus vulgaris* L.). *Journal of Agricultural and Food Chemistry*, *48*(8), 3081–3085.

McKevith, B. (2004). Nutritional aspects of cereals. *Nutrition Bulletin*, *29*(2), 111–142.

Nishita, K. D., & MM, B. (1979). *Physicochemical Properties of Rice in Relation to Rice Bread.*

Patel, M. M., & Venkateswara Rao, G. (1996). Influence of untreated, heat-treated and germinated blackgram flours on biscuit making quality of wheat flour. *Journal of Food Science and Technology*, *33*(1), 53–56.

Rasane, P., Jha, A., Sabikhi, L., Kumar, A., & Unnikrishnan, V. S. (2015). Nutritional advantages of oats and opportunities for its processing as value added foods-a review. *Journal of Food Science and Technology*, *52*, 662–675.

Reynolds, A. N., Akerman, A. P., & Mann, J. (2020). Dietary fiber and whole grains in diabetes management: Systematic review and meta-analyses. *PLoS Medicine*, *17*(3), 1003053.

Salari, M., Razavi, S. H., & Gharibzahedi, S. M. T. (2015). Characterizing the synbiotic beverages based on barley and malt flours fermented by *Lactobacillus delbrueckii* and *paracasei* strains. *Quality Assurance and Safety of Crops and Foods*, *7*(3), 355–361.

Sehgal, A., & Kwatra, A. (2003). Processing and utilization of pearl millet for nutrition security. In *Proceedings of the National Seminar on Recent Trends in Millet Processing and Utilization* (p. 16). CCS HAU, Hissar, India.

Sethi, S., Tyagi, S. K., & Anurag, R. K. (2016). Plant-based milk alternatives an emerging segment of functional beverages: A review. *Journal of Food Science and Technology*, *53*, 3408–3423.

Sharma, V., Champawat, P. S., & Mudgal, V. D. (2014). Process development for puffing of sorghum. *International Journal of Current Research and Academic Review*, *2*(1), 164–170.

Sullivan, P., Arendt, E., & Gallagher, E. (2013). The increasing use of barley and barley byproducts in the production of healthier baked goods. *Trends in Food Science and Technology*, *29*(2), 124–134.

Suszko, M. K. (2006). *Farms & Foods of Ohio: From Garden Gate to Dinner Plate*. Hippocrene Books.

Ushakumari, S. R., Latha, S., & Malleshi, N. G. (2004). The functional properties of popped, flaked, extruded and roller-dried foxtail millet (*Setaria italica*). *International Journal of Food Science & Technology*, *39*(9), 907–915.

Verma, V., & Patel, S. (2013). Value added products from nutri-cereals: Finger millet (*Eleusine coracana*). *Emirates Journal of Food and Agriculture*, 169–176.

Webster, F. (2016). *Oats: Chemistry and Technology*. Academic Press.

Wieser, H., Koehler, P., & Scherf, K. A. (2020). The two faces of wheat. *Frontiers in Nutrition*, *7*, 517313.

Yeh, A. (2004). Preparation and applications of rice flour. *Rice: Chemistry and Technology*, 495–539.

Zhu, F. (2017). Barley starch: Composition, structure, properties, and modifications. *Comprehensive Reviews in Food Science and Food Safety*, *16*(4), 558–579.

CHAPTER 5

Nutraceuticals and Bioactives in Fermented Fruit Beverages

POOJA,[1*] B. V. C. MAHAJAN,[1] PRIYANKA SUTHAR,[2] and SHUBHAM SAMKARIYA[2]

[1]*Punjab Horticultural Postharvest Technology Center, Punjab Agricultural University, Ludhiana, Punjab, India*

[2]*Department of Food Science and Technology, Dr. Y. S. Parmar University of Horticulture and Forestry, Solan, Himachal Pradesh, India*

*Corresponding author.

ABSTRACT

Fermented foods, particularly traditional or ethnic dishes, are a significant element of today's food markets. The fermented food market is seeing rapid expansion and innovation, thanks to increased demand for effective use of agricultural wastes and advancements in fermentation technology (microbial and enzyme-based processing). Furthermore, the health-promoting properties of fermented foods are gaining popularity. Many typical fermented foods contain microorganisms that can behave as "micro-factories," producing nutrients and bioactive with distinct nutritional and health benefits. The need for proper microbe and raw material selection, as well as the requirement for precise fermentation process management, are discussed. The growth and consumption of functional foods have grown in popularity because they offer health benefits with basic nutritional functions. Fruit beverages are becoming increasingly popular among consumers since they provide health benefits due to their natural properties, both in terms of disease prevention and cure. Because of the ease and ability to accommodate customer needs, beverages, as well as ease of distribution and storage for refrigerated and shelf-stable products, are currently the most active functional food category. Furthermore,

Functional Compounds and Foods of Plant Origin: Krishan Dutt Sharma, Satish Kumar, Rakesh Sharma, and Raj K. Keservani (Eds.)

they are a wonderful source of vitamins, minerals, antioxidants, omega-3 fatty acids, plant extracts, fiber, prebiotics, and probiotics, as well as other nutrients and bioactive compounds. Producing appealing foods owing to their color, texture, look, and sensory features, on the other hand, is a constant issue in the food sector because the eyes also consume. It's much more difficult to combine nutritious and appealing ingredients in a single meal. The extensive list of advantages linked with phenolic compounds, including anticancer, antioxidant, anti-aging, and anti-inflammatory capabilities, among others, completely supports their inclusion in the fortification process.

5.1 INTRODUCTION

Beverages encompass a wide range of fluid options consumed through drinking, spanning from the simplest of drinks like water to a variety of commercially available liquids such as fruit juices, artificial beverages, alcoholic drinks, dairy products, coffee, and chocolate-flavored drinks. Regardless of their diverse qualities, all beverages share a common purpose: quenching thirst. Essentially, beverages are liquids primarily intended for human consumption. While they are not typically consumed for their nutritional value, they remain vital for our survival and are deeply ingrained in our cultural practices. There are, however, certain prerequisites for beverages, including being made from food ingredients, complying with food safety regulations, and being consumed in large quantities, sometimes even surpassing the consumption of potable water.

play a significant role in our dietary intake, and their impact on health has gained increasing attention. Consumer expectations regarding their size, packaging, shelf life, and the ability to incorporate beneficial nutrients and bioactive compounds are all met by beverages. Cagno et al. (2013) have recently classified beverages as a source of nutraceuticals (Keservani et al., 2010a, b, 2017, 2020), highlighting their importance, health benefits, interactions with food, preservation of functional compounds in the human body, and their potential to contain minerals, amino acids, vitamins, crucial fatty acids, and botanical extracts. Functional beverages, such as probiotic and mineral drinks, vegetable, and fruit juices, as well as sports and energy drinks, are prime examples of this category. These functional beverages can aid in boosting the immune system (Arya et al., 2022; Keservani et al., 2015; Kumar et al., 2022; Rane et al., 2022), improving digestive and heart health, aiding in weight lost, or even acting as supplements to neutralize the aging process (Berner and O'Donnell, 1998).

5.2 CLASSIFICATION OF BEVERAGES

Beverages can be categorized in a variety of ways. The classification criteria may be influenced by several ways like (i) synthetic and natural; (ii) non-carbonated and carbonated beverages; (iii) non-alcoholic and alcoholic fermented beverages; (iv) temperature of serving; and (v) stimulants and non-stimulants based on physiological effect.

5.2.1 NATURAL AND SYNTHETIC BEVERAGES

Naturally, sourced beverages are crafted from ingredients found in nature, including fruit juices, milk, acids, flavorings, malt, sugar, and colorings. Examples of these beverages include fruit-based, malt, and dairy drinks. On the other hand, synthetic beverages resemble their natural counterparts but may contain artificial flavorings and chemicals. They are primarily created to offer affordable enjoyment to consumers. A prevalent type of synthetic beverage is soft drinks, which are typically made with flavored sugar syrup. Highly potent sweetened beverages also fall into the synthetic beverage group, utilizing artificial sweeteners to reduce calorie content.

5.2.2 NON-CARBONATED AND CARBONATED BEVERAGES

Non-carbonated and carbonated beverages differ based on the presence of dissolved carbon dioxide. Carbonated beverages, such as Coca-Cola and lemonade, are created by infusing carbon dioxide into syrup or water, resulting in effervescence and fizz when pressure is released. Soft drinks are a common example of carbonated beverages. In some cases, like beer, carbon dioxide is naturally produced during the fermentation process. Carbonation serves various purposes, offering consumers a bubbly sensation and a unique flavor due to dissolved carbonic acid. Another carbonated beverage that can be flavored is soda water. This category also encompasses hot and alcoholic beverages that lack carbonation.

5.2.3 NON-ALCOHOLIC AND ALCOHOLIC FERMENTED BEVERAGES

Ethanol is found in alcoholic beverages and can be ingested for its intoxicating effects. Fermentation, whether spontaneous or managed, produces alcoholic beverages. Alcoholic drinks can be separated into three classes depending upon the raw materials utilized and process technology used in their production.

5.2.3.1 BEER

Beer is the third most consumed beverage globally. To make beer, barley malt breakdown's soluble liquid, known as "wort," is fermented with the appropriate yeast until it contains 4–8% alcohol. Beer is characterized by its effervescence, created by carbon dioxide, and its bitterness, imparted by hops (*Humulus lupulus*) solids like iso acids and polyphenols. Various flavors of beer are available, but two prominent types are "Lager" and "Ale." Lager beer is produced using *Saccharomyces pastorianus*; a bottom fermenting yeast, at 7–12°C for a longer time, while Ale beer is fermented with *Saccharomyces cerevisiae;* a top-fermenting yeast at higher temperature of 18–25°C.

5.2.3.2 WINE

Wine is produced from a variety of fruits, with grapes, both green and red, being the most commonly used. Grape juice is extracted, that undergoes natural fermentation by yeasts like *Saccharomyces* spp., *Stellata* spp., *Pichia* spp., and lactic acid bacteria (LAB). Unlike beer, the fermentation process for wine is lengthier, and most wines are aged for months to years to develop their desired sensory qualities. Green grapes produce white wines, while red grapes are utilized for red wine production, containing anthocyanin as a coloring pigment. Red wine can also undergo secondary fermentation called "Malolactic fermentation" to mellow its harsh flavor. Wines typically contain 9–16% alcohol, and sparkling wines are known for their effervescence due to carbon dioxide, as seen in champagne.

5.2.3.3 SPIRIT

Spirits are created by distilling a fermented base, and they are not sweetened. This fermented base can be derived from sources such as molasses (a sugar industry by-product), cereal extract, fruit juices, or a combination of several fermentable substrates. Spirits have notable higher alcohol content, which can reach up to 20%. The distillation process increases alcohol concentration while reducing congeners' levels.

5.3 FRUIT WINES AS NUTRACEUTICAL DRINK

Beverages play a crucial role in supporting growth, development, and various physiological functions necessary for a healthy life. Wine, an alcoholic beverage, is produced by fermenting fruit juices using Saccharomyces cerevisiae, a common yeast that also naturally generates carbon dioxide (Combina et al., 2005). Wine possesses substantial nutritional value, accounting for roughly 10–20% of the total energy intake in adult males (Macrae et al., 1993). It contains a wide range of bioactive substances, including phenolic compounds, aromatic compounds, flavonoids (Keservani and Sharma, 2014; Keservani et al., 2016a, b), non-flavonoids, anthocyanins, ascorbic acid, tannins, stilbenes, resveratrol, and various antioxidants. The primary components of wine are water (making up 75–90% of its composition) and ethyl alcohol (comprising at least 8–14% v/v). However, the specific composition can vary depending on the source, production methods, and fermentation conditions. Key attributes used to characterize wine encompass its alcohol content, flavor profile, and aroma. The phenomenon known as the "French paradox," which gained recognition in the 1990s, brought attention to the potential positive effects of moderate wine consumption.

Despite the relatively high intake of saturated fats in the French diet compared to other nations, lower mortality rates from coronary heart disease were observed in correlation with daily wine consumption (Renaud and de Lorgeril, 1992). In moderation, alcohol can serve as a stimulant that enhances blood circulation throughout the body, while the polyphenols and other bioactive compounds present in wine exert antioxidant properties (Bors and Michel, 2002). Epidemiological investigations have demonstrated that moderate alcohol and/or wine consumption may reduce the risk of various contemporary diseases, including CVD (Bradamante et al., 2005; Cordova et al., 2005), certain types of cancers (Gronbaek et al., 1998; Jang et al., 1997), diabetes (Rimm et al., 1995), and hypertension (Gillman et al., 1995; Nevill et al., 1997). Red wines, in particular, offer heightened antioxidant benefits due to their elevated levels of phenolic and anthocyanin content, stemming from skin maceration treatment prior to fermentation. In summary, a glass of wine can be regarded as a glass of wellness, containing a plethora of advantageous bioactive compounds.

Fruits that are fresh, healthy, and ripe (excluding citrus) are still the greatest raw material for producing high-quality wines. Fruit juice has several characteristics that make it an ideal medium for winemaking, that is, the fruits have high sugar content and other nutrients that promote the opulent growth of wine yeasts. Mostly, fruits have an acidity range that is optimum enough to prevent the augmentation of unwanted microorganisms. Further, a high concentration of ethanol produced during the fermentation process also inhibits the microbial activity in the finished product. Each fruit has a distinct bouquet of flavors, as well as a high concentration of polyphenolic components that contribute to the specific flavors of the wine and its nutritional characters also. One of the most promising features is the high content of antioxidants in the form of anthocyanins, flavonoids, and phenolic compounds, which enhance the functional properties of the final product.

5.4 PHYSICOCHEMICAL PROPERTIES OF FRUITS FOR THE PRODUCTION OF QUALITY WINE

The physical and biochemical characteristics of fruits determine their fate for processing into fermented beverages. Physical parameters like firmness, texture, pulp content, and juice yield of the fruits are the characteristics that decide the eminence of the fruit for its processing into wine. Biochemical parameters *viz.*, total sugars, reducing sugars, total soluble solids, minerals, pH, phenolic content, flavonoids, acidity, water content, and vitamins are the factors that contribute to final sensory characteristics of the wine and nutritional properties like antioxidant content. The important compounds in fruits especially grape berries, from the winemaking point of view, include the total soluble solids, various organic acids viz. malic acid, tartaric acid, citric acid, benzoic acid, and so on, a broad range of phenolic compounds, nitrogenous compounds, variety of aromatic and flavoring compounds and minerals. Beyond the specific role in the physiology of the fruits, they are also associated with good nutritional properties of the fruits and various fruit products.

5.4.1 TOTAL SOLUBLE SOLIDS

Sugar content is a crucial component in influencing fruit quality and processing potential. TSS of grape berries fluctuate stuck between 19 and 25°B, depending on the cultivars, geographical factors, soil type, and meteorological conditions of the area where the grapes are grown. Sugar is a vital component of fruit growth, and it must be transported into the berry from the leaves during ripening. The grape uses the C3 (Calvin cycle) mechanism to absorb CO_2 from the air. The gluconeogenic route transports glyceraldehyde-3-phosphate (triose phosphate) into the cytoplasm, where it is converted into hexose phosphates by enzymes. The production of sucrose, which is transported and dispersed throughout the plant, consumes a large amount of these hexoses. During maturation, sucrose can be stored in the vacuoles of grape cells. As it enters the grape pulp cell vacuoles, it is hydrolyzed to glucose and fructose. To conserve the stored sugars, the grape utilizes malic acid as a respiratory substrate throughout maturation (Coombe, 1976). Sugar also serves as the carbon skeleton or foundation for numerous molecules generated in the fruit, such as organic and amino acids.

Sugars are also delivered to non-photosynthetic organs including the trunk and roots, where they are needed for development and maintenance. Fruit, however, is the primary destination for sugars produced by the leaves from the start of ripening until harvest. During the ripening process, grapes accumulate more sugar than many other fruits. Because raisin and wine quality are significantly connected with sugar concentrations in the fresh fruit, this trait makes grapes particularly well adapted for wine and dried fruit manufacturing. The sugar concentration of the berry is normally around 2% of the berry's fresh weight during the early stages of fruit growth. When veraison begins, the sugar content of grapes rises quickly, reaching 20% or more of the berry's fresh weight by harvest. The sweetness of grapes is influenced by their ripening stage, as fructose levels increase as they mature (Liu et al., 2006). At harvest, grape berries contain roughly equal levels of glucose and fructose, each ranging from 8% to 12% of fresh weight. Sucrose, along with various other sugars, is present in trace amounts (Conde et al., 2007). According to some sources, glucose and fructose concentrations range from 45.9 to 131.0 mg/mL, while sucrose accounts for less than 2% of total sugars (Liu et al., 2006).

5.4.2 ORGANIC ACIDS

Main organic acids present in the grape berry are malic and tartaric acids, accounts about 90% of total fruit acidity. Both acids and citric acid and other non-nitrogenous organic acids, are produced in the fruit. They produce a tart flavor and have a significant impact on stability, pH, and color of wine. The relative levels of acids in grape berries vary depending on cultivars, climatic situation, and viticultural practices (Liu et al., 2006; Navarro et al., 2001). Total acidity in the berry gradually increases from the onset of fruits until shortly before the onset of fruit maturation. The amounts of these acids peak near veraison and then gradually decrease throughout the ripening process (Mato et al., 2007). Because no enzyme capable of digesting tartaric acid has been discovered in the berry, it is thought to be stable once produced. Because berry volume grows while tartarate per berry remains constant, the reduction in tartaric acid concentration observed after-ripening is attributed to a dilution effect (Khan et al., 2008). During grape ripening, malic acid is stored in vacuoles and used as an energy source. The sugars produced by photosynthesis are converted to malic acid, which accumulates in the pericarp cell vacuoles during the herbaceous growth phase.

Glycolysis and the TCA cycle are involved in the conversion of sucrose to malic acid. Malate dehydrogenase, a TCA cycle enzyme, transforms phosphoenolpyruvate (PEP) to oxaloacetic acid, which is then reduced to malic acid by PEP carboxylase. In contrast to tartaric acid, the berry has many enzymes capable of metabolizing malate, and the acid is easily respired to produce CO_2 and H_2O. Malic acid concentrations decrease after veraison because of respiration, enzyme breakdown, and dilution. Due to the synthesis of acid-salts with potassium and other cations in the berry, the amount of free malic and tartaric acids in the fruit decreases throughout ripening. pH of juice is a measure of the hydrogen ion concentration in the berry and is connected to juice acidity in general. All through the early phases of berry growth, the pH of the juice remains relatively stable, hovering about 2.5, before progressively rising during ripening as the amount of malic acid in the berry decreases. Hamman and Dami (2002) found that the pH of juice from 34 distinct grape cultivars fluctuates between 3.2 and 3.7 at berry harvest.

5.4.3 TOTAL PHENOLICS

Phenolic compounds play a pivotal role in crafting high quality wine as they enhance its aroma and nutritional characteristics. Among these phenolics, anthocyanins are particularly crucial for imparting red wine with its color, while tannins contribute to both red and white wines. These phenolic substances are distributed within the grape components: the berry's skin holds pigments and tannins, the pulp have only juice, and the seeds have tannins. The quantity of phenolic compounds in grapes increases as they ripen, with variations depending on grape varieties and agro-climatic conditions. From a biological perspective, phenolic compounds play crucial roles, with insoluble cutin in the skin and lignin in the hard seed coat serving as notable examples, functioning as skin tannins, pigments, and seed tannins. However, in the context of winemaking, it's the soluble phenolics within grape berries that take center stage. Grape phenolic compounds can be broadly categorized into two groups: flavonoids and non-flavonoids.

Flavonoids, a significant portion of grape phenolic content, encompass proanthocyanidins (tannins), anthocyanins, and flavan-3-ol monomers. Among these, tannins, which are polymerized forms of flavan-3-ols, stand out as the most prevalent soluble polyphenols found in grape berries, contributing to the characteristic astringency found in red wines. The color of wine holds immense importance in terms of its visual appeal and overall quality (Bridle and Timberlake, 1997). Key anthocyanins found in grapes include cyanidin, malvidin, delphinidin, peonidin, petunidin, and pelargonidin, often in glucoside forms (Liang et al., 2008). Primary flavan-3-ol monomers present in grapes and wine include catechin, epicatechin, and epicatechin-3-O-gallate (Kennedy, 2008). Among the natural antioxidants found in red grapes and wine, catechins, tannins, and anthocyanins rank as the most concentrated (Mattivi et al., 2002). Non-flavonoid phenolics are also found in grapes and wine, albeit typically in trace amounts, except for hydroxycinnamic acids (Kennedy et al., 2006). Notably, volatile phenolics like benzaldehyde (imparting a bitter almond flavor), phenylacetaldehyde, benzyl alcohol, 2-phenyl ethanol (contributing to a rose scent), and vanillin (providing a vanilla aroma) are predominantly located in grape skin and play a significant role in the development of primary odors during ripening. Another noteworthy member of the non-flavonoid phenolic family is resveratrol, which has garnered attention for its potential health benefits. Discovered in the 1940s, resveratrol is believed to have cardioprotective properties and serves as a defense mechanism against fungi, particularly Botrytis cinerea, in grape berry skin (Conde et al., 2007). Resveratrol can exist in both free and glucoside forms and is found in relatively small quantities in red wine, typically ranging from 0.1 to 12 mg/L (Goldberg et al., 1994). It represents the sole dietary source of resveratrol, albeit in lower concentrations compared to other phenolic compounds present in wine.

5.4.4 NITROGEN COMPOUNDS

Every living organism, including grapevines, relies on nitrogen to synthesize essential components like proteins and nucleic acids, which form the building blocks of cells. Grapevines obtain their nitrogen from fertilizers and soil applications. During the winemaking process, yeast and bacteria digest some of this nitrogen, and it eventually finds its way into the grape berries, grape must, and wine. The presence of nitrogen is crucial for the development of grapevines. However, an excess of nitrogen can lead to excessive vine vigor, prolonged growth, delayed maturation, and reduced levels of anthocyanins and tannins in the fruit. Conversely, a deficiency in nitrogen can hinder vine growth and development. Excess nitrogen can also negatively impact sugar accretion in berries throughout ripening (Delgado et al., 2004). Nitrogen exists in both mineral and organic forms within grape berries and must. Grape berries lack nitrate (NO_3^-), but they do contain nitrogen in the form of ammonium cations (NH_4^+) and organic nitrogen compounds such as amino acids and proteins. In the early stages of berry development, ammonium ions make up more than half of the total nitrogen content in the fruit, while amino acids are relatively scarce.

However, as the berry ripens (veraison), the synthesis of amino acids within the berry increases significantly. This leads to a decrease in ammonium concentrations as these ions are transformed into amino acids. At the time of harvest, the primary amino acids in the fruit of most grape varieties are arginine and proline. For example, in "Thompson Seedless" grape juice at harvest, arginine is the most abundant amino acid, with concentrations ranging from 0.30 to 0.80 mg/mL. The total nitrogen content in grape must (the crushed grapes and juice mixture) can vary widely, spanning from 100 to 1200 mg/L, with red wines typically containing more nitrogen than white wines. This nitrogen, often referred to as fermentable nitrogen, is a crucial nutrient for yeast during its growth and the normal fermentation of grape must. To prevent fermentation from stopping prematurely, ammonium, in the form of phosphate, sulfate, or sulfite salts, is added to the must when the level of fermentable nitrogen falls below 150–200 mg/L (Jiranek et al., 1995). In grape must, only about seven amino acids, including proline, arginine, glutamine, alanine, glutamate, serine, and threonine, are found in concentrations exceeding 100 mg/L. Amino acids in wine have diverse origins, including native grape components that are metabolized by yeast during development, secreted by living yeast cells, generated through proteolysis during the autolysis of dead yeast cells, or produced by enzymatic breakdown of grape proteins (Cataldi and Nardiello, 2003). The total amino acid content in grape must can vary significantly, ranging from 300 to 5000 mg/L, constituting approximately 20 to 50% of the total nitrogen content in the must.

5.4.5 MINERALS

Minerals make up about 0.2 to 0.6% of the total fresh weight of the fruit. Micro-minerals like iron, copper, zinc, and manganese are split into macro-minerals like sodium, potassium, magnesium, calcium, and phosphorus. Potassium is the most significant mineral among these. It is responsible for 50 to 70% of the cations in grape juice. Potassium, calcium, and sodium are the most common cations present in grape berries. Phosphate and chloride are the most frequent mineral anions. Inorganic minerals in the soil are absorbed by the roots and either transferred directly to the fruit or remobilized via the xylem and phloem from permanent organs (the trunk or roots). The potassium content of the grape increases as it ripens. When it enters the fruit, potassium bitartrate is formed, which reduces acidity and raises the pH of the juice (Belitz et al., 2009; DiSilvestro, 2005).

5.4.5 AROMA AND FLAVOR COMPOUNDS

The wine's distinct flavor comes from aroma and flavor components. Both add to the wine's fragrance, general texture, and overall quality. Grape berries and wine are known to include a variety of aromatic and taste components, which are produced by a variety of mechanisms like (i) biochemical processes before fermentation during juice extraction; (ii) enzymatic/chemical reactions after fermentation, during aging process; and (iii) fermentation metabolism of the microorganisms.

Volatile metabolites produced by yeast fermentation and originating from sugar and amino acid metabolism contribute significantly to the vinous taste of wine. These volatile compounds include esters, higher alcohols, carbonyls, volatile fatty acids, and sulfur compounds. During the storage and aging process, oak barrel-aged wine can absorb various odoriferous chemicals, enhancing its flavor profile. However, when it comes to determining the quality and geographical characteristics of wines, the odoriferous compounds derived from grapes themselves, representing the specific grape variety, are of paramount importance. Monoterpenes, C13-nor isoprenoids, methoxypyrazines, and sulfur compounds play a pivotal role in imparting the varietal aroma to wine. Among these compounds, terpenes constitute the majority of odoriferous chemicals in *V. vinifera* grapes and wines.

These terpenes can exist in both free forms and as odorless precursors, notably glycosylated compounds. They are responsible for the distinctive aroma of Muscat grapes and wines. However, even in simple-flavored grape cultivars, terpenes can be found in lower concentrations. At present, approximately 50 monoterpene compounds have been identified. The most common monoterpene alcohols found in Muscat varietals, known for their aromatic characteristics, include linalool, geraniol, nerol, citronellol, β-terpineol, and tocotrienol. These compounds contribute to the unique and recognizable scent profile of Muscat wines, making them distinct and appealing to many wine enthusiasts.

5.5 ETHANOLIC FERMENTATION

The primary fermentation process involves the conversion of carbohydrates found in grape juice, predominantly glucose and fructose, into ethanol, which is the principal component of wine, along with the release of carbon dioxide. This fermentation can proceed through two distinct methods: induced or seeded fermentation, and indigenous or wild fermentation. In indigenous fermentation, the naturally occurring yeasts within the grape juice initiate and complete the fermentation process. Conversely, seeded fermentation involves the deliberate introduction of specific yeast strains, often Saccharomyces cerevisiae or *Saccharomyces bayanus*, at initial concentrations ranging from 10^6 to 10^7 cells/mL (Fleet and Heard, 1993). Each molecule of glucose or fructose is converted into two molecules of ethanol and two molecules of carbon dioxide.

The theoretical conversion of 180 grams of sugar yields 88 grams of carbon dioxide and 92 grams of ethanol, resulting in an ethanol yield of 51.1% on a dry weight basis. In typical fermentations starting with approximately 22 to 24% sugar content, about 95% of the sugar is transformed into ethanol and carbon dioxide, while 1% contributes to cellular material, and the remaining 4% forms other end products. These percentages may vary based on factors such as the size of the inoculum, fermentation temperature, and nutrient availability (Boulton et al., 1995). Non-*Saccharomyces* yeasts, including *Candida*, *Debaryomyces*, *Hanseniaspora*, *Hansenula*, *Kloeckera*, *Metschnikowia*, *Pichia*, *Schizosaccharomyces*, *Torulaspora*, and *Zygosaccharomyces*, are known to initiate fermentation, depending on the grape variety and associated microflora. Their growth is typically limited to the initial two or three days of fermentation, reaching ethanol concentrations of 2–5% v/v. At this point, the natural microflora becomes sensitive to ethanol and ceases to be effective.

The fermentation process is then taken over by the introduced Saccharomyces species, which are more robust fermenters and better equipped to tolerate ethanol. To ensure an appropriate inoculum size, Saccharomyces cerevisiae used for fermenting grape musts can undergo a pre-fermentative skin treatment followed by a brief aeration pulse. This procedure has been shown to enhance yeast growth rate, viability, and fermentative capability without affecting the unsaturation index of cellular fatty acids (Fleet and Heard, 1993). During the fermentation stage, which can span from 5 to 14 days for primary fermentation and an additional 5 to 14 days for secondary fermentation, there is a risk of fermentation prematurely halting and the potential development of various wine defects. The vessels used for fermentation can vary, including stainless steel tanks (commonly used for many white wines such as Riesling and Chardonnay), open wooden vats, wine barrels, or even the wine bottle itself, as seen in many sparkling wines (Kunze, 2004; Robinson, 2006).

Fermentation conditions differ for white and red wines. White wines are typically fermented at temperatures of 10–18°C for 7–14 days or longer. The lower temperature and slower fermentation rate in whites help preserve volatile flavor components. Conversely, red wines ferment for around 7 days at temperatures ranging from 20–30°C, as a higher temperature is required to extract color from the grape skins. Various factors, including fermentation temperature, yeast strain, medium composition, pH, substrate concentration, feeding method, osmotic pressure, ethanol concentration, membrane composition, and other parameters, all play a crucial role in influencing yeast fermentation performance (D'Amore, 1992).

5.6 ROLE OF MICROORGANISMS IN FERMENTED BEVERAGES

The gastrointestinal microflora forms a complex ecosystem of bacteria that plays a vital role in maintaining overall health. These bacteria reside in the digestive tract, where they interact with one another and with the host organism. Factors such as diet, medication, and environmental conditions can disrupt the balance of the normal gut microbiota. Imbalances can be addressed through two primary methods: probiotics and prebiotics.

5.6.1 PROBIOTICS

Probiotics consist of living microorganisms, predominantly bacteria and specific yeast strains, which exert a positive influence on the host organism by promoting a healthy balance of gut microbes, as defined by FAO/WHO (2001). Probiotics offer a wide array of health benefits, including improved lactose tolerance, enhanced digestive functions, reduced cholesterol levels, decreased levels of harmful compounds like ammonia, synthesis of folic acid, and restoration of the normal gut microflora, especially following antibiotic treatments. Probiotic microorganisms have been successfully incorporated into various food and beverage products. Suitable carriers for these beneficial microbes include fruits, cereals, vegetables, and soy-based beverages (Patel, 2014).

5.6.2 PREBIOTICS

Prebiotics involves the oral administration of substances that serve as bacterial stimulants, primarily targeting native microflora components. These compounds help promote the growth and activity of beneficial gut bacteria. Fruit, grains, vegetables, and soy-based beverages have all been explored as potential mediums for delivering probiotic strains, with a particular focus on fruit and vegetable juices. However, preserving the viability of these microorganisms, often aiming for 1 billion viable cells per 100 g of product, and maintaining probiotic activity throughout the product's shelf-life present significant challenges, especially in juices with low pH levels (Perricone et al., 2015). Researchers are actively exploring different approaches to develop functional beverages incorporating probiotics. Some of these approaches include:

- Optimizing the manufacturing processes of new functional beverages by harnessing the functionality of microorganisms.
- Incorporating prebiotics and symbiotics to enhance the probiotic effect.
- Utilizing and processing natural ingredients.
- Incorporating by-products from food and beverage industries for food fortification.

Numerous studies have demonstrated the use of various fruit juices as raw materials for probiotic beverages. The majority of probiotic microorganisms used in these products are strains of *Bifidobacterium* spp. and *Lactobacillus* spp., while other species, such as *Escherichia coli*, lactic bacteria, and *Saccharomyces boulardii*, are also employed as probiotics (de Vrese and Schrezenmeir, 2011). Currently, fermented dairy products dominate the probiotics market, as they provide ideal conditions for probiotic bacteria to thrive and maintain viability. However, dairy consumption can have potential drawbacks, such as lactose intolerance, allergenic proteins, and impacts on cholesterol levels (Prado, 2008).

Historically, fermentation has been utilized for the preservation of vegetables while enhancing their nutritional and sensory attributes. In most cases, vegetables are fermented with the existing microflora at room temperature, without strict control over the fermentation process. Specialized packaging has been developed to address storage challenges on store shelves, preserving the quality and functional properties of probiotic vegetable products. Many producers recommend storing these products at 4°C, although issues related to packaging abnormalities can occur due to the elevated CO_2 concentration produced during the fermentation. The production technology of wine is depicted in Figure 5.1.

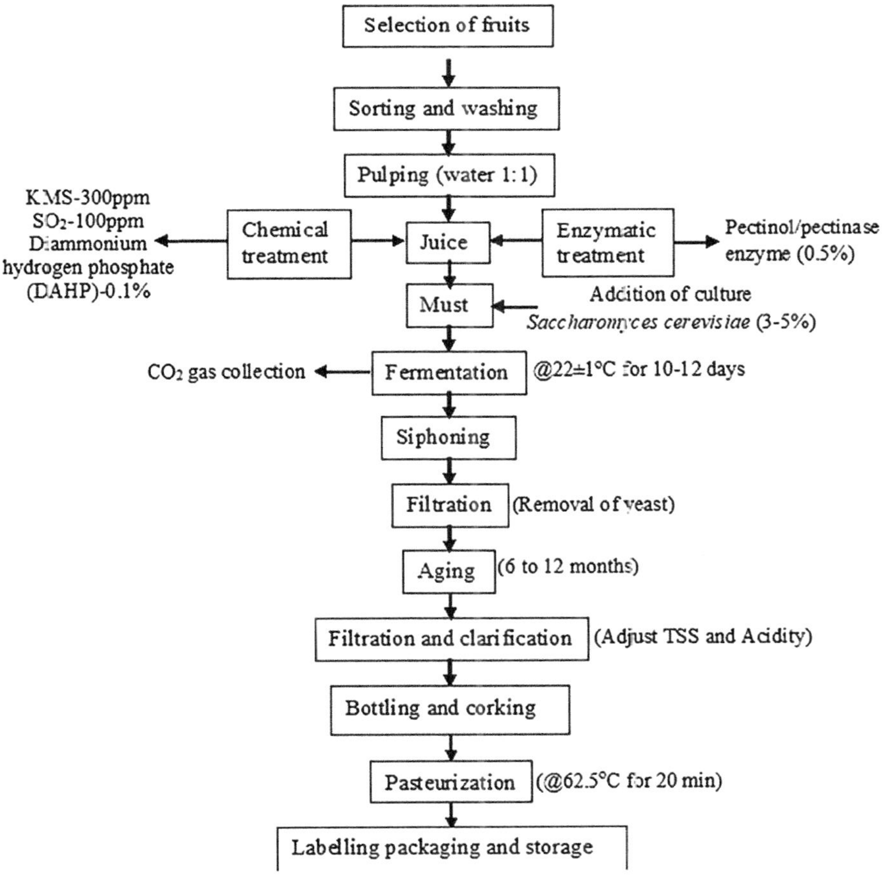

FIGURE 5.1 General flow chart for wine making.

Source: Adapted from Joshi and Ray (2021).

5.7 PROBIOTIC BEVERAGES FROM FRUIT AND VEGETABLE

The probiotic products sector is still in its infancy, with the first commercial product hitting the market in 1994. This category looks at potential as the accessibility of these products on the marketplace grows, existing technologies improve, and customer interest grows. Because of available aseptic dosing technology, probiotics can be inoculated directly into juices. Vacuum impregnation, prebiotics, and microencapsulation are utilized to keep probiotics viable throughout the product's life cycle (Prado et al., 2008). The connection between diverse probiotic cultures, particularly yeasts and bacteria, is extremely important. Although there is poor acceptability of these bacteria in many countries, genetic modification of probiotics in new food matrices could be a feasible answer to the augment in probiotic resistance in food (Ouwehand and Röytiö, 2015).

Because lactic fermentation is commonly employed to preserve vegetables, the finest strategy to produce probiotic food (Buckenhueskes, 2015). Acceptability from a sensory standpoint is an imperative consideration in the expansion of new products. Consumers desire healthy, pleasant items at a reasonable price. Traditional fermented goods serve as a foundation for producing new probiotic products that are safe and stable (Sõukand et al., 2015). Because of the high expenses of developing probiotic foodstuffs, a collaboration between industrial partners and academia may result in substantially faster product development (Khan et al., 2013).

5.8 HEALTH BENEFITS OF VEGETABLE AND FRUIT BEVERAGES

The relationship between wine/beer consumption and its potential health benefits have been expansively studied by researchers. Some key findings and insights from the studies are given in subsections.

5.8.1 HEART HEALTH BENEFITS

Moderate alcohol consumption of wine and beer has shown reduced risk of CVDs. This protective effect includes a lower threat of ischemic myocardium and overall mortality (Matsumoto et al., 2014). One of the mechanisms behind this benefit is the augment in high density lipoprotein cholesterol, which is known to have protective effect against CVD.

5.8.2 INSULIN SENSITIVITY

Moderate alcohol consumption resulted in improvement for insulin sensitivity in individuals with and without diabetes. This improvement in insulin sensitivity have role in the overall positive effects on glucose metabolism (Shai et al., 2007).

5.8.3 ADIPONECTIN LEVELS

Positive effects on the level of adiponectin, a protein hormone that regulates fatty acid oxidation and glucose uptake has been seen by moderate alcohol consumption. These effects may contribute to improved insulin sensitivity (Sierksma et al., 2004).

5.8.4 INFLAMMATION REDUCTION

Alcohol may have cardioprotective effects by influencing factors such as coagulation, endothelial functions, and inflammation (Wang et al., 2008). Additionally, moderate alcohol consumption has been inversely associated with peripheral arterial disease (PAD) (Djousse et al., 2000).

5.8.5 CONGESTIVE HEART FAILURE (CHF)

Studies have shown a lower risk of congestive heart failure (CHF) among moderate alcohol drinkers. For example, the Framingham Heart Study found that males who consumed 8 to 14 alcoholic beverages per week had a lower risk of CHF compared to nondrinkers (Bryson et al., 2006).

5.8.6 FRENCH PARADOX

'French paradox' refers to the observation that the French have a relatively low incidence of coronary heart disease despite a diet rich in saturated fat. Some studies suggest that moderate wine consumption, particularly red wine, may contribute to this phenomenon. Drinking three glasses of red wine daily has been associated with the lowest risk of cardiac diseases and mortality (Levantesi et al., 2013).

5.8.7 ANTIOXIDANTS AND POLYPHENOLS

Wine, especially red wine, contains antioxidants and polyphenols like resveratrol, quercetin, and catechins. These compounds have been linked to various health benefits, including a reduced risk of CVD. They may also have a more significant impact on platelet aggregation compared to other alcoholic beverages (Frost et al., 2004).

5.8.8 RESVERATROL

Resveratrol, a polyphenol found in red wine, has been of particular interest due to its potential health benefits. It has been associated with increased lifespan in lower organisms, mimicking the effects of dietary restriction (Baur, 2010). Resveratrol has also been shown to promote bone formation and prevent bone loss (Dai et al., 2007).

5.8.9 WEIGHT MANAGEMENT

Some studies have suggested that moderate white wine consumption, when integrated into an energy-restricted diet, may not impair the effectiveness of the diet and could be helpful for weight management in certain individuals with obesity (Flechtner-Mors et al., 2004).

5.8.10 IMMUNE SYSTEM SUPPORT

Daily wine consumption has been linked to improved immune system function. In mice, wine consumption resulted in better lymphocyte response and an increased proportion of circulating natural killer (NK) cells compared to ethanol consumption (Percival and Sims, 1999).

5.8.11 GUT HEALTH

Probiotics, found in fermented foods and beverages, can promote gut health. Consuming probiotics can help maintain a balanced gut microbiota, aid in digestion, and enhance immune function.

5.8.12 ANTIOXIDANTS

Fruits and vegetables are rich in antioxidants and phytochemicals, which act as powerful free radical scavengers and contribute to overall health and disease prevention (Oguntibeju et al., 2003).

5.8.13 BONE HEALTH

Some studies have associated moderate wine consumption with improved bone density. The presence of resveratrol in wine may contribute to bone health by promoting bone formation and preventing bone loss (Dai et al., 2007). While moderate alcohol consumption may offer certain health benefits, excessive or heavy drinking can have detrimental effects on health. The key is moderation, and individual tolerance to alcohol varies. It's advisable to consult with a healthcare professional to determine what level of alcohol consumption is appropriate for your specific health needs and circumstances. Additionally, the health benefits of alcohol should not encourage non-drinkers to start drinking solely for these potential advantages. Some of the famous fermented fruit beverage and their health benefits are listed in Table 5.1.

TABLE 5.1 Different Fermented Fruit Beverages and Their Proposed Health Benefits

Fermented Beverage (Fruits)	Functional Compound	Health Benefit	References
Pomegranate fermented juice	Phenolic compounds	Antioxidant and anti-inflammatory.	Keşa et al. (2021)
Apple cider	Epicatechin, caffeic acid, chlorogenic acid (CGA), catechin	Anticancer, hepatoprotective, anti-diabetic, anti-inflammatory, antioxidants, antimicrobial, and beneficial in a reduction of cardiotoxicity.	Tripathi and Papiya (2020)
Orange wine	Flavonoids (narirutin, hesperidin, naringin, neohesperidin, and apigenin)	Anti-inflammatory, hypolipidemic activities, and antihypertensive.	Kelebek et al. (2009)
Sea buckthorn wine	Rutin, quercetin, and myricetin	Protect hypercholesterolemia and oxidative stress.	Keşa et al. (2021); Negi et al. (2013)

TABLE 5.1 *(Continued)*

Fermented Beverage (Fruits)	Functional Compound	Health Benefit	References
Cherry wine	Biogenic amines (spermine and tyramine)	Regulation of body and stomach pH, the immune response, gastric acid secretion, and cell growth and differentiation (but at higher concentrations, biogenic amines cause nausea, oral burning, gastric intestinal problems, renal intoxication).	Ercan et al. (2013); Li et al. (2019)
Gooseberry blended bael (*Aegle marmelos* Correa) fermented beverage	Catechin, kaempferol, gallic acid and caffeic acid	Antioxidant, keeps heart healthy, helpful in losing weight and boosts immunity.	Garg et al. (2021)
Pineapple wine and vinegar	L-lysine, mellein, gallic acid, catechin, and *m*-coumaric acid	Antioxidant	Roda et al. (2017)
Persimmon wine	Gallic acid, catechin, epigallocatechin rutin, epicatechin, ferulic acid	Antioxidant, anti-carcinogenic, antimicrobial activity.	Liu et al. (2018)
Red grapes wine	Melatonin, catechins, lutein, and resveratrol	• Antioxidants; modulate cell signaling pathways, reducing harmful low-density lipoprotein (LDL) cholesterol. • Resveratrol; cardioprotective and chemopreventive.	Guilford and Pezzuto (2011)
Mango wine	Gallic acid, procyanidin, Galloyl-A-type, β-pinene, caffeoyl-quinic acid, and 2,2,6-vinyltetrahydropyran	Antioxidant, antimicrobial, anti-inflammatory, and anti-cancerous.	Patel et al. (2021)
Jamun wine	Anthocyanin (47.0 mg/100 g)	Antioxidant	Joshi et al. (2012)
Kiwi wine	Protocatechuic, ellagic, quercetin, phloretin, and gallic acid	Antioxidant, anti-aging, and anticancer activities.	Huang et al. (2021)

TABLE 5.1 *(Continued)*

Fermented Beverage (Fruits)	Functional Compound	Health Benefit	References
Pumpkin, pomegranate, and wild apricot wine (60:20:20)	Total phenols and carotenoids	Antioxidant	Sharma et al. (2018)
Strawberry wine and vinegar	Anthocyanins	Antioxidant, anti-inflammatory, weight management anticancer.	Afrin et al. (2016); Hornedo-Ortega et al. (2017)
Red dragon fruit wine	Betacynin (betanin, phyllocactin, and hylocerenin)	Antioxidant, anticancer, anti-lipidemic, and antimicrobial effects.	Jiang et al. (2020)

The challenges in discovering the nutritional properties and health effects of wine, as outlined in your statement, highlight the complexity of this area of research. Here are some additional insights into these challenges:

1. **Diverse Phenolic Compounds:** Wine contains a vast array of phenolic compounds, each with potentially unique health effects. Understanding the individual actions and bioavailability of these compounds is indeed a challenging task. Researchers must conduct in-depth analyses to identify and quantify these compounds, as well as investigate their interactions within the body.
2. **Societal and Cultural Factors:** Wine consumption is deeply embedded in societal and cultural practices. The patterns of wine consumption vary widely across different regions and communities. These variations can make it hard to draw universal conclusions about the health effects of wine, as cultural practices may influence overall lifestyle and dietary choices.
3. **Prevention of Chronic Diseases:** Wine's prospective health benefits often relate to the long-term prevention of constant diseases, such as CVD. These diseases develop over extended periods, making it challenging to measure the direct impact of wine consumption on specific health outcomes. Validated biomarkers for such diseases can be scarce.
4. **Consensus Building:** Building a scientific consensus on the health benefits of wine takes time. While there is growing evidence

supporting the potential benefits of wine, researchers continually refine their understanding. Scientific consensus typically requires rigorous and consistent findings across numerous studies.

5. **Mechanistic Understanding:** Research has shown that the molecules in wine, particularly polyphenols, can influence cellular processes and signaling pathways related to cardiovascular health. These mechanisms align with the potential reduction of arterial disease, but a more comprehensive understanding is needed.
6. **Biomarker Development:** Developing biomarkers specific to wine consumption and its health effects is crucial. Biomarkers could provide a more direct way to assess an individual's risk of developing chronic diseases based on their wine consumption habits. This requires extensive research into potential markers and their validation.
7. **Interactions with Alcohol:** Wine contains alcohol, which itself has complex effects on health. It is vital to distinguish between the latent benefits related with wine polyphenols and the effects of alcohol alone. Understanding how these factors interact, and influence health outcomes is essential.
8. **Individual Variability:** People vary in their genetic makeup, lifestyle choices, and overall health status. These individual differences can lead to variations in how wine affects different individuals. Personalized approaches to understanding wine's effects may be necessary.

5.9 CONCLUSION

Fermented foods have been well recognized for their immense health potential and the science and advancements in the field of fermented fruit-based beverages are growing increasingly popular as consumer knowledge about healthy food consumption patterns rises. In the present era, fermented fruit beverages account for one of the major segments of the functional food category. Preservation improved nutritional properties, improved taste, flavor, and bouquet, foodstuffs with elevated biological value, and superior health benefits are the most important reasons for the expansion and acceptance of fermented fruit-based beverages. Alcoholic beverages are generally free of allergens (lactose and casein), as well as cholesterol. Fermented beverages are among the most researched areas in food processing and preservation, with clinically proven and well-documented health benefits. The consumption of such beverages may be regulated within the stipulated daily intake levels to access the potential health benefits of each drink.

KEYWORDS

- **antioxidants**
- **disease prevention**
- **energy-restricted diet**
- **fermented foods**
- **free radical scavengers**
- **gut health**
- **immune system support**
- **natural killer**
- **phytochemicals**
- **probiotics**

REFERENCES

Afrin, S., Gasparrini, M., Forbes-Hernandez, T. Y., Reboredo-Rodriguez, P., Mezzetti, B., Varela-López, A., Giampieri, F., & Battino, M. (2016). Promising health benefits of the strawberry: A focus on clinical studies. *Journal of Agricultural and Food Chemistry, 64*(22), 4435–4449.

Arranz, S., Chiva-Blanch, G., Valderas-Martínez, P., Medina-Remón, A., Lamuela-Raventós, R. M., & Estruch, R. (2012). Wine, beer, alcohol and polyphenols on cardiovascular disease and cancer. *Nutrients, 4*(7), 759–781.

Arya, R. K. K., Sati, D., Bisht, D., & Keservani, R. K. (2022). Nanotechnology-based bacterial immunotherapy. In R. K. Kesharwani, R. K. Keservani, & A. K. Sharma (Eds.), *Nutraceuticals and Functional Foods in Immunomodulators* (pp. 3–19). Springer. https://doi.org/10.1007/978-981-19-2507-8_1.

Baur, J. A. (2010). Resveratrol, sirtuins, and the promise of a DR mimetic. *Mechanisms of Ageing and Development, 131*(4), 261–269.

Belitz, H. D., Grosch, W., & Schieberle, P. (2009). Coffee, tea, cocoa. *Food Chemistry*, 938–970.

Berner, L. A., & O'Donnell, J. A. (1998). Functional foods and health claims legislation: Applications to dairy foods. *International Dairy Journal, 8*(5, 6), 355–362.

Block, G., Patterson, B., & Subar, A. (1992). Fruit, vegetables, and cancer prevention: A review of the epidemiological evidence. *Nutrition and Cancer*, *18*(1), 1–29.

Bors, W., & Michel, C. (2002). Chemistry of the antioxidant effect of polyphenols. *Annals of the New York Academy of Sciences*, *957*(1), 57–69.

Boulton, R. B., Singleton, V. L., Bisson, L. F., & Kunkee, R. E. *Principles and Practices of Winemaking*. Springer Science & Business Media.

Bradamante, S., Barenghi, L., & Villa, A. (2004). Cardiovascular protective effects of resveratrol. *Cardiovascular Drug Reviews*, *22*(3), 169–188.

Bridle, P., & Timberlake, C. F. (1997). Anthocyanins as natural food colours—selected aspects. *Food Chemistry*, *58*(1, 2), 103–109.

Bryson, C. L., Mukamal, K. J., Mittleman, M. A., Fried, L. P., Hirsch, C. H., Kitzman, D. W., & Siscovick, D. S. (2006). The association of alcohol consumption and incident heart failure: The cardiovascular health study. *Journal of the American College of Cardiology*, *48*(2), 305–311.

Buckenhueskes, H. J. (2015). Quality improvement and fermentation control in vegetables. In *Advances in Fermented Foods and Beverages* (pp. 515–539). Woodhead Publishing.

Cabré, E., & Gassull, M. A. (2007). Probiotics for preventing relapse or recurrence in Crohn's disease involving the ileum: Are there reasons for failure? *Journal of Crohn's and Colitis*, *1*(1), 47–52.

Cataldi, T. R., & Nardiello, D. (2003). Determination of free proline and monosaccharides in wine samples by high-performance anion-exchange chromatography with pulsed amperometric detection (HPAEC-PAD). *Journal of Agricultural and Food Chemistry*, *51*(13), 3737–3742.

Combina, M., Elía, A., Mercado, L., Catania, C., Ganga, A., & Martinez, C. (2005). Dynamics of indigenous yeast populations during spontaneous fermentation of wines from Mendoza, Argentina. *International Journal of Food Microbiology*, *99*(3), 237–243.

Conde, C., Silva, P., Fontes, N., Dias, A. C. P., Tavares, R. M., Sousa, M. J., Agasse, A., Delrot, S., & Gerós, H. (2007). Biochemical changes throughout grape berry development and fruit and wine quality. *Structure, 1*.

Coombe, B. G. (1976). The development of fleshy fruits. *Annual Review of Plant Physiology*, *27*(1), 207–228.

Cordova, A. C., La Scienya, M. J., Berke-Schlessel, D. W., & Sumpio, B. E. (2005). The cardiovascular protective effect of red wine. *Journal of the American College of Surgeons*, *200*(3), 428–439.

Cuthbertson, W. F. J. (2002). Are the effects of dietary fruits and vegetables on human health related to those of chronic dietary restriction on animal longevity and disease? *British Journal of Nutrition*, *87*(2), 187, 188.

Dai, Z., Li, Y., Quarles, L. D., Song, T., Pan, W., Zhou, H., & Xiao, Z. (2007). Resveratrol enhances proliferation and osteoblastic differentiation in human mesenchymal stem cells via ER-dependent ERK1/2 activation. *Phytomedicine*, *14*(12), 806–814.

D'Amore, T. (1992). Cambridge prize lecture improving yeast fermentation performance. *Journal of the Institute of Brewing*, *98*(5), 375–382.

Dauchet, L., Amouyel, P., & Dallongeville, J. (2005). Fruit and vegetable consumption and risk of stroke: A meta-analysis of cohort studies. *Neurology*, *65*(8), 1193–1197.

Dauchet, L., Amouyel, P., Hercberg, S., & Dallongeville, J. (2006). Fruit and vegetable consumption and risk of coronary heart disease: A meta-analysis of cohort studies. *The Journal of Nutrition*, *136*(10), 2588–2593.

Delgado, R., Martín, P., Del Álamo, M., & González, M. R. (2004). Changes in the phenolic composition of grape berries during ripening in relation to vineyard nitrogen and potassium fertilisation rates. *Journal of the Science of Food and Agriculture*, *84*(7), 623–630.

Dembinska-Kiec, A., Mykkänen, O., Kiec-Wilk, B., & Mykkänen, H. (2008). Antioxidant phytochemicals against type 2 diabetes. *British Journal of Nutrition*, *99*(E-S1), ES109–ES117.

Di Cagno, R., Coda, R., De Angelis, M., & Gobbetti, M. (2013). Exploitation of vegetables and fruits through lactic acid fermentation. *Food Microbiology*, *33*(1), 1–10.

DiSilvestro, R. A. (2004). *Handbook of Minerals as Nutritional Supplements*. CRC press.

Djoussé, L., & Gaziano, J. M. (2007). Alcohol consumption and risk of heart failure in the Physicians' Health Study I. *Circulation*, *115*(1), 34–39.

Djoussé, L., Levy, D., Murabito, J. M., Cupples, L. A., & Ellison, R. C. (2000). Alcohol consumption and risk of intermittent claudication in the Framingham Heart Study. *Circulation*, *102*(25), 3092–3097.

Ercan, S. S., Bozkurt, H., & Soysal, Ç. (2013). Significance of biogenic amines in foods and their reduction methods. *Journal of Food Science and Engineering*, *3*(8).

Flechtner-Mors, M., Biesalski, H. K., Jenkinson, C. P., Adler, G., & Ditschuneit, H. H. (2004). Effects of moderate consumption of white wine on weight loss in overweight and obese subjects. *International Journal of Obesity*, *28*(11), 1420–1426.

Fleet, G. H. (1993). Yeast-growth during fermentation. *Wine Microbiology and Biotechnology*, 27–54.

Ford, E. S., Giles, W. H., & Mokdad, A. H. (2004). Increasing prevalence of the metabolic syndrome among US adults. *Diabetes Care*, *27*(10), 2444–2449.

Frost, L., & Vestergaard, P. (2004). Alcohol and risk of atrial fibrillation or flutter: A cohort study. *Archives of Internal Medicine*, *164*(18), 1993–1998.

Gambini, J., Inglés, M., Olaso, G., Lopez-Grueso, R., Bonet-Costa, V., Gimeno-Mallench, L., Mas-Bargues, C., Abdelaziz, K. M., Gomez-Cabrera, M. C., Vina, J., & Borras, C. (2015). Properties of resveratrol: *In vitro* and *in vivo* studies about metabolism, bioavailability, and biological effects in animal models and humans. *Oxidative Medicine and Cellular Longevity, 2015*, 837042.

Garg, N., Kumar, S., & Yadav, P. (2021). Indian goose berry fortified, antioxidant rich bael (*Aegle marmelos*) fermented beverage. *Journal of Food Science and Technology*, *58*(11), 4437–4441.

Gillman, M. W., Cook, N. R., Evans, D. A., Rosner, B., & Hennekens, C. H. (1995). Relationship of alcohol intake with blood pressure in young adults. *Hypertension*, *25*(5), 1106–1110.

Goldberg, D. M., Yan, J., Ng, E., Diamandis, E. P., Karumanchiri, A., Soleas, G., & Waterhouse, A. L. (1994). Direct injection gas chromatographic mass spectrometric assay for trans-resveratrol. *Analytical Chemistry*, *66*(22), 3959–3963.

Gordon, L. A., Percival, S. S., & Sims, C. A. (1999). Alcohol consumption as wine modifies the detrimental effects of ethanol on mouse immunity. *FASEB Journal*, *13*(4), A590. Federation of American Societies for Experimental Biology.

Grønbæk, M., Becker, U., Johansen, D., Tønnesen, H., Jensen, G., & Sørensen, T. I. (1998). Population based cohort study of the association between alcohol intake and cancer of the upper digestive tract. *Bmj*, *317*(7162), 844–848.

Guilford, J. M., & Pezzuto, J. M. (2011). Wine and health: A review. *American Journal of Enology and Viticulture*, *62*(4), 471–486.

Hamman, R. A., & Dami, I. E. (1999). *Evaluation of 35 Wine Grape Cultivars and "Chardonnay" on 4 Rootstocks Grown in Western Colorado* (Technical Report No. TR99-6). Colorado Agricultural Experiment Station.

Hickson, M., D'Souza, A. L., Muthu, N., Rogers, T. R., Want, S., Rajkumar, C., & Bulpitt, C. J. (2007). Use of probiotic Lactobacillus preparation to prevent diarrhoea associated with antibiotics: Randomized double blind placebo controlled trial. *Bmj*, *335*(7610), 80.

Homayouni, A., Azizi, A., Ehsani, M. R., Yarmand, M. S., & Razavi, S. H. (2008). Effect of microencapsulation and resistant starch on the probiotic survival and sensory properties of synbiotic ice cream. *Food Chemistry*, *111*(1), 50–55.

Hornedo-Ortega, R., Álvarez-Fernández, M. A., Cerezo, A. B., Garcia-Garcia, I., Troncoso, A. M., & Garcia-Parrilla, M. C. (2017). Influence of fermentation process on the anthocyanin composition of wine and vinegar elaborated from strawberry. *Journal of Food Science*, *82*(2), 364–372.

Huang, J., Li, H., Wang, Y., Wang, X., Ren, Y., Yue, T., & Gao, Z. (2021). Evaluation of the quality of fermented kiwi wines made from different kiwifruit cultivars. *Food Bioscience*, *42*, 101051.

Jang, M., Cai, L., Udeani, G. O., Slowing, K. V., Thomas, C. F., Beecher, C. W., Fong, H. H., Farnsworth, N. R., Kinghorn, A. D., Mehta, R. G., & Moon, R. C. (1997). Cancer chemopreventive activity of resveratrol, a natural product derived from grapes. *Science*, *275*(5297), 218–220.

Jiang, X., Lu, Y., & Liu, S. Q. (2020). Effects of different yeasts on physicochemical and oenological properties of red dragon fruit wine fermented with *Saccharomyces cerevisiae*, *Torulaspora delbrueckii* and *Lachancea thermotolerans*. *Microorganisms*, *8*(3), 315.

Jiranek, V., Langridge, P., & Henschke, P. (1995). Regulation of hydrogen sulfide liberation in wine-producing *Saccharomyces cerevisiae* strains by assimilable nitrogen. *Applied and Environmental microbiology*, *61*(2), 461–467.

Joshi, V. K., & Ramesh C. (2021). *Winemaking: Basics and Applied Aspects*.

Joshi, V. K., Sharma, R., Girdher, A., & Abrol, G. S. (2012). Effect of dilution and maturation on physico-chemical and sensory quality of jamun (Black plum) wine. *Indian Journal of Natural Products and Resources, 3*(2), 222–227.

Kausar, H., Saeed, S., Ahmad, M. M., & Salam, A. (2012). Studies on the development and storage stability of cucumber-melon functional drink. *Journal o. Agricultural Research, 50*(2), 239–248.

Kelebek, H., Selli, S., Canbas, A., & Cabaroglu, T. (2009). HPLC determination of organic acids, sugars, phenolic compositions and antioxidant capacity of orange juice and orange wine made from a Turkish cv. Kozan. *Microchemical Journal*, *91*(2), 187–192.

Kennedy, J. A. (2008). Grape and wine phenolics: Observations and recent findings. *Ciencia e Investigación Agrarian*, *35*(2), 107–120.

Kennedy, J. A., Saucier, C., & Glories, Y. (2006). Grape and wine phenolics: History and perspective. *American Journal of Enology and Viticulture*, *57*(3), 239–248.

Keşa, A. L., Pop, C. R., Mudura, E., Salanță, L. C., Pasqualone, A., Dărab, C., Burja-Udrea, C., Zhao, H., & Coldea, T. E. (2021). Strategies to improve the potential functionality of fruit-based fermented beverages. *Plants*, *10*(11), 2263.

Keservani, R. K., & Sharma, A. K. (2014). Flavonoids: Emerging trends and potential health benefits. *Journal of Chinese Pharmaceutical Sciences, 23*(12), 815.

Keservani, R. K., Kesharwani, R. K., Sharma, A. K., & Jarouliya, U. (2015). Dietary supplements, nutraceutical, and functional foods in immune response (immunomodulators).

In D. Bagchi, H. G. Preuss, & A. Swaroop (Eds.), *Nutraceutical and Functional Foods in Human Life and Disease Prevention* (Chap. 20, pp. 343–358). Boca Raton, FL: CRC Press, Taylor & Francis.

Keservani, R. K., Kesharwani, R. K., Sharma, A. K., Vyas, N., & Chadoker, A. (2010b). Nutritional Supplements: An Overview. *International Journal of Current Pharmaceutical Review and Research, 1*(1), 59–75.

Keservani, R. K., Kesharwani, R. K., Vyas, N., Jain, S., Raghuvanshi, R., & Sharma, A. K. (2010a). Nutraceutical and functional food as future food: A review. *Der Pharmacia Lettre, 2*(1), 106–116.

Keservani, R. K., Sharma, A. K., & Kesharwani, R. K. (2016a). *Nutraceutical and Functional Foods for Cardiovascular Health. In Food Process Engineering* (pp. 257–278). Apple Academic Press, CRC Press.

Keservani, R. K., Sharma, A. K., & Kesharwani, R. K. (2016b). Medicinal effect of nutraceutical fruits for the cognition and brain health. *Scientifica, 3109254*, 10. doi: 10.1155/2016/3109254.

Keservani, R. K., Sharma, A. K., & Kesharwani, R. K. (2020). *Nutraceuticals and Dietary Supplements: Applications in Health Improvement and Disease Management* (pp. 1–342). CRC Press.

Khan, R. S., Grigor, J., Winger, R., & Win, A. (2013). Functional food product development– Opportunities and challenges for food manufacturers. *Trends in Food Science and Technology. 30*(1), 27–37.

Kim, S., Khambampati, A. K., & Kim, K. Y. (2022). Mathematical concepts for image reconstruction in tomography. In *Industrial Tomography* (pp. 335–387). Woodhead Publishing.

Kumar, P., Rai, S., Verma, S. K., Prakash, P. S., & Chitara, D. (2022). Classification, mode of action and uses of various immunomodulators. In R. K. Kesharwani, R. K. Keservani, & A. K. Sharma (Eds.), *Immunomodulators and Human Health* (pp. 1–27). Springer. https://doi.org/10.1007/978-981-16-6379-6_1.

Kunze, W. (2004). Brewing malting. *Vlb, Berlin*, 18–152.

Levantesi, G., Marfisi, R., Mozaffarian, D., Franzosi, M. G., Maggioni, A., Nicolosi, G. L., Schweiger, C., Silletta, M., Tavazzi, L., Tognoni, G., & Marchioli, R. (2013). Wine consumption and risk of cardiovascular events after myocardial infarction: Results from the GISSI-Prevenzione trial. *International Journal of Cardiology, 163*(3), 282–287.

Levantesi, G., Marfisi, R., Mozaffarian, D., Franzosi, M. G., Maggioni, A., Nicolosi, G. L., Schweiger, C., Silletta, M., Tavazzi, L., Tognoni, G., & Marchioli, R. Wine consumption and risk of cardiovascular events after myocardial infarction: Results from the GISSI-Prevenzione trial. *International Journal of Cardiology, 163*(3), 282–287.

Li, H. M., Jiang, D. Q., Dai, Z. G., Zhang, Y. S., Zhang, Y., Sun, S. Y., & Zhao, Y. P. Aromatic property of cherry wine produced by malolactic fermentation of controlled and spontaneous on the bacterial evolution. *International Journal of Food Properties, 22*(1), 1270–1282.

Liang, Z., Wu, B., Fan, P., Yang, C., Duan, W., Zheng, X., Liu, C., & Li, S. (2008). Anthocyanin composition and content in grape berry skin in *Vitis* germplasm. *Food Chemistry. 111*(4), 837–844.

Lin, S. J., Ford, E., Haigis, M., Liszt, G., & Guarente, L. (2004). Calorie restriction extends yeast life span by lowering the level of NADH. *Genes and Development, 18*(1), 12–16.

Liu, C. F., & Pan, T. M. (2010). *In vitro* effects of lactic acid bacteria on cancer cell viability and antioxidant activity. *Journal of Food and Drug Analysis, 18*(2), 8.

Liu, C. T., Chu, F. J., Chou, C. C., & Yu, R. C. Antiproliferative and anticytotoxic effects of cell fractions and exopolysaccharides from *Lactobacillus casei* 01. *Mutation Research/ Genetic Toxicology and Environmental Mutagenesis*, *721*(2), 157–162.

Liu, H. F., Wu, B. H., Fan, P. G., Li, S. H., & Li, L. S. (2006). Sugar and acid concentrations in 98 grape cultivars analyzed by principal component analysis. *Journal of the Science of Food and Agriculture*, *86*(10), 1526–1536.

Liu, M., Yang, K., Qi, Y., Zhang, J., Fan, M., & Wei, X. (2018). Fermentation temperature and the phenolic and aroma profile of persimmon wine. *Journal of the Institute of Brewing*, *124*(3), 269–275.

Lollo, P. C. B., de Moura, C. S., Morato, P. N., Cruz, A. G., de Freitas Castro, W., Betim, C. B., Nisishima, L., José de Assis, F. F., Junior, M. M., Fernandes, C. O., & Amaya-Farfan, J. (2013). Probiotic yogurt offers higher immune-protection than probiotic whey beverage. *Food Research International*, *54*(1), 118–124.

Macrae, R., Robinson, R. K., & Sadler, M. J. (1993). Encyclopaedia of food science, food technology and nutrition.

Mato, I., Suárez-Luque, S., & Huidobro, J. F. (2007). Simple determination of main organic acids in grape juice and wine by using capillary zone electrophoresis with direct UV detection. *Food Chemistry*, *102*(1), 104–112.

Matsumoto, C., Miedema, M. D., Ofman, P., Gaziano, J. M., & Sesso, H. D. (2014). An expanding knowledge of the mechanisms and effects of alcohol consumption on cardiovascular disease. *Journal of Cardiopulmonary Rehabilitation and Prevention*, *34*(3), 159–171.

Mattivi, F., Zulian, C., Nicolini, G., & Valenti, L. (2002). Wine, biodiversity, technology, and antioxidants. *Annals of the New York Academy of Sciences*, *957*(1), 37–56.

Medina-Inojosa, J., Jean, N., Cortes-Bergoderi, M., & Lopez-Jimenez, F. (2014). The Hispanic paradox in cardiovascular disease and total mortality. *Progress in Cardiovascular Diseases*, *57*(3), 286–292.

Murgatroyd, P. R., Van de Ven, M. L. H. M., Goldberg, G. R., & Prentice, A. M. (1996). Alcohol and the regulation of energy balance: Overnight effects on diet-induced thermogenesis and fuel storage. *British Journal of Nutrition*, *75*(1), 33–45.

Navarro, M., Retamales, J. J., & Defilippi, B. (2000). Effect of cluster arrangement and implementation of quality CPPU table grape Sultanina treated with two gibberellin sources. *Agricultural Téchnology*, *61*, 15–21.

Negi, B., Kaur, R., & Dey, G. (2013). Protective effects of a novel sea buckthorn wine on oxidative stress and hypercholesterolemia. *Food and Function*, *4*(2), 240–248.

Nevill, A. M., Holder, R. L., Fentem, P. H., Rayson, M., Marshall, T., Cooke, C., & Tuxworth, W. (1997). Modelling the associations of BMI, physical activity and diet with arterial blood pressure: Some results from the Allied Dunbar national fitness survey. *Annals of Human Biology*, *24*(3), 229–247.

Nikhanj, P., & Kocher, G. S. (2020). Optimization of malolactic fermentation parameters with isolated and characterized lactic acid bacteria associated with grape berries. *Journal of Food Processing and Preservation*, *44*(12), p.e14905.

Oguntibeju, O. O., Truter, E. J., & Esterhuyse, A. J. (2013). The role of fruit and vegetable consumption in human health and disease prevention. *Diabetes Mellitus-Insights and Perspectives*, *3*(2), 172–180.

Patel, A. R. (2017). Probiotic fruit and vegetable juices: Recent advances and future perspective. *International Food Research Journal*, *24*(5).

Patel, V., Tripathi, A. D., Adhikari, K. S., & Srivastava, A. (2021). Screening of physicochemical and functional attributes of fermented beverage (wine) produced from local mango (*Mangifera indica*) varieties of Uttar Pradesh using novel Saccharomyces strain. *Journal of Food Science and Technology, 58*, 2206–2215.

Perricone, M., Bevilacqua, A., Altieri, C., Sinigaglia, M., & Corbo, M. R. (2015). Challenges for the production of probiotic fruit juices. *Beverages, 1*(2), 95–103.

Platiša, M. M., Gal, V., Nestorović, Z., & Gojković-Bukarica, L. (2014). Quantification of the acute effect of a low dose of red wine by nonlinear measures of RR and QT interval series in healthy subjects. *Computers in Biology and Medicine, 53*, 291–296.

Potter, J. D. (2005). Vegetables, fruit, and cancer. *The Lancet, 366*(9485), 527–530.

Prado, F. C., Parada, J. L., Pandey, A., & Soccol, C. R. (2008). Trends in non-dairy probiotic beverages. *Food Research International, 41*(2), 111–123.

Prasanna, P. H. P., Grandison, A. S., & Charalampopoulos, D. (2014). Bifidobacteria in milk products: An overview of physiological and biochemical properties, exopolysaccharide production, selection criteria of milk products, and health benefits. *Food Research International, 55*, 247–262.

Quigley, E. M. (2011). Gut microbiota and the role of probiotics in therapy. *Current Opinion in Pharmacology, 11*(6), 593–603.

Ranadheera, C., Prasanna, P., & Vidanarachchi, J. (2014). Fruit juices as probiotic carriers. In *Fruit Juices: Types, Nutritional Composition and Health Benefits* (pp. 253–267).

Rane, B. R., Patil, R. R., Jain, A. S., Keservani, R. K., & Kesharwani, R. K. (2022). Nutraceuticals as disease preventive food and immunity boosters. In Kesharwani, R. K., Keservani, R. K., & Sharma, A. K. (Eds.), *Nutraceuticals and Functional Foods in Immunomodulators* (pp. 253–267). Springer, Singapore. https://doi.org/10.1007/978-981-19-2507-8_7.

Rašić, J. L. (2003). Microflora of the intestine: Probiotics. *Probiotics*, 3911–3916.

Renaud, S. D., & de Lorgeril, M. (1992). Wine, alcohol, platelets, and the French paradox for coronary heart disease. *The Lancet, 339*(8808), 1523–1526.

Rimm, E. B., Chan, J., Stampfer, M. J., Colditz, G. A., & Willett, W. C. (1995). Prospective study of cigarette smoking, alcohol use, and the risk of diabetes in men. *BMJ, 310*(6979), 555–559.

Robinson, J., & Harding, J. (2015). *The Oxford Companion to Wine*. American Chemical Society.

Roda, A., Lucini, L., Torchio, F., Dordoni, R., De Faveri, D. M., & Lambri, M. (2017). Metabolite profiling and volatiles of pineapple wine and vinegar obtained from pineapple waste. *Food Chemistry, 229*, 734–742.

Rolle, R., & Satin, M. (2002). Basic requirements for the transfer of fermentation technologies to developing countries. *International Journal of Food Microbiology, 75*(3), 181–187.

Sanguansri, L., & Augustin, M. A. (2010). Microencapsulation in functional food product development. In *Functional Food Product Development* (pp. 1–23).

Shafer, S. L., Bartlein, P. J., & Whitlock, C. (2005). Understanding the spatial heterogeneity of global environmental change in mountain regions. In *Global Change and Mountain Regions: An Overview of Current Knowledge* (pp. 21–30).

Shai, I., Wainstein, J., Harman-Boehm, I., Raz, I., Fraser, D., Rudich, A., & Stampfer, M. J. (2007). Glycemic effects of moderate alcohol intake among patients with type 2 diabetes: A multicenter, randomized, clinical intervention trial. *Diabetes Care, 30*(12), 3011–3016.

Sharma, S., Thakur, A. D., Sharma, S., & Atanassova, M. (2018). Effect of different yeast species on the production of pumpkin-based wine. *Journal of the Institute of Brewing, 124*(2), 187–193.

Sierksma, A., Patel, H., Ouchi, N., Kihara, S., Funahashi, T., Heine, R. J., Grobbee, D. E., Kluft, C., & Hendriks, H. F. (2004). Effect of moderate alcohol consumption on adiponectin, tumor necrosis factor-α, and insulin sensitivity. *Diabetes Care, 27*(1), 184–189.

Sõukand, R., Pieroni, A., Biró, M., Dénes, A., Dogan, Y., Hajdari, A., Kalle, R., Reade, B., Mustafa, B., Nedelcheva, A., & Quave, C. L. (2015). An ethnobotanical perspective on traditional fermented plant foods and beverages in Eastern Europe. *Journal of Ethnopharmacology, 170*, 284–296.

Suter, P. M., Jequier, E., & Schutz, Y. (1994). Effect of ethanol on energy expenditure. *American Journal of Physiology-Regulatory, Integrative and Comparative Physiology, 266*(4), R1204–R1212.

Tohill, B. C., Seymour, J., Serdula, M., Kettel-Khan, L., & Rolls, B. J. (2004). What epidemiologic studies tell us about the relationship between fruit and vegetable consumption and body weight. *Nutrition Reviews, 62*(10), 365–374.

Tripathi, S., & Mazumder, P. M. (2020). Apple cider vinegar (ACV) and their pharmacological approach towards Alzheimer's disease (AD): A review. *Indian Journal of Pharmaceutical Education and Research, 54*, S67–S74.

Van Duyn, M. A. S., & Pivonka, E. (2000). Overview of the health benefits of fruit and vegetable consumption for the dietetics professional: Selected literature. *Journal of the American Dietetic Association, 100*(12), 1511–1521.

Wang, J. J., Tung, T. H., Yin, W. H., Huang, C. M., Jen, H. L., Wei, J., & Young, M. S. (2008). Effects of moderate alcohol consumption on inflammatory biomarkers. *Acta Cardiologica, 63*(1), 65–72.

Wargovich, M. J. (2000). Anticancer properties of fruits and vegetables. *HortScience, 35*(4), 573–574.

Yamagata, K. (2017). Carotenoids regulate endothelial functions and reduce the risk of cardiovascular disease. *Carotenoids, 25*(106.10), 5772.

Zhu, Y., Luo, T. M., Jobin, C., & Young, H. A. (2011). Gut microbiota and probiotics in colon tumorigenesis. *Cancer Letters, 309*(2), 119–127.

CHAPTER 6

Fermented Foods: The Mine of Functional Compounds

S. V. N. VIJAYENDRA,* SRIDHAR RACHAKONDA, D. MADHUSUDHAN, and T. JYOTHIRMAYI

CSIR–Central Food Technological Research Institute, Resource Center, Hyderabad, Telangana, India

*Corresponding author.

ABSTRACT

Foods provide nutrition for the survival of living beings. Besides providing additional nutrition than their raw form, fermented foods extend the life of the commodities by preventing the growth of spoilage microorganisms. Fermentation also contributes by giving functional attributes to the foods and reducing antinutritional factors present in the raw foods. Consumption of functional fermented foods and beverages containing probiotics, prebiotics, and synbiotics provides several health benefits to consumers. They help fight or delay chronic diseases. Functional compounds such as short-chain fatty acids (SCFAs), organic acids, bioactive peptides, carotenoids, flavonoids, vitamins, γ-aminobutyric acid, bacteriocins, conjugated linoleic acid (CLA), highly absorbable calcium, dietary fibers, phytosterols, oligosaccharides, prebiotic compounds, and so on, of fermented foods are known to improve digestion, immunity function, suppress cancers, reduces cholesterol, blood glucose level, lactose intolerance, inhibit pathogens, and reduce the risk of cardiovascular diseases (CVDs), liver diseases, chronic renal impairment, and so forth. Functional fermented foods also enhance the antioxidant activity of the foods and improve bone health and prevent osteoporosis, besides regulating hormonal, neurological, immunological, gastrointestinal, and ACE-inhibitory activities. They possess antihypertensive, anti-cholesterolemic, and antithrombotic activities. Thus, functional fermented foods are gaining

Functional Compounds and Foods of Plant Origin:Krishan Dutt Sharma, Satish Kumar, Rakesh Sharma, and Raj K. Keservani (Eds.)

attention from consumers. This chapter highlights the recent developments in the area of functional fermented foods prepared mainly using milk, fruits, vegetables, cereals, pulses, and meat products.

6.1 INTRODUCTION

The foods or food ingredients that provide therapeutic benefits in addition to providing the nutrients they possess traditionally are referred to as functional foods (Ogunremi et al., 2015). Bioactive peptides, SCFAs, conjugated linoleic acid (CLA), highly absorbable calcium, organic acids, oligosaccharides, carotenoids, flavonoids, and so on, are some examples of functional compounds (Cencic and Chingwaru, 2010). Functional foods provide special physiological and health benefits that benefit the body. Demand from consumers for more healthy foods led to the development of the concept of functional foods. Later, researchers started formulating foods supplemented with prebiotics and probiotics to convert them into functional foods. Both prebiotics and probiotics help to maintain beneficial gut microbiota, which in turn protect people from diseases and keep them healthy. Functional foods either improve health or reduce the risk of causing disease in the consumers. Functional food may provide one or more health benefits to the consumers. Not all functional foods are fermented by microbes. Functional fermented foods, in general, contribute to the functional properties of the food through fermentation by beneficial microorganisms. Functional foods are prepared either by modifying the processing method or by fortifying with selected substances that provide functional properties to the food intending to target the food to a specific disease or to improve overall health (Roberfroid, 1999). The functional food and functional beverage industry started picking up in the market in the early 1990s and now flourishing as a business worth billions of dollars. Some examples of functional foods are foods having probiotics, prebiotics, synbiotics, foods that can improve immunity function, ω-3-milk, low lactose, low cholesterol milk, and so on, Functional foods have also been targeted for cancer control, heart health, eye health, weight loss management, memory improvement, and increasing stamina.

Fermented foods have a long history, and they constitute a significant part of the diet of people across the globe. Almost every country has a unique fermented food of its own, and countries like India and other south-Asian countries have multiple numbers of fermented foods. There are several varieties of fermented foods of different commodities of the food like fermented cereals-based foods, fermented vegetable foods, fermented meat, and fish products, fermented pulse-based products, and importantly, a range of fermented milk products. In traditional methods, fermentation is carried out by microorganisms naturally present in the commodities. However, in commercial practices, selected cultures are used to ferment the foods. The microorganisms, either naturally present or added cultures, referred to as starter cultures, involved in the fermentation have a significant role in producing the fermented foods. There is a beneficial association between microorganisms, especially lactic acid bacteria (LAB), and human nutrition. Fermented foods are known to be associated with the longevity of the consumers and extend the shelf-life of perishable foods. Fermentation of foods with specific LAB removes antinutritional factors and toxic metabolites from foods. They assert health benefits (Wong et al., 2019). Many researchers have reviewed the health benefits of fermented foods (Agyei et al., 2020; Vijayendra and Halami, 2015). Fermentation of various food commodities can convert them into wellness foods (Xiang et al., 2019). Though functional fermented foods and beverages provide a range of health benefits to the consumers, their safety, in terms of plant or animal-based toxins, protein inhibitors, saponins, tannins, and so on, and microbial-derived biogenic amines, microbial toxins, and so on, is very important. In this context, the quality and safety of fermented functional foods and beverages have been reviewed very recently (Ashaolu, 2020). Hence, monitoring their quality for safety is needed. Classification of functional fermented foods based on the commodity is shown in Figure 6.1. Some of the sources of functional properties of functional fermented foods are provided in Figure 6.2, and the major functional properties of the fermented foods are depicted in Figure 6.3 and Table 6.1.

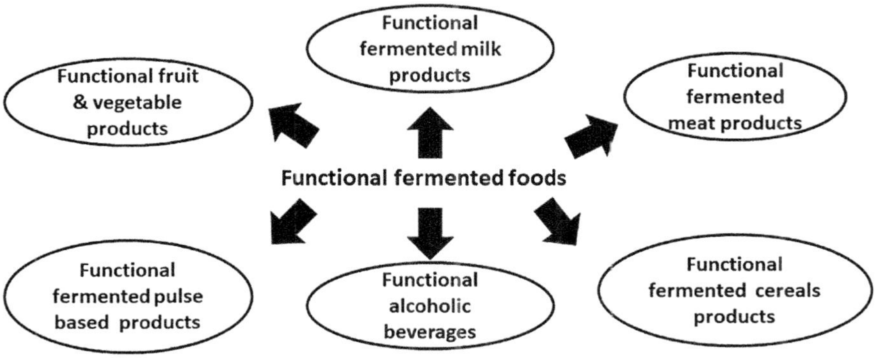

FIGURE 6.1 Classification of functional fermented foods based on commodity.

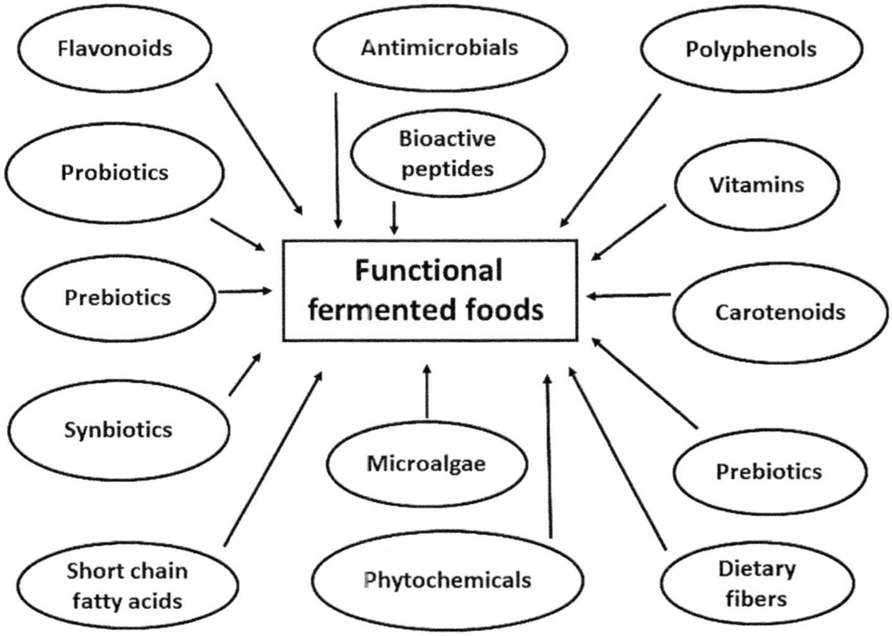

FIGURE 6.2 Functional properties of fermented foods.

TABLE 6.1 Functional Components in Fermented Foods

Fermented Food	Functional Molecules/Property
Milk-based products	Bacteriocins, anti-cancer and antihypertensive compounds, probiotics, polyphenols, poly- and monounsaturated fatty acids, phytochemicals, carotenoids, dietary fiber, vitamins, minerals, bioactive peptides, and so on.
Fruits and vegetables-based products	Antioxidants, antibacterial, antiulcerogenic, anti-inflammatory, anti-cancer, organic acids, bioactive peptides, and so on.
Cereals and pulses-based products	Phytochemicals, dietary fiber, fatty acids (EPA, DHA), vitamins, minerals, immunity boosting, antinutritional factors, antibacterial, and antifungal, and so on.
Meat-based products	Probiotics, reduced biogenic amines, nitrites, bacteriocins, prebiotics, reduced oxidative stress.

6.2 FUNCTIONAL COMPOUNDS IN FERMENTED MILK PRODUCTS

Unlike other commodities, milk has been used to produce several varieties of fermented milk products like mild acidic, acidic, and alcoholic fermented foods. Fermented milk products such as yogurt, curd, cheeses, kefir, kumis, and so on are popular in many countries and have been prepared traditionally for centuries. They play a significant role in improving the health of consumers. They inhibit the growth of several pathogens by producing antimicrobial substances like peroxides, lactic acid, vitamins, and bacteriocins. Some of the health attributes of LAB in fermented milk products include improved digestion, reduced lactose intolerance, suppression of cancers, cholesterol reduction, and inhibition of pathogens and other spoilage microbes. Several reviews are available on the technological, nutritional, and health benefits of fermented milk products (Vijayendra and Varadaraj, 2015; Vijayendra, 2021). Fermented milk significantly reduces the risk of bladder and colorectal cancers. Technological advances in the production of functional fermented milk products have been reviewed very recently.

LAB is known to produce several bacteriocins that could inhibit many Gram-positive and Gram-negative bacteria (Vijayendra et al., 2010), and reduce serum cholesterol (Vijayendra and Gupta, 2012). The probiotic cultures used in the production of fermented milk products could suppress pro-inflammatory cytokines, enhance immune response, and so forth.

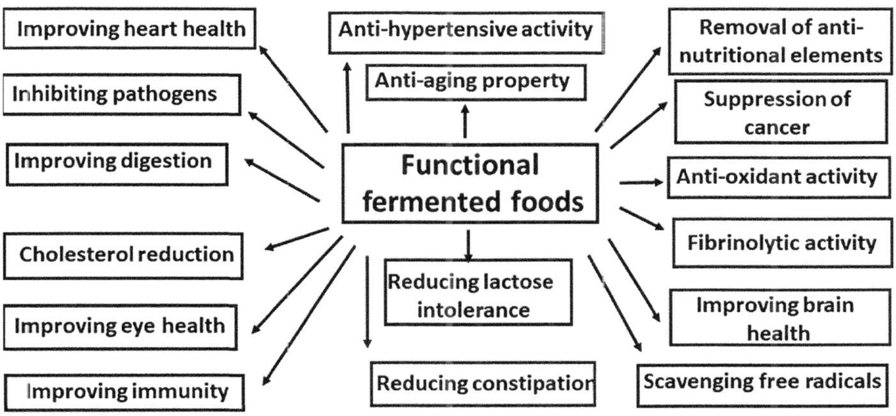

FIGURE 6.3 Functional properties of the fermented foods.

6.2.1 *BACTERIOCINS*

Bacteriocins are short-chain peptides and have antimicrobial activity. Several LAB cultures produce bacteriocins, and they are bactericidal. Bifidin, acidophilin, bulgarian, plantaricin, nisin, and pediocin produced by *Bifidobacterium*, *Lactobacillus acidophilus*, *L. bulgaricus*, *L. plantarum*,

Lactococcus lactis, and *Pediococcus* spp., respectively, are some of the examples of bacteriocins. Some bacteriocins have a broad spectrum, and some others have a narrow spectrum of activity against foodborne pathogens such as *Listeria* spp., *Staphylococcus aureus*, *Escherichia coli*, *Shigella*, *Salmonella*, and so forth. Some bacteriocins are heat stable. Some of the recent reviews provided information on the production, preservative effect, and mechanism of action of bacteriocins produced by LAB. Some bacteriocins and hydrogen peroxide produced by LAB are found to reduce or eliminate bad odor in the mouth and prevent dental caries, oral cavity-related diseases, and *Candida* infections in the mouth.

6.2.2 ANTI-CANCER AND ANTIHYPERTENSIVE COMPOUNDS

A recent review has indicated that certain SCFAs produced by some LAB prevent carcinogenesis in the colon (Górska et al., 2019). Some of the exopolysaccharides of LAB (*L. casei, L. acidophilus,* and *L. plantarum*) are known to have anti-cancer and antitumor properties (Saadat et al., 2019). Many LAB cultures are known to produce γ-aminobutyric acid (commonly referred to as GABA), a neurotransmitter, which improves health by protecting from oxidative damage caused by reactive oxygen species (ROS) and reduces blood pressure, and acts as an antihypertensive compound (Li and Cao, 2010). Many researchers concluded fermented dairy products as antihypertensive functional foods mainly due to the reduction in blood pressure and reduced heart rate because of the presence of antihypertensive peptides, and it became a topic for an exhaustive review. Several commercial fermented milk products having the antihypertensive peptides released by *L. helveticus* during the fermentation are available in the international market. Consuming such products to control hypertension could avoid the side effects posed by the use of medicines to regulate blood pressure. Hence, fermented milk products prepared with selected LAB cultures could be consumed daily to prevent developing hypertension.

6.2.3 PROBIOTICS AS FUNCTIONAL ADDITIVES

Probiotics present in fermented milk products by themselves have shown several functional properties. They are found to increase the IgA antibodies and stimulate the mucosal immune system to fight infections in the body. *L. casei* 01 decreased pro-inflammatory cytokines IL-6 leading to a reduction of inflammation in people with rheumatoid arthritis disease. Probiotics stabilized the gut microbiome by increasing the population and kind of good bacteria and decreasing harmful bacteria like some bacteroides, and improving consumer's health (González et al., 2019). Probiotics prevented antibiotic-associated diarrhea caused by *Clostridium difficile*, diarrhea induced by radiotherapy, and traveler's diarrhea (Liu et al., 2017). Probiotic LAB present in fermented milk products significantly reduced cholesterol content in the blood thus minimizing the risk of atherosclerosis and cardio-vascular diseases (CVDs) (Wang et al., 2018). The hypocholesterolemia effect of probiotics was exhaustively reviewed recently (Hassan et al., 2019). Probiotics have potential therapeutic properties in reducing non-alcoholic fatty acid liver disease, which leads to liver cancer, liver fibrosis, CVD, and chronic renal impairment if left untreated (Koopman et al., 2019). In the presence of selenium, a probiotic strain of *L. rhamnosus* reduced sunburn complications and fermented product equol produced by *L. paracasei* effectively prevented skin disorders (Kwon et al., 2019).

6.2.4 VITAMINS

Many LAB (*L. amylovorus, L. bulgaricus, L. plantarum, L. casei, Propionibacterium freudenreichii,* and *Streptococcus thermophilus*) and bifidobacteria (*Bif. animalis* and *Bif. infantis*) produce vitamins such as biotin, cobalamin, folate, riboflavin, and thiamine in fermented milk products. Biokefir was formulated by incorporating stevia syrup and hawthorn purée to enhance the product's functional properties. Stevia powder, which is rich in various glycosides, stimulates insulin secretion and improves carbohydrate metabolism. It is also high in minerals such as selenium and zinc, as well as β-carotene and vitamin C. Hawthorn can improve oxygen absorption by the heart muscle and promote blood vessel expansion, helping to reduce blood pressure. Similar to stevia, it also contains vitamin C, carotenoids, pectin, and fructose. The antimicrobial and antidiabetic potential of a synbiotic functional fermented milk product was evaluated (Shafi et al., 2019). In this study, fructooligosaccharides and isomaltooligosaccharides were used as prebiotics. Feeding this product to diabetic rats reduced blood glucose levels by 63% and inhibited the growth of *E. coli* and *Staph. aureus*.

6.2.5 POLYPHENOLS AND FATTY ACIDS

Very recently, the production of a functional symbiotic fermented probiotic milk supplemented with 4% carob pod (*Ceratonia siliqua* L.) powder has polyphenols such as gallic acid (441%) and catechin (486%), higher antioxidant activity, and improved antimicrobial activity against foodborne pathogens like *E. coli*, *Salmonella typhimurium*, *Staphylococcus aureus*, and *Listeria monocytogenes* were reported (Chait et al., 2021). A probiotic strain of *L. brevis* was used for the fermentation of this product. A recent publication reviewed the methods of enhancing functionality, nutritional value, and health properties of fermented milk products by way of supplementing bioactive molecules derived from fruits and vegetables, such as fatty acids, phenolic compounds, carotenoids, dietary fibers, phytosterols, vitamins, betacyanin, which can improve health and reduce the occurrence of non-communicable diseases (NCDs) (Prestes et al., 2021). Fruits such as apples, guava, grapes, mangosteen, oranges, cherries, pomegranate, berries, araçá (*Psidium cattleianum*), camu-camu (*Myrciaria dubia*), and guabiroba are rich in polyphenols (Fidelis et al., 2020; Keservani and Sharma, 2014; Keservani et al., 2016a, b).

The addition of strawberry juice to sheep milk enhanced the functional property by way of enhancing phenol content and antioxidant activity of the yogurt, besides extending the shelf life and inhibiting lipid oxidation. In one example, the probiotic culture itself synthesized enzymes responsible for the release of phenolic compounds in a novel fermented food. The addition of açaí fruit (*Euterpe oleracea*) pulp in yogurt increased the probiotic cells count and production of α-linoleic acid and CLA, besides mono and polyunsaturated fatty acids (PUFAs) contents. In a recent exploration, fermentation of milk cream with selected cultures of *Leuconostoc mesenteroides* and *Lactococcus lactis* inhibited *Listeria monocytogens* and increased short-chain fatty acid content and 60% CLA, besides reducing PUFAs and monounsaturated fatty acids. Thus, it increased the functional property of the cream. The antiproliferative and antioxidant activities of the yogurt were enhanced by adding an extract of camu-camu (*Myrciaria dubia*), a South American fruit seed, which has high phenolic content and vitamin C (Fidelis et al., 2020). Coconut fat is known to have several fatty acids, like capric acid and lauric acid, which are healthy for consumers. Taking this clue, coconut milk was used to prepare a functional probiotic curd for the first time, wherein coconut milk and cow milk were used in 30:70 ratio. It was accepted in terms of sensory parameters as well. Regular consumption of this product can provide health benefits.

6.2.6 CAROTENOIDS

Carotenoids have health benefits. Regular intake of carotenoids increases antioxidant and anticarcinogenic properties and helps decrease NCDs like

CVD and degeneration of eye cells; hence, it can be incorporated to increase the functional properties of fermented foods. In this direction, yogurt was prepared by adding nanoparticles (NPs) of zeaxanthin, which was derived from goji berry pulp, to enhance the functional property of the yogurt. Similarly, carotenoids extracted from cantaloupe melon (*Cucumis melo* L.) were incorporated into yogurt in the form of NPs (Medeiros et al., 2019). Microencapsulated canthaxanthin was used for the production of functional flavored fermented milk and unfermented flavored milk. It served as an antioxidant molecule as well as a coloring agent in the products.

6.2.7 PHYTOCHEMICALS

A functional fermented dairy product having gerodietetic properties was developed by adding a photocomposition consisting of extracts of medicinal plants like *Comarum*, *Crataegus,* and *Origanum vulgare*. These extracts can strengthen the joints, stimulate the cardiovascular system, and sedative effect on neurons. Phytosterols are plant-derived molecules that are similar to cholesterol and provide health benefits to consumers. Regular consumption of phytosterols helps circulate low-density cholesterol and reduce blood cholesterol content. Phytosterols are present in apples, berries, avocados, olives, seeds, nuts, and vegetable oils. The fat globules of dairy products are good carriers of phytosterols. Hence, phytosterols can be incorporated into fermented milk products to provide additional health benefits to health-conscious consumers. The addition of *Aloe* extracts which are rich in vitamins, essential oils, enzymes, and phytoncides like homonatalain, nataloin, emodin, aloin, and so on, in milk, enhanced the sensory properties and nutritional value of fermented milk. Aloe extracts have digestion-improving properties, anti-inflammatory properties, and inhibitory activity against pathogenic microorganisms. Another fermented milk with *L. rhamnose* and *Aloe* extract reduced the size of adipocytes and blood glucose levels and were effective in controlling obesity and reducing type-II diabetes.

Rosalact, a probiotic product, was prepared by enriching the pasteurized milk with extracts of rosehip and liquorice medicinal plants and fermented with a probiotic ABT-5 culture. The liquorice root extract contains several sugars, alkaloids, organic acids, alcohols, essential oils and resins, derivatives of phenol carboxylic acids, flavonoids, tannins, coumarin, and higher fatty acids, and higher aliphatic hydrocarbons. Rosalact along with Rosehips, which has several functional properties, could be used for the prevention of vitamin deficiency, CVD, colds, and kidney disease.

6.2.8 DIETARY FIBER

Pectin, lignin, cellulose, and hemicellulose are examples of dietary fiber, and they have a key role in the digestion of food we consume. Whole grains, vegetables, fruits, and root vegetables are rich in dietary fibers. Dietary fibers are used as prebiotic components in foods and promote the growth of probiotics in synbiotic foods. Enhanced survival of probiotic cells in fermented goat milk was observed with the incorporation of pomace flour of orange, passion fruit, and guava. A similar effect on the growth of *L. acidophilus* with the supplementation of date extract in a fermented milk product was reported, and the product was also found to have high antioxidant activity. Though dietary fiber present in various foods cannot be digested in the stomach or absorbed in the gastrointestinal tract, they provide several health benefits. They stimulate the growth of gut microbes, alleviate constipation, reduce levels of insulin, post-prandial blood glucose, total, and low-density cholesterol, and prevent gut cancer, heart diseases, obesity, diabetes, and so forth. They help improve viscosity and minimize whey separation in fermented dairy products. Apple pomace was incorporated in stirred yogurt to reduce syneresis, and it helped to increase the total phenolic content (TPC) of the product (Wang et al., 2020). Thus, the addition of fruits and their derivatives through their various bioactive components improves the functionality of fermented dairy products and enhances the growth and viability of probiotic cultures in these products. In some cases, they also improve the sensory properties and texture of these products.

6.2.9 BIOACTIVE PEPTIDES

Milk is known to have several bio-active peptides consisting of 2–20 amino acids, which can provide many health benefits, and they can be used as potential functional foods. The health benefits of bioactive peptides of goat, buffalo, cow, camel, and sheep milk, such as reduction of diet-related NCDs, obesity, diabetes, improved bone health, CVDs, prevention of osteoporosis, cancer, hypertension, regulation of gastrointestinal, neurological, hormonal, and immunological activities with their antioxidant, ACE-inhibitory, anticholesterolemic, antihypertensive, antithrombotic, broad-spectrum antimicrobial activity against *Bacillus, Staphylococcus, Listeria, E. coli, Candida, Diplococcus, Salmonella, Streptococci, Pseudomonas, Klebsiella*, and so on, immunomodulatory properties, have been reviewed comprehensively.

Many LAB cultures are known to produce bioactive peptides through milk fermentation by their peptidase enzymes (tripeptidases, di-peptidases, amino-peptidases, and endopeptidases). However, their real applications in a wider prospect are yet to be explored, especially at a commercial scale.

6.3 ALGAE AS A SOURCE OF FUNCTIONAL COMPOUNDS

Blue-green microalgae such as *Spirulina* sp. and *Chlorella* sp., which have several therapeutic and nutritional benefits, have been exploited as a source of functional compounds and supplemented in fermented foods, including dairy products. A comprehensive review of this aspect is available. These algae are rich in antioxidant molecules, several vitamins like provitamin-A, E, K, B_2, B_6, B_{12}, β-carotene, minerals like iron and calcium, and unsaturated fatty acids (linoleic acid and γ-linolenic acid), and pigments like phycocyanin. Many therapeutic attributes like reducing blood sugar and serum cholesterol levels and wound healing time, alleviation of hyperlipidemia, anemia, improved anti-inflammatory, antitumoral, and antiviral properties, improving vision and immunity, and so on, are assigned to these algal cultures. The addition of microalgae also stimulated the growth and viability of probiotic cultures in fermented milk products. *Chlorella* is known to prevent cellular damage and oxidative stress and reduce nephrotoxicity. It has antiviral and antibacterial properties. Supplementation of milk with *C. vulgaris* and *S. platensis* while preparing the yogurt has stimulated the growth of probiotics and lactic cultures.

6.4 PLANT PROTEIN AND PLANT INGREDIENTS AS FUNCTIONAL AGENTS

Enhancing the protein content in foods is the need of the hour to address malnutrition in several countries. Dietary proteins play a significant role to build and repair body tissues. Hence, the functional properties of non-fat fermented milk products have been enhanced by adding plant-based proteins, and the addition of soy protein isolate to the milk enhanced the essential amino acid score than other plant proteins such as pea protein, wheat gluten, and rice protein. The amino acids like leucine, lysine, methionine, isoleucine, and threonine are found to be more prevalent in these products. Supplementation of milk products with plant proteins improved the sensory and physicochemical properties of the products besides increasing the nutritional value. Assigning functional attributes to fermented dairy products by adding plant ingredients, which are rich in minerals, vitamins, and biologically active molecules from medicinal plants, was reviewed recently (Sukhikh et al., 2019). According to this review, some plants are known to strengthen blood vessels. Some stimulate the cardiovascular system, and some other has psychotropic properties. Some plants are rich in vitamins and antimicrobial compounds and extend the shelf-life of food products by preventing oxidation and photodegradation-related spoilage. Honeysuckle powder, which is rich in vitamin C, is used in dairy-based foods as an antioxidant component. Thyme extract, which has significant levels of flavonoids and anthocyanins, is used in milk products. Rosehip blooms, apple pectin, barberry fruits, and lemongrass leaves have been used in the preparation of dairy products to provide anti-inflammatory, antihypertensive, antiseptic properties, and immunomodulating effects (Yu et al., 2017). The pericarp of mangosteen (*Garcinia mangostana* L.) fruit contains high concentrations of xanthones, which have a wide range of health attributes such as antifungal, antitumor, diuretic, cardiotonic, psychotropic, and choleretic properties.

6.5 FUNCTIONAL COMPOUNDS IN FERMENTED FRUITS AND VEGETABLES

Fruits and vegetables are rich sources of various functional compounds such as phenolic acids, flavonoids, carotenoids, tannins, stilbenes, lignans, vitamins, and so on, and these have several functional and health benefits. Very recently, it became a topic for an exhaustive review. These compounds are known for their bioactive properties such as decreasing lipid oxidation, reducing the risk of cancers, asthma, diabetes, CVDs, antioxidant, antiviral, antifungal, antibacterial activities, anti-inflammatory property, inhibition of platelet aggregation, improving digestion, bioactive peptides like hypocholesterolemic peptides, antithrombotic peptides, immunomodulating peptides, antioxidant peptides and antihypertensive peptides from cheap sources like vegetables (Garcia et al., 2013). In terms of nutritional value, the fermentation of vegetables improves the digestibility of proteins and carbohydrates. Technological and functional properties of dairy and non-dairy probiotic products including that fruit and vegetables have been reviewed recently (Kumar et al., 2015). The additional advantage of vegetable-based fermented foods is the absence of issues like lactose intolerance and cholesterol.

Fermented vegetables and fruits are the sources of probiotic microorganisms, especially LAB. Gundruk, sauerkraut, sinki, khalpi, kimchi is some of the fermented vegetables. Fermentation not only increases the shelf-life but also enhances the nutritional values and sensory quality besides, providing several health benefits. An exhaustive review is available on the health benefits of fermented vegetables (Vijayendra and Prakash, 2015). Because of their health benefits, fermented vegetables can be used as functional foods. Fermented vegetables exhibited anti-constipation, anti-obesity, antioxidative, anti-cancer, antibacterial (due to the production of bacteriocins), antiaging properties, colorectal health promotion, removal of antinutritional factors, fibrinolytic activity, reduction of cholesterol, improve brain health, enhance immunity, and promotes skin health. In kimchi fermentation, production of folate and riboflavin by *L. sakei* and *Leuconostoc. mesenteroides,* respectively, have been observed by Jung et al. (2013). Fermentation of nightshade leaves with *L. plantarum* 75 improved the functionality of the product by increasing the bioavailability of vanillic acid, ferulic ellagic acids, coumaric, gallic acid, and flavonoids enhanced the functional potential, and improved the bioavailability of luteolin, quercetin, and catechin, and ascorbic acid. It also enhanced the antioxidant activity of the fermented leaves.

Natural fermentation, also known as pickling, of vegetables is very common in many countries across the globe. The total phenolic acid content significantly increased in vegetables (green beans, carrots, white cabbage, chili, pepper, cucumber, cauliflower, tomato, and sneak melon) after 30 days of pickling. A probiotic beverage prepared by fermenting drumstick (*Moringa oleifera*) leaves and beetroot (*Beta vulgaris* L.) with a probiotic LAB culture (*Lb. plantarum*) had antibacterial activity against foodborne pathogens like *Staph. aureus*, *L. monocytogenes*, *E. coli*, and *Bacillus cereus*, reduced the content of raffinose (flatulence causing sugar) and exhibited radical scavenging activity. Several fruits and vegetables (strawberry, carrot, tomato, onion, melon, and fennel) and beverages fermented with kefir grains are found to have higher antioxidative properties and an increased number of volatile compounds (Corona et al., 2016). Several bioactive compounds (organic acids, bioactive peptides, bacteriocins, exopolysaccharides, etc.) have been detected in kefir-based beverages, and they exhibited antibacterial, antiulcerogenic, anti-inflammatory, antioxidant activities (Fiorda et al., 2017). The authors of a review have enlisted various primary mechanisms related to the enhancement of the antioxidative activity of plant-based foods, including fruits and vegetables, mediated by fermentation by various microorganisms like bacteria and fungi.

LAB-mediated fermentation of fruits and vegetables changes both the type and profile of bioactive molecules. Bioavailability and bio-accessibility of food components are enhanced by fermentation. The anti-cancer effects of fermented foods, including those of fruits and vegetables, have been reviewed exhaustively. These foods help in maintaining a healthy gut microbiome and reduce the risk of cancer. Kimchi (fermented cabbage) and fermented cucumbers have shown anti-cancer properties in cell lines such as HCT-116 colon cancer cells, HeLa and HT29 cells, and reduced apoptotic cell death. The pickling process is a relatively good method to preserve phenolic acids in vegetables, and retention of most antioxidant capacities after 30 days of fermentation of chili pepper was noticed. Though sulfur-containing amino acids are lost, the higher content of R-tocopherol, riboflavin, and most individual amino acids in pickled garlic was reported (Montano et al., 2004). Similarly, fermentation significantly improved vitamin C, total flavonoid, antioxidant activities, and phenolic contents in okra seeds. Hence, fermented okra seeds have the potential to be used in the manufacture of functional foods. The total polyphenol content in apple pomace increased by solid-state fermentation with *Phanerocheate chrysosporium,* and nutraceutical properties are increased. Controlled alcoholic fermentation of orange juice for 15 days enhanced the carotenoid composition and provitamin A due to better extractability, making the juice healthier. Besides some vegetables, even tender bamboo shoots of the Himalayas are found to be a good source of LAB, and they degraded antinutritional factors, produced functional enzymes, and so forth (Tamang et al., 2009). Based on the death rates in the recent COVID-19 pandemic in different nations and regions within the same country, researchers have proposed a hypothesis that consumption of fermented vegetables-based foods containing functional molecules might be the reason for this and proposed that consumption of fermented vegetables can be considered as one of the mitigation strategies in lowering the death rates (Bousquet et al., 2021).

Fermented pickles, too, have shown their effect on tongue cancer by reducing tumor necrosis factor (TNF)-α, granulocyte–macrophage-CSF, and granulocyte colony-stimulating factor, besides increasing glutathione (GSH) peroxidase (POD) and superoxide dismutase activities, and so forth. The extract of fermented melon, tomato, and carrot exhibited higher antimicrobial activity both in foodstuff and *in vitro* against foodborne pathogens like *B. cereus, L. monocytogenes,* and *Salmonella* spp. and food spoilage microorganisms than commercial preservatives. Five times higher γ-aminobutyric acid (GABA) content, which has a role in reducing hypertension, in Som-pak, a Thai fermented vegetable product, was noted when inoculated with a fish-isolated *Lb. plantarum* L10–11 along with MSG at 1% (w/v) (Tanamool et al., 2020). A recent review on traditional Ethiopian cereal-based and fruit-based alcoholic beverages prepared like tella, booka, shamita, ogol, keribo, korefe, tej, cheka, and borde, indicated to contain significant amounts of total polyphenols and antioxidants, making them functional beverages (Fentie et al., 2020). A rice-based traditional alcoholic product of Assam is found to have phenolic and flavonoid compounds providing functional attributes to the product.

6.6 FUNCTIONAL COMPOUNDS IN FERMENTED CEREAL AND PULSE PRODUCTS

Cereals are a good source of nutrients such as carbohydrates, dietary fiber, vitamins, proteins, and minerals. Cereals consist of a major portion of food production, and cereals are consumed as a staple food in many countries. Fermented foods based on cereals are existing for centuries, and these are sources of microorganisms, which can provide functional attributes like phytochemicals, dietary fiber, fatty acids (EPA, DHA), vitamins (A, E, B, C), minerals (Zn, Fe, Mg, Ca, etc.) to the consumers (Fernandes et al., 2018). Cereals such as rice, sorghum, maize, and millets are used for the production of fermented products of cereals. Several countries in Asia, Africa, South America, and Europe have a wide variety of cereal-based, mainly based on rice, maize, sorghum, and pearl millets, fermented foods, and beverages with known functional properties (Fernandes et al., 2018). Starter cultures play a key role in the fermentation of foods. The recent trend and application of various starter cultures specifically for the production of cereal-based fermented foods have been reviewed (Ogunremi et al., 2017).

Many cereal-based probiotic foods, along with fruit and vegetable-based, meat-based, and soy-based probiotic foods are being emerged as functional foods (Min et al., 2019). A major portion of fermented cereals are prepared at the home-scale level, and with changing lifestyles, the commercialization of fermented cereal-based products has taken place in recent times. The sale of prepacked ready-to-use idli and dosa batters in India are some examples of such products. Majorly, cereals are subjected to fermentation either by lactic cultures alone or in combination with yeasts. Certain enzymes like amylases, proteases, phytases, and lipases are released by microorganisms responsible for fermentation and hydrolyzing carbohydrates, proteins, phytates, and lipids, respectively. Due to this, the fermentation of cereals improves digestibility and reduces antinutritional factors, besides providing desirable sensory attributes like texture, taste, flavor, and appearance and significantly reducing toxigenic cyanogenic glucoside content during the fermentation of cereals like cassava (Holzapfel, 2002). Fermentation also inhibits the growth of pathogenic and spoilage microorganisms in the products. Antibacterial, antitumor, anti-leukemic, and viricidal activities of fermented cereal-based products have been documented in a review a long time ago (Iwuoha and Eke, 1996). Certain lactic acid cultures are found to disintegrate indigestible oligosaccharides like raffinose, verbascose, stachyose, and so on, present in cereals and legumes during the fermentation into di- and monosaccharides and provide relief from flatulence. Cereals and legumes have several antinutritional components like tannins, phytic acid, and phenolic acids. These factors lead to malnutrition by interfering with the bioavailability of minerals like calcium, phosphorous, magnesium, iron, zinc, and molybdenum, leading to osteoporosis, anemia, impaired mental development, and recurring diarrhea, and limiting the growth rate. Hence, the fermentation of cereals and legumes by LAB and yeasts reduces these antinutritional elements and provides functional benefits to consumers (Claver et al., 2011; Frontela et al., 2008). Fermentation of cereals decreased the activity of amylase and proteinase inhibitors and increased the bioavailability of essential amino acids like methionine, isoleucine, lysine, leucine, and starch.

Mycotoxins are toxigenic secondary metabolites produced by certain fungi belonging to *Aspergillus*, *Penicillium*, *Fusarium*, and so forth. Among all mycotoxins (Fumonisins, patulin, aflatoxins, zearalenone, etc.), aflatoxin B_1 is highly toxigenic. Aflatoxins can cause cancers and other ill effects in consumers (Wagacha and Muthomi, 2008). Cereals are not an exception for the presence of mycotoxins. Several LAB cultures have a detoxification effect on mycotoxins and reduction of mycotoxin levels in fermented maize-based products.

Barley is considered a healthy cereal due to the presence of β-glucan, dietary fiber, vitamins (E), minerals, selenium, and phytochemicals such as folates, phytosterols, lignans, tocols, flavonoids, phenolic acids, and so forth. These phytochemicals exhibited anti-inflammatory, antiproliferative, antioxidant, anticarcinogenic, anti-allergic, cholesterol-reducing abilities, and gastroprotective properties. Barely supports the good growth of LAB. Hence, barley can be used to prepare various functional fermented foods. The β-glucan in barley acts as a prebiotic and promotes the growth of probiotic bacteria. Several functional aspects of barely such as reduction of the risk of inflammatory bowel disease (IBD), CVDs, colon cancer, blood cholesterol level, controlling weight gain, dissolving kidney stones, function as a skin booster, antimicrobial activity, controlling Type-II Diabetes have been reviewed recently. Hence, barely has the potential to be used in the preparation of various functional fermented food products in the form of flakes, grits, or flour. Fermented barley extract was found to attenuate the liver damage induced by alcohol consumption. A polysaccharide extracted from a fermented was barely found to activate the innate immune system and induced antitumor metastasis in experimental mice.

India is home to several rice-based traditional fermented foods (dosa, idli, dhokla, babru, uthappam, appam, chitou, anarshe, sour rice, sez, selroti, ambali, rice cake pitha, etc.) and ethnic fermented rice beverages (haria, apong, jou, rice Jann, bhati jannar, judima, rice beer, zutho, etc.) with many functional and health attributes and more details of these products are provided in a recent review (Ray et al., 2016). Along with LAB, some yeasts (*Saccharomyces*, *Saccharomycopsis*, *Hansenula*, *Candida*, etc.) and filamentous fungi (*Mucor*, *Rhizopus*, *Aspergillus*, *Penicillium*, etc.) are involved in the preparation of fermented rice products. They provide good texture and improve the flavor of the fermented products. To address the increasing malnutrition in several countries, the functional and nutritional values of pseudocereals such as quinoa, amaranth, and buckwheat, which are rich in vitamins and phytochemicals, have been enhanced by lactic fermentation and it became a topic for a recent review (Rollan et al., 2019). Beverages made from sprouted cereals and legumes are gaining popularity as non-dairy probiotic drinks (González-Amaro et al., 2015).

Similar to other fermented foods, fermentation is used as a common tool for the production of cereals-based beverages, and its popularity is increasing over some time. Fermentation improves the digestibility of cereals and enhances the bioavailability of nutrients. Very recently, the functional attributes and potential improvements in the production of non-alcoholic fermented cereal beverages, either at the commercial or home-scale level, have been reviewed thoroughly, and summarized the functional attributes of these products prepared in various countries and their health benefits. Fermented functional beverages have been produced from various cereals such as wheat (*Triticum aestivum* L.), barley (*Hordeum vulgare* L.), proso millet (*Panicum miliaceum* L.), maize (*Zea mays* L.), rice (*Oryza sativa*), sorghum (*Sorghum bicolor*), rye (*Secale cereale*), and oats (*Avena sativa* L.). The role of LAB in the fermentation of cereals-based beverages and providing functional attributes such as reduction of antinutritional factors, antibacterial, and antifungal activities in these products have been reviewed. The LAB as a source of B-vitamins to make fermented cereals-based products as functional products has been reviewed.

Not only cereals but their by-products such as bran, germ, and spent grains in the brewing industry generated during cereal processing like wet milling, dry milling, and malting have several functional attributes, and instead of using them as feed supplements or fuel, it can be fermented to derive value-added products that have application in various foods (Verni et al., 2019). These products are rich in highly soluble fiber, increased peptides, γ-linolenic acid, β carotene, decreased phytic acid and display antioxidant, anti-photoaging effect, anti-cancer and antiproliferative activities (Mabunga et al., 2015). Though several cereal-based fermented products are available in Asia and Africa, only a few fermented products made out of pseudocereals like quinoa-based beverages, and buckwheat-based products, which had higher flavonoids are available (Ludena Urquizo et al., 2017). Autochthonous LAB such as *Enterococcus durans* and *E. mundtii*, *Lb. plantarum*, having phytase degradation activity, are isolated from pseudocereals like amaranth and quinoa (Carrizo et al., 2016; Rizzello et al., 2016). Because of its several disorders' preventability, oats have been used as a good matrix for functional fermented products, and it became a topic in a recent review.

Extracts from cereals like wheat, barley, and malt significantly enhanced the acid tolerance of lactobacilli (*L. acidophilus, L. reuteri,* and *L. plantarum*) to gastric acid due to the protection received from free amino acids and free sugars. Keeping its functional attributes, fermented brown rice flour has been used as an ingredient in the preparation of bread using white rice flour, and it improved volume, texture, and rheological characteristics. Fermentation eliminated the negative effects of bran and improved softening of the bran particles and partial degradation of fiber. Fermentation of brown rice flour with yeast significantly increased soluble fiber, insoluble fiber, protein, minerals, resistant starch (RS), total ash, antioxidants, riboflavin, pyridoxine, phenolics, γ-tocotrienol, nicotinic acid, and δ-tocotrienol (Ilowefah et al., 2015).

Pulses from legume plants are a rich source of proteins, and these are the main source of proteins for vegetarians in many countries. They provide many bioactive molecules like angiotensin-converting enzyme (ACE), which can regulate blood pressure. Fermentation of pulses leads to increased nutrition, the bioavailability of phytonutrients, and the formation of bioactive compounds like bioactive peptides that can prevent potential chronic diseases (Frias et al., 2017). Hence, pulses are being explored to prepare plant-based yogurt alternates. Pulse-based fermented foods, with their nutritional and functional benefits, can help achieve food security and zero hunger, which is one of the Sustainable Development Goals of the United Nations organization, and these are affordable, viable, alternative cheap sources of proteins, especially in developing nations (Adebo et al., 2017). Germination and fermentation of pulses or flour of pulses and cereals with bacterial (both lactic or non-lactic) or fungal cultures were found to improve the nutritional quality of pulses and cereals through their endogenous enzymes (Garrido-Galand et al., 2021), ACE inhibitory activity, bioactive peptides, polyphenols, and provide antioxidant, antimicrobial, anti-cancer, and antidiabetic activities to the consumers. A recent review summarized various factors involved in these processes that affect the production of compounds that has ACE-I activity (Maleki and Razavi, 2020). Nowadays, legume-based prebiotics, probiotics, and synbiotic foods, specifically, beverages, are gaining importance, and it became a topic for a recent review. A recent publication reviewed the iron-enhancing capacity of pulse-based probiotic superfoods in women of reproductive age who are suffering from anemia (Rajagukguk et al., 2021).

6.7 FUNCTIONAL COMPOUNDS IN FERMENTED MEAT PRODUCTS

Fermentation of meat and meat products extends the shelf life besides improving the texture, color, and sensory quality of the meat. Fermentation provides a unique flavor to meat and meat products and is convenient to transport. Fermentation of meat is mediated mainly by bacteria belonging to micrococci, lactobacilli, staphylococci, and some yeasts and molds (Wang et al., 2021). Fermentation can be mediated naturally with inherent microbial cultures or artificially by adding selective microbial cultures to get the desired properties. Traditionally raw cured meats are produced through fermentation by LAB, non-lactic bacterial cultures, yeasts, and molds, and some of these are probiotics providing functional attributes to cured meats (Ojha et al., 2015). Many non-thermal technologies for the production of functional fermented meat products were reviewed (Ojha et al., 2015).

However, unlike in using starter cultures, natural fermentation may compromise the safety and quality of the product, and variations from batch to batch are possible. Although fermented dairy and vegetable products have existed for centuries, fermented meat products are relatively recent and gaining popularity due to advancements in technology, improved nutritional quality, and changes in sensory perception (Pedro et al., 2020). Very recently, the functional properties of microbial fermentation of meat and meat products, such as inhibition of undesired and pathogenic microbial populations, and reduction of biogenic amines and nitrite contents (Wang et al., 2021).

Demand for healthier meat with reduced fat and cholesterol content, low levels of salt and nitrate, and improved fatty acid profile are increasing among consumers. Hence, researchers focused on the preparation and evaluation of the quality, and functionality of meat and meat products (Kumar et al., 2017). Protein degradation during meat fermentation improves the functional properties of the meat products. Lipolysis of fat in meat releases free fatty acids, which can improve flavor and selectively encourage beneficial microbes in the gut microbiome. The antimicrobial compounds like bacteriocins produced by LAB used for the fermentation of meat enhance the shelf life and inhibit foodborne pathogens in the meat and fermented sausages. Hence, fermentation of meat and meat product can enhance its functional properties that help prevent diseases and improve the health of the consumers. Consumption of meat in large quantities regularly increases oxidative stress in consumers mainly due to the formation of oxidized compounds like oxysterols, 4-hydroxy-nominal, malondialdehyde (MDA), or protein carbonyls, which induces oxidative damage. Fatty acids, walnuts, and polyphenols can reduce the oxidative stress of meat products and make them functional meats to have a positive impact on reducing degenerative diseases. Several healthy and unhealthy components of meat products are to be considered while preparing the functional meat-based products to reduce the risk of several diseases like CVDs, cancer, bone diseases, anemia, and so on, and to improve the nutrition and health of the consumers have been reviewed.

Probiotics like in dairy products can improve the functional properties of meat and meat products. Several researchers have used selected LAB such as lactobacilli, bifidobacteria, enterococci, pneumococci, and so on, to ferment meat and the delivery strategies of these probiotics in raw, cooked, and fermented meats of chicken, beef, pork, and so on, either in the form of encapsulation or entrapment in gel dispersions have been reviewed exhaustively. Fermented probiotic meat products such as dry sausages and meat spreads have been reviewed. Bacteriocin producing *L. brevis* SCH6, *Pd. pentosaceus* BAL6 and KL14 are isolated from raw fermented meat products of Poland, such as gammon, pork roast, and sausage, and these LABs showed inhibitory activity against *L. monocyotogenes* than other pathogens tested. The presence of low and medium-chain bioactive peptides and oligopeptides due to the proteolysis by microorganisms during fermentation of cured meats has been described and reviewed. These bioactive peptides showed antihypertensive and *in vitro* antioxidant effects.

6.8 CONCLUSION

Several developments in the area of functional fermented foods have taken place in the last two decades however, it is limited mainly to laboratory studies. Not many studies used human subjects to explore the health attributes of functional fermented foods. In any case, fermented functional foods may have a better place in providing health attributes in terms of no negative side effects as compared to pharma preparations. Although functional fermented foods are gaining attention from consumers and many are being prepared on a small scale, not all products are produced in large volumes for commercialization. Hence, the production and commercialization of these products at a large scale can benefit the people at large.

KEYWORDS

- **amino acids**
- **antimicrobial activity**
- **antioxidant activity**
- **bioactive peptides**
- **cancer**
- **cardiovascular diseases**
- **diabetes**
- **fermentation**
- **functional foods**
- **gastrointestinal regulation**
- **hypertension**
- **non-communicable diseases**
- **osteoporosis**
- **prebiotics**
- **probiotics**

REFERENCES

Adebo, O. A., Njobeh, P. B., Adebiyi, J. A., Gbashi, S., Phoku, J. Z., & Kayitesi, E. (2017). Fermented pulse-based food products in developing nations as functional foods and ingredients. In Hueda, M. C. (Ed.), *Functional Food—Improve Health through Adequate Food* (pp. 77–109).

Agyei, D., Owusu-Kwarteng, J., Akabanda, F., & Akomea-Frempong, S. (2020). Indigenous African fermented dairy products: Processing technology, microbiology and health benefits. *Critical Reviews in Food Science and Nutrition, 60*(6), 991–1006.

Andrey, L., Irina, C., & Marina, N. (2020). Russian methodology for designing multicomponent foods in retrospect. *Foods and Raw Materials, 8*(1), 2–11.

Ashaolu, T. J. (2020). Safety and quality of bacterially fermented functional foods and beverages: A mini review. *Food Quality and Safety, 4*(3), 123–127.

Bousquet, J., Anto, J. M., Czarlewski, W., Haahtela, T., Fonseca, S. C., Iaccarino, G., Blain, H., Vidal, A., Sheikh, A., Akdis, C. A., & Zuberbier, T. (2021). Cabbage and fermented vegetables: From death rate heterogeneity in countries to candidates for mitigation strategies of severe COVID-19. *Allergy, 76*(3), 735–750.

Carrizo, S. L., Montes de Oca, C. E., Hebert, M. E., Saavedra, L., Vignolo, G., LeBlanc, J. G., & Rollán, G. C. (2017). Lactic acid bacteria from Andean grain amaranth: Source of functional value enzymes and vitamins. *Journal of Molecular Microbiology and Biotechnology, 27*(5), 289–298.

Cencic, A., & Chingwaru, W. (2010). The role of functional foods, nutraceuticals, and food supplements in intestinal health. *Nutrients, 2*(6), 611–625.

Čertík, M., Adamechová, Z., & Guothová, L. (2013). Simultaneous enrichment of cereals with polyunsaturated fatty acids and pigments by fungal solid-state fermentations. *Journal of Biotechnology, 168*(2), 130–134.

Chait, Y. A., Gunenc, A. G., Bendali, F., & Hosseinian, F. (2021). Functional fermented carob milk: Probiotic variability and polyphenolic profile. *Journal of Food Bioactives*.

Claver, I. P., Zhou, H. M., Zhang, H. H., Zhu, K. X., Qin, L., & Murekatete, N. (2011). The effect of soaking with wooden ash and malting upon some nutritional properties of sorghum flour used for impeke, a traditional Burundian malt-based sorghum beverage. *Agricultural Sciences in China, 10*(11), 1801–1811.

Corona, O., Randazzo, W., Miceli, A., Guarcello, R., Francesca, N., Erten, H., Moschetti, G., & Settanni, L. (2016). Characterization of kefir-like beverages produced from vegetable juices. *LWT–Food Science and Technology, 66*, 572–581.

Fentie, E. G., Emire, S. A., Demsash, H. D., Dadi, D. W., & Shin, J. H. (2020). Cereal-and fruit-based Ethiopian traditional fermented alcoholic beverages. *Foods, 9*(12), 1781.

Fernandes, C. G., Sonawane, S. K., & SS, A. (2021). Cereal-based functional beverages: A review. *Journal of Microbiology, Biotechnology and Food Sciences*, 914–919.

Fidelis, M., de Oliveira, S. M., Santos, J. S., Escher, G. B., Rocha, R. S., Cruz, A. G., do Carmo, M. A. V., Azevedo, L., Kaneshima, T., Oh, W. Y., & Shahidi, F. (2020). From byproduct to a functional ingredient: Camu-camu (*Myrciaria dubia*) seed extract as an antioxidant agent in a yogurt model. *Journal of Dairy Science, 103*(2), 1131–1140.

Frias, J., Penas, E., & Martinez-Villaluenga, C. (2017). Fermented pulses in nutrition and health promotion. In *Fermented Foods in Health and Disease Prevention* (pp. 385–416).

Frontela, C., García-Alonso, F. J., Ros, G., & Martínez, C. (2008). Phytic acid and inositol phosphates in raw flours and infant cereals: The effect of processing. *Journal of Food Composition and Analysis, 21*(4), 343–350.

Gabriel-Dănuţ, M., Gabriela, R., Elisabeta, B., Liliana, G., Doina-Georgeta, A., Oana, N., Gabriela, V., & Alina, D. (2009). Sensory evaluation and rheological behavior of probiotic dairy products with *Rosa canina* L. and *Glycyrriza glabra* L. extracts. *Innovative Romanian Food Biotechnology,* (4), 32–39.

García, M. C., Puchalska, P., Esteve, C., & Marina, M. L. (2013). Vegetable foods: A cheap source of proteins and peptides with antihypertensive, antioxidant, and other less occurrence bioactivities. *Talanta, 106*, 328–349.

Garrido-Galand, S., Asensio-Grau, A., Calvo-Lerma, J., Heredia, A., & Andrés, A. (2021). The potential of fermentation on nutritional and technological improvement of cereal and legume flours: A review. *Food Research International, 145*, 110398.

González, S., Fernández-Navarro, T., Arboleya, S., de Los Reyes-Gavilán, C. G., Salazar, N., & Gueimonde, M. (2019). Fermented dairy foods: Impact on intestinal microbiota and health-linked biomarkers. *Frontiers in Microbiology, 10*, 1046.

Górska, A., Przystupski, D., Niemczura, M. J., & Kulbacka, J. (2019). Probiotic bacteria: A promising tool in cancer prevention and therapy. *Current Microbiology, 76*, 939–949.

Hassan, A., Din, A. U., Zhu, Y., Zhang, K., Li, T., Wang, Y., Luo, Y., & Wang, G. (2019). Updates in understanding the hypocholesterolemia effect of probiotics on atherosclerosis. *Applied Microbiology and Biotechnology, 103*, 5993–6006.

Holzapfel, W., & Hornesey. (2002). Appropriate starter culture technologies for small-scale fermentation in developing countries. *International Journal of Food Microbiology, 75*(3), 197–212.

Ilowefah, M., Bakar, J., Ghazali, H. M., Mediani, A., & Muhammad, K. (2015). Physicochemical and functional properties of yeast-fermented brown rice flour. *Journal of Food Science and Technology, 52*, 5534–5545.

Iwuoha, C. I., & Eke, O. S. (1996). Nigerian indigenous fermented foods: Their traditional process operation, inherent problems, improvements, and current status. *Food Research International, 29*(5, 6), 527–540.

Jung, J. Y., Lee, S. H., Jin, H. M., Hahn, Y., Madsen, E. L., & Jeon, C. O. (2013). Metatranscriptomic analysis of lactic acid bacterial gene expression during kimchi fermentation. *International Journal of Food Microbiology, 163*(2, 3), 171–179.

Keservani, R. K., & Sharma, A. K. (2014). Flavonoids: Emerging trends and potential health benefits. *Journal of Chinese Pharmaceutical Sciences, 23*(12), 815.

Keservani, R. K., Sharma, A. K., & Kesharwani, R. K. (2016a). Nutraceutical and functional foods for cardiovascular health. In *Food Process Engineering* (pp. 257–278). Apple Academic Press, CRC Press. ISBN: 9781771884020.

Keservani, R. K., Sharma, A. K., & Kesharwani, R. K. (2016b). Medicinal effect of nutraceutical fruits for cognition and brain health. *Scientifica, 3109254*, 10. https://doi.org/10.1155/2016/3109254.

Koopman, N., Molinaro, A., Nieuwdorp, M., & Holleboom, A. G. (2019). Can bugs be drugs? The potential of probiotics and prebiotics as treatment for non-alcoholic fatty liver disease. *Alimentary Pharmacology & Therapeutics, 50*(6), 628–639.

Kumar, P., Chatli, M. K., Verma, A. K., Mehta, N., Malav, O. P., Kumar, D., & Sharma, N. (2017). Quality, functionality, and shelf life of fermented meat and meat products: A review. *Critical Reviews in Food Science and Nutrition, 57*(13), 2844–2856.

Kwon, J. E., Lim, J., Bang, I., Kim, I., Kim, D., & Kang, S. C. (2019). Fermentation product with new equal-producing *Lactobacillus paracasei* as a probiotic-like product candidate for prevention of skin and intestinal disorder. *Journal of the Science of Food and Agriculture, 99*(9), 4200–4210.

Li, H., & Cao, Y. (2010). Lactic acid bacterial cell factories for gamma-aminobutyric acid. *Amino Acids, 39*, 1107–1116.

Liu, M. M., Li, S. T., Shu, Y., & Zhan, H. Q. (2017). Probiotics for prevention of radiation-induced diarrhea: A meta-analysis of randomized controlled trials. *PLoS ONE, 12*(6), e0178870.

Liu, Q., & Yang, H. (2019). Application of atomic force microscopy in food microorganisms. *Trends in Food Science & Technology, 87*, 73–83.

Ludena Urquizo, F. E., García Torres, S. M., Tolonen, T., Jaakkola, M., Pena-Niebuhr, M. G., von Wright, A., Repo-Carrasco-Valencia, R., Korhonen, H., & Plumed-Ferrer, C. (2017). Development of a fermented quinoa-based beverage. *Food Science & Nutrition, 5*(3), 602–608.

Mabunga, D. F. N., Gonzales, E. L. T., Kim, H. J., & Choung, S. Y. (2015). Treatment of GABA from fermented rice germ ameliorates caffeine-induced sleep disturbance in mice. *Biomolecules & Therapeutics, 23*(3), 268.

Maleki, S., & Razavi, S. H. (2021). Pulses' germination and fermentation: Two bioprocessing methods against hypertension by releasing ACE inhibitory peptides. *Critical Reviews in Food Science and Nutrition, 61*(17), 2876–2893.

Medeiros, A. K. D. O. C., de Carvalho Gomes, C., de Araújo Amaral, M. L. Q., de Medeiros, L. D. G., Medeiros, I., Porto, D. L., Aragão, C. F. S., Maciel, B. L. L., de Araújo Morais, A. H., & Passos, T. S. (2019). Nanoencapsulation improved water solubility and color stability of carotenoids extracted from Cantaloupe melon (*Cucumis melo* L.). *Food Chemistry, 270*, 562–572.

Montano, A., Casado, F. J., De Castro, A., Sánchez, A. H., & Rejano, L. (2004). Vitamin content and amino acid composition of pickled garlic processed with and without fermentation. *Journal of Agricultural and Food Chemistry, 52*(24), 7324–7330.

Ogunremi, O. R., Agrawal, R., & Sanni, A. I. (2015). Development of cereal-based functional food using cereal-mix substrate fermented with probiotic strain–*Pichia kudriavzevii* OG 32. *Food Science & Nutrition, 3*(6), 486–494.

Ogunremi, O. R., Banwo, K., & Sanni, A. I. (2017). Starter-culture to improve the quality of cereal-based fermented foods: Trends in selection and application. *Current Opinion in Food Science, 13*, 38–43.

Osman, A., Goda, H. A., Abdel-Hamid, M., Badran, S. M., & Otte, J. (2016). Antibacterial peptides generated by Alcalase hydrolysis of goat whey. *LWT-Food Science and Technology, 65*, 480–486.

Pedro, D., Saldaña, E., Lorenzo, J. M., Pateiro, M., Dominguez, R., Dos Santos, B. A., Cichoski, A. J., & Campagnol, P. C. B. (2021). Low-sodium dry-cured rabbit leg: A novel meat product with healthier properties. *Meat Science, 173*, 108372.

Prestes, A. A., Vargas, M. O., Helm, C. V., Esmerino, E. A., Silva, R., & Prudencio, E. S. (2021). How to improve the functionality, nutritional value and health properties of fermented milks with added fruit bioactive compounds: A review. *Food Science and Technology, 42*.

Rajagukguk, Y. V., Arnold, M., & Gramza-Michałowska, A. (2021). Pulse probiotic superfood as an iron status improvement agent in active women—A review. *Molecules, 26*(8), 2121.

Ray, M., Ghosh, K., Singh, S., & Mondal, K. C. (2016). Folk to functional: An explorative overview of rice-based fermented foods and beverages in India. *Journal of Ethnic Foods, 3*(1), 5–18.

Roberfroid, M. B. (1999). What is beneficial for health? The concept of functional food. *Food and Chemical Toxicology, 37*(9, 10), 1039–1041.

Rollán, G. C., Gerez, C. L., & LeBlanc, J. G. (2019). Lactic fermentation as a strategy to improve the nutritional and functional values of pseudocereals. *Frontiers in Nutrition, 6*, 98.

Saadat, Y. R., Khosroushahi, A. Y., & Gargari, B. P. (2019). A comprehensive review of anti-cancer, immunomodulatory and health beneficial effects of the lactic acid bacteria exopolysaccharides. *Carbohydrate Polymers, 217*, 79–89.

Seo, Y. K., Jung, S. H., Song, K. Y., Park, J. K., & Park, C. S. (2019). Anti-photoaging effect of fermented rice bran extract on UV-induced normal skin fibroblasts. *European Food Research and Technology, 231*, 163–169.

Shafi, A., Naeem Raja, H., Farooq, U., Akram, K., Hayat, Z., Naz, A., & Nadeem, H. R. (2019). Antimicrobial and antidiabetic potential of synbiotic fermented milk: A functional dairy product. *International Journal of Dairy Technology, 72*(1), 15–22.

Siro, I., Kápolna, E., Kápolna, B., & Lugasi, A. (2008). Functional food: Product development, marketing and consumer acceptance—A review. *Appetite, 51*(3), 456–467.

Tamang, J. P., Tamang, B., Schillinger, U., Guigas, C., & Holzapfel, W. H. (2009). Functional properties of lactic acid bacteria isolated from ethnic fermented vegetables of the Himalayas. *International Journal of Food Microbiology, 135*(1), 28–33.

Tanamool, V., Hongsachart, P., & Soemphol, W. (2019). Screening and characterisation of gamma-aminobutyric acid (GABA) producing lactic acid bacteria isolated from Thai fermented fish (Plaa-som) in Nong Khai and its application in Thai fermented vegetables (Som-pak). *Food Science and Technology, 40*, 483–490.

Verni, M., Rizzello, C. G., & Coda, R. (2019). Fermentation biotechnology applied to cereal industry by-products: Nutritional and functional insights. *Frontiers in Nutrition, 6*, 42.

Vijaya Kumar, B., Vijayendra, S. V. N., & Reddy, O. V. S. (2015). Trends in dairy and non-dairy probiotic products—A review. *Journal of Food Science and Technology, 52*, 6112–6124.

Vijayendra, S. V. N. (2021). Fermented milk products and their potential health benefits. In A. K. Rai & K. A. Anu Appaiah (Eds.), *Bioactive Compounds in Fermented Foods: Health Aspects* (pp. 299–323). CRC Press.

Vijayendra, S. V. N., & Gupta, R. C. (2012). Assessment of probiotic and sensory properties of dahi and yoghurt prepared using bulk freeze dried cultures in buffalo milk. *Annals of Microbiology, 62*, 939–947.

Vijayendra, S. V. N., & Halami, P. M. (2015). Fermented vegetables—Health benefits. In *Fermented Vegetables* (pp. 327–344). CRC Press.

Vijayendra, S. V. N., & Varadaraj, M. C. (2015). Dahi, lassi and shrikhand. In A. K. Puniya (Ed.), *Fermented Milk and Dairy Products* (pp. 355–376). CRC Press.

Vijayendra, S. V. N., Rajashree, K., & Halami, P. M. (2010). Characterisation of a heat stable anti-listerial bacteriocin produced by vancomycin sensitive *Enterococcus faecium* isolated from idli batter. *Indian Journal of Microbiology, 50*(2), 243–246.

Wagacha, J. M., & Muthomi, J. W. (2008). Mycotoxin problem in Africa: Current status, implications to food safety and health, and possible management strategies. *International Journal of Food Microbiology, 124*(1), 1–12.

Wang, L., Guo, M. J., Gao, Q., Yang, J. F., Yang, L., Pang, X. L., & Jiang, X. J. (2018). The effects of probiotics on total cholesterol: A meta-analysis of randomized controlled trials. *Medicine, 97*(5).

Wang, X., Kristo, E., & LaPointe, G. (2020). Adding apple pomace as a functional ingredient in stirred-type yogurt and yogurt drinks. *Food Hydrocolloids, 100*, 105453.

Wang, Z., Wang, Z., Ji, L., Zhang, J., Zhao, Z., Zhang, R., Bai, T., Hou, B., Zhang, Y., Liu, D., & Wang, W. (2021). A review: Microbial diversity and function of fermented meat products in China. *Frontiers in Microbiology, 12*, 645435.

Wong, C. B., Odamaki, T., & Xiao, J. Z. (2019). Beneficial effects of *Bifidobacterium longum* subsp. *longum* BB536 on human health: Modulation of gut microbiome as the principal action. *Journal of Functional Foods, 54*, 506–519.

Xiang, H., Sun-Waterhouse, D., & Ruan, Z. (2019). Business fermentation-enabled wellness foods: A fresh perspective. *Published, 8*(3), 203–243.

Yu, P. E., Kaledina, M. V., Shevchenko, N. P., & Ukolova, O. V. (2017). Phytoproducts with extracts of herbal raw materials of the Crimean Peninsula on the basis of serumal-polysaccharidic fraction. *Mezhdunarodnyy studencheskiy nauchnyy vestnik [International Student Science Bulletin], 64*(10–13), 90–94.

Zieliński, H., Szawara-Nowak, D., Bączek, N., & Wronkowska, M. (2019). Effect of liquid-state fermentation on the antioxidant and functional properties of raw and roasted buckwheat flours. *Food Chemistry, 271*, 291–297.

PART II

Fruit- and Vegetable-Based Functional Foods

CHAPTER 7

Colored Vegetables – Futuristic Approach for Developing High-Value Crops

BHARTI SHREE,[1] SATISH KUMAR,[2] DEEPIKA SHARMA,[3] SHWETA SHARMA,[3*] and AYUSH KASHYAP[3]

[1]*CSK Himachal Pradesh Krishi Vishvavidyalaya, Palampur, Himachal Pradesh, India*

[2]*Department of Food Science and Technology, Dr. Y. S. P. University of Horticulture and Forestry, Nauni, Solan, Himachal Pradesh, India*

[3]*M. S. Swaminathan School of Agriculture, Shoolini University, Solan, Himachal Pradesh, India*

*Corresponding author.

ABSTRACT

Vegetables are considered protective and necessary food for balanced diets since they supply essential minerals, vitamins, carbohydrates, protein, dietary fiber, phytonutrients, and nutraceutical compounds. They also provide health protection on account of the presence of secondary metabolites exhibiting antioxidant, antimicrobial, anti-proliferative, and anti-cancerous properties. Consumption of colored/rainbow vegetables is an easy way to get a complete range of nutrients and nutraceuticals to keep us healthy. The color of vegetables represents a lot about their nutritional importance. The compounds which are vital for nutritional security must be identified. The demand is to shift the focus from food security to nutritional security; therefore, efforts must be made to breed color-rich cultivars through different breeding methods. The chapter has been devoted to a better understanding of the health and nutritional benefits of colored vegetables to address the

Functional Compounds and Foods of Plant Origin:Krishan Dutt Sharma, Satish Kumar, Rakesh Sharma, and Raj K. Keservani (Eds.)

problem of malnutrition and to assess the effects of processing on pigment bioavailability.

7.1 INTRODUCTION

Vegetables are a vital compound of healthy eating. They are important sources of carbohydrates, proteins, minerals, fiber, and vitamins (Kader et al., 2000). With the awareness of the nutritional value of vegetables, their consumption has drastically increased. They are an indispensable part of a daily diet. Vegetables play an important role in a balanced diet by supplying vital protective nutrients. According to the Indian Council of Medical Research, the minimum dietary intake of vegetables is 300 g per person per day. It contains 125 g of leafy vegetables, 75 g of other vegetables, and 100 g of root and tuber vegetables. Apart from providing nutrition, vegetables protect against many diseases. Malnutrition is defined as deficiencies, excesses, or imbalances in a person's dietary consumption of nutrients. Malnutrition, in all its forms, includes undernutrition (stunting, wasting, and underweight), insufficient minerals or vitamins, obesity, and nutrition or diet-related diseases. The ill effects of a poor diet are reflected in general in three forms of malnutrition, *viz.*, stunting, wasting, overweight, and obesity. Stunting is the staggering aftereffect of poor nutrition in early childhood. The overwhelming impacts of stunting can last a lifetime and even influence the next generation. Wasting means a child who is too thin for his or her height. Wasting in children is a dangerous consequence of poor food intake. There is also an emerging face of malnutrition, that is, childhood overweight, often called pre-obesity. This type of malnutrition results from energy intake from food and beverages that exceed children's energy requirements. Obesity is a complicated illness characterized by an excess of body fat. These build the danger of diet-related NCDs later in life (WHO, 2020). Malnutrition mostly occurs in low- and middle-income countries or in families. The developmental, community, and medical consequences of the global malnutrition dilemma are serious and eternal for individuals and families, communities, and countries. Foods derived from animals, on the other hand, are frequently out of reach for these families due to its low availability and high cost. Vegetable crops especially colored vegetables can play an important role in combating malnutrition by offering more balanced diets. A Rainbow on a platter is necessary for optimum health. Therefore, vegetable research ought to create methodologies not exclusively to deliver more sustenance, yet additionally to improve access to increasingly nutritious food (Baldermann et al., 2016). The presence of many chemical constituents, such as phytochemicals, antioxidants, and bioflavonoids in vegetables have medicinal values that have the potential to cure diseases and therefore vegetables have been used as traditional medicine since time immemorial. Edible colors present in vegetables are a vital source of antioxidants. Vegetables are irreplaceable as they form the basis of a balanced diet and help in solving nutritional deficiency, improving eye health, and reducing the risk of cardiovascular diseases (CVDs) and cancer. Vegetables are an excellent source of carbohydrates (leguminous vegetables, sweet potato, potato, onion, garlic, and fenugreek), protein (pea, bean, and sweet potato), vitamin A (carrot, tomato, pumpkin, and leafy vegetables), vitamin C (green chilies, cole crops, leafy vegetables, radish leaves, and turnip) and minerals (drumstick pods, leafy vegetables). Traditional use of medicines is recognized as a way to learn about future medicines. The terms "functional foods" and "nutraceuticals" are emerging out of benefits from food that go beyond those attributable to essential nutrients. In recent years, there is increasing interest in nutraceuticals that supply health benefits and are substitutes for modern medicine. The word Nutraceutical where "Nutra" is derived from nutrition and "critical" is from pharmaceutical refers to substances that may be considered a food or part of a food and are thought to provide medicinal and health benefits, including the treatment and prevention of disease (Keservani et al., 2010a, b, 2017, 2020). Phytochemicals are defined as bioactive constituents/compounds that sustain or promote health and occur at the intersection of the food and pharmaceutical industries. Phytochemicals (Keservani and Sharma, 2014; Keservani et al., 2016a, b) are largely considered carotenoids, polyphenols, flavonoids, phytoestrogens, terpenoids, phytosterols, limonoids, glucosinolates, and anthocyanidins. They possess health-promoting properties such as anti-inflammatory, anti-allergic, antibacterial, antifungal, chemopreventive, hypolipidemic, neuroprotective, anti-aging, diuretic, central nervous system stimulant, analgesic, protects from UVB-induced carcinogenesis, immuno-modulator, against diabetes, osteoporosis, DNA damage, and CVDs. They are powerful antioxidants and are thought to reduce the risk of chronic disease by protecting against free radicals, removing toxic carcinogens, changing metabolic activation, and further influencing processes that change the course of tumor cells. Keeping in view the present demand for nutrition-rich and colored vegetables, this chapter has been compiled in further sections.

7.2 PIGMENTS DEVELOPMENT

Based on color, vegetables come into different categories such as red (tomato, beetroot, carrot, watermelon), purple/blue (black carrot, brinjal, red cabbage), yellow/orange (tomato, orange carrot, pumpkin, and squashes, sweet potato), yellow (pumpkin, yellow capsicum, and squashes, sweet corn, yellow onion), green (broccoli, capsicum, peas, green leafy) and white/ brown vegetables like potato, cauliflower, radish, garlic, ginger, and onion. In contrast to dull-colored vegetables, vivid colors of the vegetables especially indicate their richness in numerous antioxidants (Cömert et al., 2020). The color of vegetables and fruits results from the pigments, chlorophyll, and carotenoids in the chloroplasts and chromoplasts, as well as phenolic pigments (anthocyanins, flavonols, and proanthocyanins) in the vacuole. Carotenoids, flavonoids, and chlorophylls (Chl) are the three main categories of pigments. It is known that some colors provide additional health advantages. Each class consists of subgroups of derivatives with substitution alterations to the fundamental structure for various colors and biological purposes. Each class has a distinct basic chemical structure. Physical aspects like the presence of cuticular waxes, epidermal hair, and the shape and orientation of cells in the epidermis and sub-epidermis all have an impact on how pigment is expressed. Incoming visible light is selectively absorbed and refracted by surface topography and pigments to create a reflectance spectrum unique to the skin of a specific fruit or vegetable.

Chlorophylls are the main photosynthetic pigments and the most common natural pigments in plants and are responsible for the green color. The word "chlorophyll" is derivative of the Greek words "chloros" meaning green and "phyllon" meaning leaf. There are two types of chlorophylls, namely chlorophyll a and chlorophyll b, with a typical ratio of 3:1 in fruits and vegetables, with typical blue-green and yellow-green colors. In vegetables and fruits, the degradation of chlorophyll caused by chemical changes is a physiological event occurring in different phases of growth, development, and maturation of tissues. The body of Chl consists of a porphyrin head with four pyrrole rings attached by nitrogen to magnesium ion and hydrocarbon tail. Ethylene release, pH, activation of oxidative and chlorophyllase enzyme systems are responsible for the degradation of chlorophylls. Chlorophyllide b-reductase and ferredoxin-dependent 7-hydroxymethylchlorophyllide a-reductase catalyze the conversion of chlorophyll b to chlorophyll a. Chlorophyll a is hydrolyzed to chlorophyllin and phytol by the enzyme chlorophyllase replacing the magnesium atom with hydrogen ions, leading to the formation of pheophorbide in an acidic condition. Loss of green color in crops is usually associated with the presence of pigments, that is, red, yellow, orange, and so forth.

More than 800 carotenoids are accessory pigments that are found in abundance in plants and other photoautotrophs. They are colored yellow, orange, and red. The crystalline form of carotenoids is found in chloroplasts and chromoplasts, where it acts as an auxiliary pigment for photosynthesis and is mostly linked to proteins (Sharma et al., 2021). Carotenoids are lipophilic tetraterpenoids with alternating double and single carbon bonds along their 40-carbon (C40) polyene hydrocarbon chains. To clarify their structure, the traditional (trivial) names of carotenoids—often derived from their biological source—have been replaced by a semi-systematic naming system that uses two Greek letters to indicate the two end groups. For instance, β-carotene and γ-carotene are both named based on their terminal groups. Carotenoids such as β-carotene, α-carotene, and β-cryptoxanthin can be converted to vitamin A, which is essential for immunity, growth, and vision in humans and animals. Carotenoids are dietary bioactive chemicals that protect against degenerative disorders such as heart disease, chronic liver disease, diabetes, cancer, and macular degeneration. They also reduce the risk of some cancers and function as strong antioxidants. Pumpkins are a great source of vitamin A and carotenoids. According to Solovchenko et al. (2019), the amounts of -carotene and -carotene in several species (*Cucurbita pepo*, *Cucurbita maxima*, and *Cucurbita moschata*) were determined to be 60–7400 and 0–7500 g/100 g, respectively. As a result of maturity and ripening, the content of certain carotenoids can go from zero to high levels in a matter of days. The metabolic activity of ethylene is what causes this rise in carotenoid content. The amount of carotenoids that accumulate in tomatoes throughout maturity and ripening, particularly lycopene, relies on a variety of variables, including temperature and light intensity.

Flavonoids are water-soluble phenolic pigments with various structures. The structure of phenolic compounds is an aromatic ring with hydroxyl substituents. Among them, more than 9000 polyphenol derivatives belong to flavonoids, and although they share a common 15-carbon benzo-γ-pyrone skeleton (C6–C3–C6), the combinations of substituents are different. About nine subgroups are formed from the substitutions: anthocyanins (anthocyanidins), condensed tannins (proanthocyanidins), flavonols, flavones, flavandiols, isoflavonoids, chalcones, aurones, and furovafens (Harborne, 2013). Compared to carotenoids, they have more diverse biological roles and exhibit a wider range of colors. Anthocyanins are flavonoids that give purple or red color to vegetables by masking the expression of carotenoids and chlorophylls. They are powerful pigments that give vegetables their diverse colors. Anthocyanins are flavonoids found in the cell vacuoles of vegetables. Almost all non-green visible colors in plants can be seen, including the distinctive purple and violet hues in anthocyanin-rich fruits and vegetables like the apple (*Malus pumila*), blueberry (*Vaccinium* spp.), and eggplant (*Solanum melongena*). Anthocyanins also appear to be the most researched and colorful flavonoids. The colors are altered by changes in pH levels and metal ions in addition to the combinations of substituents.

7.3 FUNCTIONAL IMPORTANT OF VEGETABLE COLORS AND THEIR INTERACTION WITH HEALTH

Vegetables' color serves as a cue for consumers to identify the food, judge its quality and ripeness, and draw conclusions about its sensory qualities. People are familiar with the archetypal colors for many vegetables, such as orange for carrots, red for tomatoes, and green for cucumbers. As a result, these colors make it simple to identify the product category and help buyers. Vegetables are regarded as a valuable component of the human diet because they diverge the color of different food products and also have positive health effects because of their high antioxidant capacity and content of various phytochemicals like flavonoids, betalains, and carotenoids.

In addition to giving vegetables their green color, chlorophylls are strong antioxidants that boost the immune system, cleanse the blood, get rid of germs, prevent cancer, normalize blood pressure, and detoxify the body. Additionally, chlorophylls have been shown to have powerful anti-mutagenic, antioxidant, anti-carcinogenic, and photosensitizing properties in photodynamic therapy through a variety of mechanisms, including the modification of genotoxic effect, induction of phase II enzymes, and apoptosis (Suparmi et al., 2016). While the enzymes chlorophyllase, PODs, and catalases help to purify the blood and increase the capacity to carry more oxygen, chlorophylls make it easier for RBCs (red blood cells), which contain hemoglobin to hold oxygen from the lungs to other body areas. Chlorophylls affect the redox environment (reactive oxygen species (ROS) and altered reduced oxidized GSH ratio) of pancreatic cancer cells, which may be the reason for their proliferative and anticancer effects. Chlorophylls have been found to inhibit heme oxygenase mRNA expression and heme oxygenase enzymatic activity. In addition to being widely consumed around the globe, green leafy vegetables also greatly increase daily intake recommendations for a number of nutrients that are vital to human health (Kim et al., 2016). Due to its sensory and health-promoting qualities, lettuce is a highly common vegetable crop used for fresh consumption and as a salad ingredient (Kim et al., 2016). When compared to spinach, another significant and popular leafy green, the total vitamin B9 and flavonoid concentrations in lettuce were greater by 16% and 22%, respectively (Wang et al., 2013). According to several studies, the amount of secondary metabolites in lettuce varied significantly between genotypes and was mainly influenced by the color of the leaves (dark red, red, green, or red and green) (Kim et al., 2016, 2018).

Chard leaves have nutritional and functional profiles (flavonoids, carotenoids) that are responsive to modern human nutrition. The Brassicaceae group includes kale, mizuna, bok choy, watercress, wild, and salad rocket includes a wide range of species with potentially beneficial traits. Cabbage leafy greens are a valuable source of phytochemicals. It contains a large amount of vitamins (C, E), K, carotenoids, and phenols. In addition to the latter phytochemicals, sulfur-containing glucosinolates and methylcysteine sulfoxide compounds characterize cruciferous leafy vegetables.

Synthesized via the isoprenoid pathway in plants, carotenoids such as lycopene, zeaxanthin, and lutein (beta-carotene – the precursor of vitamin A) come in a variety of brightly colors such as red, yellow, and orange with different health benefits (Shabbir and Nuzhat, 2018). For example, the red-colored vegetable tomato (*Solanum lycopersicum* L.) is consumed worldwide due to its high antioxidant content, taste, nutritional value, and versatility. The most common carotenoid is lycopene, followed by phytoene, phytofluene, ε-carotene, γ-carotene, β-carotene, neurospores, and lutein (Li et al., 2018).

Anthocyanins, a water-soluble vacuolar pigment, give vegetables their purple, blue, and red colors, and their expression depends on pH and structural composition. Anthocyanins show various properties and protect against ultraviolet radiation due to their activity of removing free radicals. Eggplant (*Solanum melongena* L.) is a very popular vegetable cultivated worldwide in tropical and subtropical regions. Nisha et al. (2009) reported total phenolic and anthocyanin content of purple small fruit cultivars as 106.98 mg/100 g fresh weight and 0.756 mg/100 g fresh weight, respectively, compared to other three cultivars studied such as purple moderate-large fruits containing 80.31 mg/100 g and 0.525 mg/100 g of crude total phenols and anthocyanins, green long (50.79 mg/100 g and 0.0475 mg/100 g, total phenols and anthocyanins) calculated on a fresh weight basis and purple large size (49.02

mg/100 g and 0.53 mg/100 g fresh weight of total phenolics and anthocyanins). Rising market trends in terms of consumer quality perception have resulted in colored *Brassica* leafy vegetables, for example, rich in anthocyanins, purple kale or pakis choi attracts the attention of nutritionists and horticulturists. Hao et al. (2018) reported differences in eating quality, such as cellulose, protein, sugar, and vitamin C content, between 74 red/purple and green lettuce cultivars. In addition, leaf pigmentation is significantly correlated with the composition and amount of phenolic compounds (Kim et al., 2016). In an *in vivo* study in mice fed a high-fat diet, supplementing the diet with 8% of body weight red lettuce reduced total cholesterol and low-density lipoprotein (LDL) by 9% and 123%, respectively, indicating the possible effects of red pigmented lettuces on CVDs. Swiss chard (*Beta vulgaris* var. *cicla* L.), belongs to the Chenopodiaceae family, cultivated for its green or reddish leaves and white, yellow, or red leaves promoting the health properties.

White, green, purple, and black peppercorns are packed with vitamin E, flavonoids, and capsaicinoids, as well as other carotenoid pigments such as lycopene and zeaxanthin. The genus *Capsicum* includes about 31 species, of which five species are *C. annuum*, *C. baccatum*, *C. chinense*, *C. frutescens*, and *C. pubescence*. Unripe fruits are usually colored in white, green, purple, and black, and when the fruit ripens, the color gradually changes to yellow, orange, red, and brown. The differences in fruit color are mainly due to the different accumulation of flavonoids and carotenoids. According to Liu et al. (2020), the accumulation of anthocyanins in the outer epidermis of unripe pepper fruits causes purple or black color 30 days after anthesis and turns red 50 days after anthesis and anthocyanin content, flavones, and flavonols were significantly found to be higher in purple cultivar than other cultivars and delphinidin, luteolin, chrysoeriol, and quercetin derivatives were the most abundant polyphenols.

7.4 ADVANCED BREEDING APPROACHES FOR COLOR ENHANCEMENT

Researchers across the globe are trying to develop fruits and vegetables with the greatest color intensity not only due to their better quality, taste, and appearance but also due to the beneficial impact of color pigments on human health. Prior efforts or methods to improve the quality or color of vegetables and fruits were usually based on a single step of a biochemical pathway, which turned out to be not so successful due to the complicated regulatory network at each metabolic step. Therefore, more advanced methods, such as molecular-assisted breeding, genetic engineering, and gene editing, came into the picture that are nowadays used most widely to enrich color. Molecular-assisted breeding or marker-assisted breeding uses molecular markers associated with desired traits to select a plant species for either inclusion or further breeding programs. Genetic engineering techniques allow us to improve and modify agronomic traits by changing the insertion or deletion of a particular gene at a particular site in the host plant species. Gene editing includes a group of methods and technologies that enable a particular genetic material to be added, deleted, or changed at a particular location in the genome of an individual.

Vegetables in the solanaceous family are among the most common ingredients in our diet and include tomato (red/orange), potato (brown), eggplant (purple), chili/pepper (green, red, and yellow), and so forth. Wide color spectrums are due to the presence of compounds such as anthocyanin, beta-carotene, capsanthin, chlorophyll, flavonoids, lutein, lycopene, and zeaxanthin (Gurbuz et al., 2018; Jo et al., 2017; Mezbani et al., 2019). To improve color in vegetables, genes with significant color traits are identified and isolated from other species and incorporated into the desired vegetable crop. Several genes with high lycopene and beta carotene content have been identified and *isolated*, for example, *Hp-1, Hp-2, Dg*, and *Og*^{c} in tomatoes. Cultivars with Og^c genes have approximately 25% more lycopene than normal tomato cultivars. Similarly, to produce transgenic tomato lines with enhanced carotenoids, the phytoene desaturase enzyme has been incorporated into the crtI gene, which converts phytoene to lycopene. In pepper, the capsanthin-capsorubin synthesis (*CCS*) gene determines fruit color, and removing CCS from an upstream region of the gene leads to a deep orange color in the fruit (Li et al., 2013). Three genes derived from *Erwinia*, which code for phytoene synthase (CrtB), phytoene desaturase (CrtI), and lycopene beta-cyclase (CrtY) resulted in deep yellow potato tubers with improved nutritional qualities occurred and as a result, the carotenoids content in tuber increased by almost 20-fold and beta-carotene increased by 3600-fold. In a transgenic potato when over-expressed with 3GT gene, a deeper color anthocyanin product was obtained. RNAi-mediated silencing technique can be used to regulate the expression of certain genes. Silencing of beta-carotene hydroxylase gene (bch) with RNAi-mediated silencing, resulted in beta-carotene-rich potato tubers as the product of bch gene converts beta-carotene to zeaxanthin. An antioxidant-rich blue/purple colored variety of tomato was developed by using two genes, astrviolaceum (from wild species *S. cheesemanii*) and anthocyanin fruit gene, that is, transcription factor from *S. chilense*. Similarly, genes from snapdragons were incorporated into tomatoes to increase the anthocyanins levels using gene-splicing techniques.

Vegetables belonging to Alliaceae family such as onions and garlic are particularly rich in sulfur compounds, anthocyanin, phenolic compounds, quercetin, and flavonoids. In addition to their cardio-protective effects, these phytochemicals also have anti-cancerous, antidiabetic, and antimicrobial effects and help to improve immune system health. The particular mechanisms of action include inhibition of platelet to form aggregate, blocking polycyclic hydrocarbon, mitigating cholesterol and triglyceride levels, increasing peripheral blood flow, and improving the elasticity of vessels (Marrelli et al., 2019). Onion and garlic are usually found to be a rich source of flavonoids, and these are among the few most popular, white-colored vegetables. In red onions, anthocyanins and quercetin are particularly abundant, whereas flavonoids such as quercetin are responsible for the brown and yellow color of onions. Tearless onions possess higher thiosulfate levels that might contribute to their bioactivity. Onion produces lachrymatory factor (LF) by catalyzing trans-S-1-propenyl-L-cysteine sulfoxide through alliinase and lachrymatory factor synthase (LFS) as catalyzing agents. Onion seeds irradiated with neon ion resulted in lower alliinase transcription and ultimately led to bulbs with 7.5 times lesser LF than that of normal onions (Kato et al., 2016). The tear inducing property of LF synthase gene was reduced up to more than 1500 times by silencing gene using RNA interference technique. High thiosulfinate levels are found to be related with higher cardiovascular, anti-inflammatory, and anti-carcinogenic characteristics in tearless onions. Inactivation of chalcone isomerase (CHI) resulted in disruption in flavonoid synthesis and accumulation of chalcone derivatives, might be responsible for gold-colored onions (Kim et al., 2004).

Cauliflower, cabbage, radish, kale, and broccoli are amongst few crucifer vegetables which are a rich source of lipids, flavonoids, glucosinolates, phenolics, carotenoids, and tannins. Apart from these nutrient ingredients, Cruciferae vegetables contain selenium, which is associated with inhibition of ROS accumulation, inhibits histone deactylase activity, promotes changes in estrogen metabolism, induces apoptosis, and therefore reduces tumor growth and cancer (Keck and Finley, 2004). Because of the presence of chlorophyll, flavones, and anthocyanin pigments, these vegetables are mostly green (cauliflower, broccoli, kale, cabbage,), white (radish, cauliflower), and pink/purple (red cabbage, pink radish, kale) in color. The Orange (Or) transgene of cauliflower has higher levels of beta-carotene compared to wild type (Lu et al., 2006). Similarly, another gene mutation (purple gene mutation) in cauliflower resulted in anthocyanin accumulation and deep purple color. Yadava et al. (2017) developed the first bio-fortified variety of cauliflower, named "Pusa Beta Kesari 1" rich in β-carotene (8.0–10.0 ppm). CRISPR/Cas9 mediated gene-editing technique was used for manipulating BoPDS, PDS, and FRI genes resulting in albino phenotypes in cabbage (Jansson, 2018; Ma et al., 2019). In carrots, an editing technique was used to block anthocyanin synthesis by targeting the flavanone 3-hydroxylase (DcF3H) gene. CRISPR/Cas9 gene-editing technique can be a promising technique to improve vegetables such as kale and other *Brassica* vegetables, where the carotenoid isomerase gene of Chinese kale was edited, resulting in high mutation rate, reduced carotenoid, and chlorophyll content (Sun et al., 2020).

Cucurbitaceous vegetables are commonly known as cucurbits including crops such as cucumber, melon, and gourds. These are preferred to be consumed as raw for table purposes (Sharma et al., 2021). These vegetables are a rich source of phytochemicals such as glycosides, terpenoids, tannins, steroids, carotenoids, and cucurbitacins as well these also have enormous health benefits such as anti-cancerous properties (Rajasree et al., 2016). Melons are a good source of carotenoids, beta-carotene, and lycopene which are believed to have a very vital effect on metabolic regulation. The green color of cucumber and the bitter gourd is due to the chlorophylls and flavones and these color pigments are directly associated with various health benefits. Common cucumber usually has low carotenoid content (22–48 μg/100 g) whereas *Cucumis sativus var. ishuangbannanensis* has high carotenoid content of 700 μg/100 g. This variation in the bioactive compound can be genetically manipulated to develop vegetables with enriched carotenoid content. The lycopene content of watermelon can be increased from 50–70 μg/g to 1000–2000 μg/g fresh weight suggesting that the increase in the color intensity also has the potential to improve the delivery of bioactive elements and adds health benefits too. Gene editing using the *Agrobacterium*-mediated cotyledon transformation technique resulted in virus-resistant cucumber plants by disrupting one of the genes of translation *viz.*, eIF4E (eukaryotic initiation factor 4E) gene.

Leguminous vegetables are considered a power pack of proteins, fibers, nutrients, phenolic compounds, carotenoids, protease, and saponins. These compounds possess anti-inflammatory, anti-cancerous, and antioxidant activity and reduce cholesterol. Leguminous vegetables are usually green in color and rich in chlorophyll however, their dried version has higher amounts of carotenoids and beta-carotene too. Genes such Pur, Pu, and Astr are accountable for purple color pods and stripes on pods of *Pisum fulvum* (wild type). Indigenous vegetables are those vegetables that were once grown and used more widely in the past but at present, their use is dismissed due to some unknown agronomic, economic, cultural reasons and are also not grown commercially. Chayote (*Sechium edule*) is an important underutilized vegetable that is rich in bioactive components and can provide several hidden health benefits to human (Fu et al., 2021). Flavonoids and isoflavonoid content can be manipulated at several steps using genetic engineering technique to further enhance the antioxidant properties. The leaves of *Sauropus androgynus* (Chekurmanis) are highly nutritive and are often referred as multi-mineral-packed leafy vegetable. Similarly, *Merremia tuberosa* (Kaasara kaya), an underutilized vegetable cultivated in Tamil Nadu (India) is used as medicine for diabetes, rheumatics, spleen, and liver diseases (Koneri et al., 2006). *Solanum torvum*, or Turkey berry, holds several medicinal properties, such as being used to treat tooth decay, fever, wounds, and reproductive problems. *Talinum trangulare* (Water leaf), an underutilized vegetable is soft and leafy in nature and contains carotenoids (lutein and zeaxanthin), that can work as stimulant for immune cells of the eyes (Shakuntala and Shadaksharaswamy, 1985). It also has several other health benefits such as prevent heart diseases, maintain blood pressure, cholesterol, prevent cancer and cataract. Health benefits of underutilized vegetable crops have been presented in Table 7.1.

The use of molecular markers can play a significant role in assessing the genetic diversity and nutrient potential in these underutilized vegetables. Vatanparast et al. (2016) developed SSR and SNP markers using transcriptome sequencing for winged beans and revealed the divergence of a trypsin inhibitor, which is an antinutritional factor in winged beans and also in other legumes. As a result, these underutilized vegetable crops can be used for nutritional security to feed the growing population that faces huge mal-nutrition and deficiency. Recent advancements in the OMICS (genomics, proteomics, transcriptomics, and metabolomics) technique have advanced our understanding of the structure of these underutilized crops, gene expression, and metabolism of these plants.

TABLE 7.1 Health Benefits of Functional Compounds Present in Underutilized Vegetable Crops

Family	Vegetables	Edible Part	Functional Compound	Health Benefit(s)	References
Amaranthaceae	Chenopodium (*Chenopodium* spp.)	Leaves and young shoots	Carotenoids and protein	• Cure urinary troubles. • Remove intestinal worm. • Cures constipation. • Improves appetite and hemoglobin level.	Prajapati et al. (2018)
Apiaceae	Long coriander (*Eryngium foetidum*)	Leaves and roots	Calcium, iron, carotene, and riboflavin	• Stimulate appetite. • Improve the digestion. • Soothe stomach pains. • An expectorant and aphrodisiac.	Rohini et al. (2017)
	Ashitaba (*Angelica keiskei*)	Leaves and stalks	Vitamin B_{12} and flavonoids (chalcone)	• Antibiotic activity against staphylococcus. • Rejuvenates human digestion and skin.	Koneri et al. (2006)
	Mitsuba, Japanese honewort (*Cryptotaenia japonica*)	All parts of the plant are edible from roots to seed	Beta carotene, folic acid, ascorbic acid, and iron	• Treatment of hemorrhages, colds, and fevers. • Tonic to strengthen the body.	Rohini et al. (2017)
Araceae	Elephant foot yam (*Amorphophallus campanulatus*)	Tuberous and fresh corms	Betulinic acid, β-sitisterol, stigmasterol, triacotane, lupcol, and β-sitosterol palmitate	• Antibacterial, antifungal, and cytotoxic activities due to the presence of a diterpenoid, a triterpenoid. • Possess blood purifier properties. • Treatment of piles, abdominal disorders, tumors, enlargement of the spleen, asthma, and rheumatism.	Ramalingam et al. (2010)
	Geli-geli, Bekil, Spiny Lasia (*Lasia spinosa* L.)	Young leaves and petioles, rhizomes	Polyphenols, vitamin C, and dietary fiber	• Anti-helminthic, antibacterial, anti-inflammatory, antidiabetic, anti-tumor activities.	Shefana et al. (2009)

TABLE 7.1 *(Continued)*

Family	Vegetables	Edible Part	Functional Compound	Health Benefit(s)	References
Asteraceae	Velvet plant (*Gynura bicolour*)	Succulent tender leaves and shoots	Iron	• Good for pregnant women.	Rohini et al. (2017)
	Helencha (*Enhydra fluctuans*)	Young parts and the leaves	β-carotene, saponins, cholesterol, glucoside, and enhydrin	• Cure inflammation, leukoderma, biliousness, and good in smallpox. • Possesses antioxidant, analgesic and antidiarrheal activity.	Ali et al. (2013)
	Para cress (*Spilanthes acmella* L.)	Flower heads are chewed	Unsaturated alkamid and spilanthol	• To relieve the toothache. • Treatment of skin diseases.	Seal (2012)
Athyriaceae	Vegetable fern (*Diplazium esculentum*)	Tender fronds and ferns	β-carotene, folic acid, phytic acid, and tannins		
Basellaceae	Basella (*Basella alba, Basella rubra*)	The tender shoots and purplish fruit sap	Essential amino acids like arginine, isoleucine, leucine, lysine, threonine, and tryptophan, B – sitosterol, lupeol, saponins, kaempferol, betalain, steroids, and phenolics	• Exhibit androgenic, antidiabetic, anti-inflammatory, anticancer, antiviral, CNS depressant, and wound healing properties. • Act as a cooling medicine.	Gupta et al. (2008) Lin et al. (2009); Orwa et al. (2009)
Brassicaceae	Wasabi (*Wasabia japonica* (Miq.) Matsum)	All plant parts	Isothiocyanates	• Suppression of stomach cancer cells. • Powerful digestive stimulant. • Powerful hydroxyl radical scavenging agents. • Antidiarrheal property.	Nakayama et al. (1998)
Convolvulaceae	Water spinach (*Ipomea aquatica*)	All plant parts	Carotene and flavonoids	• Has high energy value. • It is used for piles.	Umar et al. (2007)

TABLE 7.1 *(Continued)*

Family	Vegetables	Edible Part	Functional Compound	Health Benefit(s)	References
Cucurbitaceae	Wax gourd (*Benincasa hispida*)	Young leaves and flower buds	Terpenes	• Seeds exhibit anthelmintic, anti-inflammatory, demulcent, diuretic, expectorant, febrifuge, and laxative properties. • Contains anti-cancer terpenes.	IPGRI (2006)
	Gherkin (*Cucumis sativus* var. *anguria*)	Fruit	Beta-carotene	• Weight maintenance. • Reduces the risk of problems like heart disease and cancer.	Kumar et al. (2018)
	Ivy gourd (*Coccinia grandis*)	Fruits and leaves	Lycopene, β-carotene, and leaves contain protein and vitamin A	• Possess antidiabetic and antioxidant properties. • Used for skin eruption, tongue sores, and earache.	Kumar et al. (2018)
	Karchikai (*Momordica cymbalaria*)	Tubers and leaves	Beta carotene	• Having properties such as antidiarrheal, hepatoprotective, antidiabetic, antimicrobial.	Kirtikar and Bashu (1989)
	Chow-chow (*Sechium edule*)	Edible roots fruit and seeds	Aglycone flavonoid poly-phenolic, and luteolin	• Scavenge harmful oxygen-derived free radicals. • Disease prevention properties.	Prajapati et al. (2018)
	Kakrol (*Momordica cochinchinesis* and *M. dioica*)	Young leaves, flowers, and seeds, fruits	Carotenoids like lutein, tannins, saponins, and alkaloids	• Helps in the prevention of various eye diseases, cardiovascular diseases (CVDs), and even cancers. • Used in ulcers, piles, sores, and obstruction of liver and spleen. • Seeds are used to treat swelling, ulcers, and abscesses.	Kumar et al. (2012); Prajapati et al. (2018)
Cyperaceae	Keysoor (*Scirpus grossus*)	Root of keysoor	Progesterone, sugar, tannins, starch, and saponins	• Roots have astringent properties. • To check diarrhea and vomiting.	–
Fabaceae	Jack bean (*Canavalia ensiformis*)	The young leaves may be cooked	Methionine and cystine or cyanogen glycosides and oligosaccharides	• Acts as anti-inflammatory, anti-asthmatic, anti-cancer, and antiacne.	Prajapati et al. (2018)

TABLE 7.1 *(Continued)*

Family	Vegetables	Edible Part	Functional Compound	Health Benefit(s)	References
Hydrocharitaceae	Ducklettuce (*Ottelia alismoides*)	Leaves	–	• Used to cure hemorrhoids. • Has rubefacient properties.	–
Lamiaceae	Thumbai (*Leucas aspera*)	Leaves and leaves juice	Triterpenoids and oleanolic acid, ursolic acid and 3-sitosterol	• Used in rheumatism. • Bruised leaves are used for snake bites and poisonous insects.	Hughes et al. (2012)
Leguminosae	Winged bean (*Psophocarpus tetragonolobus*)	All parts of the plant and tuber-like-roots	Niacin	• Helps in reducing the formation of blood clots by reducing platelet aggregation.	Mady et al. (2009)
Malvaceae	Roselle (*Hibiscus sabdariffa*)	Young shoots, stalks, and leaves	Organic acids, minerals, amino acids, carotene, and vitamin C	• In the treatment of abscesses, bilious conditions, cancer, cough.	Verma et al. (2009)
Moringaceae	Drumstick (*Moringa oleifera*)	Leaves – eaten fresh, cooked, or stored as a powder.	Polyphenols	• Possess antioxidant activity.	Sreelatha and Padma (2009)
Polygonaceae	Water pepper (*Polygonum hydropiper*)	Leaves	Rutin	• Exhibit anti-inflammatory, astringent, carminative, diaphoretic, diuretic, emmenagogue, stimulant activities. • Helps to strengthen fragile capillaries. • Helps to prevent bleeding.	Duke and Ayensu (1985)
	Fagophyrum sp. Buckwheat (*Fagophyrum esculentum*) and buckwheat (*F. tataricum*)	Leaves and young shoots	Rutin	• Reduces hemophilia and heart attack.	Rohini et al. (2017)

TABLE 7.1 *(Continued)*

Family	Vegetables	Edible Part	Functional Compound	Health Benefit(s)	References
Portulacaceae	Waterleaf (*Talinum triangulare*)	Leaves	Lutein and zeaxanthin	• Influencing the immune cells of the eyes. • Helps to control blood pressure and cholesterol level. • Protect against cataracts and muscular degeneration.	Shakuntala and Shadaksharaswamy (1985)
	Purslane (*Portulacae oleracea*)	Tender shoots and fresh herb	Purslane shoot tissues are rich in β-carotene, lutein, neoxanthin, total carotenoids, and chlorophyll.	• Treat the bleeding of the genitor-urinary tract as well as dysentery. • Relieve sores and insect or snake bites on the skin.	Hughes et al. (2012)
Sphenocleaceae	Wedgewort (*Sphenoclea zeylanica*)	Leaves and the young shoots	B-carotene and Vit C	• Used against the stings of venomous animals and ulcers.	Rohini et al. (2017)
Zingiberaceae	Bamboo-leaved Galanga (*Alpinia allughas* (Retz.))	Rhizome	Linalool (2.86%), (–)-α-phellandrene (1.6%), and geranyl acetate (0.16%)	• Remedy for impotence.	Rastogi and Mehrotra (1990)

7.5 EFFECT OF PROCESSING ON THE STABILITY OF COLOR PIGMENTS

Color is one of the most important physical properties of raw and processed food and a natural indicator of nutritional quality that influences consumer acceptance. Natural color pigments such as anthocyanins, betalains, and chlorophylls are responsible for the different colors of fruits and vegetables. Regardless of consumers' financial considerations, vegetables are one of the most important parts of the human diet. Since the body cannot synthesize these beneficial compounds, it is important to supplement these phytochemicals with fresh products that contain these pigments. To maintain the quality of the products, consumers should be aware of the processing stage where pigments or nutrients are lost (Fabbri and Crosby, 2016). Heat treatment can promote coloring of anthocyanin pigments by converting leucoanthocyanidins to anthocyanins. Many foods containing anthocyanins are heat-treated before use, significantly affecting the anthocyanin content. Heat treatment involves heating to a temperature of 50–150°C, depending on the pH, temperature, light, oxygen, metal ions, enzymes, sugars, and the shelf life and stability of the desired anthocyanins. The degradation rate of anthocyanins increased with increasing solids content during heating. Heat treatment has shown that increased degradation of anthocyanin pigments and high-temperature processing can affect anthocyanin levels in fruit and vegetable-based foods.

Chlorophyll is the main pigment in green plants and breaks down easily during processing. Chlorophylls have been found to be sensitive to chemical and physical changes during vegetable processing. The degree of greenness of chlorophyll pigments is important to determine the final quality of such vegetables because color is one of the most important quality indicators of vegetable products. Color changes in green vegetables caused by heat treatment are due to the conversion of chlorophyll to pheophytin by replacing chlorophyll Mg with hydrogen. Cooking caused the loss of chlorophyll a and b to varying degrees, depending on the method used, as well as the type of vegetable. While loss of chlorophyll a range from 9–81% and chlorophyll b from 20–61%, pH had a significant effect on chlorophyll and betanin and a marginal effect on beta-carotene and lycopene. Heat treatment was found to affect some pigments, but no differences were usually found between heat treatment methods such as pressure cooking or steaming. The stability of chlorophyll-b during processing is higher than that of chlorophyll-a (Sledz et al., 2016). Chlorophylls are sensitive to various factors such as heat, enzymes, light, oxygen, and pH (3.5–5.0) and even during ripening. Stabilization of chlorophylls in processed foods can be achieved by short-term treatment at elevated temperature, acid neutralization, enzymatic processing of chlorophylls into chlorophyllide and metal complexes with zinc or copper ions.

Temperature, pH, water activity, oxygen, metals, and radiation are factors that affect betalain stability during processing and storage. A first order reaction for betanin degradation has been confirmed (Guldiken et al., 2016; Wootton-Beard et al., 2011). Betanin decomposes under alkaline conditions by hydrolysis of the aldimine bond, producing ferulic acid with an amine group. Betalain was found to be more stable in the presence of oxygen between pH 5.5 and 5.8. Under anaerobic conditions, betalain was stable at pH 4.0–5.0. Reduction of water activity (below 0.63) during various treatments such as spray drying and concentration improves the stability of betalains (Kearsley and Katsaboxakis, 1980). Temperature affects the stability of betalains, and an increase in temperature causes beta chains to break and polyphenol oxidase (PPO) enzymes to break down. However, temperature range, degree of heating, presence of oxygen and concentration of pigments were found to affect thermal degradation (Herbach et al., 2006). Colors oxidize and break down under the influence of light. There is an inverse relationship between light intensity between 2200 and 4400 lux. A large amount of chlorophyll, flavonoids, and carotenoids are destroyed by microwave heating, as reported by several researchers (El-Abassy et al., 2010). Chlorophyll and carotenoids are heat sensitive pigments, and microwave heating causes greater damage to carotenoid content (Singh et al., 2017).

Different physicochemical parameters affecting the stability of anthocyanins, such as pH, light, temperature, and so on, can cause the predictable color to fade or change (Sui et al., 2014). Different methods have been described to stabilize anthocyanins, such as addition of chemical reagents and modified processing (Pan et al., 2014; Santos et al., 2013). The stability of anthocyanins in food is significantly weakened by various compounds in the food matrix, such as ascorbic acid, which accelerates the degradation of anthocyanins. Among the natural red pigments, anthocyanins are thought to degrade more easily at high temperatures than carotenoids (Fernandez-Lopez et al., 2020). To ensure the stability of anthocyanins during pre-processing, handling, and storage, oxidizing conditions must be avoided by removing oxygen or creating a barrier by encapsulation, or they can be complexed with compounds such as polymers, phenolic compounds, and metals

(Fernandez-Lopez et al., 2020). Therefore, more research into the effects of breeding techniques on plant pigments is needed to investigate differences in appearance and nutritional value.

7.6 CONCLUSION

Vegetables are considered very important as they provide food and nutrition security globally, by not just being a source of energy-rich carbohydrate food, but also providing vital protective nutrients such as vitamins, minerals, and antioxidants that are necessary for overall health. It is worth noting that vegetables are also a powerhouse of micro-nutrients that has the potential to eliminate malnutrition from the population consuming them. Including deep-colored and variety of vegetables that are rich in nutrients, antioxidants, and minerals in daily diet is highly recommended by dietitians these days for a healthy and fit life. The agricultural practices for vegetable farming should be shaped in such a way that it not only enhances productivity but also enriches its nutrient constituents so that it offers both recommended food and nutrient for a healthy lifestyle. Some underutilized/indigenous vegetables are a rich source of bioactive phytochemicals that possess antioxidant activity and can provide innumerable health benefits. Despite, highlighting the significance of underutilized vegetables on human health, their consumption is limited to time, place, and people due to either lack of awareness among entire populations or due to lack of availability of planting material. However, if we continue to enhance the consumption of these underutilized vegetables, it will have the potential to fulfill all the nutritional requirements of individuals and eliminate the nutritional demand of developing nations. At present, the production and processing sector of vegetables is focused on producing colored vegetables that are packed with nutrition to fulfill the demand of people's nutrition requirements. To mitigate the harmful impact of post-harvest processes on vegetables, it is necessary to further enhance our knowledge in this specific area so that the nutrient and bioavailability of natural pigments in vegetables remain intact. The perspective of specific breeding methods is primarily to enhance the nutritional value of vegetable crops so that it improves health and is cost-effective. Several genetic engineering methods have been used by plant breeders to introduce desired traits into the elite or wild cultivars, resulting in vegetables with enhanced nutritional and commercial value. Although the concern about genetically modified crops is rising due to ethical and biosafety issues, if thoughtfully designed and developed, they can help to reduce malnutrition, diseases, and food insecurity in the country.

KEYWORDS

- **anthocyanins**
- **betalains**
- **breeding**
- **chlorophylls**
- **color pigments**
- **functional food**
- **heat treatment**
- **high-value crops**
- **leucoanthocyanidins**
- **phytochemicals**
- **raw food**

REFERENCES

Ali, R., Billah, M., Hassan, M., & Dewan, S. M. R. (2013). *Enhydra fluctuans* Lour: A review. *Research Journal of Pharmacy and Technology, 6*(9), 927–929.

Baldermann, S., Blagojević, L., Frede, K., Klopsch, R., Neugart, S., Neumann, A., Ngwene, B., Norkeweit, J., Schröter, D., Schröter, A., & Schweigert, F. J. (2016). Are neglected plants the food for the future? *Critical Reviews in Plant Sciences, 35*(2), 106–119.

Cömert, E. D., Mogol, B. A., & Gökmen, V. (2020). Relationship between color and antioxidant capacity of fruits and vegetables. *Current Research in Food Science, 2*, 1–10.

El-Abassy, R. M., Donfack, P., & Materny, A. (2010). Assessment of conventional and microwave heating induced degradation of carotenoids in olive oil by VIS Raman spectroscopy and classical methods. *Food Research International, 43*(3), 694–700.

Fabbri, A. D., & Crosby, G. A. (2016). A review of the impact of preparation and cooking on the nutritional quality of vegetables and legumes. *International Journal of Gastronomy and Food Science, 3*, 2–11.

Fasuyi, A. O. (2005). Nutritional potentials of some tropical leafy vegetable's meals: Chemical characterization and functional properties of fresh and frozen vegetables. *Food Chemistry, 62*(1), 59–64.

Fernández-López, J. A., Fernández-Lledó, V., & Angosto, J. M. (2005). New insights into red plant pigments: More than just natural colorants. *RSC Advances, 10*(41), 24669–24682.

Fu, A., Wang, Q., Mu, J., Ma, L., Wen, C., Zhao, X., Gao, L., Li, J., Shi, K., Wang, Y., & Zhang, X. (2021). Combined genomic, transcriptomic, and metabolomic analyses provide insights into chayote (*Sechium edule*) evolution and fruit development. *Horticulture Research, 8*, 35.

Guldiken, B., Toydemir, G., Nur Memis, K., Okur, S., Boyacioglu, D., & Capanoglu, E. (2016). Home-processed red beetroot (*Beta vulgaris* L.) products: Changes in antioxidant properties and bio accessibility. *International Journal of Molecular Sciences, 17*(6), 858.

Gupta, A. K. (2008). Quality standards of Indian Council of Medical Research, New Delhi.

Gürbüz, N., Uluişik, S., Frary, A., & Doğanlar, S. (2018). Health benefits and bioactive compounds of eggplant. *Food Chemistry, 268*, 602–610.

Hao, J. H., Wang, L., Liu, C. J., Zhang, X. D., Qi, Z. Y., Li, B. Y., Xiao, S., & Fan, S. X. (2018). Evaluation of nutritional quality in different varieties of green and purple leaf lettuces. In *IOP Conference Series: Earth and Environmental Science* (Vol. 185, No. 1, p. 012008). IOP Publishing.

Harborne, J. B. (2013). *The Flavonoids: Advances in Research Since 1980*. https://link.springer.com/book/10.1007/978-1-4899-2913-6.

Herbach, K. M., Stintzing, F. C., & Carle, R. (2006). Betalain stability and degradation—structural and chromatic aspects. *Journal of Food Science, 71*(4), R41–R50.

Hughes, J. D. A., & Keatinge, J. D. H. (2012). The nourished millennium: How vegetables put global goals for healthy, balanced diets within reach. In *High Value Vegetables in Southeast Asia: Production, Supply and Demand* (pp. 11–26).

Jansson, S. (2018). Gene-edited plants on the plate: The 'CRISPR cabbage story.' *Physiologia Plantarum, 164*(4), 396–405.

Jo, S. J., Kim, J. W., Choi, H. O., Kim, J. H., Kim, H. J., Woo, S. H., & Han, B. H. (2017). Capsanthin inhibits both adipogenesis in 3T3-L1 preadipocytes and weight gain in high-fat diet-induced obese mice. *Biomolecules & Therapeutics, 25*(3), 329.

Kader, A. A., Perkins-Veazie, P., & Lester, G. E. (2004). Nutritional quality of fruits, nuts, and vegetables and their importance in human health. USDA Hand Book, 66.

Kato, M., Masamura, N., Shono, J., Okamoto, D., Abe, T., & Imai, S. (2016). Production and characterization of tearless and non-pungent onion. *Scientific Reports, 6*(1), 23779.

Kearsley, M. W., & Katsaboxakis, K. Z. (1980). Stability and use of natural colors in foods: Red beet powder, copper chlorophyll powder and cochineal. *International Journal of Food Science & Technology, 15*(5), 501–514.

Keck, A. S., & Finley, J. W. (2004). Cruciferous vegetables: Cancer protective mechanisms of glucosinolate hydrolysis products and selenium. *Integrative Cancer Therapies, 3*(1), 5–12.

Keservani, R. K., & Sharma, A. K. (2014). Flavonoids: Emerging trends and potential health benefits. *Journal of Chinese Pharmaceutical Sciences, 23*(12), 815.

Keservani, R. K., Kesharwani, R. K., Sharma, A. K., Vyas, N., & Chadoker, A. (2010b). Nutritional supplements: An overview. *International Journal of Current Pharmaceutical Review and Research, 1*(1), 59–75.

Keservani, R. K., Kesharwani, R. K., Vyas, N., Jain, S., Raghuvanshi, R., & Sharma, A. K. (2010a). Nutraceutical and functional food as future food: A review. *Der Pharmacia Letter, 2*(1), 106–116.

Keservani, R. K., Sharma, A. K., & Kesharwani, R. K. (2016a). Nutraceutical and functional foods for cardiovascular health. In *Food Process Engineering* (pp. 291–312). Apple Academic Press, CRC Press.

Keservani, R. K., Sharma, A. K., & Kesharwani, R. K. (2016b). Medicinal effect of nutraceutical fruits for cognition and brain health. *Scientifica, 3109254,* 10. https://doi.org/10.1155/2016/3109254.

Keservani, R. K., Sharma, A. K., & Kesharwani, R. K. (2017). An overview and therapeutic applications of nutraceutical and functional foods. *Recent Advances in Drug Delivery Technology,* 160–201.

Keservani, R. K., Sharma, A. K., & Kesharwani, R. K. (2020). *Nutraceuticals and Dietary Supplements: Applications in Health Improvement and Disease Management.* CRC Press.

Kim, D. E., Shang, X., Assefa, A. D., Keum, Y. S., & Saini, R. K. (2018). Metabolite profiling of green, green/red, and red lettuce cultivars: Variation in health beneficial compounds and antioxidant potential. *Food Research International, 105,* 361–370.

Kim, M. J., Moon, Y., Tou, J. C., Mou, B., & Waterland, N. L. (2016). Nutritional value, bioactive compounds and health benefits of lettuce (*Lactuca sativa* L.). *Journal of Food Composition and Analysis, 49,* 19–34.

Kim, S., Jones, R., Yoo, K. S., & Pike, L. M. (2004). Gold color in onions (*Allium cepa*): A natural mutation of the chalcone isomerase gene resulting in a premature stop codon. *Molecular Genetics and Genomics, 272,* 411–419.

Koneri, R., Balaraman, R., & Saraswati, C. D. (2006). Antiovulatory and abortifacient potential of the ethanolic extract of roots of *Momordica cymbalaria* Fenzl in rats. *Indian Journal of Pharmacology, 38*(2), 111.

Kumar, N., Singh, S., & Gupta, R. (2012). *Trichosanthes dioica* Roxb.: An overview. *Pharmacognosy Reviews, 6*(11), 61.

Kumar, R., Rajasree, V., Sagar, L., Tripura, U., Karthick, K., Karthick, R., & Prasath, G. (2018). Underutilized vegetables as rich sources of medicinal value: A boon for the pharmaceutical industries and farmer income. *International Journal of Chemical Studies, 6*(4), 3320–3323.

Li, F., Song, X., Wu, L., Chen, H., Liang, Y., & Zhang, Y. (2018). Heredities on fruit color and pigment content between green and purple fruits in tomato. *Scientia Horticulturae, 235,* 391–396.

Li, Z., Wang, S., Gui, X. L., Chang, X. B., & Gong, Z. H. (2013). A further analysis of the relationship between yellow ripe-fruit color and the capsanthin-capsorubin synthase gene in pepper (*Capsicum sp.*) indicated a new mutant variant in *C. annuum* and a tandem repeat structure in the promoter region. *PLOS ONE, 8*(4), e61996.

Lin, L. J., Hsiao, Y. Y., & Kuo, C. G. (2009). *Discovering Indigenous Treasures: Promising Indigenous Vegetables From Around the World.* AVRDC-The World Vegetable Center.

Liu, Y., Lv, J., Liu, Z., Wang, J., Yang, B., Chen, W., Ou, L., Dai, X., Zhang, Z., & Zou, X. (2020). Integrative analysis of metabolome and transcriptome reveals the mechanism of color formation in pepper fruit (*Capsicum annuum* L.). *Food Chemistry, 306,* 125629.

Lu, S., Van Eck, J., Zhou, X., Lopez, A. B., O'Halloran, D. M., Cosman, K. M., Conlin, B. J., Paolillo, D. J., Garvin, D. F., Vrebalov, J., & Kochian, L. V. (2006). The cauliflower Or gene encodes a DnaJ cysteine-rich domain-containing protein that mediates high levels of β-carotene accumulation. *The Plant Cell, 18*(12), 3594–3605.

Ma, C., Liu, M., Li, Q., Si, J., Ren, X., & Song, H. (2019). Efficient BoPDS gene editing in cabbage by the CRISPR/Cas9 system. *Horticultural Plant Journal, 5*(4), 164–169.

Marrelli, M., Amodeo, V., Statti, G., & Conforti, F. (2018). Biological properties and bioactive components of *Allium cepa* L.: Focus on potential benefits in the treatment of obesity and related comorbidities. *Molecules, 24*(1), 119.

Mezbani, A., Kavan, B. P., Kiani, A., & Masouri, B. (2019). Effect of dietary lycopene supplementation on growth performance, blood parameters and antioxidant enzyme status in broiler chickens. *Livestock Research for Rural Development, 31*(1).

Nakayama, H., Suzuki, T., & Suzuki, Y. (1998). Effect of wasabi on ion secretion in guinea pig colon. *Journal of the Agricultural Chemical Society of Japan, 72, 499–507.*

Ndebia, E. J., Kamgang, R., & Nkeh-ChungagAnye, B. N. (2007). Analgesic and anti-inflammatory properties of aqueous extract from leaves of *Solanum torvum* (Solanaceae). *African Journal of Traditional, Complementary and Alternative Medicines, 4*(2), 240–244.

Nisha, P., Nazar, P. A., & Jayamurthy, P. (2009). A comparative study on antioxidant activities of different varieties of *Solanum melongena. Food and Chemical Toxicology, 47*(10), 2640–2644.

Oriola, K. O., Hussein, J. B., Oke, M. O., & Isiaka, R. D. (2022). Effects of moisture content on some physical attributes of the seeds of three accessions of winged bean (*Psophocarpus tetragonolobus* L.). *International Journal of Postharvest Technology and Innovation, 8*(2, 3), 227–242.

Orwa, C., Mutua, A., Kindt, R., Jamnadass, R., & Simons, A. (n.d.). *Agroforestree Database: A Tree Reference and Selection Guide* (Version 4).

Paciulli, M., Medina-Meza, I. G., Chiavaro, E., & Barbosa-Cánovas, G. V. (2016). Impact of thermal and high pressure processing on quality parameters of beetroot (*Beta vulgaris* L.). *LWT–Food Science and Technology, 68,* 98–104.

Pan, Y. Z., Guan, Y., Wei, Z. F., Peng, X., Li, T. T., Qi, X. L., Zu, Y. G., & Fu, Y. J. (2014). Flavonoid C-glycosides from pigeon pea leaves as color and anthocyanin stabilizing agents in blueberry juice. *Industrial Crops and Products, 58,* 142–147.

Prajapati, U., Asrey, R., Meena, N. K., & Nayak, S. L. (2018). *Underutilized Vegetables for Health and Prosperity.*

Pratap Singh, A., Singh, A., & Ramaswamy, H. S. (2017). Effect of reciprocating agitation thermal processing (RA-TP) on quality of canned tomato (*Solanum lycopersicum*) puree. *Journal of the Science of Food and Agriculture, 97*(8), 2411–2418.

Rajasree, R. S., Sibi, P. I., Francis, F., & William, H. (2016). Phytochemicals of Cucurbitaceae family: A review. *International Journal of Pharmacognosy and Phytochemical Research, 8*(1), 113–123.

Ramalingam, R., Bindu, K. H., Madhavi, B. B., Nath, A. R., & Banji, D. (2010). Phytochemical and anthelmintic evaluation of corm of *Amorphophallus campanulatus. International Journal of Pharma and Bio Sciences, 1*(2), 317–323.

Rastogi, R. P., & Mehrotra, B. N. (1990). *Compendium of Indian Medicinal Plants.* Central Drug Research Institute.

Rodriguez-Amaya, D. B. (2019). Update on natural food pigments: A mini-review on carotenoids, anthocyanins, and betalains. *Food Research International, 124,* 200–205.

Rohini, N., Shanmugasundaram, T., & Pandiarana, N. (2017). Review on underutilized vegetables for human health. *Trends in Biosciences, 10,* 5593–5603.

Santos, D. T., Albarelli, J. Q., Beppu, M. M., & Meireles, M. A. A. (2013). Stabilization of anthocyanin extract from jabuticaba skins by encapsulation using supercritical CO_2 as solvent. *Food Research International, 50*(2), 617–624.

Seal, T. (2012). Evaluation of nutritional potential of wild edible plants, traditionally used by the tribal people of Meghalaya state in India. *American Journal of Plant Nutrition and Fertilization Technology, 2*(1), 19–26.

Shabhir, W., & Nuzhat, H. (2018). Natural carotenoids: A weapon to fight lifestyle-related disorders. *Journal of Food and Nutrition Population Health, 2,* 6.

Shakuntala, M. N., & Shadaksharaswamy, M. (1985). *Foods. Facts and Principles*. Wiley Eastern Limited.

Sharma, S., Katoch, V., Kumar, S., & Chatterjee, S. (2021). Functional relationship of vegetable colors and bioactive compounds: Implications in human health. *The Journal of Nutritional Biochemistry, 92,* 108615.

Shefana, A. G., & Ekanayake, S. (2009). Some nutritional aspects of *Lasia spinosa* (kohila).

Sledz, M., Wiktor, A., Rybak, K., Nowacka, M., & Witrowa-Rajchert, D. (2016). The impact of ultrasound and steam blanching pre-treatments on the drying kinetics, energy consumption, and selected properties of parsley leaves. *Applied Acoustics, 103,* 148–156.

Sreelatha, S., & Padma, P. R. (2009). Antioxidant activity and total phenolic content of *Moringa oleifera* leaves in two stages of maturity. *Plant Foods for Human Nutrition, 64,* 303–311.

Sui, X., Dong, X., & Zhou, W. (2014). Combined effect of pH and high temperature on the stability and antioxidant capacity of two anthocyanins in aqueous solution. *Food Chemistry, 163,* 163–170.

Sun, B., Jiang, M., Zheng, H., Jian, Y., Huang, W. L., Yuan, Q., Zheng, A. H., Chen, Q., Zhang, Y. T., Lin, Y. X., & Wang, Y. (2020). Color-related chlorophyll and carotenoid concentrations of Chinese kale can be altered through CRISPR/Cas9 targeted editing of the carotenoid isomerase gene *BoaCRTISO*. *Horticulture Research, 7,* 7.

Suparmi, S., Fasitasari, M., Martosupono, M., & Mangimbulude, J. C. (2016). Comparisons of curative effects of chlorophyll from *Sauropus androgynus* (L) Merr leaf extract and Cu-chlorophyllin on sodium nitrate-induced oxidative stress in rats. *Journal of Toxicology, 8515089,* 1–7.

Umar, K. J., Hassan, L. G., Dangoggo, S. M., & Ladan, M. J. (2007). Nutritional composition of water spinach (*Ipomoea aquatica* Forsk.) leaves. *Journal of Applied Sciences, 7*(6), 803–809.

Vaňková, K., Marková, I., Jašprová, J., Dvořák, A., Subhanová, I., Zelenka, J., Novosádová, I., Rasl, J., Vomastek, T., Sobotka, R., & Muchová, L. (2018). Chlorophyll-mediated changes in the redox status of pancreatic cancer cells are associated with its anticancer effects. *Oxidative Medicine and Cellular Longevity, 4069167.*

Verma, A. R., Vijayakumar, M., Mathela, C. S., & Rao, C. V. (2009). *In vitro* and *in vivo* antioxidant properties of different fractions of *Moringa oleifera* leaves. *Food and Chemical Toxicology, 47*(9), 2196–2200.

Wang, C., Riedl, K. M., & Schwartz, S. J. (2013). Fate of folates during vegetable juice processing: Deglutamylation and interconversion. *Food Research International, 53*(1), 440–448.

Wootton-Beard, P. C., & Ryan, L. (2011). A beetroot juice shot is a significant and convenient source of bioaccessible antioxidants. *Journal of Functional Foods, 3*(4), 329–334.

World Health Organization. (2020). *UNICEF/WHO/The World Bank Group Joint Child Malnutrition Estimates: Levels and Trends in Child Malnutrition: Key Findings of the 2020 Edition.*

Yadava, D. K., Choudhury, P. R., Hossain, F., & Kumar, D. (2017). *Biofortified Varieties: Sustainable Way to Alleviate Malnutrition.* Indian Council of Agricultural Research.

Yahia, E. M., García-Solís, P., & Celis, M. E. M. (2019). Contribution of fruits and vegetables to human nutrition and health. In *Postharvest Physiology and Biochemistry of Fruits and Vegetables* (pp. 19–45). Woodhead Publishing.

CHAPTER 8

Plant-Based Milk as Potential Health Beverage

HAMID,[1*] DEEPIKA KATHURIA,[2] SHWETA SURI,[3] and SUNAKSHI GAUTAM[4]

[1]*Food Technology and Nutrition, Lovely Professional University, Phagwara, Punjab, India*

[2]*Department of Food Science and Technology, Graphic Era University, Dehradun, Uttarakhand, India*

[3]*Amity Institute of Food Technology, Amity University, Noida, Uttar Pradesh, India*

[4]*School of Bioengineering and Food Technology, Shoolini University, Solan, Himachal Pradesh, India*

*Corresponding author.

ABSTRACT

Functional foods are the most emerging food segment in the market due to their highly nutritious and functional characteristics. With an increase in hectic lifestyles and concerns about health, the beverage industry has driven up market growth by exploring alternate emerging raw, nutritious sources. Among beverages, dairy milk consists of a sufficient amount of all macro- and micro-nutrients. But still, dairy milk has limited access for its consumption around the world due to its allergenicity, lactose intolerance, and hypercholesterolemia. Non-dairy plant-based milk is considered the best alternative for health-conscious consumers. These are developed using cereals, legumes, nuts, seeds, and pseudo-cereals which consist of various bioactive components such as polyphenolics, phytosterols, unsaturated fatty acids, antioxidants, and so on, that possess health-promoting activities. Over conventional milk, these

Functional Compounds and Foods of Plant Origin: Krishan Dutt Sharma, Satish Kumar, Rakesh Sharma, and Raj K. Keservani (Eds.)

milk analogs are cheap and easy to modify the composition as per the market demand. But the only problem associated with plant analogs is low sensory acceptability as well as the presence of certain anti-nutritional components. Although, fermentation of plant-based beverages helps in maintaining their sensory profile and prevents microbial contamination by using other artificial ingredients for stability. Keeping because of the emerging health issues nowadays, these plant-based beverages have the potential to be used as an alternate source of conventional milk with higher disease-preventing bioactive components besides, a sufficient amount of nutrients. This chapter highlights different milk analogs produced from different categories of plant and their associated bioactive compounds producing health benefits.

8.1 INTRODUCTION

Milk is a liquid food produced by the mammary glands of female mammals. It is nature's complete food consisting of all the nutrients which are important to sustain a healthy life. The presence of both macro- and micro-nutrients (calcium, selenium, and riboflavin), and bioactive peptides in milk makes it the key contributor to wholesome growth and maintenance of the body system (Iid et al., 2020; Kumar et al., 2022; Paul et al., 2020; Sharma et al., 2022; Shree et al., 2022). Besides being a nutritionally rich food, its consumption is decreasing progressively due to some negative health impacts. The major reasons for this drastic shift are milk protein allergy, lactose intolerance, calorie concern, hypercholesterolemia, and cultural ethics. Also, veganism and other types of diet (veganism, Paleo, flexitarian diet) play a major role in the shift towards the elimination of animal-origin foods. Nowadays, people have become very concerned about their health and are easily influenced by popular trends, social media, food blogs, and research or scientific-based advice. These above-mentioned factors are responsible for the change in the eating habits of people. To overcome such problems, there is a new concept of plant-based milk, which gives a healthy, animal welfare-friendly, and sustainable alternative to dairy milk. Plant milk is the extracted result after reducing the size of plant materials such as cereals, legumes, nuts, oil seeds, and pseudo-cereals in water. This fluid is homogenized properly which resembles dairy milk in appearance and consistency (Jaglan et al., 2020; Kumar et al., 2022; Sethi et al., 2016; Sharma et al., 2022). Various types of particulate matter are suspended in an aqueous solution of plant-based milk, which is present in the form of a complex colloidal dispersion. These are divided into five broader categories *viz*., cereal-based, legume-based, oil seed-based, nut-based, and pseudo-cereals-based (Jeske et al., 2018a; Sethi et al., 2016). Crops such as soybean, coconut, oat, almond, rice, hemp, and quinoa are majorly utilized for the preparation of milk.

Mild-flavored soya milk, the first commercial plant-based milk, was marketed in Hong Kong in 1940, and afterward, this trend spread rapidly in Western countries (Jeske et al., 2018). In 1999, soy products were approved by the FDA with the health claim to reduce the risk of heart disease. Since then, more than 2700 new soy products have been launched in the market (Mäkinen et al., 2015). Afterward, different types of plant-based dairy substitutes were explored, and these products are now available in the form of beverages and functional products in the market. The newer research is more focused on the development of functional products, especially beverages from non-dairy sources. However, these plant-based dairy substitutes have been developed and consumed for centuries as traditional dishes in various cultures. But now, the increasing popularity of these substitutes and their development is the need of the hour because they have become a part of the lifestyle and not just a thirst quencher (Chandel et al., 2023; Kumar et al., 2020; Sethi et al., 2016; Suthar et al., 2022). Calcium is very important for regulating body functions and maintaining teeth and bone health. Plant-based milk has variable calcium content and is lower as compared to dairy milk but can be enhanced by fortification. Whereas, concerning vitamin E, dairy milk has low levels as compared to non-dairy plant-based milk such as almond milk. The calories served by these alternatives are also less than 140–150 calories which represent only 7–7.50% of total calorie intake in a day (Craig et al., 2021). These substitutes are a better option for lactose intolerant or dairy allergic people as they do not contain lactose or dairy allergens and also have low cholesterol content. Although the protein content in dairy milk is higher than in other plant-based milk alternatives, soymilk (2.5–3.16%) can fulfill this gap up to a certain limit. These plant-based beverages have the potential to contribute to the health and nutrition of the population as well as contribute to food safety. As consumers become more comfortable with milk alternatives, the competition between dairy and dairy alternative beverages intensifies. This chapter covers all aspects either positive or negative related to plant-based milk to enhance knowledge as well as its practical implications.

8.2 ROLE OF MILK IN THE HUMAN DIET

Milk is the first food for mammals that acts as an essential element for health and a balanced diet. It is a composite physiological fluid produced by the female mammary gland to nourish her newborn baby. Milk is an important nutrient food source that contains all the vital macro- and micro-nutrients required to meet the needs of the human body, especially for vulnerable groups such as infants, school-going children, and elderly people. The protein present in milk comprises highly biological active peptides which are divided into two fractions, soluble plus insoluble. The soluble protein includes whey protein accounting for 20% while the insoluble protein is casein which constitutes 80% of the total milk protein. Milk protein is also enriched with essential amino acids and has a greater protein digestibility corrected amino acid score (Pereira and Paula, 2014). These proteins acquire biological and nutritional properties, especially whey protein consisting of α-lactalbumin, β-lactoglobulin, lactoperoxidase, lactoferrin, and serum albumin exert antioxidant, anticarcinogenic, and antiviral effects. Milk fat is considered a good source of both energy as well as lipid-soluble micronutrients. It contains bioactive lipids such as triacylglycerides, diacylglycerides, and phosphorus-containing lipids. Additionally, milk predominantly consists of 70% saturated fatty acids and 2% polyunsaturated fatty acids (PUFAs). Around, 12% of milk fat significantly reduces the total cholesterol/high-density lipoproteins (HDLs) ratio which would reduce the risk factor associated with cardiovascular disease (CVD). The major trans-fatty acid found in milk is conjugated linoleic acid (CLA). It has been of great importance because of its health-promoting effects on the cardiovascular system, and immune function as well as prevention of cancer (Benjamin et al., 2009). Furthermore, milk contains fat-soluble vitamins such as vitamin A and D plus water-soluble vitamins such as vitamin B complex. Vitamin A helps with growth and development along with maintaining eye health, while vitamin D is crucial for calcium absorption and improves bone health. On the other hand, the vitamin B complex participates in metabolic pathways for the generation of energy, neurotransmitters as well as hormonal synthesis. Among micronutrients, milk also delivers minerals, particularly calcium, phosphorus, zinc, potassium, magnesium, and selenium (Muehlhoff et al., 2013). Calcium is the major mineral found in large (1200 mg/L) amounts in milk which is distributed among the micellar as well as watery phases. Apart from calcium content, milk is additionally perceived as a promising phosphorus source with a concentration of around 950 mg/L. These nutrients are necessary for bone formation in childhood and its maintenance through adulthood, to reduce the chances of osteoporosis and bone fractures in old age (Rizzoli, 2014).

The complex mixture of folate, vitamin D, and calcium available in milk has protective effects against cancer and prevents metabolic syndrome, glucose tolerance, and insulin sensitivity. These micronutrients regulate cell proliferation, thereby affecting the carcinogenesis process. Bioactive peptides produced from casein digestion as well as calcium manage blood pressure by inhibiting angiotensin-converting enzyme-II activity (ACE-II), thus exerting antihypertensive effects (Del Mar Contreras et al., 2013). Also, milk plays a major role in reducing the risk of non-insulin-dependent diabetes mellitus (DM). The advantage of milk in forestalling infectious diseases has been perceived for millennia. Its action against infectious microorganisms has been ascribed to antibodies present in milk. It comprises significant amounts of immunoglobulins and certain physiologically dynamic mixtures for warding off disease in the infant (Séverin et al., 2005). Hence, milk plays an important role in human development in underdeveloped and developing countries. It has been recommended that the consumption of milk and milk products prevents various chronic diseases like obesity, cancer, hypertension, and CVD.

8.3 PUBLIC HEALTH CONCERN: WHY PLANT-BASED MILK?

Although, milk is rich in nutrients such as protein, fat, vitamins, and minerals which are essential in the human diet but still the consumption of milk is still declining around the world especially, in western countries. The type of fatty acids present in milk is a controversial subject because saturated fats (palmitic, myristic, and lauric) present in milk have opposing effects. Further, people find difficulty in digesting milk sugars (lactose intolerance) or might be allergic to milk protein. Lactose intolerance is a non-immunological response causing indigestion of lactose available while consuming milk. This problem is found among 75% of the world population, who face difficulty in digestion and absorption of milk due to the absence of the β-D-galactosidase enzyme. Therefore, lactose gets accumulated and fermented by intestinal microbiota in the colon which releases certain organic acids such as acetic, butyric, and propionic acid as well as gases like CO_2, methane, hydrogen, and so forth. These compounds further cause flatulence, diarrhea, nausea, vomiting, metabolic acidosis, bloating, and constipation (Fassio et al., 2018). Another adverse immune response to milk dietary antigen (protein) causes allergy. The main allergen-causing proteins are casein and whey protein. It starts with casein (80% of milk protein) which when converted into opioid compounds known as β-casomorphines, binds to the A1 allele of β-casein, thus causing allergy to the human body

(El-Agamy, 2007). The allergy caused by milk protein mainly affects the gastrointestinal tract causing vomiting, diarrhea, and respiratory tract that leads to sneezing, wheezing, and coughing as well as showing the symptoms of itching, rash, hives, and edema. In addition to this, another factor that eliminates milk from the diet is religious ethics. As per the report of the World Atlas, India ranked first with 38% of the population being considered vegetarian, followed by Israel, Taiwan, Italy, Austria, Germany, the UK, and Australia with 5% in the year 2021. In India, the highest percentage of vegetarians was found in Rajasthan (74.9%), Haryana (69.25%), Punjab (66.75%), and Gujarat (60.95%) in 2018 (Reddit, 2017). This proportion of the population considered that milk originated from animal sources, neither milk nor its products are consumed by the vegan population. Therefore, the demand for plant-based milk has driven up in the world market, as the water-soluble extract of cereals, legumes, pseudo-cereals, oil seeds, and nuts, resembles dairy milk.

8.4 TYPES AND NUTRITIONAL PROFILE OF PLANT-BASED MILK

Plant-based milk is a water-soluble extract derived from various plant sources. It is formed by extracting the raw material in water by following a process of maceration, grinding, and filtration with a particle size distribution of 5–20 μm that appears similar to dairy milk. Prepared milk also consists of additives, stabilizers, and emulsifiers, and is subjected to proper treatment for improved shelf-life. Although no specific definition and classification are cited in the literature, in general, plant-based milk is divided into five categories based on the type of raw material being used. It includes cereal-based (rice, corn, oat, barley, and millets), legume-based (soybean, peanut, lupin, cowpea, and chickpea), nut-based (almond, coconut, tiger nut, hazelnut, and walnut), oil seed-based (sesame, flaxseed, and hemp), and pseudo-cereal based (quinoa, amaranth, and buckwheat) (Sethi et al., 2016). The nutritional quality of plant-based milk is directly related to the raw material from which it is prepared and the processing methodology. For example, cereal-based milk is rich in carbohydrates, while legume-based milk possesses a higher concentration of protein with low saturated fat. Unlike dairy milk, plant-based milk does not contain lactose or cholesterol as well as has a lower concentration of calcium and vitamins. Hence, it is sold after fortifying with calcium and vitamins, especially vitamin B_{12}. The nutritional characteristics of different plant-based kinds of milk are shown in Table 8.1.

8.4.1 CEREAL-BASED MILK SUBSTITUTE

8.4.1.1 RICE MILK

Rice is the major cereal crop that is used as a staple food by more than half of the world population, mainly in Asian countries. It is rich in starch content but deficient in lysine and threonine (Jiang et al., 2016). It also consists of an appreciable amount of sinapic acid, ferulic acid, and p-coumaric acid. The composition of the components may vary depending upon the variety as well as on fractions obtained during milling, such as broken, head, and brown rice obtained from similar varieties. Besides, the rice also contains a high amount of carbohydrates majorly starch which gives a characteristic sweet taste to rice milk beverages. However, escalation in the production and consumption of rice has been reported in various countries around the world due to some alternate products with high nutritional value. Milk is a good alternative to the consumption of rice. Traditionally, rice milk is developed using whole kernels of brown and milled rice. Koyama and Kitamura (2014) described milk from rice as a mixture of the rice slurry developed by grinding rice with water in a 1:9 (solid: liquid) ratio. Although, rice milk is deficient in nutritional composition various studies have claimed that rice milk fortified with calcium and B_{12} vitamin when consumed reduces the percent of malnutrition among infants (Winston, 2009). Rice milk is nowadays generally used as a highly nutritional beverage due to its high amount of calcium (245.5 mg), carbohydrates (25.28%), energy (133 Kcal), and low amount of saturated fatty acids (0.16 mg) with zero cholesterol. Therefore, milk prepared from rice grains has a similar nutritive value to that of dairy milk as given in Table 8.1 and is free from milk sugar lactose. Further, rice milk also acts as an alternative source of milk for patients who are allergic to almonds and soybeans.

8.4.1.2 CORN MILK

Maize is also one of the major cereal crops in India, known as Indian corn, sweet corn, sugar corn, pole corn, or corn. Maize has high nutritional values. Maize contains various B complex vitamins, starch, protein, carbohydrate, moisture, fat, fiber, and ash. Corn harvested at a milky stage is low in protein therefore, it can be supplemented with high-grade protein like soybean for the development of milk. Trisnawati et al. (2013) developed corn milk by blending it with soymilk in 30:70 ratios. Corn milk has a pleasant taste and nutritive value. Corn milk attracts health-conscious consumers of different age groups as it contains 1.87% protein, 1.09% fat, 6.25% sugar, 0.54% ash content, and very low cholesterol and saturated fat compared to dairy milk (Rahman et al., 2019).

8.4.1.3 OAT MILK

Cereal grains are consumed by a large population around the world. Oats have received significant concentration due to their high nutritional value as a part of the consumer's daily diet in the form of ready-to-eat (RTE) cereal. Oats have a good amount of dietary fibers such as β-glucan, functional protein, lipid, and starch components, therefore, act as a potential source for the development of milk (Sethi et al., 2016). β-glucan is a soluble fiber that increases the viscosity of milk and also raises the gastric emptying time and gastrointestinal transit time, which thereby decreases blood glucose level. Thus, oat milk is highly recommended for people who are suffering from irritable bowel syndrome (IBS) and inflammatory bowel disease (IBD). The enzymatic method for the development of milk has been optimized by Deswal et al. (2014) where the slurry is treated with 77.78 mg/Kg of enzyme for liquefaction at 75°C for 49 min. It has been observed that oat milk consists of a sufficient amount of nutrients (Table 8.1) except calcium which is necessary for growth and development. Consequently, it has to be fortified with calcium before consumption to be used as an alternate source of dairy milk. Nowadays, UHT-treated oat milk has been commercially available in the market in the form of tetra packs.

8.4.1.4 BARLEY MILK

Barley is one of the important food grains in the category of underutilized cereal grains since ancient times. This cereal grain consumption has an important nutritional benefit for human health. It consists of a high amount of dietary fiber, vitamin E, thiamin, niacin, and minerals such as molybdenum, selenium, manganese, copper, and so forth (Suman and Sreeja, 2019). Although barley is majorly used for feed and malting purposes, recently it has been often used for milk production as well. The conversion of barley grains into barley milk leads to various changes in the nutritional content of milk as compared to fresh grains. Table 8.1 represents the nutritional composition of barley milk.

8.4.1.5 MILLETS MILK

Millets are also a type of cereal grain that is smaller in size. It has been a staple food for thousands of years for humans who live in dry and hot areas of the world. It includes crops like pearl, finger, foxtail, proso, kodo, barnyard millet, and so forth. Millets are a rich source of protein, calcium, iron, dietary fiber, phenolics, thiamine, riboflavin, and nicotinic acid. These crops are not effectively utilized in the world but need to be diversified through value addition. Millet consists of high protein and mineral content but is low in calories; hence, milk developed from it is preferred over other sources of plant milk (Raajeswari and Nithya, 2016). Therefore, millet milk acts as an ideal source for replacing dairy milk where high nutrition and low-calorie foods are preferred. The nutritional characteristics of different types of millet milk are given in Table 8.2.

8.4.2 MILK LEGUME-BASED SUBSTITUTE

8.4.2.1 SOYBEAN MILK

Soybeans are the highest-yielding protein source, with around 40% protein (dry basis). It is also abundant in fat (18–22%), dietary fiber, and minerals. With the advancement of technologies, a variety of products are being produced throughout the world, either by following modern processing techniques or using traditional methods. Among these products, soymilk acts as an inexpensive, rich source of protein that can be processed into yogurt, tofu, and a variety of other products. Soy milk is a creamy, milk-like product made by soaking and grinding soybean in water. It is an off-white emulsion that resembles dairy milk in appearance. The direct utilization of soybean in human foods is very limited due to its offensive off-flavor. Kinney (2003) stated that the negative flavors associated with soybean and soy products include painty, grassy, beany, metallic, bitter, or chalky which occur due to lipoxygenase catalyzed oxidation reaction. Therefore, to improve the quality of soy milk, several treatments such as alkaline soaking, hot grinding, cold grinding, and gallic acid-aided grinding have been used by various researchers from time to time to reduce beany flavor through the inactivation or inhibition of lipoxygenases (Kinney, 2017). After the removal of the beany flavor, different methods have been developed by various investigators for the production of soy milk. Kumar et al. (2019) prepared soymilk by blanching seeds at 98°C for 20 min then mixing in distilled water in the ratio of 1:5 and heating at 121C for 20 min. The nutritional composition of soymilk is affected by the processing method, especially the seed-to-water ratio used for blending, which is increased with the increased concentration of soybean. The nutritional value of soymilk observed by different researchers is given in Table 8.1.

8.4.2.2 Peanut milk

Energy and protein deficiencies are common among low-income populations in developing countries. Peanut milk has significant potential as a substitute for dairy beverages due to its high nutritional value. Peanuts are rich in quality protein, essential fatty acids (e.g., linolenic and linoleic acid), and minerals. Peanut milk is a health-promoting beverage that can be prepared from raw peanuts through processing steps that include soaking in water, followed by grinding and filtration.

Several researchers have explored the development of peanut-based milk. For example, Siddeeg et al. (2020) formulated peanut milk using roasted seeds ground in water at a 1:4 ratio. Similar to dairy milk, peanut milk contains valuable components such as proteins (2.8%), ash (0.2%), carbohydrates (2.0%) including glucose (0.05%) and other sugars like sucrose (0.63%) and pentosans (0.20%), as well as starch (0.20%) and fats (4.4%) such as palmitic (0.40%), stearic (0.07%), oleic (2.30%), linoleic (1.30%), and arachidic (0.07%) acids, which are desirable for human nutrition (Diarra et al., 2005). Children who are allergic to cow's milk can consume peanut milk and alternatives like peanut butter and curd (Diarra et al., 2005).

One challenge in the development of peanut milk is its poor stability during processing, transportation, and storage due to high fat content. This stability can be improved by adding emulsifying agents, such as alginates, gelatin, or vegetable gums at 1% (w/w).

8.4.2.3 Chickpea milk

Chickpea is the third most essential legume crop grown in the world. It is recognized as a valuable source of complex carbohydrate, dietary fiber (3.82%), protein (23.64%), vitamins (riboflavin (0.173 mg), thiamin (0.453 mg), niacin (1.602 mg), pyridoxine (0.466 mg), folate, and vitamin A) and minerals like Na (121 mg), K (870 mg), Ca (176 mg), Mg (176 mg), P (226 mg), Mn (2.11 mg), Zn (4.32 mg), Cu (1.10 mg), and Fe (7.72 mg) which is an important constituent of daily diets for humans (Aguilera et al., 2009). Like other legumes, chickpea also consists of anti-nutritional and beany flavor components which could be reduced by using soaking and cooking techniques or by addition of seasonings like turmeric, and black pepper (Kishor et al., 2016). It is evident from the study that chickpea milk is highly effective in managing persistent diarrhea.

8.4.2.4 Cowpea milk

Cowpea is also a cheap legume crop whose consumption pattern is very encouraging in Nigeria. It is a rich source of protein (26.61%), carbohydrate (56.24%), and ash content in the human diets in many developing countries. The lower amount of lipids (3.99%) and fiber (1.38%) are also present in cowpea (Owolabi et al., 2012). The protein significantly possessed sulfur-containing amino acids, which are deficient in cereal crops. Hence, in most of the studies, cowpea milk has been developed as protein fortification in maize milk. Cowpea milk prepared by boiling seeds followed by grinding in water at 1:10 consists of 92.5% moisture, 5.5% carbohydrate, 1.5% protein, 0.4% fat, and 0.2% ash content (Aduol et al., 2020). In-addition minerals are also present in variable quantity in cowpea milk includes magnesium (17.62 mg) and iron (0.49 mg), zinc (0.13 mg) and calcium (0.01 mg) (Aduol et al., 2020).

8.4.2.5 Lupin milk

After soybean, lupin is another plant-based protein-rich legume. It is one of the important economic crops not only because of its nutritional content but also due to its higher adaptability to human and animal nutrition. Besides protein, lupin is also a good source of dietary fiber, and essential amino acids, while lower in lipids, and starch content. The seeds of lupin consist of protein (37.4%), lipid (11.34%), dietary fiber (34.44%), ash content (3.79%), carbohydrate (39.8%), vitamins like thiamin (0.34 mg), riboflavin (0.61 mg), α-tocopherol (0.47 mg), γ-tocopherol (51.6 mg) and δ-tocopherol (0.25 mg). It also provides an appreciable amount of minerals like Na (31.94 mg), Mg (128.98 mg), Fe (4.58 mg), and Zn (3.22 mg) (Atudorei et al., 2021). Therefore, milk derived from lupin is nutritionally rich with high functional values (Table 8.2). Like soy milk, lupin milk also possesses anti-nutritional components and has a beany flavor which can be removed by heat inactivation or fermentation. Different authors developed milk analogs of lupin by using protein isolate solution along with the addition of fat, sucrose, and emulsifiers.

8.4.3 TREE NUT-BASED MILK SUBSTITUTE

8.4.3.1 ALMOND MILK

Almonds are considered a brain food that improves mental alertness, and concentration, as well as helps with getting good sleep at night. It is one of the major sources of vitamin E and B complex. It also consists of a sufficient amount of protein, dietary fiber, polyunsaturated fats, polyphenols, phytosterols, vitamins, and minerals (calcium, potassium, phosphorus, iron, and zinc) that make it a good non-dairy supplement (Bernat et al., 2015). Due to the presence of trace elements, almond milk has great significance in developing countries where iron and zinc deficiency are at their maximum. Almond milk is an oil-in-water emulsion where the dispersed phase includes complex proteins and oil droplets. This emulsion gets breakdown forming flocculation, coalescence, therefore proper stirring and homogenization are required to form milk of smooth consistency. The major problem associated with almond milk is its highly nutritious content (Table 8.1) and the presence of unsaturated fatty acids. This makes milk highly susceptible to oxidation and the growth of spoilage microorganisms. Therefore, proper thermal treatments must be applied to milk to extend its shelf-life as the lipoxidase enzyme gets inactivated at pasteurization temperature. Further, almond milk is also found to be allergic in most people in the United States (US) which may even cause fatal anaphylaxis (Paul et al., 2020).

8.4.3.2 COCONUT MILK

Coconut is a versatile fruit, nut, or seed which is widely cultivated in several countries around the world. Milk developed from coconut is the most well-known product in the world. It is oil in water emulsion which is generally produced by grating coconut meat followed by squeezing it with or without the addition of water. It is the most widely used dairy milk substitute owing to its ease of preparation, nutritional value, and digestibility. The major component of coconut milk is coconut oil (38% by weight) which includes medium-chain fatty acids that are easily absorbed as well as metabolized in the liver to form ketone compounds. These compounds are effective in improving memory in people suffering from Alzheimer's disease (Ekanayaka et al., 2013). Coconut milk has good nutritional value (Table 8.1) in the context of the presence of proteins, amino acids (arginine, glutamic acid, and aspartic acid), minerals (potassium, calcium, and phosphorus), vitamins (B-complex, E, and C), and saturated fats. Lauric acid and oleic acids are fatty acids that are present in large quantities in coconut milk (Yuliana and Rangga, 2010).

8.4.3.3 TIGER NUT MIL

It is an underutilized perennial edible tuberous rhizome that is known by different names *viz.*, chufa, earth almond, rush nut, yellow nut sedge, atadwe, and edible galingale. Tiger nut tubers contain double the amount of starch as potato or yam tubers. Tiger nut also consists of protein, minerals (potassium and phosphorus), ascorbic acid, and tocopherols. This plant can be eaten raw, roasted, or pressed for juice extraction to form a beverage or milk. The obtained milk is healthy, having 8.19% protein, 25.5% fat, 1.8% ash, and 58.01% carbohydrate (Ogbonna et al., 2013). It has been reported that tiger nut milk consists of greater amounts of minerals such as iron and magnesium as well as carbohydrates. Unlike soy milk or almond milk, tiger nut milk does not cause any allergies. A milky drink called *horchata de chufa* is commercially produced from tiger nuts in Spain. The main challenge associated with tiger nut milk is its starch content, which causes gelatinization and retrogradation during pasteurization and cooling. Therefore, in tiger nut milk production, either the pasteurization step is eliminated, or the starch is converted into free sugars before processing (Okorie and Nwanekezi, 2014).

8.4.3.4 HAZELNUT MILK

Hazelnut is a hard-shelled fruit that is consumed as a whole nut. It is a potentially rich source of energy (600–650 Kcal/100 g). It also consists of protein, carbohydrate, lipid (oleic acid), vitamins, dietary fibers, total phenolics, tocopherols, minerals, phytosterols, and squalene. Hazelnut is highly appreciated for its taste, aroma, and crunchy texture and therefore is extensively utilized in snacks and confectionary items like chocolates, pastries, and so forth. Milk can also be developed from hazelnuts by grinding the nuts in water, followed by homogenization and filtration (Bernat et al., 2014).

8.4.3.5 WALNUT MILK

Walnut is described as the "doctrine of signatures" which resembles an organ of the body. The walnut kernel looks like a brain which is also known as the glands of Jupiter. It is one of the most promising functional foods consisting of healthy fatty acid composition (omega-3, omega-6 fatty acids, and oleic acid) and also possesses higher protein, dietary fiber, vitamin E, vitamin K and minerals like potassium, phosphorus, iron, zinc, and selenium (Pereira et al., 2008). The milk developed from walnut is an emulsion of oil in water where the dispersed phase is oil, and the rest of the soluble solids play an active role in product stability. Based on nut composition and processing method, the nutritional value in walnut milk ranged from 1.30–1.70% for protein, 7.37–18% for lipid, 0.20–0.41% for ash while around 4.76% carbohydrate content (Bolarinwa et al., 2018). Milk does possess iron (0.64 ppm), magnesium (19.85 ppm), calcium (23.63 ppm), sodium (1.64 ppm), and zinc (2.02 ppm) content as well (Bolarinwa et al., 2018).

8.4.4 OILSEED-BASED MILK SUBSTITUTE

8.4.4.1 SESAME MILK

Sesame also known as the queen of oilseeds is a tiny seed crop cultivated in tropical and sub-tropical climates around the world. It possesses 50–60% lipid, 18–25% protein, 13–14% carbohydrate and 5% ash content (Nagemdra et al., 2012). The seeds also have a significant amount of trace minerals such as iron, zinc, calcium, magnesium, manganese, phosphorus, and copper. Lipid is the major energy source of sesame seeds, majorly comprising of unsaturated fats such as oleic acid (39.10%), and linoleic acid (40.00%) with minor amounts of stearic acid, palmitic acid, and trace quantities of linolenic acid (Namiki, 2007). Similar to soymilk, sesame milk also has its beany flavor for which seeds are subjected to the process of roasting and blanching before preparation of milk to reduce enzymatic off-flavor. The milk developed had 2.64–2.97% protein, 6.43–7.82% fat, 0.22–0.48% ash, and 4.04% carbohydrate content (Ahmadian-Kouchaksaraei et al., 2014). Sesame protein available in seeds causes a major problem during the preparation of sesame milk. As it does not get solubilized in water but is susceptible to heat denaturation. Therefore, various processing steps such as soaking, germination, fermentation, and microwave treatment are used to modify the functionality of the protein. Sesame milk is marketed in countries like Spain under the brand name Ecomil and in South Korea as Vegemil (Sethi et al., 2016).

8.4.4.2 FLAXSEED MILK

Flaxseed popularly known as linseed is a rich source of fat (41% with majorly PUFAs) followed by total dietary fiber (28%), protein (20%), and ash (3.4%) content. The utilization of flaxseed as a value-added product is a new prospect due to increased awareness of functional food among consumers. The presence of bioactive components in flaxseeds has increased its use in the form of salad dressings, food additives, and substitutes for milk (Rabetafika et al., 2011).

8.4.4.3 HEMP MILK

Hemp is an Asian crop whose seeds are relatively underutilized. It consists of an equal amount of fat and protein of about 25 to 30%. In addition, seeds are rich in dietary fiber and tocopherol content. Although, hemp seeds have been incorporated into various bakery products beverage/milk is the most emerging product. Other than dairy, soy, and nut-based milk, hemp milk also delivers high nutritional value (Table 8.1) with low allergenicity, but it also tends to flocculate due to unstable oil-in-water emulsion. Cold-pressured-without-heat-pasteurized hemp milk can also be developed to preserve the nutrients and original clear flavor of the seeds (Wang et al., 2018).

8.4.5 PSEUDOCEREAL-BASED MILK SUBSTITUTE

8.4.5.1 QUINOA MILK

Quinoa is an underutilized fruit of South America that has gained importance nowadays due to its nutritional characteristics. It is a good source of protein (12–16.5%) whose quality is equivalent to casein. The vitamins and minerals, especially vitamin C, E, folate, iron as well as calcium are higher in amount when compared with true cereals. Traditionally, quinoa has been used in soups, porridges, stews, breakfast cereals, granola bars, and beer. It could be consumed in popped or extruded form. To magnify the traditional use of quinoa, beverages/milk has also been developed. This wonder grain is gluten-free; therefore, it can be consumed by those who are allergic to gluten protein. Kaur and Tanwar (2016) reported 0.68–2.90% protein, 0.23–0.93% fat, 0.11–0.28% ash, and 14.7–16.20%

carbohydrate in quinoa milk developed from soaked, germinated, and malted seeds. Moreover, milk containing quinoa pericarp imparts a bitter and astringent taste due to the presence of 5% saponins (Vega-Gálvez et al., 2010). Hence, before preparation, these components have to be removed.

8.4.5.2 AMARANTH (AMARANTHUS SP.) MILK

Amaranth is one of the world's oldest pseudo-cereals which is considered a super food. It consists of high-quality protein, unsaturated fatty acids (omega-6), dietary fiber, vitamin E, B_2, C, and minerals (Aderibigbe et al., 2020). The seeds also possess a higher amount of lysine and tryptophan content. Although, amaranth is also an underutilized crop but its ability to improve the nutritional quality of staple food products has made it a promising solution for combating malnutrition. An instant functional beverage has been developed using extruded amaranth grains. The incorporation of amaranth into the beverage provides higher acceptability than without amaranth. For example, a non-alcoholic refreshing drink called Kunu is consumed in Nigeria and prepared by fermenting amaranth seeds was much better than developed from sorghum (Isaac-Bamgboye et al., 2019). Arguelles-Lopez et al. (2018) formulated an amaranth chai-based beverage (70:30 amaranths to chai flour ratio). Based on the potential contribution towards nutrients offered by amaranth, its associated functional drinks can be utilized as an alternative to energy-dense beverages that are dominating the market.

8.4.5.3 BUCKWHEAT MILK

Similar to quinoa and amaranth, buckwheat is also considered a health-promoting pseudo-cereal. Buckwheat constitutes a gluten-free protein with balanced amino acid content. Buckwheat flour also consists of minerals, such as zinc, manganese, copper, potassium, magnesium, and phosphorus as well as vitamins, including thiamine, riboflavin, niacin, pantothenic acid, and pyridoxine (Kim et al., 2002). Buckwheat behaves like a prebiotic and can be used for the development of probiotic drinks. Kowalska and Ziarno (2020) developed a buckwheat fermented beverage using lactic acid bacterial culture and bifidobacterium.

TABLE 8.1 Nutritional Composition of Plant-based Milk

Nutrient (per 100 g)	Type of Non-Dairy Milk						
	Rice	**Oat**	**Barley**	**Soy**	**Almond**	**Coconut**	**Hemp**
Energy (Kcal)	50–133	45–91	–	43–95	24–55.40	27	19.00–21.06
Protein (g)	0.85–1.12	1.0–3.70	2.90	3.1–8.71	0.5–1.70	1.3–4.56	0.83–4.00
Fat (g)	1.0–0.33	0.69–1.5	0.47	2.4–4.35	1.82–3.40	2.0–5.8	1.25–3.00
Ash (g)	0.10	0.49	0.82	–	0.11–3.04	1.1	–
Carbohydrate (g)	10.1–25.28	6.30–7.50	9.91	2.2–5	0.68–4.50	1.9	2.50–20.00
Fiber (g)	–	0.9–3.2	0.57	0.5–0.96	1.25–1.6	0.1	–
Calcium (mg)	88–245.5	34.52–120	47.50	130–205.86	13.10–120	120	12.03–20.25
Phosphorus (mg)	34.4–63	31.3–160.11	301.19	89–108	28–75.20	73	54.84–63.98
Potassium (mg)	50	30.6–240.20	182.88	119–364.29	59–65.33	26	110.00–126.58
Magnesium (mg)	35	9.2	–	18–49	10–42.05	2.3	–
Sodium (mg)	33.3–72	49	–	56–65	6.38–50	40	85.69–140.01
Zinc (mg)	0.75	1.23	2.44	0.75	4.58	–	–
Iron (mg)	–	1.56	2.31	–	1.40	–	4.89–6.58
Vitamin A (μg)	25	–	–	32.57	–	–	–
Vitamin D (μg)	0.3	0.8	–	0.8–1.86	0.8	0.8	–
Vitamin E (μg)	–	0.1	–	0.3–4.00	1.8	0.03	–
Vitamin B_2 (μg)	302.5	0.2	–	0.2–0.24	0.2	–	–
Vitamin B_{12} (μg)	0.1–1	0.4	–	0.4–0.68	0.4	0.4	–
References	Padma et al. (2019); Vanga and Raghavan (2018)	Salama et al. (2011); Syed et al. (2020)	Salama et al. (2011)	Vanga et al. (2018)	Alozie and Ukpong (2015); Hasan (2012)	Nadeeshani et al. (2015)	Vahanvaty (2009)

TABLE 8.2 Nutritional Component of Millets Milk in g/100 g (Sheela et al., 2018)

Nutrient	Foxtail Millet	Kodo Millet	Proso Millet	Barnyard Millet
Moisture (%)	42.29	38.35	41.25	43.15
Protein (%)	6.85	7.93	7.05	8.12
Fat (%)	0.50	0.56	0.70	0.53
Carbohydrate (%)	56.04	60.09	57.85	56.98
Reducing sugars (%)	2.02	0.50	2.08	1.09
Fiber (%)	2.43	2.36	1.7	1.87
Ash (%)	0.14	0.43	0.42	0.41
Phytic acid (%)	15.31	21.97	13.97	10.01

8.5 TRADITIONAL AND MODERN METHODS OF MILK PRODUCTION

Plant-based milk is produced through the extraction of coarse material from plants and breaking it down to reduce its particle size in water combined with homogenization (Reyes-Jurado et al., 2021). Dry and wet milling of cereals, legumes, and nuts is required before the separation of solids. The proper production technique is a very important factor for the nature of the product. The arrangement of components, stability, viscosity, particle size, as well as sensory characteristics like color and texture all, depending on the processing conditions in the case of plant-based milk beverages.

8.5.1 TRADITIONAL METHOD

The coarse particles present in plant-based milk are separated by filtration, decantation, and centrifugation (Reyes-Jurado et al., 2021). Heating and boiling are the main step in the production of plant-based milk which can destroy the heat-sensitive compounds in the product. Most plant-based beverages are traditionally produced which starts by milling pre-soaked seeds (soybean, oats, peanuts, etc.) followed by extracting the seed material in water, separating coarse material from it, homogenization of the colloidal suspension, and heat treatment for the stability of the product (Jeske et al., 2018b). The raw material is changed into slurry by grinding the seeds and then straining them to get desirable particle size. Ultra-high-temperature processing is used for the pasteurization of these milk alternatives (Reyes-Jurado et al., 2021). UHT in combination with aseptic packaging is used to attain the utmost stability.

8.5.2 MODERN METHODS

The modern methods of processing, such as high-pressure processing, high-pressure homogenization (HPH), ultrasound, and pulse electric field have been in trend now a day. The main reason behind this is the presence of thermo-labile and functional components in plant-based milk which is prone to potential loss during conventional processing methods (Munekata et al., 2020). These methods help lower the quality changes in the products and enhance their stability and shelf-life. Various researchers have reported the production and processing of plant-based milk like soy milk, almond milk, oat milk, hazelnut milk, and legume-based milk by using modern technology.

8.5.2.1 HIGH-HYDROSTATIC-PRESSURE PROCESSING

The HPP technique is very effective in reducing allergenic characteristics present in plant-based milk (Munekata et al., 2020). HPP has been reported to modify the characteristics and nature of protein, carbohydrate, and fat as it affects the non-covalent bonds of compounds. A range of 400–600 MPa is effective for the processing of plant-based milk (Dhakal et al., 2016).

8.5.2.2 HIGH-PRESSURE HOMOGENIZATION (HPH) AND ULTRA-HIGH-PRESSURE HOMOGENIZATION (UHPH)

For the production of plant-based milk, this technique is the most preferred as it reduces the microbial load similar to pasteurization. This method is of utmost importance to acquire emulsion stability as well as minimize the loss of volatile components of plant-based (Munekata et al., 2020). UHPH techniques help reduce the size of fat globules and other dispersed particles present in plant-based milk and thus produce a stable emulsion of fine texture (Toro-Funes et al., 2014a). The beany flavor of soymilk can also be reduced by this process and give the result better than the conventional pasteurization process (Reyes-Jurado et al., 2021).

8.5.2.3 PULSE ELECTRIC FIELD

A pulse electric field is applied to inactivate microorganisms and thus preserve the flavor of plant-based milk for a longer time (Munekata et al.,

2020). PEF is very helpful in stabilizing the fat globules in plant-based milk, aggregation of protein and fats as well as excludes phase separation. The inactivation of the lipoxygenase enzyme in some milk alternatives can also be done by using pulse electric fields (Reyes-Jurado et al., 2021).

8.5.2.4 HIGH-INTENSITY PULSED ELECTRIC FIELDS

The main reason for opting HIPEF is due to the positive effects such as enhanced shelf-life and minimum loss of quality including the flavor of the product. This technique can also eliminate the microbial load from the product more effectively than traditional heat treatment (Morales-De et al., 2010).

8.5.2.6 ULTRASONICATION

Ultrasound is used in plant-based milk for the stability of emulsion and uniform dispersion of particles. Ultrasonication also acts as a potential technology for the homogenization of dairy milk, but its function related to plant-based milk is not explored too much extent (Munekata et al., 2020).

8.5.2.7 OHMIC HEATING

Resistive heating or Ohmic heating is known for uniform heating, less processing time, optimum sterilization effect, and high flavor retention. This technique is very helpful in the production of high-protein dairy alternatives, as it does not change the structural configuration and functional properties of the protein. This method of production is also effective in reducing chymotrypsin activity in soymilk (Reyes-Jurado et al., 2021).

8.6 FUNCTIONAL COMPONENTS IN PLANT-BASED MILK AND THEIR HEALTH BENEFITS

8.6.1 CEREAL-BASED MILK

Cereal grains are a valuable source of dietary fiber, vitamins, minerals, and other phytochemicals. Milk developed from cereal grains varies remarkably in its protein, fat, and carbohydrate content. The amount of protein in cereal-based milk is lower, and the amount of carbohydrates is higher than that in dairy milk. Among cereals, rice is the major crop utilized worldwide that possesses various free phenolic acids like protocatechuic acid (38.43–1519 µg/g), caffeic acid (16.8–311.67 µg/g), vanillic acid (9.0–189.83 µg/g), p-coumaric acid (2.57–77.1 µg/g), ferulic acid (8.17–184.45 µg/g) and sinapic acid (6.03–134.93 µg/g) (Biswas et al., 2011). These phenolic compounds present in rice possess broad antioxidant activity. Functional components like phytosterols, especially β-sitosterol and γ-oryzanol, present in rice and its milk lower cholesterol, control hypertension, and act as an antidiabetic, anti-inflammatory, and antioxidant (Biswas et al., 2011). Rice milk is free of lactose content, which makes it a good option for lactose-intolerant individuals. In oats, the presence of β-glucan is associated with a decreased level of blood glucose. Oat milk (both fermented and non-fermented) substantially helps in the reduction of the level of glucose, serum total cholesterol, and total glycerides along with LDL cholesterol in diabetic rats (feeding Alloxan + standard diet + 5 mL oat milk given by epigastric tube for 4 weeks) with a significant increase in HDLs (Gohari et al., 2019). The existence of β-glucan, which promotes fecal cholesterol and bile acid excretion while slowing lipid absorption, might explain oat milk's hypolipidemic action. Furthermore, β-glucan inhibits endogenous cholesterol production by suppressing insulin release. However, the cholesterol-lowering effect of oats may be due to their lipid and protein composition (Gohari et al., 2019). Furthermore, they concluded that treatment with oat milk (both fermented and non-fermented types) led to the cure of hyperlipidemia and hyperglycemia in diabetic rats.

Similar to oats, barley also consists of β-glucans that have health advantages such as reducing blood cholesterol and lowering the glycemic index (GI). Barley grains are also rich in phytoestrogens, vitamin E, and phenolic antioxidants, all of which have biological activity and can boost the nutritive value of products derived from these cereals substantially. Only barley grain contained γ-tocotrienol (one of the isomers of vitamin E), among all cereal crops. The combination of these and many other bioactive substances in oats and barley makes them essential items for both patients and healthy people suffering from a variety of metabolic diseases (Shvachko et al., 2021). Among cereals, sweet corn exhibits higher antioxidant activity due to the presence of many bioactive components like lutein, frolic acid, and phenolic compounds including flavonoids in its grains (Kljak et al., 2015). Corn milk is also reported to contain phenolic acids, such as 7.07 mg GAE/100 g of milk (Ateteallah et al., 2019). Corn milk also provides a lower amount of energy with no cholesterols which help in controlling obesity and related complications. Millet milk can also be chosen above other plant milk because of its nutritional advantages. It has high protein content and low-calorie content (Raajeswari and Nithya, 2016). Millets are high in phytochemicals such as phenolics, sterols, lignans, resistant starch (RS), β-glucan, and phytates. Polyphenols are phenolic acids, tannins, and flavonoids (Keservani and Sharma, 2014; Keservani et al., 2016a, b) that serve as antioxidants in tiny amounts (Ambati and Sucharitha, 2019). Millets lower α-glucosidase and pancreatic amylase levels, lowering postprandial hyperglycemia and decreasing complex carbohydrate enzymatic hydrolysis. Due to the activities of these phytochemicals, millet and its drinks have functional and health-promoting benefits, including antidiabetic, anti-obesity, and CVD, and play a function in the immunological system of humans (Ambati and Sucharitha, 2019). In streptozotocin (STZ)-induced diabetic rats, finger millet-enriched probiotic fermented milk has a significant antidiabetic and hypolipidemic effect. It was also successful in treating other diabetes problems, including polyphagia and polydipsia, and had a glucose-lowering impact similar to the antidiabetic medication metformin (Chaudhary et al., 2020).

8.6.2 LEGUME-BASED MILK

Legume seeds in the human diet are also acknowledged for their nutritional and health-beneficial properties due to the presence of various forms of bioactive compounds. These components include phenolic compounds, phytosterols, RS, oligosaccharides, saponins, and so forth. Soybean is one of the major cheap sources of quality protein and has many health benefits. The benefits of soy milk are attributed to its components, which include soybean protein content, fiber, saponins, protein peptides, and majorly due to isoflavones content. Soy milk is reported to affect blood serum lipid levels, as given in Table 8.3. The effect of feeding soymilk to weanling male albino rats was shown to decrease serum lipids such as total cholesterol, phospholipids, and triacylglycerols while the increase in HDL-cholesterol which might be due to low levels of sterols, and high amount of proteins in soy milk. Soy milk may help to reduce the risk of diabetes and its consequences, such as weight reduction, reduced blood glucose, serum lipids level, oxidative stress level as well as control of inflammation (Mohammadi et al., 2015). In addition, fermentation improves the efficiency of soy milk. Soy milk and fermented soymilk reduce blood glucose (in soy milk-fed rats from 149.73 to 143.64 and 153.08 to 133.92 in fermented soymilk-fed rats), total cholesterol, triglyceride, and malondialdehyde (MDA) concentrations. In the fermented soy milk-fed rats, red blood cells, body weight, hemoglobin, hematocrit, and superoxide dismutase all increased significantly (Mohammadi et al., 2015).

Peanuts are also regarded as functional foods because they include a variety of bioactive components that are beneficial to human nutrition, including peanut protein, unsaturated fatty acids (80% of total fat content) like oleic and linoleic acids, fat-soluble vitamins, and antioxidants known as tocopherol (Settaluri et al., 2012). Moreover, peanut milk is a vital source of plant protein, and it provides all the vital amino acids required for body development. Tryptophan, phenylalanine, tyrosine, and methionine are all essential amino acids present in peanuts which are essential in the prevention of protein-calorie mal-nutrition disorders (Kwashiorkor and Marasmus). Peanut milk also contains antioxidants, phytosterols (which prevent cholesterol absorption from the diet), and some important polyphenolic compounds such as phenolic acids, flavonoids, and resveratrol which protect against oxidative stress, and illnesses such as stroke, coronary artery disease cancer (Settaluri et al., 2012).

As compared to other legume crops, lupin seeds contain substantial quantities of phytochemicals such as polyphenols (19.4 mg GAE/100 g), phytosterols (53.6 mg), tocopherols (0.47 mg), and 1.74 mg of squalene or triterpene (Kalogeropoulos et al., 2010). Lutein, apigenin, and diosmetin are the most often discovered flavones, whereas genistein is the most important isoflavone. They act as antioxidants by trapping singlet oxygen and also exhibit non-steroidal phytoestrogenic activity (Khan et al., 2015). On the other hand, high fiber and protein content aid in the prevention of obesity and diabetes linked to a rise in BMI (body mass index), whilst antioxidant, anti-hyperlipidemic, and anti-inflammatory properties of lupin phytochemicals protect against a variety of chronic illnesses (Khan et al., 2015). Chickpeas are a low-cost protein source with an essential amino acid profile that is well-balanced. As a result, specific amino acids play a crucial role in the peptide fraction's antioxidant properties. This legume is an economically vital source of dietary fiber along with bioactive substances (e.g., saponins, phytosterols, oligosaccharides) when combined with their low GI, chickpea might help in the risk reduction of CVDs. Polyphenols in chickpea seeds like ferulic and p-coumaric acids are present in small amounts that have already been demonstrated to lower lipid levels (Jukanti et al., 2012). The two significant phenolic compounds found in chickpeas are isoflavones biochanin A [5,7-dihydroxy-4'-methoxyisoflavone] and formononetin [7-hydroxy-4'-methoxyisoflavone]. Furthermore, cowpea is a good source of proteins, dietary fibers, minerals, and polyphenolics including B group vitamins. The effect of RS with dietary fiber, lowers appetite and diminishes insulin reactions, which is associated with cowpea's low GI. Phenolic acids and flavonol glycosides are the primary polyphenols found in all cowpea cultivars (Awika and Duodu, 2017). Anthocyanins and/or flavan-3-ols are found in some cultivars. Ferulic acid with coumaric acid is the most profuse phenolic acid in cowpea besides gallic acid. Extracts of cowpea polyphenols have already been shown to have anti-inflammatory activity, with potential benefits in the treatment of diabetes, cancer, and CVDs (Awika and Duodu, 2017). Besides this, the hypolipidemic effects of the hydrolysis products of cowpea protein are mediated by the destruction of cholesterol micelles in the gastrointestinal tract and binding of bile acid, along with the change in hepatic and adipocytic enzyme activity and gene expression of lipogenic proteins, all of which can amend lipid profiles and delay the onset of CVD (Marques et al., 2015).

8.6.3 TREE NUT-BASED MILK

Besides cereals and legumes, nuts are also a predominant source of essential nutrients, especially fatty acids (mono and poly-unsaturated fatty acid) except coconut which is rich in saturated fatty acids and fat-soluble constituents such as tocotrienols, phytosterols, alkyl phenols, carotenoids, and so forth. Other than these, nuts are a relatively good source of phenolic acids, flavonoids, coumarins, lignans, stilbenes (resveratrol), tannins, and so forth. Walnuts, almonds, and hazelnuts possess a diverse group of polyphenolic compounds, especially proanthocyanidins (condensed tannins) and hydrolyzable tannins. Among them, hazelnuts had the highest value of 491 mg/100 g for proanthocyanidins (Lainas et al., 2016). A few compounds like isoflavones are only recorded in pistachios, whereas dihydrochalcones are in hazelnuts. These nuts ameliorate the effects of neuroinflammation and apoptosis and help improve memory, learning ability, and anxiety. Supplementation with walnuts and almonds at 400 mg/kg/day for 28 days significantly reduced cadmium- and scopolamine-induced memory impairment in rats, attributed to their antioxidant activity (Haider et al., 2018). Almonds and walnuts are also essential sources of tocopherol and antioxidants. Even coconuts and hazelnuts do possess antioxidant activity that helps from free radical damage. However, walnuts consisting of omega-3 fatty acids show radical scavenging activity towards Colo-205 human cancer cells (Anjum et al., 2017). Most of the epidemiological research done using nuts is allied with CVD due to their fatty acid composition. As PUFAs are effective in reducing the risk associated with coronary disease, nitric acid synthase gene expression, and blocking inflammation. Ellagic acid present in walnut protects against inflammation by lowering inflammatory molecules like nitric oxide, MDA, tumor necrosis alpha-factor, nuclear factor –κB expression, and cyclooxygenase (El-Shitany et al., 2014).

Almond, walnut, and coconut can increase HDL levels while decreasing LDL cholesterol levels in the bloodstream (Ekanayaka et al., 2018). Almonds and walnut consist of phytosterol and α-linolenic acid while coconut contains lauric acid which reflects a cholesterol-lowering effect (Ekanayaka et al., 2018). Jenkins et al. (2008) reported that supplementation with almonds at 73 g/day significantly decreased serum concentrations of MDA and creatine-adjusted urinary isoprostane. The almond has also been acknowledged for reducing post-prandial glycemia, insulinemia, and oxidative stress. It possesses a protective effect against virus infection due to the production of cytokines at a higher concentration. In addition, evidence suggests that consumption of nuts helps in reducing blood pressure levels and the risk associated with type-2 diabetes. Certain clinical trials by Domènech et al. (2019) were performed using walnut among 236 hypertensive patients and recorded that individuals were able to reduce systolic BP (–4.61 mm Hg) in the whole cohort and 24 h ambulatory systolic BP (–8.5 mm Hg). Coconut milk consists of monolaurin which has antimicrobial activity as well as monoglyceride that destroys lipid-coated viruses like HIV (Suyitno, 2003). On the other hand, cashew nuts are related to the presence of alkaloids, saponins, phenolic acid, and tannins which are responsible for antimicrobial activity. Cashew nuts are also used in managing colitis by reducing neutrophil infiltration, pro-inflammatory cytokines, and so forth (Siracusa et al., 2020). The presence of bioactive components in nuts possesses various therapeutic properties and therefore, milk prepared from these nuts also reflects health-promoting effects which are highlighted in Table 8.3.

TABLE 8.3 Therapeutic Properties of Plant-based Milk with Their Associated Bioactive Components

Product	Methodology	Bioactive Compound	Activity	Remark	References
Oat milk (fermented + non-fermented).	Diabetic rats (Alloxan + Standard diet + 5 mL oat milk by epi gastric tube for 4 weeks).	β-Glucan, functional lipids, and proteins.	Curing of hyperglycemia and hyperlipidemia.	Reduction in level of glucose, serum total cholesterol, triglyceride, low-density lipoproteins (LDLs) cholesterol.	Gohari et al. (2019)
Millet milk	Weanling male Albino rats with supplemented millet milk for 6 weeks.	Calcium	Substitute milk for lactose-intolerant infants.	Calcium content was 80 mg/100 ml in millet milk.	Ambati et al. (2019)
Soy milk (fermented + non-fermented).	Sprague-Dawley diabetic rats 1 mL/day of soymilk for 28 days.	Soy protein and isoflavones.	Hypoglycemic and antidiabetic effects.	Decrease blood glucose level, total cholesterol, triglyceride, malondialdehyde (MDA).	Mohammadi et al. (2015)
Fermented almond milk tea.	Diabetes mellitus (DM)-induced rat with 50 mg/Kg STZ dose.	Saponins, flavonoids, phenols (373 μg/mL).	Antidiabetic effect.	Decrease in plasma glucose, triglyceride, and cholesterol level while the increase in serum insulin.	Sukumaran et al. (2021)
Almond milk.	Biofilm formation by *Streptococcus mutans*.	Lauric acid, capric acid, polyphenols.	The anti-carcinogenic potential of unsweetened almond milk.	Soymilk is highly carcinogenic.	Lee et al. (2018)
Cashew, almond, and coconut milk.	Biofilm formation by *Streptococcus mutans*.	Phytochemicals.	Anti-carcinogenic activity.	Sweetened milk was more effective.	Huang et al. (2019)
Coconut milk kefir.	Different concentrations of kefir were tested against breast cancer cell lines and gram-positive and negative bacteria.	Organic acids, peptides (bacteriocins), ethanol, and diacetyl.	Antibacterial, antifungal, anti-carcinogenic.	Activation of the immune response, delay of early-stage tumor production, and enzyme activity.	Lakshmi et al. (2017)

TABLE 8.3 *(Continued)*

Product	Methodology	Bioactive Compound	Activity	Remark	References
Coconut milk.	FRAP, ORAC, and DPPH analysis.	Phenolic compound.	Antioxidant activity.	Highest activity in comparison to goat and cow milk.	Alyaqoubi et al. (2015)
Coconut milk.	Administration of milk in ethanol-induced gastric ulcer rat.	Tanin, saponin, flavonoids.	Anti-ulcerative potential.	Milk caused alteration of the gastric pylorus.	Tolulope et al. (2019)
Quinoa milk.	Determination of glycemic index.	Phytoecdysteriod.	Antidiabetic activity.	Higher protein and lower glycemic index.	Pineli et al. (2015)

8.6.4 OIL SEED-BASED MILK

With the increasing demand for functional products, edible oil seeds have also become a target for researchers for quality and utilizing them for the preparation of food products. The milk developed from these edible oil seeds requires emulsifiers and stabilizers for its stability. Likewise, nuts and oil seeds are also a rich source of essential fatty acids, which play a major role in brain development and spatial memory. Other than these components, oil seeds do possess flavonols, phenolic acids, procyanidins, phytosterols, and so on, which are associated with lowering the risk of certain degenerative diseases like CVD, cancer, and diabetes. Sesame is a good source of PUFAs along with natural antioxidants such as lignin (sesamin and sesamolin) and tocopherol. Even flaxseed is also considered to be a valuable source of α-linolenic acid and lignin. It is evident from studies that lignan has antioxidant and anti-carcinogenic activities and increases hepatic mitochondrial and peroxisomal fatty acid oxidation rates. Furthermore, tocopherol lowers cholesterol levels which have tremendously increased the application of sesame seeds in various healthy product forms as asserted for tumor prevention as well as liver and heart protection (Cheng et al., 2006). According to the investigation by Qin et al. (2019), ingestion of sesamol at 100 mg/kg body weight exerts an anti-obesity effect and improves metabolic disorders associated with obesity. Hamza and Mahmoud (2013) also observed that the presence of phytoestrogens in sesame seeds helps ameliorate cardiotoxicity. On the other hand, the administration of 30 g of flax seed for 6 months continuously helps in reducing systolic and diastolic blood pressure due to its high lignin concentration (Rodriguez-Leyva et al., 2013). In addition, flax seeds act as an anti-depressant and show a protective effect on cortical neurons against NMDA-induced neutron toxicity (Li et al., 2012). Moreover, flax seeds, as well as hemp seeds, are also responsible for protecting women against breast cancer and also reducing the risk of mortality women. Certain clinical studies revealed that postmenopausal breast cancer women possessed a higher concentration of enterolactone in the blood, which lowered the risk of breast cancer. β-sitosterol in hemp and flax seeds plays an important role in reducing hypercholesterolemia, the occurrence of colon cancer as well as exhibiting antimicrobial and anti-inflammatory effects.

On the other hand, sunflower seeds consist of an adequate amount of selenium that has free radical scavenging activity and prevents cancer development by inhibiting the growth of cancer cells and inducing apoptosis. Sunflower seed peptide helps in increasing the bioavailability of calcium by transporting it into Caco-2 cells and hence can be used as a calcium supplement (Bao et al., 2021). Certain phenolic acids (caffeic acid, gallic acid, ferulic acid, and chlorogenic acid (CGA)) and flavonoids (heliannone, quercetin, and kaempferol) compound identified in sunflower seeds prevent diabetes, lower hypertension, and inflammatory activities as well as help with wound healing (Gad et al., 2018). Although no literature has been cited related to oil seed-based milk acting against chronic diseases using hemp milk. It can be said that the associated milk consumed will have a health-promoting effect.

8.6.5 PSEUDOCEREAL-BASED MILK

Pseudo-cereal act as an important resource exceptionally for higher nutritional as well as phytochemical characteristics with good quality proteins. The major crops it includes are quinoa, amaranth, and buck-wheat which are gluten-free and hence, can easily be incorporated into the diets of people suffering from celiac disease. The consumption of 50 g of quinoa for 6 weeks was effective in tolerating celiac disease among patients (Zevallos et al., 2014). These grains possess various bioactive ingredients such as polyphenols, phytosterols, sapogenins (bioavailability form of saponins), phytoecdysteroids, betalains, bioactive peptides, and so forth. A diverse group of saponins is majorly present in quinoa and amaranth seeds that reveal antifungal, antiviral, anti-proliferative, and hepatoprotective activities. The phytoecdysteriod available in quinoa significantly improves the anabolic/catabolic hormone status (testos-terone and cortisol) as well as helps in treating metabolic syndrome and post-menopausal disorder (Foucault et al., 2014). The polysaccharides (PS) present in quinoa show antioxidant activity and significantly have immunomodulatory capacity. In quinoa, certain polysaccharide fractions also contain galacturonic acid that suppresses nitric acid production and has a cytotoxic effect on the human liver and breast cancer (Hu et al., 2017). The oral administration of amaranth seeds was recorded to have high ACE inhibiting activity in spontaneously hypertensive rats. Further, the antidiabetic effect of amaranth seeds is associated with the inhibition of dipeptidyl peptidase IV while the antithrombotic effect is linked with the hydrolysis of Albs, Globs, and Gluts that inhibit blood clot forma-tion (Tovar-Pérez et al., 2019). Furthermore, buckwheat is also rich in phenolic acids, flavonoids (rutin, vitexin, orientin, quercetin, isovitexin, and isoorientin), phytosterols, fagopyrins, and so on, which have strong anti-inflammatory, anticancer, and protective effects. As per the report by Zhou et al. (2018) administration of Tartary buckwheat protein for 6 weeks modulates the intestinal microbiota and also exerts hypocholesterolemic and anti-inflammatory effects. Buckwheat protein also prevents gallstone formation and the occurrence of mammary and colon cancer by lowering cell proliferation and serum estradiol concentration. Nowadays, chia seeds have also become very popular among the population due to their phytochemical characteristics which exhibit protection from arteriosclerosis, inflammation, hypertension, diabetes, and cancer. Consumption of chia during pregnancy also develops the brain and retina of the fetus (2016). No work has been documented about milk hence, studies have to be undertaken in association with milk developed from these pseudocereals for protection from health-related disorders.

8.7 PLANT-BASED MILK: AN EMERGING PROBIOTIC CARRIER

According to WHO, probiotics are live microorganisms that when administered in sufficient amounts confer health benefits on the host. The microorganisms it includes are lactic acid bacteria (LAB) of the species *Lactobacillus*, *Bifidobacterium*, *Streptococcus*, and *Enterococcus*. In most cases, these probiotic bacteria are incorporated into the human diet in the form of fermented products like yogurt, cheese, kefir, and so on, developed from dairy milk. Moreover, plant-based milk extracted from cereals, pseudocereals, legumes, oil seeds, nuts, and so on, has now also been accounted for its use as a probiotic carrier. Cereals and legumes naturally harbor microorganisms that exhibit spontaneous fermentation for the production of traditional fermented products. However, when milk is formed from these plant sources, the existing microorganisms cause interference with the added probiotic bacteria. This leads to high microbial contamination while preparing probiotic drinks. Hence, probiotic fermentation provides a way of obtaining live probiotic bacteria, which not only improves the quality but also enhances the nutritional characteristics. Although dairy milk and its associated products are the major sources of probiotic bacteria, these microbes do not indicate favorable conditions for their growth in plant-based milk. However, a few recent studies have indicated some of the major plant-based milks, such as rice milk, soy milk, and coconut milk, act as a carrier matrix for probiotic fermentation. It is known that cereals and legumes are an inexpensive source of non-digestible oligosaccharides that act as prebiotics that break easily during fermentation and provide a symbiotic effect. Other than this, maize, oats, buckwheat, peanuts, hemp, cashew milk, and so on are now also being employed for the development of probiotic drinks. *Lactobacillus* and *Bifidobacterium* are the major probiotic strains that are used in plant-based milk; for example, *Bifidobacterium* is effectively grown in soy milk and, thereby, increases the protein content. The technique, type of bacteria, and the viable count of the bacteria used for the development of different plant-based probiotic milk are given in Table 8.4.

Probiotic fermentation also increases the concentration of bioactive compounds (polyphenols) as well as their bio-accessibility by converting the insoluble form of phenols into a more soluble form. Further, probiotic fermentation helps in the synthesis of vitamins such as vitamin K and B complex and reduces anti-nutritional compounds like phytate, cyanide, tannins, and so on, which is beneficial for human health. The LAB fermentation of soymilk releases bioactive peptides, which possess antihypertensive effects by inhibiting angiotensin-converting enzymes (ACEs) (Myagmardorj et al., 2018). The consumption of probiotic soymilk at 200 mL/day helps in modulating blood lipoprotein levels and is thus considered quite effective in managing diabetes complications and atherosclerotic risks (Feizollahzadeh et al., 2017). The investigation done by Veerichetty et al. (2018) revealed that fermented soya and flax seed milk exhibits an antidiabetic effect on alloxan-induced diabetic rats with an increase in insulin production in both normoglycemic and hyperglycemic rats. Hence, fermented soya and flax seeds can be replaced with medicines to reduce Type-1 diabetes. On the other hand, fermentation of almond milk using *L. rhamnosus* and *B. bifidum* revealed the energetic metabolism of enterocytes and increased the iron uptake by Caco-2 cells (Bernat et al., 2014a). Wansutha et al. (2018) developed probiotic almond milk using the LAB strain while Ravindran and Radhai Sri (2020) prepared probiotic oat milk using microencapsulated *L. plantarum* and reported that the milk exhibited antioxidant activity by inhibiting DPPH activity. In addition, the quinoa yoghurt beverage formulated by Obaroakpo et al. (2019) indicates maximum α-glucosidase and ACE inhibiting activity which thereby prevents diabetes and hypertension. Hence, it can be concluded that probiotic plant-based milk can also be incorporated into the human diet to manage metabolic imbalances.

TABLE 8.4 Techniques Used for the Development of Plant-based Probiotic Milk

Product	Probiotic Strain	Incubation Condition	Viability During Storage	Remark	References
Rice milk	*L. casei*, *B. longum*, *L. bulgaricus*, *S. thermophilus*, and *L. acidophilus*	37°C for 16 h	10^4–10^9 CFU/mL	The viable count of *S. thermophilus* increased on storage HDPE and LDPE bottles.	Padma et al. (2019)
Maize milk	*L. paracasei* and *S. cerevisiae*	37°C for 24 h	Around 10^6 CFU/mL during storage for 28 days at 4°C	Lactic and acetic acid production provides safety.	Menezes et al. (2018)
Oat milk	*L. plantarum*	30°C for 12 h	More than 10^7 CFU/mL on storage at 4°C for 7 weeks	Dietary fiber in oat possessed probiotic effect.	Wang et al. (2018)
Cowpea milk	*Lactobacillus acidophilus*, *Bifidobacterium animalis*, *Streptococcus thermophilus*.	45°C for 14 h	More than 10^7 CFU/mL	Use of *Bifidobacterium* genus leads to rejection of the product due to higher acidity and more viscous texture of product.	Aduol et al. (2020)
Soybean milk	*Lactobacillus*, *Lactococcus*, and yeast	12 h	More than 10^7 CFU/mL	The addition of inulin improved the microorganism efficiency.	dos Santos et al. (2019)
Cashew nut milk	*B. animalis*, *L. acidophilus* and *L. plantarum*	30°C	10^7 CFU/mL on storage for 30 days	*B. animalis* had non-significant effect on sensory characteristics.	Bruno et al. (2019)
Coconut milk	*L. acidophilus* and *L. fermentum*	37°C for 6 h	10^7–10^9 CFU/mL	Coconut milk blended with banana pulp provides prebiotic effect due to oligosaccharides.	Manoj (2018)
Almond milk	*L. reuteri* and *S. thermophilus*	37°C for 24 h	10^7 CFU/mL	Glucose generates prebiotic effect and inulin increase functional properties.	Bernat et al. (2014)
Hemp milk	*L. fermentum*, *L. plantarum* and *B. bifidu*	37°C for 24 h	10^7 CFU/mL	Prebiotic effect of hemp seeds.	Nissen et al. (2020)

TABLE 8.4 *(Continued)*

Product	Probiotic Strain	Incubation Condition	Viability During Storage	Remark	References
Flaxseed milk	*L. plantarum*	37°C for 4 h	–	Bioactive peptide (10 kDa) exhibits higher antioxidant, ACE inhibition, antimicrobial activity.	Sharma et al. (2020)
Sesame milk	*L. plantarum*	37°C for 18 h	10^9 CFU/mL	Antioxidant activity in milk without added sugar.	Fitrotin et al. (2015)
Quinoa milk	*L. plantarum*	30 for 48 h	Around 10^7 CFU/mL	Higher amount of starter culture inhibits the growth of enterococci present in native quinoa.	Paz et al. (2020)
Quinoa milk	*L. casei*	30°C for 24 h	10^6 CFU/mL	Germination of quinoa seed prior to milk production improves functional characteristics.	Obaroakpo et al. (2019)

8.7 CHALLENGES IN PRODUCTION, SHELF-LIFE, AND MARKETABILITY

Rapid growth and popularity of these plant-based dairy substitutes have been observed throughout the world. Presently, plant-based milk shares 7.4% of total dairy milk production and the proportion could increase to 18.50% by 2023 (Schiano et al., 2020). In particular, about 5% market value has been reported for non-dairy plant-based beverages (Munekata et al., 2020). In 2019, the global plant-based milk market was valued at USD 12.40 billion and estimated at USD 19.67 billion by 2023 with a 12% CAGR (Market Analysis Report, 2020). As per the report of transparency market research (TMR) the market for plant-based milk was valued at approximately US $14 billion in 2019 and forecasted for approximately 8% of its annual growth from 2019 to 2029 (TMR, 2019). The major reason behind this drastic shift from dairy to plant-based milk is the health of individuals or families, changing lifestyles and consumption patterns as well as an increasing preference for sustainable, organic, and plant-based foods. Plant-based milk includes various bioactive components such as omega-3 fatty acids, vitamins, polyphenols, and so on, which help in boosting the market. International companies like White Wave Food Company (US), Blue Diamond Growers (US), Coca-Cola Company (US), and Pureharvest (Australia) are working efficiently on the development of plant-based milk beverages (Market Analysis Report, 2020). However, some plant-based milk is costly and has low calcium, protein, and vitamin D content. To maintain the nutritional quality of milk, Palsgaard developed a calcium (0.20 to 0.30%) rich almond, or rice milk called Palsgaard® RecMilk 131 using emulsion stability that ranges between 0.18 and 0.20% to achieve a good consistency.

Plant-based milk has wide application in the food and beverage industries. But still, milk developed from plant sources brings about processing challenges due to its physiological composition and structure which differ significantly from raw dairy milk. It is observed that consumers expect that plant-based milk should be identical in terms of physicochemical characteristics and sensory attributes including creaminess, mouth feel, and color to dairy milk. During the production of plant-based milk, colloidal instability, off-flavor, and the presence of anti-nutritional factors are the major problems, along with lower shelf-life. Colloidal instability is dependent on dispersed phase particle size, processing method, and storage conditions. The dispersed particles include protein, fat, and starch granules which make the product unstable resulting in a sandy, chalky, or gritty mouth feel as well as causing sedimentation during storage. Traditionally, colloidal milk was used for size reduction, but this method resulted in microbial contamination. The most effective method, that is, ultra-high-pressure homogenization (UHPH) not only results in size reduction but also maintains the safety of milk from microbial spoilage (Valencia-Flores et al., 2013). Other than this, beverage industries have also found a suitable means of blending plant-based milk with plant oils, stabilizers, thickeners, emulsifiers, and other additives. The addition of stabilizers and thickeners creates a network required to suspend particles, increases viscosity, and improves mouth feel. However, in the case of milk obtained from cereals and pseudo-cereals, the presence of large amounts of starch gets gelatinized and forms a viscous paste when heated at 55–65°C which can be avoided by treating milk with α-amylase enzyme for liquefaction (Tano-Debrah et al., 2005).

Milk developed from plant sources develops an off flavor during storage, which is inherently available due to the presence of lipoxygenase enzyme and unsaturated fatty acid. The best example is soy milk, which is not acceptable due to its unpleasant beany flavor. The lipoxygenase enzyme catalyzes the lipid oxidation reaction that leads to the production of certain non-volatile hydroperoxide compounds. This compound further gets decomposed into n-hexanal and n-hexanol (Sethi et al., 2016). Different methods have been developed by researchers to improve the flavor quality of milk, such as high-temperature processing, alkali treatment, rapid hydration grinding, β-cyclodextrin entrapment, and so forth. Certain other novel techniques have also been employed like Ohmic heating, pulse electric field, high-pressure processing, cold plasma, and so on, that are effective in removing off-flavor from plant-based milk (Paul et al., 2020). In addition, blending legume milk with other cereals and nut milk was also quite effective in enhancing the nutritional quality and overall acceptability of the milk. Apart from the removal of off-flavor, legumes are also responsible for containing some anti-nutritional components such as trypsin and protease inhibitors, phytic acid, saponins, lecithin, and so on, which get incorporated into milk as well. The destruction of enzyme inhibitors such as trypsin and protease inhibitors are essential as it results in certain health issues such as pancreatic hypertrophy as well as decreased protein absorption. Hence, the processing techniques must be selected carefully so that the trypsin inhibitor gets denatured and other anti-nutritional compounds get removed completely with minimum destruction of nutrients. Various treatments, such as blanching, soaking, sprouting, steam injection, β-cyclodextrin entrapment, fermentation, and so on, have been employed for the removal of anti-nutritional factors. Besides, the shelf-life of plant-based milk is also an important factor because of its higher nutritional value that serves as a vehicle for the rapid growth of microorganisms. Although, thermal treatment has been followed to extend the shelf-life of the milk excessive heating has caused a detrimental effect on nutrients. Various techniques have been developed by using time-temperature combinations to obtain the desired product with less or no pathogenic bacteria. Heat treatments like pasteurization, sterilization, and UHT are being followed to increase the shelf-life of plant-based milk. In addition, novel techniques like high-pressure processing, ultrasonic treatment, and pulsed electric fields are also being used for the preservation of plant-based milk. Some of the important innovative techniques associated with overcoming the challenges of colloidal stability, removal of beany flavor and anti-nutrients as well as shelf-life extension are highlighted in Table 8.5.

8.8 CONSUMER PREFERENCES AS POTENTIAL HEALTH BEVERAGES

Over the years, the plant-based milk substitute has become very popular among consumers and various commercial products are available on the market. Consumers are willing to pay for plant-based milk due to the presence of nutritive components in it. In most studies, the consumption of plant-based milk by consumers was based on its health benefits. Consumers are shifting toward these alternatives but have not entirely substituted dairy consumption. According to the survey conducted by the International Food Information Council Foundation and Consumer Report, there is still confusion among a large population about the nutritional value of plant-based beverages (Stewart et al., 2020). Due to a lack of awareness, consumers are having difficulties understanding this concept. Most of the population consumes dairy and dairy products because they do so out of habit and like the flavor. And those who drink non-dairy beverages find them appealing as long as they like their flavor. Overall, many people opted for milk avoidance due to some health issues but didn't completely exclude it from the diet.

8.9 FUTURE OF PLANT-BASED MILK

In recent years, plant-based milk has shown solid growth and promise for the future. In the past 12 months, around 40% of the consumers have utilized more than one type of plant-based milk, which reflects that consumers need more

TABLE 8.5 Innovative Techniques Used to Overcome the Challenges During Milk Production

Challenge	Sample	Technique	Remark	References
Colloidal stability	Lentil milk	Dynamic-high pressure treatment (180 and 900 bar) followed by heat at 65°C/30 min or 85°C for 2 min.	Increased protein stability, reduction in particle size, more whitish appearance of milk at 900 bars.	Jeske et al. (2019)
	Peanut milk	High-pressure homogenization (HPH-150 and 300 MPa) and sterilization (121°C for 15 min).	Pressure homogenization had a significant effect on reducing particle size and consistency index while increasing flow behavior index. The use of roasted seed further enhances the stability of HPH and thermal treated milk.	Zaaboul et al. (2019)
	Soy milk	UHPH at 207 and 276 MPa, 121 and 145°C.	Absence of lipoxygenase enzyme, no aggregation of the particle during storage for 28 days.	Sidhu and Singh (2016)
	Soy milk	UHPH at 200 MPa, 55 and 75C.	Non-significant effect on isoflavone profile on UHPH treatment, minimum conversion to aglycone form during storage for 21 days at 4°C.	Toro-Funes et al. (2014b)
	Tiger nut milk	UHPH (200 and 300 MPa at 40°C) and homogenized-pasteurized (HP) at 18 MPa, 80°C for 15 s.	Maximum colloidal stability at 300 MPa but minimum fat oxidation at 200 MPa.	Codina-Torrella et al. (2017)
	Almond and hazelnut milk	UHPH at 172 MPa along with low heating (LH) at 85°C for 30 min or high heating (HH) at 121°C for 15 min.	UHPH treatment along with LH exhibit improved stability, and a significant decrease in particle size; HH caused aggregation of fat and protein molecule.	Bernat et al. (2014)
	Almond milk	UHPH at 300 MPa and 65°C, Pasteurization (90°C for 90 s), UHT (142°C for 6 s).	UHPH revealed maximum stability with decreased particle size, no off-flavor, and microbial growth was recorded.	Valencia-Flores et al. (2013)
Removal of any flavor	Soy milk	Soaking (4, 20, 30, 40, and 50°C) at 3 and 9 pH.	The pH of 9 and 25°C of temperature possessed minimum beany flavored components.	Li et al. (2019)
	Peanut milk	Alkali treatment (1% $NaHCO_3$), pressure blanching (121°C and 15 psi for 2, 3, and 5 min).	Pressure blanching for 3 min was more effective in improving the sensory quality of the milk.	Jain et al. (2013)

TABLE 8.5 *(Continued)*

Challenge	Sample	Technique	Remark	References
	Soy milk	Pulse electric field treatment at 0 to 15 kV for 547 μs.	Inactivation of lipoxygenase enzyme and 5.20 and 3.51 $\log_{10}$ reduction in *E. coli* and *Staphylococcus aureus*.	Li et al. (2013)
Anti-nutritional factor	Soy milk	Ultrasonic (400 W, 25 kHz for 1–16 min) and microwave (2450 MHz, 70 to 100°C for 2 to 10 min) treatment.	Around 52 and 84% reduction in TIA using ultrasonic and microwave (100°/10 min) treatment, respectively.	Vanga et al. (2020)
	Soy milk	Microwave-assisted heating at 540–810 W and 80–90°C.	Maximum inactivation occurred at 810 W and 90°C.	Varghese and Pare (2019)
	Soy milk	Dielectric-barrier discharge plasma treatment at 25 V for 21 min.	Inactivation of trypsin inhibitor due to conformational changes and oxidative modification.	Li et al. (2017)
	Soy milk	HPP (500–700 MPa) treatment at 45–65°C.	Significant effect on TIA, complete inactivation of lipoxygenase enzyme.	Manassero et al. (2016)
	Soy milk	Ohmic (220 V and 50 Hz), induction cooker, and electric strove heating.	More effective inactivation of trypsin and chymotrypsin inhibitor activity in ohmic heating but it caused slight protein aggregation.	Lu et al. (2015)
Shelf-life	Peanut milk	Ultrasonic (200, 300, and 400 W) and hydrodynamic cavitations (6, 8, and 10 bar).	Ultrasonic treatment at 500 W was much effect with maximum inactivation of yeast and mold as well as improved stability.	Salve et al. (2019)
	Soy milk	UHPH (70, 140, and 210 MPa) and storage at 4°C).	Maintain colloidal stability for 3 weeks.	Mukherjee et al. (2017)
	Almond milk	Pulse electric field (7, 14, 21, and 28 kV/cm) and thermal treatment (90°C for 60 s).	PEF treatment at 28 kV provides more colloidal and microbial stability for 28 days with higher retention of bioactive compounds.	Manzoor et al. (2020)
	Tiger nut milk	UHPH (200 and 300 MPa at 40°C) and HP treatment.	Increase in shelf-life up to 57 days at 300 MPa UHPH with minimum microbial growth and higher sensory characteristics.	Codina-Torrella et al. (2018)

variety. Thereby, plant-based milk offers the same with great potential. The increased driven force for the purchase, usage patterns, and growth potential of plant-based milk has created a great opportunity for the milk industry to the development of innovated products. It has also been observed that plant-based milk is not always the healthier beverage (low protein content) thus, manufacturers developed milk by addition of functional ingredients as well as protein for strengthening the health segment. As consumers look for milk products that are rich in protein and bioactive compounds. Oat milk has been commonly marketed due to its heart benefits properties. Furthermore, to achieve consumer acceptability progressive efforts have been made by the researchers are indeed expected from the scientific community to improve the product quality with a continuously changing behavior toward novel food products. Hence, it can be concluded that plant-based milk is a great topic for research in novel product development in food science and technology.

KEYWORDS

- **consumer preferences**
- **dairy consumption**
- **dairy products**
- **functional food**
- **health benefits**
- **health beverages**
- **milk alternates**
- **non-dairy beverages**
- **nutritional value**

REFERENCES

Aderibigbe, O. R., Ezekiel, O. O., Owolade, S. O., Korese, J. K., Sturm, B., & Hensel, O. (2022). Exploring the potentials of underutilized grain amaranth (*Amaranthus* spp.) along the value chain for food and nutrition security: A review. *Critical Reviews in Food Science and Nutrition*, *62*(3), 656–669.

Aduol, K. O., Onyango, A. N., & Imathiu, S. M. (2020). Proximate, microbial and sensory characteristics of cowpea milk fermented with probiotic starter cultures. *European Journal of Agriculture and Food Sciences*, *2*, 1–7.

Aguilera, Y., Martín-Cabrejas, M. A., Benítez, V., Mollá, E., López-Andréu, F. J., & Esteban, R. M. (2009). Changes in carbohydrate fraction during dehydration process of common legumes. *Journal of Food Composition and Analysis*, *22*(7, 8), 678–683.

Ahmadian-Kouchaksaraei, Z., Varidi, M., Varidi, M. J., & Pourazarang, H. (2014). Influence of processing conditions on the physicochemical and sensory properties of sesame milk: A novel nutritional beverage. *LWT–Food Science and Technology*, *57*(1), 299–305.

Alozie Yetunde, E., & Udofia, U. S. (2015). Nutritional and sensory properties of almond (*Prunus amygdalu* Var. *Dulcis*) seed milk. *World Journal of Dairy and Food Sciences*, *10*(2), 117–121.

Alyaqoubi, S., Abdullah, A., Samudi, M., Abdullah, N., Addai, Z. R., & Musa, K. H. (2015). Study of antioxidant activity and physicochemical properties of coconut milk (*Pati santan*) in Malaysia. *Journal of Chemical and Pharmaceutical Research*, *7*(4), 967–973.

Ambati, K., & Sucharitha, K. V. (2019). Millets: Review on nutritional profiles and health benefits. *International Journal of Recent Scientific Research*, *10*(7), 33943–33948.

Anjum, S., Gani, A., Ahmad, M., Shah, A., Masoodi, F. A., Shah, Y., & Gani, A. (2017). Antioxidant and antiproliferative activity of walnut extract (*Juglans regia* L.) processed by different methods and identification of compounds using GC/MS and LC/MS techniques. *Journal of Food Processing and Preservation*, *41*(1), e12756.

Argüelles-López, O. D., Reyes-Moreno, C., Gutiérrez-Dorado, R., Sánchez-Osuna, M. F., López-Cervantes, J., Cuevas-Rodríguez, E. O., Milán-Carrillo, J., & Perales-Sánchez, J. X. K. (2018). Functional beverages elaborated from amaranth and chia flours processed by germination and extrusion. *Biotecnia*, *20*(3), 135–145.

Ateteallah, A. H., & Osman, A. A. K. A. (2019). Antioxidant, some flavor components, microbiological and microstructure characteristics of corn milk yoghurt. *Food and Nutrition Sciences*, *10*(5), 551.

Atudorei, D., Stroe, S. G., & Codină, G. G. (2021). Impact of germination on the microstructural and physicochemical properties of different legume types. *Plants*, *10*(3), 592.

Awika, J. M., & Duodu, K. G. (2017). Bioactive polyphenols and peptides in cowpea (*Vigna unguiculata*) and their health promoting properties: A review. *Journal of Functional Foods*, *38*, 686–697.

Bao, X., Yuan, X., Feng, G., Zhang, M., & Ma, S. (2021). Structural characterization of calcium-binding sunflower seed and peanut peptides and enhanced calcium transport by calcium complexes in Caco-2 cells. *Journal of the Science of Food and Agriculture*, *101*(2), 794–804.

Benjamin, S., & Spener, F. (2009). Conjugated linoleic acids as functional food: An insight into their health benefits. *Nutrition & Metabolism*, *6*, 1–13.

Bernat, N., Cháfer, M., Chiralt, A., & González-Martínez, C. (2015). Development of a non-dairy probiotic fermented product based on almond milk and inulin. *Food Science and Technology International*, *21*(6), 440–453.

Bernat, N., Cháfer, M., Chiralt, A., & González-Martínez, C. (2014). Hazelnut milk fermentation using probiotic *Lactobacillus rhamnosus* GG and inulin. *International Journal of Food Science & Technology*, *49*(12), 2553–2562.

Bernat, N., Chafer, M., Rodríguez-García, J., Chiralt, A., & González-Martínez, C. (2015). Effect of high pressure homogenisation and heat treatment on physical properties and stability of almond and hazelnut milks. *LWT–Food Science and Technology*, *62*(1), 488–496.

Biswas, S., Sircar, D., Mitra, A., & De, B. (2011). Phenolic constituents and antioxidant properties of some varieties of Indian rice. *Nutrition & Food Science, 41(2), 123–135.*

Bolarinwa, I. F., Aruna, T. E., Adejuyitan, J. A., Akintayo, O. A., & Lawal, O. K. (2018). Development and quality evaluation of soy-walnut milk drinks. *International Food Research Journal, 25*(5), 2033–2041.

Bruno, L. M., Lima, J. R., Wurlitzer, N. J., & Rodrigues, T. C. (2019). Non-dairy cashew nut milk as a matrix to deliver probiotic bacteria. *Food Science and Technology, 40*, 604–607.

Chandel, R., Kumar, V., Kaur, R., Kumar, S., Kumar, A., Kumar, D., & Kapoor, S. (2023). Bioactive compounds, health benefits and valorization of *Pyrus pyrifolia* (sand pear): A review. *Nutrition & Food Science, 53(6)*.

Chaudhary, J. K., & Mudgal, S. (2020). Antidiabetic and hypolipidaemic action of finger millet (*Eleusine coracana*)-enriched probiotic fermented milk: An *in vivo* rat study. *Food Technology and Biotechnology, 58*(2), 192.

Cheng, F. C., Jinn, T. R., Hou, R. C., & Tzen, J. T. (2006). Neuroprotective effects of sesamin and sesamolin on gerbil brain in cerebral ischemia. *International Journal of Biomedical Science: IJBS, 2*(3), 284.

Codina-Torrella, I., Guamis, B., Ferragut, V., & Trujillo, A. J. (2017). Potential application of ultra-high pressure homogenization in the physico-chemical stabilization of tiger nuts' milk beverage. *Innovative Food Science & Emerging Technologies, 40*, 42–51.

Codina-Torrella, I., Guamis, B., Zamora, A., Quevedo, J. M., & Trujillo, A. J. (2018). Microbiological stabilization of tiger nuts' milk beverage using ultra-high pressure homogenization: A preliminary study on microbial shelf-life extension. *Food Microbiology, 69*, 143–150.

Craig, W. J., & Fresán, U. (2021). International analysis of the nutritional content and a review of health benefits of non-dairy plant-based beverages. *Nutrients, 13*(3), 842.

del Mar Contreras, M., Sanchez, D., Sevilla, M. Á., Recio, I., & Amigo, L. (2013). Resistance of casein-derived bioactive peptides to simulated gastrointestinal digestion. *International Dairy Journal, 32*(2), 71–78.

Deswal, A., Deora, N. S., & Mishra, H. N. (2014). Optimization of enzymatic production process of oat milk using response surface methodology. *Food and Bioprocess Technology, 7*, 610–618.

Dhakal, S., Giusti, M. M., & Balasubramaniam, V. M. (2016). Effect of high pressure processing on dispersive and aggregative properties of almond milk. *Journal of the Science of Food and Agriculture, 96*(11), 3821–3830.

Diarra, K., Nong, Z. G., & Jie, C. (2005). Peanut milk and peanut milk based products production: A review. *Critical Reviews in Food Science and Nutrition, 45*(5), 405–423.

Domènech, M., Serra-Mir, M., Roth, I., Freitas-Simoes, T., Valls-Pedret, C., Cofán, M., López, A., Sala-Vila, A., Calvo, C., Rajaram, S., & Sabaté, J. (2019). Effect of a walnut diet on office and 24-hour ambulatory blood pressure in elderly individuals: Findings from the WAHA randomized trial. *Hypertension, 73*(5), 1049–1057.

dos Santos, D. C., de Oliveira Filho, J. G., Santana, A. C. A., de Freitas, B. S. M., Silva, F. G., Takeuchi, K. P., & Egea, M. B. (2019). Optimization of soymilk fermentation with kefir and the addition of inulin: Physicochemical, sensory and technological characteristics. *LWT–Food Science and Technology, 104*, 30–37.

Ekanayaka, R. A. I., Ekanayaka, N. K., Perera, B., & De Silva, P. G. S. M. (2013). Impact of a traditional dietary supplement with coconut milk and soya milk on the lipid profile in normal free living subjects. *Journal of Nutrition and Metabolism, 2013*, 481068.

El-Agamy, E. I. (2007). The challenge of cow milk protein allergy. *Small Ruminant Research, 68*(1, 2), 64–72.

El-Shitany, N. A., El-Bastawissy, E. A., & El-Desoky, K. (2014). Ellagic acid protects against carrageenan-induced acute inflammation through inhibition of nuclear factor kappa B, inducible cyclooxygenase and proinflammatory cytokines and enhancement of interleukin-10 via an antioxidant mechanism. *International Immunopharmacology*, *19*(2), 290–299.

Fassio, F., Facioni, M. S., & Guagnini, F. (2018). Lactose maldigestion, malabsorption, and intolerance: A comprehensive review with a focus on current management and future perspectives. *Nutrients*, *10*(11), 1599.

Feizollahzadeh, S., Ghiasvand, R., Rezaei, A., Khanahmad, H., Sadeghi, A., & Hariri, M. (2017). Effect of probiotic soy milk on serum levels of adiponectin, inflammatory mediators, lipid profile, and fasting blood glucose among patients with type II diabetes mellitus. *Probiotics and Antimicrobial Proteins*, *9*, 41–47.

Fitrotin, U., Utami, T., Hastuti, P., & Santoso, U. (2015). Antioxidant properties of fermented sesame milk using *Lactobacillus plantarum* Dad 13. *International Research Journal of Biological Sciences*, *4*(6), 56–61.

Foucault, A. S., Even, P., Lafont, R., Dioh, W., Veillet, S., Tomé, D., Huneau, J. F., Hermier, D., & Quignard-Boulangé, A. (2014). Quinoa extract enriched in 20-hydroxyecdysone affects energy homeostasis and intestinal fat absorption in mice fed a high-fat diet. *Physiology & Behavior*, *128*, 226–231.

Gad, H. A., & El-Ahmady, S. H. (2018). Prediction of thymoquinone content in black seed oil using multivariate analysis: An efficient model for its quality assessment. *Industrial Crops and Products*, *124*, 626–632.

Gohari, S. T., El-Batawy, O. I., & Mahdy, S. M. (2019). Biological attributes of unfermented and fermented oat milk compared to cow milk in alloxan-induced diabetic rats. *Journal of Food and Dairy Sciences*, *10*(3), 61–69.

Haider, S., Batool, Z., Ahmad, S., Siddiqui, R. A., & Haleem, D. J. (2018). Walnut supplementation reverses the scopolamine-induced memory impairment by restoration of cholinergic function via mitigating oxidative stress in rats: A potential therapeutic intervention for age-related neurodegenerative disorders. *Metabolic Brain Disease*, *33*, 39–51.

Hamza, A. H., & Mahmoud, R. H. (2013). Soy milk and sesame seeds (phytoestrogens) ameliorate cardiotoxicity induced by adriamycin in experimental animals. *American Journal of Research Communication*, *1*(10), 1–15.

Hasan, N. A. (2012). Almond milk production and study of quality characteristics. *Journal of Academia*, *2*(1), 1–8.

Hu, Y., Zhang, J., Zou, L., Fu, C., Li, P., & Zhao, G. (2017). Chemical characterization, antioxidant, immune-regulating, and anticancer activities of a novel bioactive polysaccharide from *Chenopodium quinoa* seeds. *International Journal of Biological Macromolecules*, *99*, 622–629.

Huang, Y., Thompson, T., Wang, Y., Yu, Q., Zhu, L., Xu, X., Wen, Z. T., & Townsend, J. A. (2019). Analysis of cariogenic potential of alternative milk beverages by *in vitro Streptococcus mutans* biofilm model and *ex vivo* caries model. *Archives of Oral Biology*, *105*, 52–58.

Iid, I. I., Kumar, S., Shukla, S., Kumar, V., & Sharma, R. (2020). Putative antidiabetic herbal food ingredients: Nutra/functional properties, bioavailability and effect on metabolic pathways. *Trends in Food Science & Technology*, *97*, 317–340.

Isaac-Bamgboye, F. J., Edema, M. O., & Oshundahunsi, O. (2019). Nutritional quality, physico-chemical properties and sensory evaluation of Amaranth-Kunu produced from fermented grain amaranth (*Amaranthus hybridus*). *Annals of Food Science and Technology, 20*(2), 322–331.

Jaglan, P., Kumar, V., Suthar, P., Paul, A. A., & Kumar, S. (2020). Recent advances in the profiling and characterization of antioxidants. In P. Kaur (Ed.), *Oxidative Stress and Natural Antioxidants* (pp. 172–208). Bentham Science.

Jain, P., Yadav, D. N., Rajput, H., & Bhatt, D. K. (2013). Effect of pressure blanching on sensory and proximate composition of peanut milk. *Journal of Food Science and Technology, 50*, 605–608.

Jenkins, D. J., Kendall, C. W., Marchie, A., Josse, A. R., Nguyen, T. H., Faulkner, D. A., Lapsley, K. G., & Blumberg, J. (2008). Almonds reduce biomarkers of lipid peroxidation in older hyperlipidemic subjects. *The Journal of Nutrition, 138*(5), 908–913.

Jeske, S., Bez, J., Arendt, E. K., & Zannini, E. (2019). Formation, stability, and sensory characteristics of a lentil-based milk substitute as affected by homogenization and pasteurization. *European Food Research and Technology, 245*, 1519–1531.

Jeske, S., Zannini, E., & Arendt, E. K. (2018). Past, present and future: The strength of plant-based dairy substitutes based on gluten-free raw materials. *Food Research International, 110*, 42–51.

Jeske, S., Zannini, E., Cronin, M. F., & Arendt, E. K. (2018b). Impact of protease and amylase treatment on proteins and the product quality of a quinoa-based milk substitute. *Food & Function, 9*(6), 3500–3508.

Jiang, S. Y., Ma, A., Xie, L., & Ramachandran, S. (2016). Improving protein content and quality by over-expressing artificially synthetic fusion proteins with high lysine and threonine constituent in rice plants. *Scientific Reports, 6*(1), 34427.

Jukanti, A. K., Gaur, P. M., Gowda, C. L. L., & Chibbar, R. N. (2012). Nutritional quality and health benefits of chickpea (*Cicer arietinum* L.): A review. *British Journal of Nutrition, 108*(S1), S11–S26.

Kalogeropoulos, N., Chiou, A., Ioannou, M., Karathanos, V. T., Hassapidou, M., & Andrikopoulos, N. K. (2010). Nutritional evaluation and bioactive microconstituents (phytosterols, tocopherols, polyphenols, triterpenic acids) in cooked dry legumes usually consumed in the Mediterranean countries. *Food Chemistry, 121*(3), 682–690.

Kaur, I., & Tanwar, B. (2016). Quinoa beverages: Formulation, processing and potential health benefits. *Romanian Journal of Diabetes Nutrition and Metabolic Diseases, 23*(2), 215–225.

Keservani, R. K., & Sharma, A. K. (2014). Flavonoids: Emerging trends and potential health benefits. *Journal of Chinese Pharmaceutical Sciences, 23*(12), 815.

Keservani, R. K., Sharma, A. K., & Kesharwani, R. K. (2016a). Nutraceutical and functional foods for cardiovascular health. In *Food Process Engineering* (pp. 291–312). Apple Academic Press, CRC Press.

Keservani, R. K., Sharma, A. K., & Kesharwani, R. K. (2016b). Medicinal effect of nutraceutical fruits for cognition and brain health. *Scientifica, 2016*, 3109254. https://doi.org/10.1155/2016/3109254.

Khan, M. K., Karnpanit, W., Nasar-Abbas, S. M., Huma, Z. E., & Jayasena, V. (2015). Phytochemical composition and bioactivities of lupin: A review. *International Journal of Food Science & Technology, 50*(9), 2004–2012.

Kim, S. L., Kim, S. K., & Park, C. H. (2002). Comparisons of lipid, fatty acids, and tocopherols of different buckwheat species. *Food Science and Biotechnology, 11*(4), 332–336.

Kinney, A. J. (2003). Engineering soybeans for food and health. *AgBio Forum, 6*(1, 2), 18–22.

Kishor, K., & David, J. (2016). Sensory evaluation of value-added cottage cheese with different levels of chickpea milk and turmeric powder. *Advances in Life Sciences*, *5*(23), 10909–10911.

Kljak, K., & Grbeša, D. (2015). Carotenoid content and antioxidant activity of hexane extracts from selected Croatian corn hybrids. *Food Chemistry*, *167*, 402–408.

Kowalska, E., & Ziarno, M. (2020). Characterization of buckwheat beverages fermented with lactic acid bacterial cultures and bifidobacteria. *Foods*, *9*(12), 1771. https://doi.org/10.3390/foods9121771.

Koyama, M., & Kitamura, Y. (2014). Development of a new rice beverage by improving the physical stability of rice slurry. *Journal of Food Engineering*, *131*, 89–95.

Kumar Mishra, B., Hati, S., Das, S., & Prajapati, J. B. (2019). Biofunctional attributes and storage study of soy milk fermented by *Lactobacillus rhamnosus* and *Lactobacillus helveticus*. *Food Technology and Biotechnology*, *57*(3), 399–407.

Kumar, S., Kumar, V., Sharma, R., Paul, A. A., Suthar, P., & Saini, R. (2020). Plant proteins as healthy, sustainable, and integrative meat alternates. *IntechOpen*. https://doi.org/10.5772/intechopen.94094.

Kumar, S., Sharma, S., Kumar, V., Sharma, R., Minhas, A., & Boddu, R. (2022). Cruciferous vegetables: A mine of phytonutrients for functional and nutraceutical enrichment. In *Current Advances for Development of Functional Foods Modulating Inflammation and Oxidative Stress* (pp. 401–426). Academic Press.

Kumar, V., Kaur, J., Sharma, K., Kumar, S., & Sharma, R. (2022). Role of analytical techniques in food quality control and safety. In P. Chandra & P. S. Panesar (Eds.), *Nanosensing and Bioanalytical Technologies in Food Quality Control* (pp. 1–17). Springer. https://doi.org/10.1007/978-981-16-7029-9_12.

Lainas, K., Alasalvar, C., & Bolling, B. W. (2016). Effects of roasting on proanthocyanidin contents of Turkish Tombul hazelnut and its skin. *Journal of Functional Foods*, *23*, 647–653.

Lakshmi, T. S., MaryPramela, A., & Iyer, P. (2017). Anti-microbial, anti-fungal, and anti-carcinogenic properties of coconut milk kefir. *International Journal of Home Science*, *3*, 365–369.

Lee, J., Townsend, J. A., Thompson, T., Garitty, T., De, A., Yu, Q., Peters, B. M., & Wen, Z. T. (2018). Analysis of the cariogenic potential of various almond milk beverages using a *Streptococcus mutans* biofilm model *in vitro*. *Caries Research*, *52*(1, 2), 51–57.

Li, J., Xiang, Q., Liu, X., Ding, T., Zhang, X., Zhai, Y., & Bai, Y. (2017). Inactivation of soybean trypsin inhibitor by dielectric-barrier discharge (DBD) plasma. *Food Chemistry*, *232*, 515–522.

Li, X. B., Yang, Z. X., Yang, L., Chen, X. L., Zhang, K., Yang, Q., Wu, Y. M., Liu, S. B., Tao, K. S., & Zhao, M. G. (2012). Neuroprotective effects of flax lignan against NMDA-induced neurotoxicity *in vitro*. *CNS Neuroscience & Therapeutics*, *18*(11), 927–933.

Li, X., Liu, X., Hua, Y., Chen, Y., Kong, X., & Zhang, C. (2019). Effects of water absorption of soybean seed on the quality of soymilk and the release of flavor compounds. *RSC Advances*, *9*(6), 2906–2918.

Li, Y. Q., Tian, W. L., Mo, H. Z., Zhang, Y. L., & Zhao, X. Z. (2013). Effects of pulsed electric field processing on quality characteristics and microbial inactivation of soymilk. *Food and Bioprocess Technology*, *6*, 1907–1916.

Lu, L., Zhao, L., Zhang, C., Kong, X., Hua, Y., & Chen, Y. (2015). Comparative effects of ohmic, induction cooker, and electric stove heating on soymilk trypsin inhibitor inactivation. *Journal of Food Science*, *80*(3), C495–C503.

Mäkinen, O. E., Uniacke-Lowe, T., O'Mahony, J. A., & Arendt, E. K. (2015). Physicochemical and acid gelation properties of commercial UHT-treated plant-based milk substitutes and lactose-free bovine milk. *Food Chemistry*, *168*, 630–638.

Manassero, C. A., Vaudagna, S. R., Sancho, A. M., Añón, M. C., & Speroni, F. (2016). Combined high hydrostatic pressure and thermal treatments fully inactivate trypsin inhibitors and lipoxygenase and improve protein solubility and physical stability of calcium-added soymilk. *Innovative Food Science & Emerging Technologies*, *35*, 86–95.

Manoj Kumar, C. T., Chauhan, O. P., Rajani, C. S., & Sabikhi, L. (2018). Effect of coconut milk, skim milk powder, and banana pulp on sensory and functional properties of coconut curd and its applicability as a carrier for probiotic microorganisms. *Journal of Food Processing and Preservation*, *42*(2), e13460.

Manzoor, M. F., Zeng, X. A., Ahmad, N., Ahmed, Z., Rehman, A., Aadil, R. M., Roobab, U., Siddique, R., & Rahaman, A. (2020). Effect of pulsed electric field and thermal treatments on the bioactive compounds, enzymes, microbial, and physical stability of almond milk during storage. *Journal of Food Processing and Preservation*, *44*(7), e14541.

Market analysis report. (2020). *Plant-Based Beverages Market Size, Share & Trends Analysis Report by Product (Soy Base, Coconut Base, Almond Base, Rice Base, Oats Base, Hemp Base), by Flavor, by Region, and Segment Forecasts, 2020–2027*. Markets and Markets. https://www.marketsandmarkets.com/Market-Reports/plant-based-beverage-market-34414144.html (accessed on 5 November 2024).

Marques, M. R., Freitas, R. A. M. S., Carlos, A. C. C., Siguemoto, É. S., Fontanari, G. G., & Arêas, J. A. G. (2015). Peptides from cowpea present antioxidant activity, inhibit cholesterol synthesis, and its solubilization into micelles. *Food Chemistry*, *168*, 288–293.

Martínez-Villaluenga, C., Frías, J., & Vidal-Valverde, C. (2006). Functional lupin seeds (*Lupinus albus* L. and *Lupinus luteus* L.) after extraction of α-galactosides. *Food Chemistry*, *98*(2), 291–299.

Menezes, A. G. T., Ramos, C. L., Dias, D. R., & Schwan, R. F. (2018). Combination of probiotic yeast and lactic acid bacteria as starter culture to produce maize-based beverages. *Food Research International*, *111*, 187–197.

Mohammadi Sartang, M., Mazloomi, S. M., Tanideh, N., & Rezaian Zadeh, A. (2015). The effects of probiotic soymilk fortified with omega-3 on blood glucose, lipid profile, hematological and oxidative stress, and inflammatory parameters in streptozotocin nicotinamide-induced diabetic rats. *Journal of Diabetes Research*, *2015*, 696372. https://doi.org/10.1155/2015/696372.

Morales-De La Peña, M., Salvia-Trujillo, L., Rojas-Graü, M. A., & Martín-Belloso, O. (2010). Isoflavone profile of a high-intensity pulsed electric field or thermally treated fruit juice-soymilk beverage stored under refrigeration. *Innovative Food Science & Emerging Technologies*, *11*(4), 604–610.

Muehlhoff, E., Bennett, A., & McMahon, D. (2013). *Milk and Dairy Products in Human Nutrition*. Food and Agriculture Organization of the United Nations (FAO).

Mukherjee, D., Chang, S. K., Zhang, Y., & Mukherjee, S. (2017). Effects of ultra-high pressure homogenization and hydrocolloids on physicochemical and storage properties of soymilk. *Journal of Food Science*, *82*(10), 2313–2320.

Munekata, P. E., Domínguez, R., Budaraju, S., Roselló-Soto, E., Barba, F. J., Mallikarjunan, K., Roohinejad, S., & Lorenzo, J. M. (2020). Effect of innovative food processing technologies on the physicochemical and nutritional properties and quality of non-dairy plant-based beverages. *Foods*, *9*(3), 288. https://doi.org/10.3390/foods9030288.

Myagmardorj, B., Purev, M. E., & Batdorj, B. (2018). Functional properties of fermented soymilk by *Lactobacillus fermentum* BM-325. *Mongolian Journal of Chemistry, 19*(45), 32–37.

Nadeeshani, R., Wijayaratna, U. N., Prasadani, W. C., Ekanayake, S., Seneviratne, K. N., & Jayathilaka, N. (2015). Comparison of the basic nutritional characteristics of the first extract and second extract of coconut milk. *International Journal of Food Science, 0410003.*

Nagendra Prasad, M. N., Sanjay, K. R., Prasad, D. S., Vijay, N., Kothari, R., & Nanjunda Swamy, S. (2012). A review on nutritional and nutraceutical properties of sesame. *Journal of Nutrition and Food Science, 2*(2), 1–6.

Namiki, M. (2007). Nutraceutical functions of sesame: A review. *Critical Reviews in Food Science and Nutrition, 47*(7), 651–673.

Nissen, L., di Carlo, E., & Gianotti, A. (2020). Prebiotic potential of hemp blended drinks fermented by probiotics. *Food Research International, 131*, 109029. https://doi.org/10.1016/j.foodres.2020.109029.

Ogbonna, A. C., Abuajah, C. I., & Utuk, A. R. (2013). Tigernut milk: A nutritious but underutilized food ingredient. *Food Biology, 2*(2), 14–17.

Okorie, S. U., & Nwanekezi, E. C. (2014). Evaluation of proximate composition and antinutritional factors of *Cyperus esculentus* (tigernut) as influenced by boiling. *IOSR Journal of Environmental Science, Toxicology and Food Technology, 2*, 70–73.

Owolabi, A. O., Ndidi, U. S., James, B. D., & Amune, F. A. (2012). Proximate, antinutrient and mineral composition of five varieties (improved and local) of cowpea, *Vigna unguiculata*, commonly consumed in Samaru community, Zaria-Nigeria. *Asian Journal of Food Science and Technology, 4*(2), 70–72.

Padma, M., Jagannadha Rao, P. V. K., Edukondalu, L., Aparna, K., & Ravi Babu, G. (2019). Storage studies of probiotic rice milk during refrigerated conditions. *International Journal of Chemical Studies, 7*, 1114–1117.

Paul, A. A., Kumar, S., Kumar, V., & Sharma, R. (2020). Milk analog: Plant-based alternatives to conventional milk, production, potential and health concerns. *Critical Reviews in Food Science and Nutrition, 60*(18), 3005–3023. https://doi.org/10.1080/10408398.2019.1647631.

Paz, P. C., Janny, R. J., & Håkansson, Å. (2020). Safeguarding of quinoa beverage production by fermentation with *Lactobacillus plantarum* DSM 9843. *International Journal of Food Microbiology, 324*, 108630. https://doi.org/10.1016/j.ijfoodmicro.2020.108630.

Pereira, J. A., Oliveira, I., Sousa, A., Ferreira, I. C., Bento, A., & Estevinho, L. (2008). Bioactive properties and chemical composition of six walnut (*Juglans regia* L.) cultivars. *Food and Chemical Toxicology, 46*(6), 2103–2111. https://doi.org/10.1016/j.fct.2008.01.022.

Pereira, P. C. (2014). Milk nutritional composition and its role in human health. *Nutrition, 30*(6), 619–627. https://doi.org/10.1016/j.nut.2013.10.011.

Pineli, L. D. L. D. O., Botelho, R. B., Zandonadi, R. P., Solorzano, J. L., de Oliveira, G. T., Reis, C. E. G., & Teixeira, D. D. S. (2015). Low glycemic index and increased protein content in a novel quinoa milk. *LWT–Food Science and Technology, 63*(2), 1261–1267. https://doi.org/10.1016/j.lwt.2015.03.022.

Qin, H., Xu, H., Yu, L., Yang, L., Lin, C., & Chen, J. (2019). Sesamol intervention ameliorates obesity-associated metabolic disorders by regulating hepatic lipid metabolism in high-fat diet-induced obese mice. *Food & Nutrition Research, 63*. https://doi.org/10.29219/fnr.v63.3016.

Raajeswari, P. A., & Nithya, N. (2006). Standardization of value added recipes with millet milk powder. *Research & Reviews: Journal of Food Science and Technology*, *5*(2), 10–17.

Rabetafika, H. N., Van Remoortel, V., Danthine, S., Paquot, M., & Blecker, C. (2011). Flaxseed proteins: Food uses and health benefits. *International Journal of Food Science & Technology*, *46*(2), 221–228. https://doi.org/10.1111/j.1365-2621.2010.02492.x.

Rahman, M. M., Miaruddin, M., Chowdhury, M. G. F., Khan, M. H. H., & Rahman, M. E. (2019). Development of processing method for sweetened condensed corn milk. *Bangladesh Journal of Agricultural Research*, *44*(2), 377–385.

Ravindran, S. (2021). Probiotic oats milk drink with microencapsulated *Lactobacillus plantarum*—An alternative to dairy products. *Nutrition & Food Science*, *51*(3), 471–482.

Reddit, Map Porn. (2017). Vegetarianism by country. *Wikipedia*. Retrieved from: https://en.wikipedia.org/wiki/Vegetarianism_by_country#cite_note-82 (accessed on 5 November 2024).

Reyes-Jurado, F., Soto-Reyes, N., Dávila-Rodríguez, M., Lorenzo-Leal, A. C., Jiménez-Munguía, M. T., Mani-López, E., & López-Malo, A. (2021). Plant-based milk alternatives: Types, processes, benefits, and characteristics. *Food Reviews International*, 1–32.

Rizzoli, R. (2014). Dairy products, yogurts, and bone health. *The American Journal of Clinical Nutrition*, *99*(5), 1256S–1262S.

Rodriguez-Leyva, D., Weighell, W., Edel, A. L., LaVallee, R., Dibrov, E., Pinneker, R., Maddaford, T. G., Ramjiawan, B., Aliani, M., Guzman, R., & Pierce, G. N. (2013). Potent antihypertensive action of dietary flaxseed in hypertensive patients. *Hypertension*, *62*(6), 1081–1089.

Salama, F. M., Azzam, M. A., Abdl-Rahman, M. A., Naga, A. E., & Abdl-Hameed, M. S. (2011). Optimization of processing techniques for production of oat and barley milks. *Journal of Food and Dairy Sciences*, *2*(10), 577–591.

Salve, A. R., Pegu, K., & Arya, S. S. (2019). Comparative assessment of high-intensity ultrasound and hydrodynamic cavitation processing on physico-chemical properties and microbial inactivation of peanut milk. *Ultrasonics Sonochemistry*, *59*, 104728.

Schiano, A. N., Harwood, W. S., Gerard, P. D., & Drake, M. A. (2020). Consumer perception of the sustainability of dairy products and plant-based dairy alternatives. *Journal of Dairy Science*, *103*(12), 11228–11243.

Sethi, S., Tyagi, S. K., & Anurag, R. K. (2016). Plant-based milk alternatives: An emerging segment of functional beverages: A review. *Journal of Food Science and Technology*, *53*, 3408–3423.

Settaluri, V. S., Kandala, C. V. K., Puppala, N., & Sundaram, J. (2012). Peanuts and their nutritional aspects—A review. *Food Nutrition and Science*, *3*, 1644–1650.

Séverin, S., & Wenshui, X. (2005). Milk biologically active components as nutraceuticals. *Critical Reviews in Food Science and Nutrition*, *45*(7, 8), 645–656.

Sharma, K., Kaur, R., Kumar, V., Kumar, S., Singh, A., & Gautam, N. (2022). Prebiotic and probiotic potential of cereals. In *Functional Cereals and Cereal Foods* (pp. 163–188). Springer, Cham.

Sharma, P., Sharma, D., Kaur, S., & Borah, A. (2021). Optimization of flaxseed milk fermentation for the production of functional peptides and estimation of their bioactivities. *Food Science and Technology International*, *27*(7), 585–597.

Sharma, S., Shree, B., Sharma, D., Kumar, S., Kumar, V., Sharma, R., & Saini, R. (2022). Vegetable microgreens: The gleam of next generation super foods, their genetic enhancement, health benefits and processing approaches. *Food Research International*, *155*, 111038.

Sheela, P., Kanchana, S., Maheswari, T. U., & Hemalatha, G. (2018). Optimization of parameters for the extraction of millet milk for product development. *Research Journal of Agricultural Sciences*, *9*(6), 1345–1349.

Shree, B., Kumar, S., Sharma, S., & Katoch, V. (2022). Functional significance of underutilized high value cruciferous vegetables—An exotic gleam in the gloomy guise of their functional importance. *South African Journal of Botany*, *145*, 420–437.

Shvachko, N. A., Loskutov, I. G., Semilet, T. V., Popov, V. S., Kovaleva, O. N., & Konarev, A. V. (2021). Bioactive components in oat and barley grain as a promising breeding trend for functional food production. *Molecules*, *26*(8), 2260.

Siddeeg, A., Salih, Z. A., Ammar, A. F., Saeed, N. S. M., Howladar, S. M., & Alzahrani, F. O. (2020). Production of Peanut Milk and Its Functional, Physiochemical, Nutritional and Sensory Characteristics. *Annual Research & Review in Biology*, 79–88.

Sidhu, J. S., & Singh, R. K. (2016). Ultra-high pressure homogenization of soy milk: Effect on quality attributes during storage. *Beverages*, *2*(2), 15.

Siracusa, R., Fusco, R., Peritore, A. F., Cordaro, M., D'Amico, R., Genovese, T., Gugliandolo, E., Crupi, R., Smeriglio, A., Mandalari, G., & Cuzzocrea, S. (2020). The antioxidant and anti-inflammatory properties of *Anacardium occidentale* L. cashew nuts in a mouse model of colitis. *Nutrients*, *12*(3), 834.

Sukumaran, P., Venkatesan, S., Kumaran, A., & Kannan, D. (2021). Development of novel fermented almond milk tea and its evaluation as an antidiabetic drink. *Journal of Pharmaceutical Research International*, *33*(2), 75–87.

Suman, D., & Sreeja, V. (2019). Barley: A cereal with potential for development of functional fermented foods. *International Journal of Fermented Foods*, *8*(1), 1–13.

Suthar, P., Kumar, S., Kumar, V., Vaidya, D., Sharma, A., & Sharma, A. (2022). *Murraya koenigii* (L.) Spreng: Speculative ethnobotanical perspectives of ubiquitous herb with versatile nutra/functional properties. *South African Journal of Botany*, *145*, 111–134.

Suyitno, T. (2003). Health benefit of coconut milk. *Indonesian Food and Nutrition Progress*, *10*(2), 106–112.

Syed, S. J., Gadhe, K. S., & Shaikh, R. P. (2020). Studies on quality evaluation of OAT milk. *Journal of Pharmacognosy and Phytochemistry*, *9*(1), 2275–2277.

Tano-Debrah, K., Asiamah, K., Sakyi-Dawson, E., & Budu, A. S. (2005). Effect of malt enzyme treatment on the nutritional and physicochemical characteristics of cowpea-peanut milk. In *Proceedings of the 1st International Edible Legume Conference in conjunction with the IVth World Cowpea Congress*, Durban, South Africa, 17-21 April, University of Pretoria, 1–7.

Tolulope, O. A., David, O. A., Sylvester, E. S., & Ugochukwu, U. (2019). The histochemical and biochemical effects of coconutmilk in wistar rats with ethanol-induced gastric ulcer. *Anatomy Journal of Africa*, *8*(2), 1558–1569.

Toro-Funes, N., Bosch-Fusté, J., Veciana-Nogués, M. T., & Vidal-Carou, M. C. (2014). Effect of ultra high pressure homogenization treatment on the bioactive compounds of soya milk. *Food Chemistry*, *152*, 597–602.

Toro-Funes, N., Bosch-Fusté, J., Veciana-Nogués, M. T., & Vidal-Carou, M. C. (2014). Changes of isoflavones and protein quality in soymilk pasteurised by ultra-high-pressure homogenisation throughout storage. *Food Chemistry*, *162*, 47–53.

Tovar-Pérez, E. G., Lugo-Radillo, A., & Aguilera-Aguirre, S. (2019). Amaranth grain as a potential source of biologically active peptides: A review of their identification, production, bioactivity, and characterization. *Food Reviews International*, *35*(3), 221–245.

Transparency Market Research [TMR]. (2019). Global industry analysis, size, share, growth, trends, and forecast, 2019–2029. TMR, Albany, NY, USA.

Trisnawati, C. Y., Srianta, I., & Marsono, Y. (2013). Effect of corn varieties on the characteristics of soycorn milk. *International Food Research Journal*, *20*(3), 1187–1190.

Ullah, R., Nadeem, M., Khalique, A., Imran, M., Mehmood, S., Javid, A., & Hussain, J. (2016). Nutritional and therapeutic perspectives of Chia (*Salvia hispanica* L.): A review. *Journal of Food Science and Technology*, *53*(4), 1750–1758.

Vahanvaty, U. S. (2009). Hemp seed and hemp milk: the new super foods? *ICAN: Infant, Child, & Adolescent Nutrition*, *1*(4), 232–234.

Valencia-Flores, D. C., Hernández-Herrero, M., Guamis, B., & Ferragut, V. (2013). Comparing the effects of ultra-high-pressure homogenization and conventional thermal treatments on the microbiological, physical, and chemical quality of almond beverages. *Journal of Food Science*, *78*(2), E199–E205.

Vanga, S. K., & Raghavan, V. (2018). How well do plant based alternatives fare nutritionally compared to cow's milk? *Journal of Food Science and Technology*, *55*(1), 10–20.

Vanga, S. K., Wang, J., & Raghavan, V. (2020). Effect of ultrasound and microwave processing on the structure, *in-vitro* digestibility and trypsin inhibitor activity of soymilk proteins. *LWT-Food Science and Technology*, *131*, 109708.

Varghese, T., & Pare, A. (2019). Effect of microwave assisted extraction on yield and protein characteristics of soymilk. *Journal of Food Engineering*, *262*, 92–99.

Veerichetty, V. (2018). Antidiabetic potential of the combination of fermented soy milk and flaxseed milk in alloxan-induced diabetic rats. *International Journal of Green Pharmacy (IJGP)*, *12*(03).

Vega-Gálvez, A., Miranda, M., Vergara, J., Uribe, E., Puente, L., & Martínez, E. A. (2010). Nutrition facts and functional potential of quinoa (*Chenopodium quinoa* Willd.), an ancient Andean grain: A review. *Journal of the Science of Food and Agriculture*, *90*(15), 2541–2547.

Wang, C., Liang, S., Wang, H., & Guo, M. (2018). Physiochemical properties and probiotic survivability of symbiotic oat-based beverage. *Food Science and Biotechnology*, *27*, 735–743.

Wang, Q., Jiang, J., & Xiong, Y. L. (2018). High pressure homogenization combined with pH shift treatment: A process to produce physically and oxidatively stable hemp milk. *Food Research International*, *106*, 487–494.

Wansutha, S., Yuenyaow, L., Jantama, K., & Jantama, S. S. (2018). Antioxidant activities of almond milk fermented with lactic acid bacteria. *Thaiphesatchasan*, *42*.

Winston, J. C. (2009). Health effects of vegan diets. *American Journal of Clinical Nutrition*, *89*, 1627S–1633S.

Yuliana, N., & Rangga, A. (2010). Manufacture of fermented coco milk-drink containing lactic acid bacteria cultures. *African Journal of Food Science*, *4*(9), 558–562.

Zaaboul, F., Raza, H., Cao, C., & Yuanfa, L. (2019). The impact of roasting, high pressure homogenization and sterilization on peanut milk and its oil bodies. *Food Chemistry*, *280*, 270–277.

Zevallos, V. F., Herencia, I. L., Chang, F., Donnelly, S., Ellis, J. H., & Ciclitira, P. J. (2014). Gastrointestinal effects of eating quinoa (*Chenopodium quinoa* Willd.) in celiac patients. *Official Journal of the American College of Gastroenterology*, *109*(2), 270–278.

Zhou, X. L., Yan, B. B., Xiao, Y., Zhou, Y. M., & Liu, T. Y. (2018). Tartary buckwheat protein prevented dyslipidemia in high-fat diet-fed mice associated with gut microbiota changes. *Food and Chemical Toxicology*, *119*, 296–301.

CHAPTER 9

Micro-Green: A Concentrated Source of Nutritional Bioactives

SANJEEV KUMAR,[1] DEV RAJ,[2*] ASHOK K. SENAPATI,[2] and AVNISH K. PANDEY[3]

[1]*Department of Vegetable Science, Navsari Agricultural University, Navsari, Gujarat, India*

[2]*Department of Postharvest Technology, Navsari Agricultural University, Navsari, Gujarat, India*

[3]*Department of Fruit Science, Navsari Agricultural University, Navsari, Gujarat, India*

*Corresponding author.

ABSTRACT

Increased health awareness in people about general well-being through improved food habits has resulted in novel developments in food production and processing. Micro-greens, the young greens with delicate textures, distinctive flavors, and higher nutrition have emerged as potential functional foods. They have gained popularity as they contain exceptionally higher volumes of several bioactive compounds including phenolics, alkaloids, vitamins, minerals, and so on, as compared to their derivatives especially, in certain groups of population owing to their higher nutraceutical value and unique taste. They have been reported effective against several chronic health diseases. With the rising global issues of continuous reduction in land resources, sedentary lifestyle coupled with unhealthy dietary habits, and the role and importance of the foods in the overall management of health, the specific interest in the field of micro-greens has increased. The nutritional and functional significance of micro-greens, their classification, growth conditions, and postharvest management are discussed in detail in this chapter.

Functional Compounds and Foods of Plant Origin:Krishan Dutt Sharma, Satish Kumar, Rakesh Sharma, and Raj K. Keservani (Eds.)

9.1 INTRODUCTION

With the advancement in medical sciences, many diseases and ailments prevalent earlier, like infectious and communicable diseases, have been eliminated. But at the same time, the spectrum of life in terms of rapid urbanization, industrialization, increase in per capita income, and consumer expenditure is changing rapidly with the continuous economic growth. Further, unhealthy food consumption patterns are contributing to the rise of lifestyle-linked diseases such as diabetes, obesity, cardiovascular (CVD) diseases, stroke, hypertension, and any type of cancer in both developed and developing nations (Kumar et al., 2020, 2022; Sharma et al., 2022). The UN Decade of Action on Nutrition, with a scheduled execution from 2016 to 2025, was established by the United Nations general assembly (UNGA) as an enormous opportunity to address all types of malnutrition. Malnutrition comprises undernutrition (wasting, stunting, and underweight), overnutrition (overweight, obesity), vitamin or minerals deficiency, and non-communicable diseases (NCDs), which are all linked to dietary consumption patterns of the individual. Nearly 41 million people (71% of global deaths) die every year due to NCDs, such as cardiovascular diseases, diabetes, sleep disorders, stress, obesity, chronic respiratory diseases, and some forms of cancer. More than 15 million annual deaths are reported in the age group of 30 to 69 due to NCDs (WHO, 2021). On a global level, unhealthy dietary patterns and inadequate nutrition are among the main risk factors for chronic diseases and altogether they are constituting a threat to the socio-economic facet of the countries (WHO, 2021).

Among 88 countries, India is identified with the highest rates of domestic inequalities in malnutrition and is likely to miss global nutrition targets by the year 2025. According to the Global Hunger Index 2021, India holds 101st position among a total of 116 nations, and 14% of the Indian population has been reported to be under-nourished (www.globalhungerindex.org). Despite the widespread global challenges like malnutrition, poverty, mental health issues, climate change, economic crisis, and so on, the unforeseen sporadic COVID-19 pandemic has obstructed the overall health of the population, food systems, and working styles. Therefore, in order to maximize life expectancy and quality, a balanced diet must be maintained by identifying food components or ingredients capable of strengthening an individual's overall immunity. Nutrition has been a developmental goal for many years, and ensuring nutritional security has long been a primary concern for developing nations like India. Food-based strategies are known to be a sustainable approach as they empower individuals and households to have complete control over the quality of food in their diet via the production and consumption of nutritionally rich foods. These strategies are considered ideal to achieve long-term goals like assured access to an adequate nutritional diet, rational consumer selection, appropriate preparation methods, and substantial feeding (Perez-Escamilla, 2017).

Micro-green is a novel category of edibles withstanding huge potential in curing various nutritional deficiency-related diseases or ailments (Pinto et al., 2015) and presents a feasible residentially grown choice for nutritive security. Micro-green is commercially gaining popularity at a global level and is popularly called "Vegetable Confetti" (www.koppertcress.com/en). Microgreens can be cultivated or produced easily in rural as well as urban areas, with limited land resources and skilled vegetable farmers or even interested consumers themselves. In addition, they give an opportunity to the rural residents to improve the nutritional status of the food they produce and consume. Microgreens can be cultivated organically with or without soil near or inside residential or household areas within a short period of 10–15 days. Further, no degradation or loss of micronutrients is observed in micro-greens as it is typically consumed in raw form. Phytonutrient profile differs according to the developmental stage of the micro-greens and the content oftentimes decline from the stage of seedling to the fully mature stage (Ebert et al., 2014). Immature micro-greens are approximately 6 times nutritionally richer when compared to their mature or fully developed counterparts (Xiao et al., 2012). So, micro-greens can be recognized as a functional food that has disease preventing or health-promoting properties. Subsequent increase in the consumption of microgreens can be observed lately, with rising consumer awareness and liking of microgreens for its tender and soft texture, distinct refreshing flavor and concentrated biologically active (bio-active) compounds such as vitamins, antioxidants, and minerals, and so forth. All the more, NASA scientists are exploring the obstacles and utilities of cultivating micro-greens in space. Microgreens are an important crop in accordance with their cultivation in space as they offer oxygen regeneration, nitrogen fixation, availability of fresh ingredients, and vital nutrients that enhance the morale of astronauts for extended stays in space (Kyriacou et al., 2017).

9.2 HISTORICAL DEVELOPMENT

Both Indian and Persian festivities and rites have long utilized wheatgrass. Wheatgrass is traditionally used in both Indian and Persian festivals and rituals. On the occasion of *Navratri pooja* in India, wheat or barley seeds are sown on the first day of pooja and on the last or final day, the saplings grown are offered to the mother goddess as a part of the ceremony. Wheatgrass is being utilized for its rejuvenating and detoxifying properties from ancient times in *Ayurveda*. In traditional Chinese medicine, it is used to notify the spleen, help support digestion and drain dampness in the body and is sometimes used interchangeably with barley grass. However, Charles Schnabel's experiments in the 1930s led to the introduction of wheatgrass in the western world (Jaglan et al., 2020; Sean, 2002; Sharma et al., 2021, 2022; Shree et al., 2022). By the year 1940, canned Schnabel's powdered grass was available in major drug stores throughout the US and Canada (Meyerowitz, 1999). Also, sunflower, radish, and buckwheat were cultivated in winter season in the 1960s and healthy locally produced or home grown "grasses" were publicized for its benefits for health and well-being in the 1970s. Ann Wigmore, the founder of the Hippocrates Health Institute – Ann Wigmore, was a well-known advocate of wheatgrass. He emphasized on consumption of raw wheatgrass in the diet and opined that wheatgrass cleanses the body toxins and also provides an optimal nutritional balance so also referred as whole food. In the 1980s, local cooks or chefs in restaurants started cultivating seedlings also known as "cresses" for garnishing. The use of term "micro-green" firstly started in 1998 in the USA and local producers or farmers started offering fresh microgreens to the local retailers throughout North America in 2000s. Micro-greens were seen in local market or grocery stores for food enthusiasts (Pinto et al., 2015). Presently, the well-known chefs of India, namely; Vikas Khanna, Kunal Kapoor, Ranveer Brar and Sanjeev are preparing various blends of micro-greens with Indian food items and are increasing the popularity of these items amongst the people highlighting their importance.

Micro-green shares many characteristics with sprouts. The market for micro-green is rapidly growing (Riggio et al., 2019a; Wood, 2019), along with facing challenges of marketing or bio-safety concerns. Micro-green is an emerging new category of edible veggies (Pinto et al., 2015). It includes small sized crops ranging from 5 cm–10 cm (Xiao et al., 2012). Micro-green comprises of stem, leaves (cotyledon) and uppermost young leaves (first pair). Based on the plant growth stages, micro-greens fall in the stage younger than baby greens and older stage than the sprouts. This novel class of food have a short life span even under refrigeration and is utilized in small amounts for garnishing or as toppings or seasonings (Riggio et al., 2019). Micro-green is cultivated from a variety of seeds (aromatic herbs, vegetables, and wild herbaceous plants), identified by a variety of colors, tastes, and textures, and are fresh and tender. Micro-greens are picked off after a few days or a week of germination after formation of cotyledons and appearance of first true leaves (Sharma et al., 2020). Globally, there are approximately 70–80 plant species identified as micro-greens. However, Table 9.1 enlists the popular micro-greens for homestead and commercial utility over the years.

TABLE 9.1 Important Micro-Green for Homestead and Commercial Purposes

Micro-Green	Family	Micro-Green Color
• Chive (*Allium schoenoprasum*)	Alliaceae	Green
• Onion/Scallion (*Allium cepa*)		
• Shallot (*Allium cepa* var. *aggregatum*)		
• Garlic (*Allium sativum*)		
• Garnet amaranth (*Amaranthus hypochondriacus*)	Amaranthaceae	Red
• Cilantro (*Coriandrum sativum*)	Apiaceae	Green
• Celery (*Apium graveolens*)		
• Sunflower (*Helianthus annuus*)	Asteraceae	Green/variegated
• Chrysanthemum (*Glebionis coronaria*)		
• Marigold (*Tagetes* sp.)		
• Arugula (*Eruca sativa* Mill.)	Brassicaceae	Green
• Green daikon Radish (*Raphanus sativus* L.)	Brassicaceae	Green
• Opal Radish (*Raphanus sativus* L.)	Brassicaceae	Greenish purple
• Mizuna (*Brassica rapa* L.)	Brassicaceae	Green
• Pepper cress (*Lepidium bonariense* L.)	Brassicaceae	Green
• Purple/red mustard (*Brassica juncea* L.)	Brassicaceae	Purplish green/red
• Red cabbage (*Brassica oleracea* var. *capitata* f. *rubra*)	Brassicaceae	Purplish green
• Broccoli (*Brassica oleracea* var. *italica*)	Brassicaceae	Purplish green
• Cauliflower (*Brassica oleracea* var. *botrytis*)		
• Kohlrabi (*Brassica oleracea* var. *gongylodes* L.)		
• Kale (*Brassica oleracea* var. *acephala* DC)		
• Pak-choi (*Brassica rapa* subsp. *chinensis*)		
• Collard (*Brassica oleracea* var. *viridis* L.)		
• Brussels sprouts (*Brassica oleracea* var. *gemmifera* DC)		
• Rutabaga (*Brassica napus* var. *napobrassica* Peterm)		

TABLE 9.1 *(Continued)*

Micro-Green	Family	Micro-Green Color
• Bull's blood beet (*Beta vulgaris* L.)	Chenopodiaceae	Reddish green
• Magenta spinach (*Spinacia oleracea* L.)	Chenopodiaceae	Red
• Red beet (*Beta vulgaris* L.)	Chenopodiaceae	Reddish green
• Orach (*Atriplex hortensis* L.)	Chenopodiaceae	Red
• Cucumber (*Cucumis sativus*)	Cucurbitaceae	Green
• Golden pea (*Pisum sativum* L.)	Fabaceae	Yellow
• Pea tendrils (*Pisum sativum* L.)	Fabaceae	Green
• Alfalfa (*Medicago sativa*) and fenugreek (*Trigonella foenum-graecum*)	Fabaceae	Green
• Lentil (*Lens culinaris* Medik.)	Fabaceae	Green
• Green basil (*Ocimum basilicum* L.)	Lamiaceae	Green
• Opal basil (*Ocimum basilicum* L.)	Lamiaceae	Greenish purple
• Mint (*Mentha piperita* L. and *Mentha spicata* L.), chia (*Salvia hispanica*) and lemon balm (*Melissa officinalis*)	Lamiaceae	Green
• Wood sorrels (*Oxalis acetosella*) and clover (*Trifolium pretense* and *T. repens*)	Oxalidaceae	Green
• Sweet corn (*Zea mays* L.)	Poaceae	Yellow
• Buckwheat (*Fagopyrum tataricum* L.)	Poaceae	Green
• Lemon grass (*Cymbopogon citratus*)	Poaceae	Green
• Wheat (*Triticum aestivum*) and barley (*Hordeum vulgare*)	Poaceae	Green
• Sorrel (*Rumex acetosa* L.)	Polygonaceae	Reddish green
• Claytonia (*Claytonia perfoliata*) and purslane (*Portulaca oleracea*)	Portulacaceae	Green
• Dill (*Anethum graveolens*)	Apiaceae	Green
• Fennel (*Foeniculum vulgare*)	Umbelliferae	Green
• Linseed/flaxseed (*Linum usitatissimum*)	Linaceae	Green
• Carrot (*Daucus carota*)	Umbelliferae	Green

Source: Turner et al. (2020); Xiao et al. (2012).

9.3 MICRO-GREEN AS A FUNCTIONAL FOOD

Micro-greens have gained interest and is becoming popular as a functional food that has significant richness in micronutrients and bio-active compounds including carotenoids (Ghoora et al., 2020; Paradiso et al., 2018; Sun et al., 2013; Xiao et al., 2012, 2016). They are also known for their variety of attractive colors, flavors, and textures (Renna et al., 2017). USDA reports claims that microgreens have approximately five times more vitamin and carotenoid value in comparison to mature vegetables. Microgreens comprises of vitamin E, vitamin C, minerals like calcium (Ca) and magnesium (Mg), β-carotene and different microgreens contain different levels of phytonutrients (Kyriacou et al., 2016). Many studies have revealed increase in the concentration of bioactive compounds during the stage of sprouting (Gan et al., 2017). Plant secondary metabolites such as redox-active compounds like vitamins, glucosinolates, minerals as well as polyphenols (Keservani and Sharma, 2014; Keservani et al., 2016a, b) and anthocyanins are also found in microgreens (Kyriacou et al., 2019).

Micro-greens largely contain phenolics, minerals, vitamins, and antioxidants when compared to fully formed greens or seeds. Therefore, also referred to as functional foods that consists of various health promoting or disease preventing properties in addition to their nutritive richness. Microgreens are also recognized as well-established carriers of bio-active components (Mir et al., 2017), as micro-green is full of nutrients like K, Fe, Zn, Mg, and Cu, concentrations of which may vary from one to another micro-green (Xiao et al., 2016). The affluent nutritional profile of micro-greens comprises phytochemicals in the same amount as that of fully-formed greens (Xiao et al., 2012). Research studies have proclaimed that the nutritional richness of the microgreen is up to nine times higher when compared to those established in mature fully developed greens (Pinto et al., 2015). Carotenoids are a major category of phytochemical, and their significance is not limited to their role as a retinol precursor. They also offer antioxidant and anti-tumor properties that have importance in the communication of gap-junction, regulation of gene function and hormone, as well as immunity modulation (Adadi et al., 2018; Rao and Rao, 2007). Many herbaceous plants as well as vegetables are well recognized for its health benefits. In particular, the vegetables from *Brassicaceae* family, contains glucosinolates (Fuentes et al., 2015), carotenoids (Niranjana et al., 2015; Nishino et al., 2009), and selenium (Donaldson, 2004) that offers protection against cancer (Herr and Buchler, 2010). The end products obtained from the hydrolysis of glucosinolates have antimicrobial properties (Delaquis and Mazza, 1995; Gonzalez-Lamothe et al., 2009). Other nutrients like omega-3 fatty acids, vitamin B_{12}, iron, calcium, and zinc found in these functional foods have significant importance in growth and development (Morrison and Regnault, 2016).

Huang et al. (2016) reviewed and concluded that micro-green supplementation of red cabbage had health-promising properties in the mice fed with a diet with high fat value. Micro-greens incorporation attenuated body weight gain with reduced level of low-density lipoproteins (LDL) cholesterol level, and inhibited hepatic cholesterol ester, inflammatory cytokines, and triglyceride levels. Interestingly, raw cabbage supplementation offering a low-fat diet raised LDL and HDL cholesterol levels (Huang et al., 2016). Choe et al. (2018) studied the use of micro-green as a functional food and its role in the prevention of dietary diseases such as cardiovascular disease (CVD), obesity, Type-2 diabetes mellitus (DM), and cancer. Earlier researchers have authenticated the utility of different micro-greens as a functional food for the availability of various promising compounds (Table 9.2).

TABLE 9.2 Health-Promoting Compounds in Micro-Green

Crop	Bioactive	References
Lettuce	Lactucaxanthin	Niizu and Rodriguez-Amaya (2005)
Tartary buckwheat	High DPPH activity	–
Cabbage	High TAA	Xiao et al. (2012)
Garnet amaranths	High phylloquinone and violaxanthin	Xiao et al. (2012)
Red sorrel	β-carotene and lutein	Xiao et al. (2012)
Brassica sp.	High polyphenols	Sun et al. (2013)
Amaranth	Violaxanthin, neoxanthin, lutein, α and β-carotene, chlorogenic acid (CGA)	Ebert et al. (2014)
Lettuce	Ca, Mg, Zn, Fe, Mn Cu, Mo, and Se	Pinto et al. (2015)
Cauliflower	Ca, Mg	Xiao et al. (2016)
Broccoli	P	Xiao et al. (2016)
Arugula	K	Xiao et al. (2016)
Kohlrabi	Fe	Xiao et al. (2016)
Pea and lupin	High flavonoid levels	Klopsch et al. (2018)
Wheat and barley	Higher carotenoid content	Niroula et al. (2019)
Broccoli and cauliflower	Phylloquinone, vitamin C, carotenoids, glucosinolates, and polyphenols	Xiao et al. (2019)
Cauliflower, rapini, radish (red ruby and China rose)	Higher contents of vitamin C, phylloquinone, glucosinolates, tocopherols, and TPC	Xiao et al. (2019)
Ruby radish	Higher antioxidant value	Xiao et al. (2019)
Radish and mustard	Ascorbic acid, carotenoids, and isothiocyanates	De la Fuente et al. (2019)
Broccoli, kale, and radish	Polyphenols	De la Fuente et al. (2019)

TABLE 9.2 *(Continued)*

Crop	Bioactive	References
Kale	Ca, Fe, Mg, Mn, P, K, and Vitamin K	De la Fuente et al. (2019)
Radish and mustard	Maximum bio-accessibility of bioactive compounds, namely total soluble polyphenols, total isothiocyanates, and macro-minerals (K, Ca, Mg)	De la Fuente et al. (2019)
Basil and Swiss chard	K and Mg	Kyriacou et al. (2019)
Green basil and coriander	β-carotene and total polyphenols	Kyriacou et al. (2019)
Fenugreek	Ascorbic acid, N, and Ca	Kumar et al. (2020)
Beetroot	β-carotene and K conte	Kumar et al. (2020)
Red cabbage, beetroot	Total antioxidant activity	Kumar et al. (2020)
Alfalfa	High agmatine	–
Fenugreek	High spermine	–
Lentil	High spermidine (SPD)	–
Alfalfa	Highest levels of polyamines agmatine (AGM)	–
Fenugreek	Highest level of cadaverine (CAD), putrescine (PUT) and spermine (SPM)	–
Lentil	SPD	–
Lentil	Higher total chlorophyll, carotenoid, and ascorbic acid	Kowitcharoen et al. (2021)
Buckwheat	High phenolic content and DPPH scavenging activity	Kowitcharoen et al. (2021)
Purple radish and red cabbage	Higher Anthocyanin, vitamin C	Kowitcharoen et al. (2021)

Since micro-greens are blessed with vital nutrients and contain high amounts of vitamins, minerals, and beneficial plant compounds, their consumption reduces the threat of various diseases (Bazzano et al., 2002).

9.4 POSTHARVEST MANAGEMENT OF MICRO-GREENS

Micro-greens are characterized by rapid respiration and relatively short shelf-life, leading to limited industrial production and marketing (Chandra et al., 2012; Kou et al., 2013). The demand and supply cycle of micro-greens is greatly influenced by emerging gastronomic trends and consumer preference for sensory characteristics (Kyriacou et al., 2016). Micro-green respires during different stages, such as during the germination process and metabolization of carbohydrates (CHO) in the cotyledonary leaves. After the complete depletion of CHO sources, degradation of micro-green occurs. Therefore, some needed modifications in the metabolic activity of plants and storage life extension by a few days could be considered advantageous. Micro-greens have a similar bearing as that of fresh produce since they are subjected to pursue the patterns of stress induced conditions rather than following natural senescence, mechanical or physical trauma caused during cutting, handling, processing, excessive heat, desiccation, and packaging headspace composition, all of which may result in the acceleration of the deteriorating characteristics of the produce and reduces the storage stability of the product (Hodges and Toivonen, 2008; Kou et al., 2014). The blunt blades used for cutting procedure, reduces the storage life span of raw and fresh-cut leafy plants, thus, harvesting micro-greens must be done with sharp blade to avoid mechanical injuries, bruising or damage to the cells adjoining the cut (Portella and Cantwell, 2001).

Time interval of harvesting the micro-greens have crucial implications in the bioactives composition and the overall shelf-life of harvested produce (Clarkson et al., 2005; Garrido et al., 2015; Hasperue et al., 2016). The effect is specific to the species harvested and accelerates in the season of spring or summer, due to increase in the light intensity and photoperiodism. The lifespan of lollo rosso lettuce (*Lactuca sativa* L. "Ravita"), baby red chard (*Beta vulgaris* L. var. *flavescens*), and leaf roquette (*Eruca vesicaria*) was extended by 2–6 days following the final day of harvest. This extension was associated with diurnal changes in sucrose levels and starch content in the leaves (Clarkson et al., 2005). Passive MAP with films having relatively high OTR (oxygen transmission rate) or perforated packaging technique is reported to enhance the value of micro-greens after harvest (Kyriacou et al., 2016).

Although the thermal conditions and packaging environment are unquestionably the most important elements for shelf-life extension, calcium (Ca) treatments before harvest and storage may improve the quality of the produce during storage period of microgreens. Pre-harvest application (200 mL) of calcium lactate (1–20 mM), calcium amino acid chelates (1–20 mM) and particularly calcium chloride (10 Mm; 6.5 pH level) improves the overall postharvest quality and storage stability broccoli characterized by a significant decline in the electrolyte's concentration at storage time at 5°C (Kou et al., 2015). Temperature is one of the most critical elements that influences the deterioration rate of micro-greens after harvest, as it interacts with ethylene, reduced pO_2 and elevated pCO_2 (Kader, 2002; Kou et al., 2014). Temperature directly affects the storage stability of micro-greens via regulation of respiration and metabolic changes leading to senescence (Xiao et al., 2014).

Edible coatings are regarded as novel packaging strategy for postharvest management of the fresh and raw produce. They form a modified atmospheric conditions around the plant produce and act as a barrier for gases and water vapor. It helps in retarding the food deterioration rates and enhances the overall quality of the produce. Edible coatings are developed from natural polymers like carbohydrates (CHO), proteins and lipids. Edible coatings are considered efficient to enhance the postharvest quality of fresh-cut raw produce, such as minimally processed (MP) lettuce and celery sticks (Fortunati et al., 2017; Tay and Perera, 2004). In the last decade, the increasing interest in utilization of *Aloe vera* gel in the form of edible coating for fruits and vegetables, for its film-forming, biodegradable, antimicrobial as well as biochemical properties (Misir et al., 2014). The work carried out on the shelf-life extension of micro-green is presented in Table 9.3.

TABLE 9.3 Extension of Shelf-Life of Micro-Green

Crop	Treatment and Its Effect	References
Buckwheat and spinach	Shelf-life of 2–4 days under ambient conditions and may extend to 10–14 days at 5°C when packed in 16.6 pmol/(m^2 s Pa) high OTR film.	Kou et al. (2013, 2014)
Radish	OTR polyethylene bags for longer shelf-life.	Xiao et al. (2014)
Daikon radish (*Raphanus sativus* var. *longipinnatus*)	14 days storage in OTR polyethylene bags, but optimum storage temperature was 1°C.	Xiao et al. (2014)
Sunflower	Packing in polystyrene trays for shelf-life extension and quality maintenance compared to LDPE bags.	Dalal et al. (2019)
Radish (*Raphanus sativus* L.) Roselle (*Hibiscus sabdariffa*)	Pre-harvest spray treatment with *Aloe vera* gel as an ergonomic pre-harvest treatment followed by packaging in PET clamshell for improving the postharvest quality and shelf-life extension.	Ghoora and Srividya (2020)

Micro-green is highly perishable, hence it cannot be stored or preserved over a long period in ambient as well as under refrigerated conditions. To overcome this, and further fulfillment of commercial utility of micro-greens, these are available in the market in the form of living micro-greens. Living micro-greens are the most nutritious and fresh greens stored under refrigerated conditions for up to 14 days or under ambient conditions (room temperature) for 4–6 days by fulfilling complete water requirements.

9.5 BIO-SAFETY OF MICRO-GREEN

Several postharvest factors that interact with the microbial load on micro-greens including proximity to the media at harvest, residual humidity after pre-packaging washing treatments and temperature during storage. It has been reported that the total aerobic mesophilic bacteria (AMB) plate count initially for microgreens (radish, buckwheat, and Chinese cabbage) that were not washed were 7.1, 7.2, and 7.8 log CFU/g, respectively. This AMB count is considered high when compared to cilantro and baby spinach (Allende et al., 2004; Chandra et al., 2012; Kou et al., 2013; Wang et al., 2004). According to Chandra et al. (2012), the soft-textured, delicate hypocotyls of micro-greens may favor microbial growth in comparison to their mature counterparts.

Washing of micro-greens particularly with chlorinated water prior to packaging, effectively reduces AMB population (Chandra et al., 2012). Initial AMB counts on the buckwheat micro-green before storage were reduced by a range of 0.3, 0.9, and 1.3 log CFU/g following water; 50 mg/L, and 100 mg/L (chlorinated wash treatments), respectively (Kou et al., 2013), whereas similar chlorinated water treatments on radish micro-greens did not resulted in effective outcomes (Xiao et al., 2014). Pre-harvest applications (200 mL) of calcium lactate, calcium amino acid chelates and particularly calcium chloride (10 mM at pH 6.5) inhibited proliferation of AMB, yeast, and mold in broccoli micro-green populations (Kou et al., 2015). Sanitation and expansion of industrial micro-green production still remains crucial procedures to work upon for the preparing packed and RTE micro-greens. Chandra et al. (2012) have illustrated a 2 min dip treatment in 0.5% (w/v) citric acid solution in combination with a 50% ethanol spray, that is counted as effective for standard industrial sodium hypochlorite disinfection (2 minutes dip in 100 mL^{-1} at 7.0 pH level) to control the proliferation of AMB as well as coliform population on the micro-greens of Chinese cabbage stored for 9 days at 5°C temperature under darkness.

Wild herbaceous plants are becoming increasingly popular for their nutritive profile and are favored and promoted by health-conscious individuals as part of a healthy lifestyle, particularly in wild, green-centered cuisines (Łuczaj and Pieroni, 2016). Wild greens are referred as an enriched source of proteins, fats, vitamins, sugars, and minerals (Garcia-Herrera et al., 2016;

Guil-Guerrero et al., 1999; Sanchez-Mata et al., 2016). Wide range of phytochemicals that has potential antioxidant activity are also reported in many of these specified species (Barros et al., 2016) and some studies have demonstrated the presence of molecules in wild species exhibiting antimicrobial potential (Di Venere et al., 2016) and other biological-pharmacological activities (Heinrich et al., 2016). Hence, some of the wild green plants have gained considerable importance as a of functional food, fortified food or additive powders.

Simultaneously, most of these plants are grown in humanly disturbed areas such as regions of human habitation (ruderals), farmlands (weeds), borders of pathways, and soil on sides of constructed roads rich in nitrate or contaminated by metallic trace elements (Sanchez-Mata and Morales, 2016) whose detrimental effects on humans are well established (Bryan et al., 2012; Jaishankar et al., 2014). Due to efficient root-to-shoot translocation paths, the plants accumulate high level of pollutants in the environmentally exposed aerial parts (Bandiera et al., 2016; Brunetti et al., 2009) and their utilization in food may entail health risks as well. Lenzi et al. (2019) studied and compared three leafy species (*Sinapis arvensis* L., *Sanguisorba minor* Scop. and *Taraxacum officinale* Weber ex F. H. Wigg.) from wild regions at micro-green as well as baby green stages and obtained competitive yields that enhances the dietary (Keservani et al., 2010a, b, 2017, 2020) intake of macro-elements, micro-elements, and bioactive compounds. In contrast, the wild small sized greens presented high amount of nitrates and metal traces which are potentially detrimental for health, thereby emphasizing the urgent need for cautious actions towards the utilization of wild species plants for developing micro-greens and baby leaves.

9.6 CONCLUSION

Diseases like cardiovascular, diabetes, sleep disorders, obesity, chronic respiratory diseases, and some forms of cancers are escalating largely because of rapid changes in lifestyle, and malnutrition is still affecting many countries, including India. Healthy food-based strategies are a reliable approach that empowers individuals, especially women and local households, to take over complete control of diet quality via the production and consumption of foods with better nutritional profiles. Micro-green, a new class of edibles, are setting up new trends in addressing diet and nutritional challenges because of the presence of 4–6 times higher health-promoting compounds like polyphenols, anthocyanins, vitamins, glucosinolates, minerals, phenolics, and carotenoids in them. The advancement of research on micro-green has now presented the potentiality of such edibles not only for homestead utility for securing nutrition but also for commercial exploration. There are many plant species currently being recommended and grown successfully worldwide as micro-green, and people are still investigating and exploring a wide variety of wild species for their utility as micro-green. No doubt, the cultivation of micro-green is very easy, however, they require a different set of environmental conditions when grown as a component of rural, urban/or peri-urban horticulture. The pre-and postharvest handling factors for better quality and shelf-life of micro-greens need due consideration owing to their highly perishable nature. Simultaneously, different wild species are being explored for their utility as micro-green. However, the presence of high amounts of nitrates and heavy metals suggests some precautions while using and recommending new crop species. Therefore, it is imperative to take some fundamental research on the inclusion of new edibles in the diet. Such efforts should also be supplemented with extension activities to sensitize and create awareness among the masses about the use of micro-green as excellent nutritional and healthy food choices.

KEYWORDS

- **aerobic mesophilic bacteria**
- **cultivation**
- **heavy metals**
- **microbial growth**
- **micro-greens**
- **postharvest factors**
- **soft-textured hypocotyls**

REFERENCES

Adadi, P., Barakova, N. V., & Krivoshapkina, E. F. (2018). Selected methods of extracting carotenoids, characterization, and health concerns: A review. *Journal of Agricultural and Food Chemistry*, *66*(24), 5925–5947.

Allende, A., Luo, Y., McEvoy, J. L., Artés, F., & Wang, C. Y. (2004). Microbial and quality changes in minimally processed baby spinach leaves stored under super atmospheric oxygen and modified atmosphere conditions. *Postharvest Biology and Technology*, *33*(1), 51–59.

Bandiera, M., Dal Cortivo, C., Barion, G., Mosca, G., & Vamerali, T. (2016). Phytoremediation opportunities with alimurgic species in metal-contaminated environments. *Sustainability*, *8*(4), 357.

Barros, L., Morales, P., Carvalho, A. M., & Ferreira, I. C. (2016). Antioxidant potential of wild plant foods. In *Mediterranean Wild Edible Plants: Ethnobotany and Food Composition Tables* (pp. 209–232).

Bazzano, L. A., He, J., Ogden, L. G., Loria, C. M., Vupputuri, S., Myers, L., & Whelton, P. K. (2002). Fruit and vegetable intake and risk of cardiovascular disease in US adults: The first National Health and Nutrition Examination Survey Epidemiologic Follow-up Study. *The American Journal of Clinical Nutrition*, *76*(1), 93–99.

Brunetti, G., Soler-Rovira, P., Farrag, K., & Senesi, N. (2009). Tolerance and accumulation of heavy metals by wild plant species grown in contaminated soils in Apulia region, Southern Italy. *Plant and Soil*, *318*, 285–298.

Bryan, N. S., Alexander, D. D., Coughlin, J. R., Milkowski, A. L., & Boffetta, P. (2012). Ingested nitrate and nitrite and stomach cancer risk: An updated review. *Food and Chemical Toxicology*, *50*(10), 3646–3665.

Butkutė, B., Taujenis, L., & Norkevičienė, E. (2018). Small-seeded legumes as a novel food source. Variation of nutritional, mineral, and phytochemical profiles in the chain: Raw seeds-sprouted seeds-microgreens. *Molecules*, *24*(1), 133.

Chandra, D., Kim, J. G., & Kim, Y. P. (2012). Changes in microbial population and quality of microgreens treated with different sanitizers and packaging films. *Horticulture, Environment, and Biotechnology*, *53*, 32–40.

Choe, U., Yu, L. L., & Wang, T. T. (2018). The science behind microgreens as an exciting new food for the 21st century. *Journal of Agricultural and Food Chemistry*, *66*(44), 11519–11530.

Clarkson, G. J., Rothwell, S. D., & Taylor, G. (2005). End of day harvest extends shelf life. *HortScience*, *40*(5), 1431–1435.

Dalal, N., & Siddiqui, S. (2019). Effect of chemical treatment, storage, and packaging on physico-chemical properties of sunflower microgreens. *International Journal of Chemical Studies*, *7*(5), 1046–1050.

de la Fuente, B., López-García, G., Máñez, V., Alegría, A., Barberá, R., & Cilla, A. (2019). Evaluation of the bioaccessibility of antioxidant bioactive compounds and minerals of four genotypes of Brassicaceae microgreens. *Foods*, *8*(7), 250.

Delaquis, P., & Mazza, G. (1998). Functional vegetable products. In *Functional Foods: Biochemical and Processing Aspects* (pp. 193–233). Lancaster: Technomic Publishing Company, Inc.

Di Venere, D., Gatto, M. A., Ippolito, A., & Bianco, V. V. (2016). Antimicrobial potential of wild edible herbaceous species. In *Mediterranean Wild Edible Plants: Ethnobotany and Food Composition Tables* (pp. 233–252).

Donaldson, M. S. (2004). Nutrition and cancer: A review of the evidence for an anti-cancer diet. *Nutrition Journal*, *3*(1), 1–21.

Ebert, A. W., Wu, T. H., & Yang, R. Y. (2014). Amaranth sprouts and microgreens–a homestead vegetable production option to enhance food and nutrition security in the rural-urban continuum. In *Proceedings of the Regional Symposium on Sustaining Small-Scale*

Vegetable Production and Marketing Systems for Food and Nutrition Security (SEAVEG 2014), Bangkok, Thailand, 25–27.

Fortunati, E., Giovanale, G., Luzi, F., Mazzaglia, A., Kenny, J. M., Torre, L., & Balestra, G. M. (2017). Effective postharvest preservation of kiwifruit and romaine lettuce with a chitosan hydrochloride coating. *Coatings*, *7*(11), 196.

Fuentes, F., Paredes-Gonzalez, X., & Kong, A. N. T. (2015). Dietary glucosinolates sulforaphane, phenethyl isothiocyanate, indole-3-carbinol/3, 3′-diindolylmethane: Antioxidative stress/inflammation, Nrf2, epigenetics/epigenomics and *in vivo* cancer chemopreventive efficacy. *Current Pharmacology Reports*, *1*, 179–196.

Gan, R. Y., Lui, W. Y., Wu, K., Chan, C. L., Dai, S. H., Sui, Z. Q., & Corke, H. (2017). Bioactive compounds and bioactivities of germinated edible seeds and sprouts: An updated review. *Trends in Food Science & Technology*, *59*, 1–14.

García-Herrera, P., & Sánchez-Mata, M. D. C. (2016). The contribution of wild plants to dietary intakes of micronutrients (II): Mineral elements. In *Mediterranean Wild Edible Plants: Ethnobotany and Food Composition Tables* (pp. 141–171).

Garrido, Y., Tudela, J. A., & Gil, M. I. (2015). Time of day for harvest and delay before processing affect the quality of minimally processed baby spinach. *Postharvest Biology and Technology*, *110*, 9–17.

Ghoora, M. D., & Srividya, N. (2020). Effect of packaging and coating technique on postharvest quality and shelf life of *Raphanus sativus L.* and *Hibiscus sabdariffa L.* microgreens. *Foods*, *9*(5), 653.

González-Lamothe, R., Mitchell, G., Gattuso, M., Diarra, M. S., Malouin, F., & Bouarab, K. (2009). Plant antimicrobial agents and their effects on plant and human pathogens. *International Journal of Molecular Sciences*, *10*(8), 3400–3419.

Guerrero, J. G., Madrid, P. C., & Isasa, M. T. (1999). Mineral elements determination in wild edible plants. *Ecology of Food and Nutrition*, *38*(3), 209–222.

Hasperué, J. H., Guardianelli, L., Rodoni, L. M., Chaves, A. R., & Martínez, G. A. (2016). Continuous white–blue LED light exposition delays postharvest senescence of broccoli. *LWT-Food Science and Technology*, *65*, 495–502.

Heinrich, M., Kerrouche, S., & Bharij, K. S. (2016). Recent advances in research on wild food plants and their biological–pharmacological activity. In *Mediterranean Wild Edible Plants: Ethnobotany and Food Composition Tables* (pp. 253–269).

Herr, I., & Büchler, M. W. (2010). Dietary constituents of broccoli and other cruciferous vegetables: Implications for prevention and therapy of cancer. *Cancer Treatment Reviews*, *36*(5), 377–383.

Hodges, D. M., & Toivonen, P. M. (2008). Quality of fresh-cut fruits and vegetables as affected by exposure to abiotic stress. *Postharvest Biology and Technology*, *48*(2), 155–162.

Huang, H., Jiang, X., Xiao, Z., Yu, L., Pham, Q., Sun, J., Chen, P., Yokoyama, W., Yu, L. L., Luo, Y. S., & Wang, T. T. (2016). Red cabbage microgreens lower circulating low-density lipoprotein (LDL), liver cholesterol, and inflammatory cytokines in mice fed a high-fat diet. *Journal of Agricultural and Food Chemistry*, *64*(48), 9161–9171.

Jaglan, P., Kumar, V., Suthar, P., Paul, A. A., & Kumar, S. (2020). Recent advances in the profiling and characterization of antioxidants. In P. Kaur (Ed.), *Oxidative Stress and Natural Antioxidants*. BenthamBriefs in Biomedicine and Pharmacotherapy, *1*, 172–208.

Jaishankar, M., Tseten, T., Anbalagan, N., Mathew, B. B., & Beeregowda, K. N. (2014). Toxicity, mechanism, and health effects of some heavy metals. *Interdisciplinary Toxicology*, *7*(2), 60.

Kader, A. A. (2002). US grade standards. In *Postharvest Technology of Horticultural Crop* (p. 287–300). University of California Agriculture and Natural Resources, *3311*(287).

Keservani, R. K., & Sharma, A. K. (2014). Flavonoids: Emerging trends and potential health benefits. *Journal of Chinese Pharmaceutical Sciences*, *23*(12), 815.

Keservani, R. K., Kesharwani, R. K., Sharma, A. K., Vyas, N., & Chadoker, A. (2010b). Nutritional supplements: An overview. *International Journal of Current Pharmaceutical Review and Research*, *1*(1), 59–75.

Keservani, R. K., Kesharwani, R. K., Vyas, N., Jain, S., Raghuvanshi, R., & Sharma, A. K. (2010a). Nutraceutical and functional food as future food: A review. *Der Pharmacia Lettre*, *2*(1), 106–116.

Keservani, R. K., Sharma, A. K., & Kesharwani, R. K. (2016a). Nutraceutical and functional foods for cardiovascular health. In *Food Process Engineering* (pp. 257–278). Apple Academic Press, CRC Press.

Keservani, R. K., Sharma, A. K., & Kesharwani, R. K. (2016b). Medicinal effect of nutraceutical fruits for cognition and brain health. *Scientifica*, *3109254*, 10. https://doi.org/10.1155/2016/3109254.

Keservani, R. K., Sharma, A. K., & Kesharwani, R. K. (2017). An overview and therapeutic applications of nutraceutical and functional foods. In *Recent Advances in Drug Delivery Technology* (pp. 160–201).

Keservani, R. K., Sharma, A. K., & Kesharwani, R. K. (2020). *Nutraceuticals and Dietary Supplements: Applications in Health Improvement and Disease Management*. CRC Press, 1–342.

Klopsch, R., Baldermann, S., Voss, A., Rohn, S., Schreiner, M., & Neugart, S. (2018). Bread enriched with legume microgreens and leaves—Ontogenetic and baking-driven changes in the profile of secondary plant metabolites. *Frontiers in Chemistry*, *6*, 322.

Kou, L., Luo, Y., Park, E., Turner, E. R., Barczak, A., & Jurick II, W. M. (2014). Temperature abuse timing affects the rate of quality deterioration of commercially packaged ready-to-eat baby spinach. Part I: Sensory analysis and selected quality attributes. *Postharvest Biology and Technology*, *91*, 96–103.

Kou, L., Luo, Y., Yang, T., Xiao, Z., Turner, E. R., Lester, G. E., Wang, Q., & Camp, M. J. (2013). Postharvest biology, quality, and shelf life of buckwheat microgreens. *LWT-Food Science and Technology*, *51*(1), 73–78.

Kou, L., Yang, T., Liu, X., & Luo, Y. (2015). Effects of pre- and postharvest calcium treatments on shelf life and postharvest quality of broccoli microgreens. *HortScience*, *50*(12), 1801–1808.

Kowitcharoen, L., Phornvillay, S., Lekkham, P., Pongprasert, N., & Srilaong, V. (2021). Bioactive composition and nutritional profile of microgreens cultivated in Thailand. *Applied Sciences*, *11*(17), 7981.

Kumar, S., Patel, N., Singh, S., & Saravaiya, N. S. (2020). Effect of different light sources on growth and quality of microgreens. *Research Report—Horticulture & Agro-Forestry Sub-Committee of Navsari Agricultural University*, Navsari, India, 348–371.

Kumar, S., Sharma, S., Revathi, R., & Minhas, A. (2020). Cruciferous vegetables: A mine of phytonutrients for functional and nutraceutical enrichment. In S. K. Gupta, S. Sharma, & A. Minhas (Eds.), *Current Advances for Development of Functional Foods Modulating Inflammation and Oxidative Stress* (pp. 1–22). Elsevier.

Kumar, V., Kaur, J., Sharma, K., Kumar, S., & Sharma, R. (2022). Role of analytical techniques in food quality control and safety. In P. Chandra & P. S. Panesar (Eds.), *Nanosensing and*

Bioanalytical Technologies in Food Quality Control (pp. 1–18). Springer. https://doi.org/10.1007/978-981-16-7029-9_12.

Kyriacou, M. C., De Pascale, S., Kyratzis, A., & Rouphael, Y. (2017). Microgreens as a component of space life support systems: A cornucopia of functional food. *Frontiers in Plant Science*, *1587*.

Kyriacou, M. C., El-Nakhel, C., Graziani, G., Pannico, A., Soteriou, G. A., Giordano, M., Ritieni, A., De Pascale, S., & Rouphael, Y. (2019). Functional quality in novel food sources: Genotypic variation in the nutritive and phytochemical composition of thirteen microgreens species. *Food Chemistry*, *277*, 107–118.

Kyriacou, M. C., Rouphael, Y., Di Gioia, F., Kyratzis, A., Serio, F., Renna, M., De Pascale, S., & Santamaria, P. (2016). Micro-scale vegetable production and the rise of microgreens. *Trends in Food Science & Technology*, *57*, 103–115.

Lenzi, A., Orlandini, A., Bulgari, R., Ferrante, A., & Bruschi, P. (2019). Antioxidant and mineral composition of three wild leafy species: A comparison between microgreens and baby greens. *Foods*, *8*(10), 487.

Łuczaj, Ł., & Pieroni, A. (2016). Nutritional ethnobotany in Europe: From emergency foods to healthy folk cuisines and contemporary foraging trends. In *Mediterranean Wild Edible Plants: Ethnobotany and Food Composition Tables* (pp. 33–56).

Meyerowitz, S. (2006). *Wheatgrass: Nature's Finest Medicine: The Complete Guide to Using Grass Foods & Juices to Help Your Health*. Sproutman Publications.

Mir, S. A., Shah, M. A., & Mir, M. M. (2017). Microgreens: Production, shelf life, and bioactive components. *Critical Reviews in Food Science and Nutrition*, *57*(12), 2730–2736.

Misir, J., Brishti, F. H., & Hoque, M. M. (2014). Aloe vera gel as a novel edible coating for fresh fruits: A review. *American Journal of Food Science and Technology*, *2*(3), 93–97.

Morrison, J. L., & Regnault, T. R. (2016). Nutrition in pregnancy: Optimising maternal diet and fetal adaptations to altered nutrient supply. *Nutrients*, *8*(6), 342.

Niizu, P. Y., & Rodriguez-Amaya, D. B. (2005). New data on the carotenoid composition of raw salad vegetables. *Journal of Food Composition and Analysis*, *18*(8), 739–749.

Niranjana, R., Gayathri, R., Mol, S. N., Sugawara, T., Hirata, T., Miyashita, K., & Ganesan, P. (2015). Carotenoids modulate the hallmarks of cancer cells. *Journal of Functional Foods*, *18*, 968–985.

Niroula, A., Khatri, S., Timilsina, R., Khadka, D., Khadka, A., & Ojha, P. (2019). Profile of chlorophylls and carotenoids of wheat (*Triticum aestivum* L.) and barley (*Hordeum vulgare* L.) microgreens. *Journal of Food Science and Technology*, *56*, 2758–2763.

Nishino, H., Murakoshi, M., Tokuda, H., & Satomi, Y. (2009). Cancer prevention by carotenoids. *Archives of Biochemistry and Biophysics*, *483*(2), 165–168.

Paradiso, V. M., Castellino, M., Renna, M., Gattullo, C. E., Calasso, M., Terzano, R., Allegretta, I., Leoni, B., Caponio, F., & Santamaria, P. (2018). Nutritional characterization and shelf-life of packaged microgreens. *Food & Function*, *9*(11), 5629–5640.

Pérez-Escamilla, R. (2017). Food security and the 2015–2030 sustainable development goals: From human to planetary health: Perspectives and opinions. *Current Developments in Nutrition*, *1*(7), e000513.

Pinto, E., Almeida, A. A., Aguiar, A. A., & Ferreira, I. M. (2015). Comparison between the mineral profile and nitrate content of microgreens and mature lettuces. *Journal of Food Composition and Analysis*, *37*, 38–43.

Portela, S. I., & Cantwell, M. I. (2001). Cutting blade sharpness affects appearance and other quality attributes of fresh-cut cantaloupe melon. *Journal of Food Science*, *66*(9), 1265–1270.
Rao, A. V., & Rao, L. G. (2007). Carotenoids and human health. *Pharmacological Research*, *55*(3), 207–216.
Renna, M., Di Gioia, F., Leoni, B., Mininni, C., & Santamaria, P. (2017). Culinary assessment of self-produced microgreens as basic ingredients in sweet and savory dishes. *Journal of Culinary Science & Technology*, *15*(2), 126–142.
Riggio, G. M., Jones, S. L., & Gibson, K. E. (2019). Risk of human pathogen internalization in leafy vegetables during lab-scale hydroponic cultivation. *Horticulturae*, *5*(1), 25.
Riggio, G. M., Wang, Q., Kniel, K. E., & Gibson, K. E. (2019a). Microgreens—A review of food safety considerations along the farm to fork continuum. *International Journal of Food Microbiology*, *290*, 76–85.
Sánchez-Mata, D., & Morales, R. (2016). The Mediterranean landscape and wild edible plants. In *Mediterranean Wild Edible Plants: Ethnobotany and Food Composition Tables* (pp. 15–31).
Sánchez-Mata, M. D. C., Matallana-González, M. C., & Morales, P. (2016). The contribution of wild plants to dietary intakes of micronutrients (I): Vitamins. In *Mediterranean Wild Edible Plants: Ethnobotany and Food Composition Tables* (pp. 111–139).
Sean, M. (2002). Wheatgrass, healthy for the body and the bank account. *Australia's National Rural Affairs Weekly-Landline*, *1310*, 2006–1006.
Sharma, D., Shree, B., Kumar, S., Kumar, V., Sharma, S., & Sharma, S. (2020). Stress induced production of plant secondary metabolites in vegetables: Functional approach for designing next generation super foods. *Plant Physiology and Biochemistry*. https://doi.org/10.1016/j.plaphy.2022.111038.
Sharma, K., Kaur, R., Kumar, V., Kumar, S., Singh, A., & Gautam, N. (2022). Prebiotic and probiotic potential of cereals. In K. Kaur & S. Kumar (Eds.), *Functional Cereals and Cereal Foods* (pp. 163–188). Springer.
Sharma, S., Dhingra, P., & Koranne, S. (2020). Microgreens: Exciting new food for 21st Century. *Ecology, Environment and Conservation*, *26*, 2020.
Sharma, S., Katoch, V., Kumar, S., & Chatterjee, S. (2021). Functional relationship of vegetable colors and bioactive compounds: Implications in human health. *Journal of Nutritional Biochemistry*. https://doi.org/10.1016/j.jnutbio.2021.108634.
Sharma, S., Shree, B., Sharma, D., Kumar, S., Kumar, V., Sharma, R., & Saini, R. (2022). Vegetable microgreens: The gleam of next generation super foods, their genetic enhancement, health benefits and processing approaches. *Food Research International*, *111038*.
Shree, B., Kumar, S., Sharma, S., & Katoch, V. (2022). Functional significance of underutilized high value cruciferous vegetables-an exotic gleam in the gloomy guise of their functional importance. *South African Journal of Botany*. https://doi.org/10.1016/j.sajb.2022.111038.
Sun, J., Xiao, Z., Lin, L. Z., Lester, G. E., Wang, Q., Harnly, J. M., & Chen, P. (2013). Profiling polyphenols in five Brassica species microgreens by UHPLC-PDA-ESI/HRMSn. *Journal of Agricultural and Food Chemistry*, *61*(46), 10960–10970.
Tay, S. L., & Perera, C. O. (2004). Effect of 1-methylcyclopropene treatment and edible coatings on the quality of minimally processed lettuce. *Journal of Food Science*, *69*(2), fct131–fct135.

Turner, E. R., Luo, Y., & Buchanan, R. L. (2020). Microgreen nutrition, food safety, and shelf life: A review. *Journal of Food Science*, *85*(4), 870–882.

Wang, H., Feng, H., & Luo, Y. (2004). Microbial reduction and storage quality of fresh-cut cilantro washed with acidic electrolyzed water and aqueous ozone. *Food Research International*, *37*(10), 949–956.

Wood. (2021). Worldwide indoor farming market outlook 2019–2024—The decrease in cultivable land is driving growth. *Research and Markets*.

World Health Organization (WHO). (2021). Malnutrition. https://www.who.int/news-room/fact-sheets/detail/mal-nutrition (accessed on 5 November 2024).

Xiao, Z., Codling, E. E., Luo, Y., Nou, X., Lester, G. E., & Wang, Q. (2016). Microgreens of Brassicaceae: Mineral composition and content of 30 varieties. *Journal of Food Composition and Analysis*, *49*, 87–93.

Xiao, Z., Lester, G. E., Luo, Y., & Wang, Q. (2012). Assessment of vitamin and carotenoid concentrations of emerging food products: Edible microgreens. *Journal of Agricultural and Food Chemistry*, *60*(31), 7644–7651.

Xiao, Z., Lester, G. E., Luo, Y., Xie, Z. K., Yu, L. L., & Wang, Q. (2014). Effect of light exposure on sensorial quality, concentrations of bioactive compounds and antioxidant capacity of radish microgreens during low temperature storage. *Food Chemistry*, *151*, 472–479.

Xiao, Z., Luo, Y., Lester, G. E., Kou, L., Yang, T., & Wang, Q. (2014). Postharvest quality and shelf life of radish microgreens as impacted by storage temperature, packaging film, and chlorine wash treatment. *LWT-Food Science and Technology*, *55*(2), 551–558.

Xiao, Z., Rausch, S. R., Luo, Y., Sun, J., Yu, L., Wang, Q., Chen, P., Yu, L., & Stommel, J. R. (2019). Microgreens of Brassicaceae: Genetic diversity of phytochemical concentrations and antioxidant capacity. *LWT*, *101*, 731–737.

CHAPTER 10

Carrot and Carrot Products: A Superb Functional Food

SAKSHI SHARMA,[1*] GHANSHYAM ABROL,[2] RAKESH SHARMA,[1] and KRISHAN DUTT SHARMA[1]

[1]*Department of Food Science and Technology, Dr. Y. S. Parmar University of Horticulture and Forestry, Solan, Himachal Pradesh, India*

[2]*Rani Lakshmi Bai Central Agricultural University, Jhansi, Uttar Pradesh, India*

*Corresponding author.

ABSTRACT

Carrot is one of the major root vegetables grown extensively in about 125 countries in the world. China is the largest producer of carrots, Russia at second position followed by the USA, Uzbekistan, Poland, Ukraine, the UK, Turkey, Japan, and France. Carrot root has a significant amount of phenolic acids and carotenoids, as well as a number of other bioactive components with numerous health advantages. As a result of its increasing popularity as a significant source of natural antioxidants, consumption of fresh carrot and its products are rising significantly. Carotenoids helps in lowering blood sugar level, avoiding heart disease, improving vision, and lowering colon cancer. However, due to the seasonal and perishable nature of carrots, it is not possible to make it readily available throughout the year. Improved postharvest handling practices and value-addition by processing in the growing season is a most important alternative. Because of the nutritional and therapeutic importance and its abundant availability particularly during the winter season, the postharvest researchers must suggest viable alternatives for its profitable use. Efforts have been made to review the information available on this very important root vegetable for the benefit of the researchers and the consumers.

Functional Compounds and Foods of Plant Origin:Krishan Dutt Sharma, Satish Kumar, Rakesh Sharma, and Raj K. Keservani (Eds.)

10.1 INTRODUCTION

Carrot (*Daucus carota* L.) is well-known root vegetable that has been cultivated throughout the world and is a good source of various bioactive compounds such as carotenoids and phenolic acids, and other phytochemicals containing valuable health-promoting properties, are found in higher amounts. The carrot is considered a primary vegetable in many countries. In recent years, carrot is becoming popular due to increasing in awareness of its health-improving and nutritional significance besides, flavor and texture enhancing properties in many diets across the world. There are numerous types of carrots, including orange, yellow, red, white, and purple. This genetic diversity coupled with growing circumstances and exposure to ethylene impacts the sensory (volatiles and polyacetylenes), nutritional (carotenoids, vitamins, and minerals), and quality characteristics.

Yellow and purple-fleshed cultivars were the first cultivated carrots. However, the orange carrot which is popular today was introduced in the 15th century in Central Europe. A boosting in the popularity of this cultivar was due to its high pro-vitamin A content (Nowacka and Wedzik, 2016). The major antioxidant pigments that present in carrots are carotenoids and anthocyanins (Sharma, 2021). The type of pigment present in the carrot is responsible for the cultivar differences. Carotenoids, the phytochemicals responsible for the yellow, orange, or red color in many yellow and orange-fleshed cultivars, are present in significant quantities. The popular orange carrot is high in α and β-carotene (Sharma and Sharma, 2020). The carotene concentration in orange and yellow-fleshed varieties rises as they mature, and the outer layer contains a higher amount of carotenes compared to the inner core. In the last 40 years, conventional breeding methods have significantly boosted the overall carotenoid levels, reaching 1000 ppm carotenoids per unit of fresh weight. Additionally, carrots can offer a substantial vitamin A content in the human diet because of the enhanced bioavailability of carotenoids in comparison to other vegetables. The yellow carrot color is due to lutein a carotenoid, which plays a major role in preventing of macular degeneration (Sun et al., 2009). The red-colored carrots are rich in lycopene. Whereas purple carrot possesses a high amount of anthocyanin (Sun et al., 2009), and white flesh cultivars contain very few pigments.

Carrots were initially being used for medicinal purposes and then slowly started using as food. Carrot is rich in pro-vitamin A, ascorbic acid, vitamin K, dietary fiber, potassium, and other bioactive compounds. The phenolic acid present in raw carrot is chlorogenic acid (CGA), along with caffeic acid as the second most prevalent. The consumption of carrots and carrot-based products is on the rise, driven by their recognition as valuable sources of natural antioxidants with potential anti-cancer properties. Carotenoids found in carrots contribute to heart disease prevention, vision improvement, blood sugar reduction, and the mitigation of colon cancer risks (Saini, 2020). Additionally, carotenoids function as effective scavengers of free radicals. Phenolic compounds have received significant interest due to their physiological roles, which encompass antioxidant, antimutagenic, and antitumor properties (Bendokas et al., 2020). Antioxidants are naturally present compounds in food that play a crucial role in mitigating the harmful impacts of reactive oxygen species (ROS) and free radicals, which are implicated in the development of degenerative human diseases like cancer and heart diseases, among others.

However, because carrots are perishable and subject to seasonal availability, it is not feasible to ensure their year-round availability. One vital approach is processing them during the primary growing season. Carrots and their various derivatives encompass a wide range, each necessitating distinct processing and production methods. Similar to other vegetables, the processing carrots deteriorate the final quality of carrots due to various factors, including exposure to heat, handling, mechanical damage, and accelerated biochemical reactions. Consequently, the primary focus for food processors is to maintain the sensory attributes as well as nutritional content of processed carrot. Fortunately, there are several feasible processing technologies accessible, whether on a household or at industrial scale. A thorough grasp of carrot production, its nutritional content, and applicable processing techniques is invaluable for effectively utilizing and preparing carrots and their associated products. This affordable root vegetable has the versatility to be transformed into a multitude of value-added products, including juice, beverages, candies, preserves, soups, stews, curries, pies, halwa, ice cream, and various cooked vegetable dishes (Iorizzo et al., 2020; Sharma et al., 2021, 2022).

10.2 DISTRIBUTION AND PRODUCTION

Carrots are believed to originate in Afghanistan, with evidence of cultivation dating back 5000 years. They were grown in Egypt around 2000 BC and brought to southern Europe by Moorish invaders. By the 13th century, they were cultivated in China, Germany, France, and England. Orange carrots, the most common type, appeared in the Netherlands in the 16th century. They were introduced to the USA about 250 years ago and to Canada about 150 years later. Carrots also reached India from Persia. China leads as the world's largest carrot producer, followed by Russia, the USA, Uzbekistan, Poland, Ukraine, the UK, Turkey, Japan, and France. However, India is ranked at 14th in its production amongst 125 countries that cultivate carrots. During the year 2016–2017, total carrot production in India was estimated to tune at 1,379,000 tons from an 88,000-hectare area (NHB, 2017). In India, the major carrot-growing states include Andhra Pradesh, Haryana, Uttar Pradesh, Assam, Punjab, Tamil Nadu, and Karnataka.

Carrots are grown worldwide, with planting seasons varying by climate. Tropical areas cultivate them in winter, while temperate regions grow them in spring, summer, and autumn. Carrots exhibit diverse colors: red due to lycopene, orange from β-carotene (a precursor of vitamin A), and purple or yellow due to anthocyanin and xanthophylls pigments, respectively. Carrots are grown on a large scale in many countries, including China, Russia, the USA, Uzbekistan, Poland, England, India, Malaysia, Indonesia, the Philippines, Central and West Africa, South America, and the Caribbean. In England, they rank as the second most important vegetable after potatoes. In Australia, carrot is the third most important vegetable after potato and tomato. China is the major carrot-producing country in the world. In India, key carrot-producing states include Uttar Pradesh, Assam, Karnataka, Andhra Pradesh, Punjab, and Haryana. Carrot yield depends on plant characteristics and environmental factors, while subpar farming practices and inadequate cultivars can lead to lower yields.

10.3 NUTRITIONAL COMPOSITION OF CARROT

Carrot plants are composed of two primary components: the storage root and the fragile leaves. Although the root is the most frequently consumed in both Japan and China, however; young tender foliage is sometimes incorporated into stir-fries or salads. Carrots do not significantly contribute to calorie intake in human diets, but they do provide essential dietary fiber and a range of nutrients in the form of phytochemicals, such as carotenoids, anthocyanins, and various phenolic compounds (Clarkson et al., 2021). The nutritional composition of carrots, as shown in Table 10.1, can vary based on factors like cultivar, maturity, season, and environmental conditions.

TABLE 10.1 Approximate Composition of Carrot Root

Characteristic	Amount	References
Moisture	86.0–92.4%	Karki (2009); Mannozzi et al. (2019); Upadhyay et al. (2008)
Total soluble solids	8–16.1% (Asiatic cvs.)	
	14.3–15.6% (European cvs.)	
Total sugars	2.71–6.36%	
Ascorbic acid	3–9 mg/100 g	
Total carotenoid	6–54.8 mg/100 g	
Total polyphenolics	0.10–0.26 mg/g	
Dietary fiber	37–55.8%	
Crude fiber	1.2–2.8 g/100 g	
Protein	0.7–0.93 g/100 g	
Fat	0.2–0.5 g/100 g	

The moisture content of fresh carrots ranging from 86–92.4% has been observed by various investigators. Karki (2009) have reported 8% total soluble solids in fresh carrot. However, Total soluble solids in 30 varieties of carrot ranging from 6.57–8.69% have been observed by Anand (2001). Whereas Sra et al. (2011) reported the total sugars of 4.3, 4.2, 4.4, and 4.4%, in SEL-21, PC-34, Ambala local, and Nantes varieties of carrot, respectively. Karki (2009) has recorded higher (6.84%) total sugar content in fresh carrots. The total phenolic content (TPC) in carrots as 2.66 and 12.40 mg/100 g has been reported by Ferreira et al. (2015). Whereas 19.20 and 55.00 mg/100 g total phenolics in fresh carrots have been reported by Karki (2009). Bystricka et al. (2015) have reported a range of 8.13–11.37 mg/100 g of total phenols in five varieties of carrot. The fiber content 1.2% by Gopalan et al. (1991), 1.87% by Upadhyay et al. (2008) and 0.67% in fresh carrot has been reported by Karki (2009). On a dry weight basis, crude fiber as 2.33% has been reported by Tadesse et al. (2015).

Carrots have been considered one of the best sources of carotenoids. Carotenoids contain mainly β-carotene, that is, about 80%. The carotenoid concentration in carrots typically falls within the range of 60 to 120 mg/100 grams, although certain varieties can exhibit levels as high as 300 mg/100 grams. Karki (2009) recorded 5.69 mg/100 g. Sra et al. (2011) have recorded very high total carotenoids as 30.70, 35.80, 34.60, and 38.20 mg/100 g in "SEL-21," "PC-34," "Ambala local," and "Nantes" varieties of carrot, respectively. The protein in fresh carrots as 0.70% has been reported by Gopalan et al. (1991). A slightly higher protein content of 1.65% in fresh carrots has been reported by Al-Amin et al. (2015) and 1.70% by Karmoker et al. (2011). Gopalan et al. (1991) have reported 0.20% fat in fresh carrots, whereas, Karmoker et al. (2011); and Al-Amin et al. (2015) have recorded 0.30% fat content in fresh carrots. Crude fat on a dry weight basis in carrots has been reported as 1.92% by Tadesse et al. (2015).

Carrots are rich in carbohydrates and contain minerals, including calcium, phosphorus, iron, magnesium, sodium, potassium, copper, and zinc. According to Gopalan et al. (1991), their mineral content per 100 grams is as follows: Ca (34.00 mg), Fe (0.40 mg), P (25.00 mg), Na (40.00 mg), K (240.00 mg), Mg (9.00 mg), Cu (0.02 mg), and Zn (0.20 mg). Furthermore, carrots contain polyacetylenes as a secondary metabolite, ranging from 20.0 to 100.0 mg per kilogram, as reported by Baranska et al. (2005) and Christensen and Kreutzmann (2007). These polyacetylenes play a role in protecting the plant from pathogens and environmental stresses. Elevated levels of polyacetylenes display toxicity towards bacteria, fungi, and mammalian cells. They can also trigger allergic skin responses (Christensen and Brandt, 2006). Furthermore, polyacetylenes are among the compounds accountable for the development of the bitter taste that may occur in stored carrots.

10.4 FUNCTIONAL PROPERTIES

Carrot root and its products have been used for medicinal purposes for well over a century. For example, an infusion of carrot has been used as a folk remedy for the treatment of threadworms. Carrot juice is reported to be a diuretic, and it facilitates the elimination of uric acid as well as favorably affects nitrogen balance. Carrots offer several nutritional advantages, with the fresh product being low in energy value and a good source of fiber, potassium, and other minerals. Carrots are rich in carotene, especially β-carotene, making them a top source among vegetables. Carotenes not only support vision but may also provide protection against cancer and heart disease (Seregelj et al., 2021). In addition, it has recently been demonstrated that the juices extracted from raw carrots have anti-microbial properties, particularly against Listeria. Carrots are a significant source of various phytonutrients, including phenolics, polyacetylenes (as noted by Kidmose et al., 2004), and carotenoids. The substantial presence of β-carotene is believed to play a significant role in the biological and medicinal attributes of carrots, as emphasized by Seregelj et al. in their 2021 study. Carotenoids have been linked to fortifying the immune system and lowering the risk of degenerative conditions like cancer, cardiovascular disease (CVD), age-related macular degeneration, and cataracts. Additionally, carotenoids have been identified as potential inhibitors of Alzheimer's disease (López-Gámez et al., 2021). A multitude of animal experiments and epidemiological investigations have suggested that carotenoids can impede carcinogenesis in rodents and may exert anti-carcinogenic effects in humans. Within biological systems, β-carotene functions as a free radical scavenger and a quencher of singlet oxygen, displaying properties such as anti-mutagenic, chemo-preventive, photo-protective, and immune-enhancing, as elaborated upon by Ghorbani et al. in their 2021 discussion. Over the last 10 years, carotenoids like β-carotene have garnered significant attention due to their protective effects to prevent the prevalence of certain types of cancers. Furthermore, carrots are rich in ascorbic acid (Vitamin C), β-carotene, and tocopherol, classifying them as vitamin-enriched foods (Iid et al., 2020; Kumar et al., 2022; Sharma et al., 2022).

10.5 HARVESTING AND HANDLING OF FRESH CARROT

Harvesting of carrots is done less than three months after sowing when its crown diameter reaches 2.50 cm. In the fresh market, carrots are typically harvested when they reach a diameter of about 1.80 cm or larger at the upper end, which is considered partially mature. For fresh-cut processing, carrots are harvested when they are tender and sweet. Delayed harvesting can enhance their ability to withstand extended storage by reducing the risk of decay. A high-quality carrot is characterized by being firm, straight from top to bottom, smooth with minimal hairiness, sweet without any bitterness or harshness in taste, and free from cracks or sprouting. Delaying the harvest of carrots can lead to woody roots and a decline in quality. The timing of harvest significantly impacts the chemical composition of carrots. Carrots harvested later tend to score higher in sensory attributes such as juiciness, crispness, sweetness, and overall flavor, but they may have lower bitterness scores, making them more preferred by consumers. In controlled environments like climate chambers, temperature plays a more crucial role than light in altering carrot sensory qualities. Carrots grown at lower, consistent temperatures (9–12°C) exhibit increased sweetness, acidity, crispness, juiciness, fructose and glucose content, weight, and root length were compared to those cultivated at higher temperatures (18°C to 21°C). Carrots are typically ready for harvesting within 100 to 120 days from sowing, with the desired size being a primary consideration. Marketable carrots should have a diameter of at least 2.5 to 4 cm at the upper end, and smaller to medium-sized carrots are generally preferred for the fresh market. For processing purposes, carrots with a thickness of about 3.0 cm at the crown and a length of about

10 cm are preferred, while those intended for the fresh market are typically harvested when they reach a thickness of 2–3 cm at the crown.

Carrots are typically harvested using either manual labor or mechanical systems like digger elevators or top-lifting machines. The harvesting of carrot from hand results in approximately 10% damage to the roots once they are lifted, whereas mechanical harvesting can cause damage ranging from 20% to 30% of the harvested roots. After the leaves are removed, the roots undergo a vigorous rubbing and washing process with ample water to eliminate hairiness and even outer skin, enhancing their visual appeal. On a smaller scale, washing is done manually, but for commercial operations, washing machines equipped with eccentrically mounted drums and running water are employed to expedite and economize the process. Asian carrot varieties generally yield more roots per hectare compared to European varieties. Carrot processing varieties, known for their larger root sizes, also yield more than those destined for the fresh market. Asian carrot varieties typically yield between 300 and 400 quintals per hectare, while European processing varieties yield around 300–350 quintals per hectare, and fresh market varieties yield approximately 200–250 quintals per hectare. The size of the carrots serves as the basis for harvesting. Early carrots intended for the market are pulled out when they are partially developed. They are usually dug out from sufficiently moist soil using a spade, after which the roots are trimmed and washed before being sent to the market. If sold without tops, the petioles should be trimmed to leave about a half-inch green "stem." To prolong the shelf life of carrots, it is crucial to rapidly cool them after harvest. Prior to being brought to market, the leaves are removed, and the carrots are sorted based on their size. Mechanical grading systems are also in use. Any misshapen, cracked, forked, or poorly developed roots are identified and separated for rejection. Carrot roots have a relatively low respiration rate compared to other vegetables. When stored at room temperature, carrots can be kept for about 3–4 days. However, under controlled conditions with temperatures ranging from 0–4°C and humidity levels of 93–96%, they can be stored for an extended period of 3–4 months.

10.6 PROCESSED PRODUCTS

Carrots are processed into various products, including canned goods, dehydrated items, juices, candies, preserves, and more. Thermal processing and chemical treatments have been used to improve the color and quality of canned carrots, while also increasing carotenoid levels. Carrot candy and preserves are made by coating carrots with sugar or syrup to boost sweetness and preservation. Additionally, carrots can be processed into intermediate moisture foods with approximately 55% moisture content for longer shelf life and versatility.

Carrot beverages are created by combining skim milk or fruit juices with carrot juices. Carrot juice is known for its high levels of α and β-carotene. It's often used in blends with other fruit juices and is particularly popular in Germany as a non-alcoholic beverage. The yield and quality of carrot juice obtained through pressing depend on factors like pretreatment conditions, including pH, temperature, and processing time.

The production and consumption of fermented products have been steadily increasing due to their therapeutic properties and high nutritional value, as highlighted by Karagul et al. in 2004. Live LAB found in yogurt offer several health benefits, including protection against gastrointestinal issues, improved digestion of lactose for those with lactose intolerance, reduced cancer risk, lower cholesterol levels, enhanced immune response (Arya et al., 2022; Keservani et al., 2015; Kumar et al., 2022; Rane et al., 2022), and better absorption of protein, calcium, and iron. Carrot juice is also incorporated into the yogurt industry, resulting in nutritionally rich food products. Carrot yogurt, for instance, is made by blending milk with carrot juice at various concentrations before fermentation, as described by Salwa et al. in 2004. Additionally, various sweet products made from carrots have been developed and preserved. Carrot halwa, a beloved dessert in northern India, is made by cooking finely grated carrots with sugar and frying them in hydrogenated oil, often accompanied by grated dehydrated milk known as khoa. Additionally, there's another sweet treat called kheer mix, which is created using dehydrated carrots, skim milk powder, sugar, and various other ingredients. Drying carrot products, including carrot powder, shreds, and chops, are employed in a range of recipes, such as carrot curry, carrot halwa, and biscuits, as detailed in a report by Suman and Kumari in 2002.

10.6.1 DEHYDRATION OF CARROT

Traditionally, carrots are dried in the open Sun, but this method is time-consuming, weather-dependent, and affects the uniformity and quality of the final product (Sagar et al., 2004). In Sun drying, carrots are spread in open yards and dehydrated using solar energy. To address these limitations, industrial driers, such as convective or solar driers, are used to ensure hygiene, quality maintenance, and product uniformity (Doymaz, 2004). Upadhyay et al. (2008) reported that drying carrot pomace using a conventional drier yielded superior results in terms of β-carotene and ascorbic acid retention.

The moisture content in dried carrots is reduced to about 10%, and they are usually transferred to portable bins at 44.4°C to complete the dehydration process. Various methods have been developed to enhance the sensory characteristics of dehydrated carrots, such as taste, color, and flavor. Upadhyay et al. (2008) also studied the drying kinetics of carrot pomace and suggested that drying at 65°C retained maximum ascorbic acid and β-carotene.

A comparison of different drying techniques (vacuum drying, hot air drying, and hot air + vacuum drying) revealed that combining hot air and vacuum drying retained the highest carotenoid content due to reduced drying time compared to other methods. Products made from osmo-convectively dehydrated carrots had moderate to excellent acceptability (Singh et al., 2006).

The degradation of carotenoids during carrot processing have been successfully retard by using high-temperature short time (HTST). During the processing of carrot, carotenoids undergo isomerization and oxidation which increases the total carotenoid content in the final product. Sharma et al. (2000) studied the effect of the dehydration method on the retention of carotenoids in the carrot sample and found that losses of carotenoids ranged from 6–21% in cabinet-dried samples as compared to 8–22% in fluidized-bed-dried samples. Authors further revealed that un-blanched freeze drying had highest losses as compared to blanched cabinet drying of carrot samples. Kumar et al. (2001) compared the total energy requirement, drying rate and physicochemical properties of dehydrated carrots using hot air- and freeze-drying methods. They found that the carotenoid content was significantly affected by drying processes and the combination different drying techniques of dehydrated vegetables obtained superior quality. Sagar et al. (2004) prepared the carrot slices which were spread on aluminum trays @ 0.20 Kg/m^2. Drying in hot air was carried out with a crossflow at 56–60°C in a cabinet dryer with air at a flow rate of 1.20–1.80 m per sec. They recorded carotenoid content in different cultivars namely "Pusa Kesar," "Sel-14" "Pusa Meghali" and "Sel-16" as 40.86, 38.46, 38.31, and 40.72%, respectively on a dry weight basis. Reducing the water activity (a_w) results in a longer shelf-life of carrots, though carotenoids degrade faster in dehydrated systems through autocatalytic oxidation.

The effect of pretreatment conditions on the physicochemical parameters of carrot juice and the effect of different blanching solutions and blanching times (1–5 min) on the quality of carrot juice has been studied. Over the last year, many authors used high pressure as a substitute for conventional blanching. In 2011, Smith used high-pressure processing (HPP) technique to preserve processed food without the application of high temperature, resulting in higher retention of various quality parameters of processed products. Similarly, another technique, Ultrasound has been studies extensively in past decade to evaluate the quality of food products. In this technique, cavitation is produced at power level 10 to 1000 W/cm^{-2} and sound frequency between 20 and 100 kHz via vibrations. The formation of cavitation is phenomena in which bubbles are formed and then eventually collapse which generates localized pressure of 100 MPa and temperature about 5000 K possessing force enough to cause cell wall damages of food components. Therefore, ultrasound treatment reduces the application of heat requirements along with enhanced heat transfer (Smith, 2011).

In previous studies, it was reported that steam blanching significantly enhances the total carotenoids due to the isomerization of carotene (Dadan et al., 2021; Puuponen-Pimia et al., 2003; Sulaeman et al., 2001). The degradation of β-carotene is reportedly associated with the development of off-flavor in dehydrated carrots and sweet potato flakes. Carotenoids are readily degraded by enzyme lipoxygenase. Blanching can lower the activity of these enzymes. The stability of plant pigments such as anthocyanins and carotenoids in papaya and pineapple is enhanced when blanched and pre-dried with sodium metabisulfite by preventing the oxidation of carotenoids whereas orthophosphoric acid showed no effect on the carotenoid oxidation. The study further showed that both plant pigments significantly decrease as the time and temperature of blanching increases. Glycerol and sugar solutions are hygroscopic and therefore carotenoids are better preserved in such systems. The retention of β-carotene content in blanched and un-blanched samples was increased while ascorbic acid content in the sample decreased (Banga and Bawa, 2002). Blanching of carrots before juice extraction resulted in better color retention in carrot products (Ribas-Agustí et al., 2018). After drying, blanched carrots showed lower ascorbic acid content whereas, non-enzymatic browning remains unaffected. In all the batches of blanched carrots, the carotenoids degradation rate was least within water activity 0.31 to 0.54 ranges however; water activity above and below increases the degradation rate of carotenoids. Blanched carrots possess higher carotenoids whereas, its concentration decreases significantly during storage. It was suggested that the leaching of total soluble solids during blanching led to the increasing carotenoids content which also beneficial for reducing non-enzymatic browning, but also encourages carotenoids destruction enzymes. In this regard, effect of various pretreatment such as freezing, blanching, sulfiting, sodium chloride and sucrose dipping, and starching on carotenoids present in carrots has been reported in researches.

Ma et al. (2015) observed that the carotene (β and α) and lutein content in un-blanched carrot juice was 42.4, 16.3, and 1.7 mg/mL, respectively which is higher or close to blanched treatments which further suggests that blanching treatment can promote the preservation of carotenoids in the prepared carrot juice. Previously, Miebach and Spieb (2003) investigated and suggested that blanching at 90°C significantly enhances both stability and yield of carotenoids in carrot juice. Authors also reported that this mainly due to softening of plant tissues which increases the carrot cell permeability; thereby assists the easy dissolution of carotenoids in the carrot juice. Sezer and Demirdoven (2015) studied the effect of microwave blanching conditions on carrot slices and observed that the peroxidase (POD) activities of all samples were inhibited and the pectin methyl esterase (PME) of carrot slices were at least 97% inactivated in all optimum points and conventional blanching conditions. They concluded that the thermostable form of PME in carrots was inactivated by microwave blanching treatment at the 360 and 630 W optimum points. The carotenoid content increased by 2–25% in carrots and was converted to a stable form by heat treatment. Additionally, carotenoid retentions in microwave blanching carrots were determined to be 35.73% higher than the conventional blanching counterparts (Ramesh et al., 2002). Zou and Jiang (2016) observed a significant increase in total carotenoids and ascorbic acid contents in the sample sonicated for 40 min compared to the sample sonicated for 20 min and the control (non-sonicated) sample. Pretreatment of sulfur dioxide of dehydrated carrots of both un-blanched and blanched samples, significantly reduces the carotenoids content. Mohamed and Hussein (1994) investigated the carrots before dehydration which was treated with sodium metabisulfite and showed that the carotenoids retention was two time higher than the sample without pretreatment of sodium metabisulfite. In 1995, Zhao and Chang reported that 2.5% of sulfite led to higher retention of carotenoids present in carrot. Pretreatment of carrot metabisulfite reduces the browning during both drying and storage period, whereas encapsulation for β carotene by using maltodextrin provide protection from oxidation. In microcrystalline cellulose powder, β-carotene stability is enhanced.

10.6.2 MINIMALLY PROCESSED (MP) CARROT

In order to meet the demand of consumer for both convenient and fresh products, minimally processed (MP) products are introduced as a new category under fresh products. In this process, washing, peeling, abrasion, shredding, slicing, and chopping is involved to make products in fresh like, RTE, and convenient for consumers (Kumar et al., 2022; Varming et al., 2004). The handling of MP products resulted in mechanical damages along with physiological stresses which ultimately causes the deterioration of carrots than the whole roots. The final product sometimes become slimy, produce off odors and lose color and firmness, mainly due to the higher respiration rate and microbial spoilage (Barry-Ryan et al., 2000). Also, washing of rooted crops before peeling/chopping can restrict the cross-contamination among inner parts and cortex of carrot. Washing of crops using chlorine in warm water was reported as most effective method to reduce aerobic mesophilic bacteria (AMB) during storage period (Klaiber et al., 2015). Several studies have been focused to extend the shelf-life of MP products with primary focus on nutritional implications, physiological parameters, quality control and packaging conditions (Rocha et al., 2007). In a study, the storage at 4°C for 7 days had no significant difference was observed between air and vacuum packaged carrots concerning pH, total soluble solids, carotenoid content and whiteness index (Rocha et al., 2007). In comparison with air packed samples, the weight loss in the vacuumed packed samples was noticeably lower (Rocha et al., 2007). The growth of spoilage causing pathogenic microorganisms can be control during the storage period of MP carrots by using anti-microbial packaging. For example, edible-film prepared from chitosan has been used to as effective method for the extension of shelf-life of carrot. Combination of chitosan and sodium casein showed significant antimicrobial properties. In another study, coating film prepared by using 4% yam starch, 2% glycerol, and 1.5% chitosan has been reported as most effective in restricting the growth of coliforms and LAB in sliced MP-carrots stored for 15 days at 10°C. MP carrot packages stored at 6°C for 14 days retains the color, taste, and vacuum by heating carrot sticks for less than 2 min with 100 ppm chlorine water at 60°C in a parallel plate radiofrequency applicator (Orsat et al., 2001). The methodology for the preparation of MP carrots is given in Table 10.2.

TABLE 10.2 Methodology for Carrot Minimal Processing

Unit Operation	Purpose
Peeling	Higher yield and pectin content in steam peeling than hand or mechanical peeling.
Blanching	Blanching for 6 min at 80°C.
Steaming	Ideal operation for the inactivation of enzymes which is responsible for discoloration from surface without affecting texture, flavor, and color of the product.

10.6.3 CARROT PUREE

As consumer's demand for quality and diversified products is very high, carrot roots are readily transformed into carrot puree and is used as base ingredient various food products (Zademowski et al., 2003). However, pomace obtained from preparation of carrot puree is rich source of dietary fibers, antioxidants, and natural carotenoids which have been shown to possess health-promoting effects. As per the estimates, 80% of the carotenoids remain in the produced carrot pomace during the preparation of juice due to deposition of crystalline carotenoids in the chloroplast which further, stabilized the carotene crystals by the protein or residual membranes. This crystal form of carotene serves high stability of carotenoids in carrot pomace. Various food processing methods such as heat treatment or mechanical homogenization led to the disintegration of carrot tissues and thereby, improve bioavailability of carotene. Cooking involves heating of plant materials which break down the cell structures and hence increase the bioavailability (Hedren et al., 2002). However, the severity of treatment may induce the degradation of carotenoids after being released from the plant cell matrix.

10.6.4 CARROT PICKLE AND PRESERVE

Carrot pickles are prepared by lactic acid fermentation, resulting in an excellent product with improved health-promoting constituents. By applying acidified brine and potassium metabisulfite, carrots can be preserved in good condition for about 6 months in non-airtight at room temperature. Also, candies were also prepared from carrots by coating it with high sugar syrup of total soluble solids of about 70–75°B. For the preparation of intermediate moisture foods, carrots have been used having moisture content of 55%. The processed products usually have good color, texture, and flavor.

Ready-to-serve products remain acceptable for 6 months at low temperature (1 to 3°C) in glass container with retention of β-carotene of about 40%.

10.6.5 CARROT JUICE

Like most other vegetables and fruit juices, carrot juice is essentially prepared by using vegetable water (<90%), besides other soluble solids (sugars, vitamins (B and C), protein, and minerals), insoluble solids – mainly fat soluble β-carotene, dietary fibers, and a very small fraction of fat. Carrot juice has characteristics bright orange color and is rich in of β-carotene which is protected properly from oxidation by inactivation of enzymes. Therefore, carrot juice has been marketed as health drink owing to its high nutritional value. Traditional methods yield 0.5 to 0.55 kg carrot juice from 1 kg pretreated carrots further depends on both carrot variety and process used for carrot juice production. In 2006, Sun et al. used enzymatic maceration as pretreatment and reported that the yield can be improved up to a reasonable percentage of 6.5%, however, this method considered to as contamination. This yield was still not as high as compared to that of oranges and apples due to the difficulties in the separation of solid–liquid during juicing of carrots. In 2003, Cassano et al. (2003) suggested that reverse osmosis (RO) can be used as alternative methods to prepare concentrated carrot juice. In this study, author demonstrated that higher yield can be achieved combing RO technology with traditional carrot juice production process by recovering most of the soluble solids from carrot. This method also offers higher yield without contamination from either valuable by-product like insoluble fiber-rich fractions or from liquid produce.

Blends prepared from carrot juice are most popular under alcoholic beverages and its rate of consumption has been reported in different countries, particularly, Germany. Carrot juice is frequently blended with other frits also. The processing of carrot juice includes extraction, canning, and followed by its storage, as suggested by different operators. However, due to yield of carrot juice, industries are adopting novel technologies for the extraction of juices, for example, mash heating, depolymerizing enzymes, and decanter technology. The bitterness of "Kinnow" mandarin juices has been successfully reduced by blending it with carrot juice. The quality and yield of extracted carrot juice vary with different pretreatment conditions like temperature, pH, and time. The application of carrot juice was also utilized in the yoghurt industry (Simova et al., 2004). The blending of carrot juice with yoghurt offers nutritionally rich food by Salwa et al. (2004) and reported that the best quality carrot yoghurt could be develop before fermentation by adding carrot juice at 5 to 20%.

10.7 EFFECTS OF STORAGE ON QUALITY OF PROCESSED PRODUCTS

Changes occur in the physical, chemical, and microbiological quality of value-added products during storage. Changes in the metabolic activity during storage significantly affect the phenolic and protein content. These changes are directly related to quality deterioration and elevated during storage temperature. Physical changes involve color, texture, or flavor, while chemical changes include browning, fat oxidation, and so forth. Molds, yeasts, and bacteria are the main causes of biological changes. Next to the microorganism oxygen is the most destructive agent resulting sometimes in severe oxidative changes in food containing little or no fat. Quality of food changes during storage depending upon the food sensitivity, atmosphere, and water content. These changes are often deteriorative and affect the texture, color, and flavor during drying. Such deterioration increases with the time and temperature of storage. High temperature and high humidity conditions can cause problems with quality changes of dehydrated products (Sagar, 2009). The most troublesome irreversible change that leads to the loss of various desirable attributes is non-enzymatic browning. Non-enzymatic browning (Maillard reactions) is one of the most important chemical phenomena that may affect food quality during processing and storage. Its effect on color, flavor, nutritional value, textural properties, and stability of foods has been demonstrated. Packaging of products significantly increases the shelf-life of processed products. Selection of suitable package used during storage particularly depends on relative humidity and temperature. For example, permeability of gases and moisture during storage in polyethylene deteriorate the food product quality however; glass, tin, and aluminum foils are considered to retain the food quality during storage.

According to Suman and Kumari (2002), chemical evaluation of dried carrots revealed the carrot powder retained maximum β-carotene during the storage for over 3 months. The stability of carotenoid in carrot was highest during long term storage at frozen conditions (Puuponen-Pimia et al., 2003). Several other vegetables also have high quantity of carotenoids during refrigerated storage than the fresh one (Kidmose et al., 2004). MP carrots does not cause the reduction of carotenoids during storage however, such products are readily decomposed by accelerated metabolic activity and microbial spoilage. In carrot pickle, acidity and reducing sugars increases significantly from 1.57 to 2.61% and 84.5 to 86.5 mg/g, respectively over storage of 4 months (Chawla et al., 2005). Further, authors also reported that 31.71% β-carotene was retained during storage period of 4 months whereas; ascorbic acid content has been reduced to 3.96 mg/100 g from 5.76 mg/100 g during the storage.

The storage studies on extracted carotenoids from carrot waste suggested that under extended light exposure or high temperature (45°C), carotenoids present in carotenoid powder undergo isomerization. According to Singh et al. (2006), pigments in dried carrot pulp (spray-dried) waste also prone to degradation depending on the storage time and temperature. Whereas Sagar (2009), revealed that temperature about 7°C was found to retain most of the β-carotene and total carotenoids in RTE dehydrated carrot shreds. During storage, both β-carotene and ascorbic acid content decreases and non-enzymatic browning value increases (Negi and Roy, 2001). On exposure to air, oxidation of highly unsaturated molecules occurs during storage unde-hydrated vegetables which resulted in loss of color. This results in the development of off-flavors in dehydrated carrots due to degradation β-carotene. Carotenes are susceptible to oxidation when exposed to light. During air drying, β-carotene rapidly degraded in carrots however; pretreatment with sodium metabisulfite found to be beneficial in the better retention. Many workers have attributed the reduction of total phenols due to their utilization in the browning reactions during storage.

10.8 CONCLUSION

With changes in the dietary habits of consumers and the continuously transforming food industry with the changes in food consumption patterns, it is anticipated that the market for fresh commodities shall increase in the near future. Carrots, beyond the normal food constituents, contain different bioactive ingredients that can sustain health. Furthermore, new introductions like black carrots have enriched anthocyanin's content, which can assist consumers in overall health management. Research shall focus on how to effectively retain the biological macro- and micro-molecules during the processing, as the loss may lead to reduced quality and bioactivity of the carrots. Efforts shall also be made to create novel functional foods for various age groups to explore the maximum health benefits.

KEYWORDS

- **abrasion**
- **anthocyanin**
- **consumer demand**
- **cross-contamination**
- **microbial spoilage**
- **ready-to-eat**
- **respiration rate**
- **shredding**

REFERENCES

Al-Amin, M., Hossain, M. S., & Iqbal, A. (2015). Effect of pretreatments and drying methods on dehydration and rehydration characteristics of carrot. *Universal Journal of Food and Nutrition Science*, *3*(2), 23–28.

Anand, V. (2001). Genetic evaluation and correlation studies in some cultivars of European carrot (Master's thesis, Department of Vegetable Science, Dr. YS Parmar University of Horticulture and Forestry, Solan, India), 70–76.

Arya, R. K. K., Sati, D., Bisht, D., & Keservani, R. K. (2022). Nanotechnology-based bacterial immunotherapy. In R. K. Kesharwani, R. K. Keservani, & A. K. Sharma (Eds.), *Nutraceuticals and Functional Foods in Immunomodulators* (pp. 3–19). Springer, Singapore. https://doi.org/10.1007/978-981-19-2507-8_1.

Banga, R., & Bawa, A. S. (2002). Studies on dehydration of grated carrots. *Journal of Food Science and Technology*, *39*(3), 268–271.

Baranska, M., Schulz, H., Baranski, R., Nothnagel, T., & Christensen, L. P. (2005). *In situ* simultaneous analysis of polyacetylenes, carotenoids and polysaccharides in carrot roots. *Journal of Agricultural and Food Chemistry*, *53*(17), 6565–6571.

Barry-Ryan, C., Pacussi, J. M., & O'Beirne, D. (2000). Quality of shredded carrots as affected by packaging film and storage temperature. *Journal of Food Science*, *65*(4), 726–730.

Bendokas, V., Stanys, V., Mažeikienė, I., Trumbeckaite, S., Baniene, R., & Liobikas, J. (2020). Anthocyanins: From the field to the antioxidants in the body. *Antioxidants*, *9*(9), 819.

Bystrická, J., Kavalcová, P., Musilová, J., Vollmannová, A., Tóth, T., & Lenková, M. (2015). Carrot (*Daucus carota* L. ssp. sativus (Hoffm.) Arcang.) as source of antioxidants. *Acta Agriculturae Slovenica*, *105*(2), 303–311.

Cassano, A., Drioli, E., Galaverna, G., Marchelli, R., Di Silvestro, G., & Cagnasso, P. (2003). Clarification and concentration of citrus and carrot juices by integrated membrane processes. *Journal of Food Engineering*, *57*(2), 153–163.

Chawla, P., Ghai, S., & Sandhu, K. S. (2005). Studies on the nutritional and organoleptic characteristics of carrot pickle during storage. *Journal of Food Science and Technology, 42*(4), 358–360.

Christensen, L. P., & Brandt, K. (2006). Bioactive polyacetylenes in food plants of the Apiaceae family: Occurrence, bioactivity, and analysis. *Journal of Pharmaceutical and Biomedical Analysis, 41*(3), 683–693.

Christensen, L. P., & Kreutzmann, S. (2007). Determination of polyacetylenes in carrot roots (*Daucus carota* L.) by high-performance liquid chromatography coupled with diode array detection. *Journal of Separation Science, 30*(4), 483–490.

Clarkson, J. J., Zuntini, A. R., Maurin, O., Downie, S. R., Plunkett, G. M., Nicolas, A. N., Smith, J. F., Feist, M. A. E., Gutierrez, K., Malakasi, P., & Bailey, P. (2021). A higher-level nuclear phylogenomic study of the carrot family (Apiaceae). *American Journal of Botany, 108*(7), 1252–1269.

Dadan, M., & Nowacka, M. (2021). The assessment of the possibility of using ethanol and ultrasound to design the properties of dried carrot tissue. *Applied Sciences, 11*(2), 689.

Ferreira, V. B., da Silva, T. T. C., Couto, S. R. M., & Srur, A. U. O. S. (2015). Total phenolic compounds and antioxidant activity of organic vegetables consumed in Brazil. *Food and Nutrition Sciences, 6*(9), 798.

Ghorbani, Z., Zamindar, N., Baghersad, S., Paidari, S., Jafari, S. M., & Khazdooz, L. (2021). Evaluation of quality attributes of grated carrot packaged within polypropylene-clay nanocomposites. *Journal of Food Measurement and Characterization, 15*(4), 770–3781.

Gopalan, C., Ramasastry, B. V., & Balasubramanian, S. C. (1991). *Nutritive Value of Indian Foods*. National Institute of Nutrition, Hyderabad.

Iid, I. I., Kumar, S., Shukla, S., Kumar, V., & Sharma, R. (2020). Putative antidiabetic herbal food ingredients: Nutra/functional properties, bioavailability, and effect on metabolic pathways. *Trends in Food Science & Technology, 97*, 317–340.

Iorizzo, M., Curaba, J., Pottorff, M., Ferruzzi, M. G., Simon, P., & Cavagnaro, P. F. (2020). Carrot anthocyanins genetics and genomics: Status and perspectives to improve its application for the food colorant industry. *Genes, 11*(8), 906.

Karagül-Yüceer, Y., Wilson, J. C., & White, C. H. (2001). Formulation and processing of yogurt affect the microbial quality of carbonated yogurt. *Journal of Dairy Science, 84*(3), 543–550.

Karki, S. (2009). *Development and Evaluation of Functional Food Products from Carrot Pomace* (Master's thesis, Department of Food Science and Technology, Dr. YS Parmar University of Horticulture and Forestry, Solan, India), 118–121.

Karmoker, P., Saha, T., & Shams-Ud-Din, M. (2011). Processing and preservation of mixed chips from potato, papaya, and carrot. *Bangladesh Journal of Agricultural Engineering, 22*(1, 2), 53–60.

Keservani, R. K., Kesharwani, R. K., Sharma, A. K., & Jarouliya, U. (2015). Dietary supplements, nutraceuticals, and functional foods in immune response (Immunomodulators). In D. Bagchi, H. G. Preuss, & A. Swaroop (Eds.), *Nutraceutical and Functional Foods in Human Life and Disease Prevention* (pp. 343–358). Boca Raton, FL: CRC Press, Taylor and Francis.

Kidmose, U., Hansen, S. L., Christensen, L. P., Edelenbos, M., Larsen, E., & Nørbæk, R. (2004). Effects of genotype, root size, storage, and processing on bioactive compounds in organically grown carrots (*Daucus carota* L.). *Journal of Food Science, 69*(9), S388–S394.

Klaiber, R. G., Baur, S., Wolf, G., Hammes, W. P., & Carle, R. (2005). Quality of minimally processed carrots as affected by warm water washing and chlorination. *Innovative Food Science & Emerging Technologies*, *6*(3), 351–362.

Kumar, H. P., Radhakrishna, K., Nagaraju, P. K., & Rao, D. V. (2001). Effect of combination drying on the physicochemical characteristics of carrot and pumpkin. *Journal of Food Processing and Preservation*, *25*(6), 447–460.

Kumar, P., Rai, S., Verma, S. K., Prakash, P. S., & Chitara, D. (2022). Classification, mode of action, and uses of various immunomodulators. In R. K. Kesharwani, R. K. Keservani, & A. K. Sharma (Eds.), *Immunomodulators and Human Health* (pp. 3–19). Springer, Singapore. https://doi.org/10.1007/978-981-16-6379-6_1.

Kumar, S., Saini, R., Suthar, P., Kumar, V., & Sharma, R. (2022). Plant secondary metabolites: Their food and therapeutic importance. In *Plant Secondary Metabolites: Physico-Chemical Properties and Therapeutic Applications* (pp. 371–413). Singapore: Springer Nature Singapore.

Kumar, S., Sharma, S., Kumar, V., Sharma, R., Minhas, A., & Boddu, R. (2022). Cruciferous vegetables: A mine of phytonutrients for functional and nutraceutical enrichment. In *Current Advances for Development of Functional Foods Modulating Inflammation and Oxidative Stress* (pp. 401–426). Academic Press.

Ma, T., Luo, J., Tian, C., Sun, X., Quan, M., Zheng, C., Kang, L., & Zhan, J. (2015). Influence of technical processing units on chemical composition and antimicrobial activity of carrot (*Daucus carota* L.) juice essential oil. *Food Chemistry*, *170*, 394–400.

Mannozzi, C., Rompoonpol, K., Fauster, T., Tylewicz, U., Romani, S., Dalla Rosa, M., & Jaeger, H. (2019). Influence of pulsed electric field and ohmic heating pretreatments on enzyme and antioxidant activity of fruit and vegetable juices. *Foods*, *8*(7), 247.

Mayer-Miebach, E., & Spieß, W. E. L. (2003). Influence of cold storage and blanching on the carotenoid content of Kintoki carrots. *Journal of Food Engineering*, *56*(2, 3), 211–213.

National Horticulture Board (NHB). (2017). *Indian Horticulture Database*. Ministry of Agriculture, Government of India, Gurgaon, India.

Nowacka, M., & Wedzik, M. (2016). Effect of ultrasound treatment on microstructure, color, and carotenoid content in fresh and dried carrot tissue. *Applied Acoustics*, *103*, 163–171.

Orsat, V., Gariepy, Y., Raghavan, G. S. V., & Lyew, D. (2001). Radio-frequency treatment for ready-to-eat fresh carrots. *Food Research International*, *34*(6), 527–536.

Puupponen-Pimiä, R., Häkkinen, S. T., Aarni, M., Suortti, T., Lampi, A. M., Eurola, M., Piironen, V., Nuutila, A. M., & Oksman-Caldentey, K. M. (2003). Blanching and long-term freezing affect various bioactive compounds of vegetables in different ways. *Journal of the Science of Food and Agriculture*, *83*(14), 1389–1402.

Ramesh, M. N., Wolf, W., Tevini, D., & Bognar, A. (2002). Microwave blanching of vegetables. *Journal of Food Science*, *67*(1), 390–398.

Rane, B. R., Patil, R. R., Jain, A. S., Keservani, R. K., & Kesharwani, R. K. (2022). Nutraceuticals as disease preventive food and immunity boosters. In R. K. Kesharwani, R. K. Keservani, & A. K. Sharma (Eds.), *Nutraceuticals and Functional Foods in Immunomodulators* (pp. 3–19). Springer, Singapore. https://doi.org/10.1007/978-981-19-2507-8_7.

Ribas-Agustí, A., Martín-Belloso, O., Soliva-Fortuny, R., & Elez-Martínez, P. (2018). Food processing strategies to enhance phenolic compounds bioaccessibility and bioavailability in plant-based foods. *Critical Reviews in Food Science and Nutrition*, *58*(15), 2531–2548.

Rocha, A. M., Mota, C. C., & Morais, A. M. (2017). Physico-chemical qualities of minimally processed carrot stored under vacuum. *Journal of Foodservice*, *18*(1), 23–30.

Rodriguez-Saona, L. E. (1999). *The Potato: Composition, Non-Enzymatic Browning and Anthocyanins*. Oregon State University.

Sagar, V. R. (2009). Changes in quality of ready-to-eat dehydrated carrot shreds during storage. *Journal of Food Science and Technology*, *46*(2), 177–178.

Sagar, V. R., Pal, N., & Kumar, R. (2004). Quality characteristics of carrot varieties for dehydration. *Journal of Food Science and Technology*, *41*(4), 417–420.

Saini, H. K. (2020). *Pseudocereals-Based Carrot Pomace Flour Enriched Bakery Products* (Master's thesis, Department of Food Science and Technology, Dr. YSP University of Horticulture and Forestry, Nauni, Solan), 1–157.

Salwa, A. A., Galal, E. A., & Neimat, A. E. (2004). Carrot yogurt: Sensory, chemical, microbiological properties and consumer acceptance. *Pakistan Journal of Nutrition*, *3*(6), 322–330.

Senthil, A., Mamatha, B. S., Vishwanath, P., Bhat, K. K., & Ravishankar, G. A. (2011). Studies on development and storage stability of instant spice adjunct mix from seaweed (*Eucheuma*). *Journal of Food Science and Technology*, *48*, 712–717.

Sharma, A. K., Sarkar, B. C., & Sharma, H. K. (2005). Optimization of enzymatic process parameters for increased juice yield from carrot (*Daucus carota* L.) using response surface methodology. *European Food Research and Technology*, *221*, 106–112.

Sharma, D., Shree, B., Kumar, S., Kumar, V., Sharma, S., & Sharma, S. (2022). Stress-induced production of plant secondary metabolites in vegetables: Functional approach for designing next generation superfoods. *Plant Physiology and Biochemistry, 192, 252–272.*

Sharma, G. K., Semwal, A. D., & Arya, S. S. (2000). Effect of processing treatments on the carotenoids composition of dehydrated carrots. *Journal of Food Science and Technology*, *37*(2), 196–200.

Sharma, S., & Sharma, K. D. (2020). Development of carrot juice concentrate enriched functional cookies. *International Journal of Current Microbiology and Applied Sciences*, *9*(12), 3129–3135.

Sharma, S., Katoch, V., Kumar, S., & Chatterjee, S. (2021). Functional relationship of vegetable colors and bioactive compounds: Implications in human health. *The Journal of Nutritional Biochemistry*, *92*, 108615.

Sharma, S., Shree, B., Sharma, D., Kumar, S., Kumar, V., Sharma, R., & Saini, R. (2022). Vegetable microgreens: The gleam of next generation superfoods, their genetic enhancement, health benefits, and processing approaches. *Food Research International*, *155*, 111038.

Simova, E. D., Frengova, G. I., & Beshkova, D. M. (2004). Synthesis of carotenoids by *Rhodotorula rubra* GED8 co-cultured with yogurt starter cultures in whey ultrafiltrate. *Journal of Industrial Microbiology and Biotechnology*, *31*(3), 115–121.

Singh, B., Panesar, P. S., & Nanda, V. (2006). Utilization of carrot pomace for the preparation of a value-added product. *World Journal of Dairy & Food Sciences*, *1*(1), 22–27.

Sra, S. K., Sandhu, K. S., & Ahluwalia, P. (2011). Effect of processing parameters on physico-chemical and culinary quality of dried carrot slices. *Journal of Food Science and Technology*, *48*, 159–166.

Sulaeman, A., Keeler, L., Giraud, D. W., Taylor, S. L., Wehling, R. L., & Driskell, J. A. (2001). Carotenoid content and physicochemical and sensory characteristics of carrot chips deep-fried in different oils at several temperatures. *Journal of Food Science*, *66*(9), 1257–1264.

Suman, M., & Krishna Kumari, K. (2002). A study on sensory evaluation, β-carotene retention, and shelf-life of dehydrated carrot products. *Journal of Food Science and Technology*, *39*(6), 677–681.

Sun, T., Simon, P. W., & Tanumihardjo, S. A. (2009). Antioxidant phytochemicals and antioxidant capacity of biofortified carrots (*Daucus carota* L.) of various colors. *Journal of Agricultural and Food Chemistry*, *57*(10), 4142–4147.

Sun, Y., Wang, Z., Wu, J., Chen, F., Liao, X., & Hu, X. (2006). Optimising enzymatic maceration in pretreatment of carrot juice concentrate by response surface methodology. *International Journal of Food Science & Technology*, *41*(9), 1082–1089.

Tadesse, T. F., Abera, S., & Worku, S. (2015). Nutritional and sensory properties of solar-dried carrot slices as affected by blanching and osmotic pretreatments. *International Journal of Food Science and Nutrition Engineering*, *5*(1), 24–32.

Upadhyay, A., Sharma, H. K., & Sarkar, B. C. (2008). Characterization and dehydration kinetics of carrot pomace. *Agricultural Engineering International: CIGR Journal*.

Varming, C., Jensen, K., Møller, S., Brockhoff, P. B., Christiansen, T., Edelenbos, M., Bjørn, G. K., & Poll, L. (2004). Eating quality of raw carrots: Correlations between flavour compounds, sensory profiling analysis, and consumer liking test. *Food Quality and Preference*, *15*(6), 531–540.

Zadernowski, R. (2003). Quality of carrot juice as conditioned by raw material and technology. *Fruit Processing*, *3*, 183–195.

Zhao, Y. P., & Chang, K. C. (1995). Sulfite and starch affect color and carotenoids of dehydrated carrots (*Daucus carota*) during storage. *Journal of Food Science*, *60*(2), 324–326.

Zou, Y., & Jiang, A. (2016). Effect of ultrasound treatment on quality and microbial load of carrot juice. *Food Science and Technology*, *36*, 111–115.

CHAPTER 11

Nutritional Profile, Bioactives, and Health Benefits of Quinoa

KARTIK SHARMA,[1] JASPREET KAUR,[2] VIKAS KUMAR,[3*] ANTIMA GUPTA,[3] SURESH SHARMA,[1] and VISHESH BHADARIYA[4]

[1]*CSIR–Institute of Himalayan Bioresource Technology, Palampur, Himachal Pradesh, India*

[2]*Food Technology and Nutrition, Lovely Professional University, Phagwara, Punjab, India*

[3]*Food Science and Technology, Punjab Agricultural University, Ludhiana, Punjab, India*

[4]*Chemical Engineering, Lovely Professional University, Phagwara, Punjab, India*

*Corresponding author.

ABSTRACT

Quinoa, a pseudo-cereal, is a strategic crop that is considered a complete food in itself, with numerous health benefits due to the presence of various micro- and macro-nutrients and phytochemicals. These phytochemicals are responsible for possessing various bioactivities such as antioxidant activity, hypocholesterolemic activity, anti-obesity, anti-diabetic activity, and many more. The absence of gluten makes Quinoa a special crop for patients suffering from celiac disease. The crop is used traditionally for several purposes and is being explored by researchers to attain maximum benefits from it. The chapter focuses on nutritional and biological properties, health apprehensions of Quinoa along with its different methods of processing and traditional uses.

Functional Compounds and Foods of Plant Origin: Krishan Dutt Sharma, Satish Kumar, Rakesh Sharma, and Raj K. Keservani (Eds.)

11.1 INTRODUCTION

Quinoa (*Chenopodium quinoa* Willd.) belongs to the *Chenopodiaceae* family and is a pseudo-cereal having its origin in South America (Andes region), where it is consumed as a dietary staple and is known as mother grain (Arneja et al., 2015; Simnadis et al., 2015; Wu, 2015). The genus *Chenopodium* includes 250 species and is distributed worldwide. The family *Chenopodiaceae* has the characteristic to survive and grow well in adverse extreme conditions where no other crops can survive. It is one of those plants that are found in the salt level of the USA (James, 2009). It is known by various names *viz.*, tupapasupha in Aymara, ayara in Quechua, Suba in Chibcha, Dawe in Mapudungun, and so forth (Vega-Gálvez et al., 2010). Quinoa, being a starchy dicotyledonous seed, not cereal, is hence called pseudo-cereal. The crops of this particular family have been used widely for food and feed purposes for decades (Bonifacio, 2003). The history of Quinoa's consumption by humans' dates back to 5000 years. Earlier, Quinoa was the staple food crop of the Andean region, which was later on replaced, by other imported foods such as pasta and rice, which were comparatively cheaper (Nowak and Charrondiere, 2016). Seeds, tender stems, leaves, and inflorescences of the plant can be consumed either in the human diet or given to animals in their feed or as forage (Reguera et al., 2018). The seeds have a variable shape, varying from flat to round; however, their color may vary from grey to white and black (Nowak and Charrondiere, 2016). Although the plant (Quinoa) is not well known that much, due to its superior quality in terms of nutritional composition as compared to other cereals, the demand for such crops is increasing considerably (González et al., 2014). Today, researchers are getting attracted to this miracle plant owing to its nutritive and functional quality (amino acids with high contents of methionine and lysine along with antioxidants (polyphenols) (Nowak and Charrondiere, 2016).

Quinoa is said to possess bioactive compounds in much higher concentrations than any other pseudo cereals and cereals (Arneja et al., 2015). The protein and lipids in Quinoa are higher compared to other cereals (Wu, 2015). Being rich in various micro- and macro-nutrients along with essential vitamins and minerals, the crop has gained the status of "complete food" and is considered a golden grain (Arneja et al., 2015; Tang and Tsao, 2017; Wu, 2015). These amino acids and minerals lower anxiety and aid memory during stress conditions, thereby making Quinoa to fall under the category of functional foods. Quinoa can be an attractive option for functional food development for human nutrition, especially for various cellular mechanics that require membrane protection utilizing antioxidants like neuronal activity and so forth. Iron solubility of Quinoa increases up to 5 times post-fermentation, due to potential reduction in phytates, whereas cooking doesn't alter the soluble iron content (Bianchi et al., 2015). Quinoa is suitable for people suffering from celiac disease as the seeds are gluten free (Nowak and Charrondiere, 2016). In addition, the plant also contains toxic compounds and bitter and anti-nutritional factors such as saponins. These toxic compounds are, however, for their protection from getting attacked by birds, pests, insects, and so on, and are present majorly in hulls, which are removed during dehulling, whereas the level of saponins can be reduced during processing conditions such as polishing, washing, and so forth. Few sweet varieties of Quinoa are also available, which possess about one-tenth of saponin level as compared to normal varieties of Quinoa (Gee et al., 1993; Nowak and Charrondiere, 2016). The plant has the potential to enhance the livelihood conditions of the modern era (González et al., 2014). FAO categorizes Quinoa as a major crop for food security in the 21st century. Quinoa has significant potential to combat malnutrition and hunger, leading to the declaration of this crop as a crop of interest, and 2013 was declared the International Year of Quinoa by the FAO of the United Nations (Tang and Tsao, 2017; Vilcacundo and Hernández-Ledesma, 2017). Recent studies suggest that few varieties of Quinoa possess celiac toxic epitopes onto them which are responsible for the activation of immune response in a person suffering from celiac disease. As there seems much need for exploring studies for a proper understanding of the crop quinoa, this review focuses on chemical composition, nutritional value, health benefits, anti-nutritional factors, human studies, animal studies, adverse effects, and traditional uses.

11.2 HISTORY AND DISTRIBUTION

Quinoa was domesticated over 7000 years ago in Andean countries (Bazile et al., 2016). Being native to the Andes, its presence can be seen more than 5000 years back. It was introduced in England in the 1970s (Bonifacio, 2003). In the USA, it was first introduced in Southern Colorado which then extended to different states and then to Canada (Ruiz et al., 2014). It is widely distributed in the regions of the Andes, covering Peru, Colombia, Bolivia, Ecuador, Chile, and North Argentina (Bonifacio, 2003). During pre-Columbian cultures, the seed was one of the major crops of Latin America

(James, 2009). It is historically under commercial cultivation in the South American plateau and was then introduced in North America, Africa, Asia, and Europe (Dallagnol et al., 2013). After centuries of ignorance of post-Spanish conquest, its potential recognized during second half of the 20th century which resulted in increased import among the countries and Quinoa started spreading its roots crossing the borders of Andean countries. The commercial cultivation of Quinoa increased from 8 countries in 1980 to 75 in 2014. In 2015, 20 more countries introduced Quinoa for cultivation on a large scale owing to its nutritional benefits (Bazile et al., 2016). Quinoa is grown in large amounts in South America, China, Canada, Europe, the USA (California and Colorado), and India. Experimentally, it is cultivated in the UK and Finland (James, 2009). Various government initiatives have significantly increased the production of Quinoa across the globe, especially in Bolivia. The annual per-person consumption of Quinoa in the US, Peru, and Bolivia is 0.03 kg, 1.15 kg, and 2.37 kg, respectively (Vilcacundo and Hernandez, 2017).

11.3 PROXIMATE AND NUTRITIONAL COMPOSITION

Quinoa seeds are known as complete food which provides nutrients especially high-quality proteins. Quinoa seeds alco contains good volume of starch, lipids, and saponins along with various vitamins and minerals (James, 2009). The proximate comparative composition along with the energy value of quinoa grain is enlisted in Table 11.1. The moisture content of quinoa seeds ranges from 8.2 to 13.1. However, the composition may vary depending upon the soil in which the crop is cultivated, climatic conditions, and irrigation conditions. Various researchers found that there exists the highest proximate composition in quinoa bran among its different anatomical parts. Table 11.1 shows the proximate composition of Quinoa.

TABLE 11.1 Proximate Composition of Quinoa Seeds and Other Cereals (g/100 g of Fresh Weight)

Parameter	Quinoa	Wheat	Barley	Corn	Rice
Protein	12.5–16.7	13.68	9.91	9.42	6.81
Fat	5.5–8.5	2.47	1.3	4.74	0.55
Ash	2.7–3.8	1.13	0.62	0.67	0.19
Carbohydrate	60–74.7	71.13	77.72	74.26	81.68
Crude fiber	1.92–10.5	10.7	15.6	7.3	2.8
Energy (Kcal)	368	339	352	365	370

TABLE 11.1 *(Continued)*

Composition of Anatomical Parts of Quinoa				
	Ash	Lipid	Crude Protein	Total Dietary Fiber
Bran	5.7–6.8	14.2–17.8	22.3–32.2	1.2–1.7
Whole	2.6–3	6–6.6	11–13.7	1.1–1.8
Perisperm	0.8–1.4	1–2.8	4.8–7.4	0.6–1
Milled Grain	2.7	6.7	13.3	12.7
Embryo	4.3	10.2	23.5	18.9

Source: Ando et al. (2002); Arneja et al. (2015); Vega-Gálvez et al. (2010); Wright et al. (2002); Nowak and Charrondiere (2016); Arneja et al. (2015).

11.3.1 PROTEIN

Protein is considered the major structural and functional component of the body as it provides energy for various bodybuilding and other activities. The protein content varies from 12–23%, which is comparatively higher than rice (7.5–9.1%), corn (10.2–13.4%), barley (10.8–11%), but comparable to that of wheat (14.3–15.4%). Of the total protein, 57% exists in embryos, 39% in perisperm, and 4% in bran (including pericarp and seed coat) of quinoa seed (Wu, 2015). The protein of quinoa seed is a rich source of methionine (0.4–1%) and lysine (5.1–6.4%), the amino acids which are rarely present in cereal grains. Quinoa proteins also contain a higher content of histidine (28.8 mg/g) which is higher than that found in soy (27.6 mg/g), barley (22.5 mg/g), and 23.5 mg/g wheat proteins (James, 2009). Among the non-essential amino acids, it contains cysteine (14.4%), alanine (41.6%), arginine (77.3%), aspartate (80.3%), glutamate (132.1%), glycine (49.2%), serine (40.2%) and 18.9% tyrosine (James, 2009; Koziol, 1992; USDA, 2005). The protein of any cereal is categorized into four groups, *viz*., globulin, albumin, glutelins, and prolamins. The first two form the major proteins (76.6%) in Quinoa, the values of which are quite higher as compared to other cereals like rice (19.62%), maize (38.3), wheat (17.1), whereas, prolamins and glutelins are present at much lower level, that is, 7.2% and 12.7%, respectively, which is comparatively lower (Wu, 2015).

The food is said to have good nutritional value if it has good protein quality and its efficiency ratio, depending upon amino acids content, anti-nutritional factors, digestibility, and bioavailability. The cooked protein quality of Quinoa and the value of protein efficiency ratio (PR) are similar to that of casein found in milk. Also, PR for cooked quinoa seeds is 30%

higher than uncooked grains. The quinoa proteins contain adequate essential amino acids which are sufficient for almost all age groups as per the requirements suggested by WHO and FAO as given in Table 11.2. Quinoa proteins supply more than 180% of the recommended essential amino acid intake for adults (Graf et al., 2015). The digestibility was found to be near 83–95%, which is 91% in the case of casein however, it can be further increased by the removal of saponins (James, 2009). Apart from this, 91.6% of raw quinoa protein is absorbable and has a biological value of 71 with a net protein utilization (NPU) of 68 (Wu, 2015). Besides, offering high digestibility, absorbability, good protein quality, and quantity, it also offers gluten-free properties and thereby is a good choice for those suffering from gluten intolerance or celiac disease. The high bioavailability of quinoa protein is another factor that makes it an interesting crop, which is contributed by the low level of trypsin inhibitors, which in turn reduces the enzymatic digestion and absorption of protein (Graf et al., 2015). The comparative overview of essential amino acids of the quinoa seeds is given in Table 11.2.

TABLE 11.2 Essential Amino Acids of Some Seeds (mg/g)

Amino Acid	Quinoa	Rice	Wheat	Maize	Requirements Suggested by WHO/FAO		
					2–5 yrs. Old	10–12 yrs. Old	Adults
Tryptophan	11.4	11.4	12.2	7.3	11	9	5
Histidine	28.8	21.3	20.6	26.4	19	19	16
Threonine	29.8	38.5	28.8	38.8	34	28	9
Isoleucine	35.7	41.2	42.6	40.5	28	28	13
Methionine + cystine	36.2	36.8	37.6	40.2	25	22	17
Valine	42.1	61.1	44.1	50.1	35	25	13
Lysine	54.2	38.7	26.8	299	58	44	16
Leucine	59.5	82.3	68.1	125	66	44	19
Phenylalanine + Tyrosine	60.9	105	82.2	86.6	63	22	19

Source: James (2009); Kaziol (1992); USDA (2005).

11.3.2 CARBOHYDRATE

Carbohydrate plays an important role in providing basic nutrition to the body. It has also some physiological health effects on the body such as, it provides energy to the body and controlling insulin metabolism and the level of blood glucose in the body. Quinoa seeds contain a considerable amount of carbohydrates, which is comparable to carbohydrates present in rice and barley. Starch is the principal component of carbohydrates in quinoa seeds (James, 2009). Starch represents 52–60.4% of fresh grain weight, of which 77.5% is contributed by amylopectin while 11–12.4% is contributed by amylose, which is less than the amylose content of rice (17%) and 28% of wheat (Kaziol, 1992). The existence of most of the starch granules is there in cells of perisperm, where the starch granules are densely packed (Wu, 2015). The starch of Quinoa possesses good viscosity, swelling capacity and water absorption. It possesses good stability during retrogradation and freezing (Navruz-Varli and Sanlier, 2016). Sugars form 3% of carbohydrates in quinoa seeds. Sugars present in quinoa seeds (Figure 11.1) are mainly maltose, fructose, glucose, D-ribose, and D-maltose (James, 2009). Along with this, it contains 10% of total dietary fiber, 2.5% crude fiber, 3.2% pentosans, 2% reducing sugars, and 2.5% non-reducing sugars. The detailed pictorial representation of carbohydrates is given in Figure 11.2 (Arneja et al., 2015; James, 2009; USDA, 2005; Wright et al., 2002).

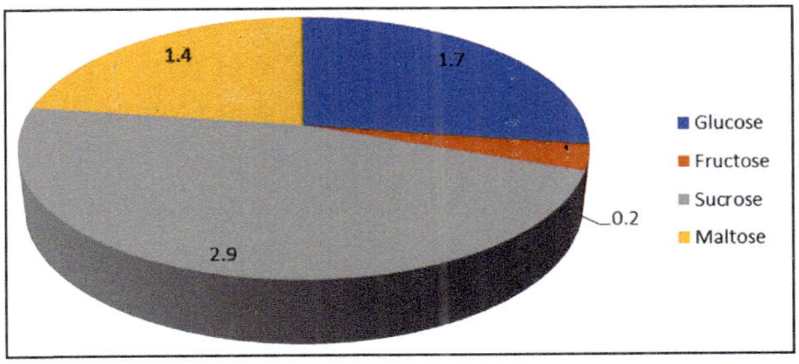

FIGURE 11.1 Sugar content in Quinoa (g/100 g dry matter).

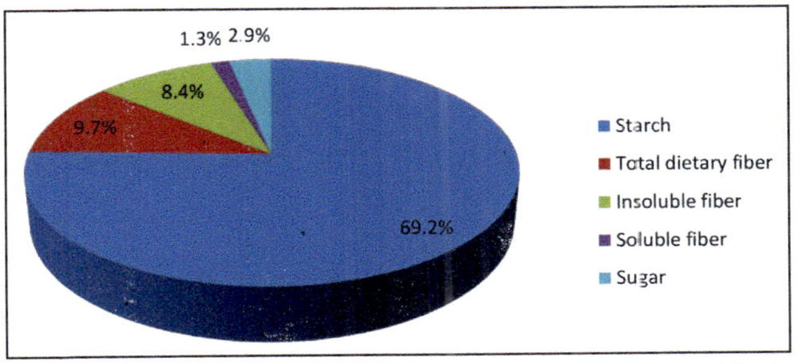

FIGURE 11.2 Carbohydrate composition of quinoa seeds (percentage (%) dry basis).

11.3.3 LIPID

Lipids are considered the concentrated source of energy, which is utilized by the body for the formation of various cell membrane components and other normal functions. Quinoa seeds possess good lipid composition constituting almost 1.8–10% of lipids with oil content on a dry basis of 7%, higher than corn oil (4.9%) but lower than that of soy (20.9%). The iodine value of quinoa seeds is 54%, acid value is 0.5% and peroxide value is 2.44%, which states that the lipid is quite stable and does not undergo oxidation if subjected to oxygen and heat. The saponification value (192%) of its lipid is comparable to that of 194% of soybean oil (Wu, 2015). The lipids of quinoa seeds are high in neutral lipids, containing triglycerides as the major fraction which accounts for more than 50% of the total neutral lipids, whereas diglycerides contribute only 20% of it. Among the polar lipids, phosphatidylcholine, and lysophosphatidyl ethanolamine are the major lipids contributing almost 57% of the total polar lipid. The fatty acid composition is categorized as total monounsaturated fatty acid contributing 25–28.7% with oleic acid as the major one, polyunsaturated fatty acids (PUFAs) contributing 58.3% with linoleic acid as the major one, and total saturated fatty acids contributing 1.9–12.3% with palmitic acid as the major one (James, 2009; Navruz-Varli and Sanlier, 2016). The lipids of Quinoa are almost similar to that of soy and corn and its comparison with soy and corn is given in Table 11.3, however, the complete description of the fatty acid composition of Quinoa in milled grain, whole grain, perisperm, and embryo is given in Table 11.4.

TABLE 11.3 Fatty Acids Composition of Some Foods as Crude Fat (Percentage of Lipid Fraction)

Fatty Acid	Quinoa	Corn	Soy	Peanut	Olive
	1. Saturated Fatty Acids				
Myristic acid	0.1–2.4	Traces	Traces	NR	NR
Stearic acid	0.6–1.1	2.8	3.6	2.0	2.8
Palmitic acid	9.2–11.1	10.7	10.7	9.3	9.6
	2. Monosaturated Fatty Acid (MUFA)				
Myristoleic acid	1	–	–	–	–
Palmitoleic acid	0.2–1.2	Traces	0.2	NR	NR
Oleic acid	22.8–29.5	26.1	22	44.7	79.4
	3. Polyunsaturated Fatty Acid (PUFA)				
Linoleic acid (C18:2)	48.1–52.3	57.7	56	35.8	7.6
Lenolenic acid (C18:3)	4.6–8	2.2	7	NR	0.6

Source: James (2009); Kaziol (1992); USDA (2005).

TABLE 11.4 Fatty Acid Composition of Quinoa (on Percent Dry Basis)

Fatty Acid	Milled Grain	Whole Grain	Perisperm	Embryo
Total fatty acid	6.7	6.5	5	10.2
Myristic acid	0.2	0.1–2.4	0.1	0.2
Palmitic acid	10.3	9.2–11.1	10.8	9.5
Stearic acid	0.8	0.6–1.1	0.7	0.9
Oleic acid	25.6	22.8–29.5	29.5	19.7
Linoleic acid	52.0	48.1–52.3	49	56.4
Lenolenic acid	9.8	4.6–8	8.7	11.7
Others	1.3	1.4	1.2	1.6

Source: Ando et al. (2002).

The quality of lipids in food or product has its significance, where some lipids are known for causing various cholesterol and heart-related problems and at the same time, some lipids prove to be a boon for the body by nature due to their protective functions. For instance, omega-6 and omega-3 are important fatty acids, the former stimulates the inflammatory activity while the latter deals with anti-inflammatory processes in the body. Its lower ratio is desirable in lowering the risk of cancer, auto-immune diseases, cardiovascular diseases (CVDs), and inflammation (Bastidas et al., 2016). Essential fatty acids are the ones, which are not synthesized in the body but are required by the body from outer sources, the deficiency of which may lead to various disorders. Linoleic acid, an essential PUFA is present in quinoa seeds in abundance. Along with the presence of fatty acids in quinoa seeds, it also contains vitamin E, which protects the fatty acids along with other antioxidants (Arneja et al., 2015). In various studies, PUFAs are shown to have improved insulin sensitivity, develop the brain, maintain immunity and prostaglandin metabolism, serve membrane function, and prevent CVDs. The oil of quinoa seeds is highly nutritive and unsaturated and has a polyunsaturation index ranging from 3.9 to 4.7 (Graf et al., 2015; James, 2009). The fatty acid ratio of Quinoa, that is, n–6/n–3 at 6.2, lies under the recommended value (Arneja et al., 2015). In nutshell, the Quinoa has a balanced quality of lipids (oil) which are beneficial for human health in one or the other form. The seeds also contain α-tocopherol (0.59–2.6 mg/100 g), a natural antioxidant that possesses a defense mechanism against lipid oxidation, which makes the oil shelf-stable (James, 2009).

11.3.4 VITAMINS AND MINERALS

They are the micronutrients as are required in minute amounts for various functions of the body such as regulation of cell growth, its development, protection from oxidative damage, and improvement of vision. Cereals are known to be the major source of vitamin B-complex. Quinoa seeds are rich source of vitamins and minerals (Graf et al., 2015). It contains a high level of total folate and vitamin B_6, 100 g of which is sufficient to fulfill the requirements in children as well as adults as suggested by the National Academy of Sciences. In terms of an edible portion (100 g), riboflavin fulfills almost 40% of the daily needs of adults and 80% of children. Each 100 g seeds supply vitamin B_6 (0.20 mg), pantothenic acid (0.61 mg), folic acid (23.5 g), and 7.1 g of biotin (Bhargava et al., 2006). Some cereal-based foods such as bread and rice contribute 14%, 19%, 28%, 30%, and 45% of the daily intake of vitamin B_6, folate, niacin, riboflavin, and thiamin (Wu, 2015). Thiamin content of Quinoa is lower than those present in oats, however, niacin, total folate, vitamin B_6, and riboflavin is comparatively higher. Folates quinoa are 10 times the content present in wheat, while niacin in Quinoa is insufficient to fulfill the daily need but its presence in the diet is beneficial (Arneja et al., 2015; Bhargava et al., 2006; Tang et al., 2015; Vega-Galvez et al., 2010). B-carotene content in quinoa seeds is 0.39 mg/100 g of dry weight, which is higher than that of barley (0.01 mg/100 g) and 0.02 mg/100 g of wheat (Arneja et al., 2015). The Vitamin composition of Quinoa along with its comparison with different cereals and recommended dietary intake (RDI) is given in Table 11.5. Some vitamins act as natural antioxidants which prevent the cell membranes from getting damaged by free radicals. Quinoa also contains one such vitamin *viz.*, vitamin E, which possesses a protective effect at the cell membrane level. It also protects lipids from oxidation, thereby making the oil of quinoa seed more stable during processing conditions and storage (Vega-Gálvez et al., 2010). Folic acid is an essential factor for the proper development of embryos and the production of RBCs. The folic acid is not synthesized in the human body and is thus required from the outer source. Quinoa is, therefore, a wonderful source of folic acid as it contains more amount of folic acid than the daily requirement, therefore, it can be recommended for pregnant women (Wu, 2015).

TABLE 11.5 Comparison of Vitamin Contents of Some Cereals (mg/100 g of Dry Weight)

	Quinoa	Barley	Oats	Rice	Wheat	RDI Value
Thiamin (B_1)	0.29–0.38	0.191–0.49	0.763	0.47	0.55	1.5
Riboflavin (B_2)	0.30–0.39	0.114–0.20	0.139	0.10	0.16	1.7

TABLE 11.5 *(Continued)*

	Quinoa	Barley	Oats	Rice	Wheat	RDI Value
Niacin (B_3)	1.24–1.52	4.604–5.44	0.961	5.98	5.88	20
Pyridoxine (B_6)	0.487	0.260	0.119	–	0.43	2.0
Total folate (B_9)	0.184–0.781	0.023	0.056	–	0.054	0.4
Ascorbic acid (C)	4	0	0	0	0	60
β-carotene	0.39	0.01	–	–	0.02	–
α-Tocopherol (E)	5.37	0.35	–	0.18	1.15	30

Source: Kaziol (1992); USDA (2005); Bhargava et al. (2006); Vega-Gálvez et al. (2010); Vega-Galvez et al. (2010); Graf et al. (2015); Tang et al. (2015); Wu (2015).

The minerals present in quinoa seeds include potassium, magnesium, phosphorus, iron, and other minerals, which as a whole contribute to the ash content of quinoa seeds. As far as the distribution of minerals is concerned (Table 11.6), minerals Mg, K, P, and so on, are present in the embryo, while calcium (Ca) along with some phosphorus is found in the pericarp (Arneja et al., 2015; Vega-Gálvez et al., 2010). K, Ca, and Mg are present in abundance than those present in other cereals like oats, barley, and so forth. The mineral composition of Quinoa along with its comparison with different cereals is given in Table 11.7. Few minerals such as Mg, Cu, Mn, and Fe in 100 mg of quinoa seeds cover the daily requirement of infants as well as adults. Zinc and phosphorus content in 100 g of quinoa seeds fulfills only 40 to 60% of the daily requirement of adults, while it is sufficient for children. The potassium fulfills only 18–22%, while calcium contributes around 10% of the daily requirements of adults and infants (James, 2009).

TABLE 11.6 Mineral Composition of the Whole, Dehulled, Flour, Bran, Embryo, and Perisperm of Quinoa Seed (mg/100 g)

Mineral	Whole	Dehulled	Flour	Bran	Embryo	Perisperm
Potassium	732–825.7	656	639.3–855	2908.5	1125.4	387.9
Magnesium	452.6–502	467.9	161–415.2	958.31	750.2	215.2
Calcium	86.3–121.3	55.1	70–91.8	481.3	139.7	71.8
Phosphorus	359.5–411	404.9	22–462	350.8	482.6	286.6
Iron	9.5–15	14.2	2.6–9.2	14.3	11.3	7.2
Manganese	3.7	NR	3.5	10.8	5.1	2.4
Copper	0.7	NR	0.6–7.6	1.4	0.8	0.5
Zinc	0.8–4	4	0.8–3.8	0.7	1.1	0.6
Sodium	1.3	NR	1.2–93	3.2	1.5	0.5

Source: Ando et al. (2002); James (2009); USDA (2005).

TABLE 11.7 Comparison of Mineral Content of Different Cereals (mg/100 g of Dry Weight)

Mineral	Quinoa	Wheat	Rice	Oats	Barley	Maize	RDI Value
Calcium	32.9–121.3	50.3	6.9	58	29	17.1	1000
Phosphorus	411–468.9	467.7	137.8	734	221	292.6	1000
Magnesium	178–502	169.4	73.5	235	79	137.1	400
Iron	5.5–15	3.8	0.7	5.4	2.5	2.1	18
Zinc	2.75–4.4	4.7	0.6	3.11	2.1	2.9	15
Potassium	732–1193	578.3	118.3	566	280	377.1	NR
Copper	1.0–5.1	7	0.2	0.4	0.4	NR	2
Sodium	1.2–42.5	8.9	6.9	4	9	6.9	NR
Manganese	1.9–17.8	3.9	2.3	5.6	1.3	0.5	NR
Chlorine	110–230	63.3	NR	NR	26	13.7	NR
Sulfur	150–220	NR	NR	NR	NR	NR	NR
Silicon dioxide	115	19.7	NR	NR	0.5	NR	NR
Aluminum	6–17	NR	NR	NR	NR	NR	NR
Boron	0.8–1.3	NR	0.03	NR	0.5	0.2	NR
Cobalt	0.004–0.005	0.002	0.001	NR	0.008	NR	NR

Source: Arneja et al. (2015); Bhargava et al. (2006); James (2009); Kaziol (1992); USDA (2005, 2011); Vega-Gálvez et al. (2010).

It also contains some carotenoids such as zeaxanthins and lutein in the range of 1.2–1.8 mg/100 g (Tang et al., 2015). The phytate or phytic acid content in any cereal is of major concern as it reduces the bioavailability of minerals by binding with them. The level of phytic acid in Quinoa is (1.18 g/100 g or 10.5–13.5 mg/g of seeds), which is quite low as compared to other cereal grains like corn (720 mg/g), rice (60 mg/g) and wheat (390 mg/g). However, this can further be reduced by various processing conditions like soaking, cooking, fermentation, and so forth. Fermentation not only reduces the phytic acid content but also enhances iron solubility (Graf et al., 2015). Also, the iron content of quinoa seeds is reported to be highly soluble, and therefore, it is easily made available to persons suffering from iron deficiency (Vega-Gálvez et al., 2010).

11.3.5 DIETARY FIBER

Dietary fibers are essential for maintaining proper health as it imparts proper digestive health and various other functional benefits. They give a feeling of satiety and reduce the absorption of lipids and cholesterol, enhance intestinal microbiota, and thereby reduce the risk of inflammation and gastrointestinal infection (Graf et al., 2015). The fiber content is almost similar to cereals ranging between 7 and 10% on a dry basis, whereas the soluble fiber ranges between 1.3 and 6.1 (James, 2009). Some studies revealed the negative interaction of the fiber consumption with development of diseases such as obesity, CVD, and diabetes (Type-2). In Quinoa around 78% of total fiber is insoluble in nature in the form of arabinose, galacturonic acid, xylose, glucose, and galactose subunits; whereas soluble fiber represents 22% of total fibers and comprises arabinose, galacturonic acid, and glucose subunits (Graf et al., 2015).

11.4 PHYTOCHEMICAL COMPOSITION

11.4.1 PHENOLICS

Even though all plant sources are good sources of polyphenols, quinoa seeds possess an excellent amount of bioactive polyphenols which prevent oxidative stress, help in proper metabolism and improved cell signaling, and possess anti-diabetic, anti-inflammatory, anti-cancerous, and anti-obesity activities (Graf et al., 2015). It contains the potential to alter the antioxidant status of an organism (Arneja et al., 2015). Phenolics containing a single ring are termed phenolic acids, while the ones with multi-rings are termed polyphenols. Phenolic acids exist either in free or bound form and are known to be associated with the pectin of the cell wall (Graf et al., 2015; Tang and Tsao, 2017). As per the study conducted by Pasko et al. (2008), quinoa seeds (490.2 mg per kg) contain a higher amount of phenolics than amaranth (464 mg per kg). Among the major prevailing acids, gallic acid and p-hydroxy-benzoic acid were the major ones, contributing almost 320 mg per kg and 76.8 mg per kg of dry weight, respectively (Pasko et al., 2008). Rosmaric acid and gallic acid are present in quinoa seeds, which possess great anti-oxidant activity (Wu, 2015). Apart from these acids, Quinoa contains a wide diversity of phenolic acids, such as vanillic acid (43.4 mg per kg), cinnamic acid (10 mg per kg), caffeic acid (40 mg per kg), hydroxybenzoic acid, and hydroxycinnamic acid and their derivatives (Pasko et al., 2008; Tang and Tsao, 2017). The content of phenolic acid was higher in sprouted seeds, that is, 516.3 mg per kg. However, the amount of gallic acid in sprouted seed was very less (70 mg per kg). Few phenolic acids were additionally present during the sprouting stage, which was not normally present in seeds namely, ferulic acid (440 mg per kg), and at the same time, there exists no p-hydroxybenzoic acid or vanillic acid during sprouting (Pasko et al., 2008).

11.4.2 FLAVONOIDS

Flavonoids (Keservani and Sharma, 2014; Keservani et al., 2016a, b), mainly epicatechin and quercetin exhibit great antioxidant activity, thereby reducing the risk of cervical cancer, heart-related diseases, leukemia, breast cancer, prostate cancer, colorectal cancer, and type II diabetes (Wu, 2015). The total flavonoid content (TFC) of quinoa seed is 2238 mg per kg of dry weight. The major flavonoids include sorientin, vitexin, rutin with a trace amount of morin, hesperidin, and noehesperidin (Pasko et al., 2008). It contains more than dozen flavonol glycosides constituted mainly by kaempferol (36.7 μmol/100 g) and quercetin (43.4 μmol/100 g) derivatives that have concentration levels up to 839 μg/g of dry weight (Graf et al., 2015; Wu, 2015). As per the study conducted by Pasko et al. (2008), light has a significant effect on the flavonoids of sprouts of Quinoa. Total flavonoids were found to be more in sprouts grown in light conditions than the ones grown in dark conditions. Rutin is the dominant flavonoid in the sprouts of Quinoa grown in light conditions whereas it was absent in sprouts grown in dark conditions. Sprouts grown in dark conditions were characterized by the presence of vitexin and isovitexin flavonoids, whereas the sprouts grown in light conditions were characterized by the presence of isovitexin only, which is completely absent in quinoa seeds.

11.4.3 ISOFLAVONES

These are a group of flavonoids that acts as phytoestrogen in humans and animals. They help in improving bone strength by increasing its density, reducing arterial resistance, possessing antioxidant activities, and stimulating the secretion of osteoprotegerin. Genistein (0.05–0.41 mg/100 g) and daidzein (0.70–2.05 mg/100 g) are the major isoflavones present in the seeds of Quinoa, which is comparatively low to that of genistein (83.8 mg/100 g) and daidzein (58.3 mg/100 g) of soybean (Graf et al., 2015). These hormones help in protection of plant from pathogens and UV rays (Vega-Gálvez et al., 2010).

11.4.4 ANTHOCYANINS

Anthocyanins are well known for their property to reduce the risk of CVDs, which is relatively due to the protection imparted by anthocyanins against oxidative stress. They also protect estrogenic activity, DNA cleavage, enzyme inhibition, increased production of cytokines, reduced capillary permeability or fragility, and enhanced membrane strengthening. Few studies suggest that a low dose of it can lower the lipid level and blood pressure and reduce ischemia and inflammation in patients suffering from CVDs. Quinoa possesses a high level of anthocyanins (102.4 mg CGE/100 g) as compared to soybean, amaranth, and jasmine rice (Wu, 2015).

11.4.5 PHYTOSTEROLS

Phytosterols are plant sterols, and lipophilic compounds found in the cell membranes of plants. They have almost similar activity in the plants as is anticipated from the cholesterol in humans. The phytosterol content of quinoa seeds is 118 mg/100 g of seeds, which is contributed by the presence of constituents such as campesterol, β-sitosterol, stigmasterol, and brassicasterol (Graf et al., 2015). The content of stigmasterol (3.2 mg/100 g), campesterol (15.6 mg/100 g), and β-sitosterol (63.7 mg/100 g) in quinoa seeds higher than many other cereals and pulses (Ryan et al., 2007). These phytosterols play an important role in human nutrition and can reduce the cholesterol serum and production of atherogenic lipoprotein in the liver. These phytosterols are said to possess anticarcinogenic and anti-inflammatory effects (Graf et al., 2015).

11.4.6 PHYTOECDYSTEROIDS

Phytoecdysteroids are plant-derived ecdysteroids that are synthesized by plants to protect themselves from insects. Quinoa seeds are said to have the highest amount of phytoecdysteroids among various edible crops, which is in the range of 138–570 μg/g of total phytoecdysteroids (Graf et al., 2015). More than 13 phytoecdysteroids have been documented till now from the quinoa seeds, among which 20-HE (20-hydroxyecdysone) is the most prevailing one, contributing about 62–90% of the total phytoecdysteroids. Following 20-HE, mastertone A, 24,28-dehydromakisterone, and 24-epimakisterone are the set of 2nd most abundant phytoecdysteroids in Quinoa (Graf et al., 2014, 2015b).

11.4.7 BETALAINS

Betalains are nitrogen-containing molecules that impart red, yellow, and black color to seeds and other vegetative parts of Quinoa (Bhargava et al., 2006). Isobetanin and betanin are the most abundant betalains in quinoa seeds which possess health relating effects including anti-inflammatory effects and antioxidant effects. Few studies suggest that some betalains show higher antioxidant potential as compared to those polyphenols. Besides, betalains can be used as natural dye due to their stability at a wide range of pH, that is, 3 to 7. Betalains are approved by the European Union (E-162) and the U.S. FDA to be used as colorants in sauces, cosmetics, pharmaceuticals, soups, and other dairy products (Graf et al., 2015). The total betanin content of Quinoa ranges from 1.5- to 61 mg/kg (Tang and Tsao, 2017).

11.4.8 GLYCINE BETAINE

Betaine, or glycine betaine, is also known as N-trimethylated amino acid (N,N,N-trimethylglycine). Cereal crops are the best source for the contribution of betaine to the diet. The betaine content of Quinoa ranges from 3930–6000 μg/g which is comparatively higher than the other cereal crops such as wheat (174–706 μg/g), millet (50–150 μg/g), buckwheat (<20 μg/g), amaranth (646 μg/g). Betaine, along with its precursor (choline), plays an important role in the regulation of homocysteine and is employed in the treatment of diabetes, CVDs, and obesity (Graf et al., 2015).

11.4.9 CAROTENOIDS

Carotenoids are known to be the precursor of vitamin A, which acts as antioxidants and is responsible for imparting color to the plant (Bastidas et al., 2016). The content of carotenoids in the sweet variety of Quinoa is 0.4 mg/10 g, whereas, in the bitter one, it is 0.08 mg/10 g, which is higher than the carotenoid content of einkorn (0.1 mg/10 g) and 0.07 mg/10 g of corn (Wu, 2015). However, the total carotenoid content of quinoa seed oil ranges from 1.69–17.61 mg/Kg, which is contributed by individual carotenoid components viz. lutein (3.96–12.01 mg/Kg), zeaxanthin (0.31–5.37 mg/Kg) and 0.26–1.07 mg/Kg of β-carotene (Tang and Tsao, 2017). Besides being the precursor of vitamin A, carotenoids are responsible for several other functions, the major one being to quench the singlet oxygen (one of the major ROS). It has anti-cancerous and anti-inflammatory activity and acts as a chemo-protective agent. They also prevent age-related macular degeneration, improve the health of the eyes, reduce heart-related problems, protect skin from UV damage, and ischemic strokes, and many more (Wu, 2015).

11.4.10 TOCOPHEROL

Tocotrienols and tocopherols are a class of organic chemical compounds, many of which possess an activity similar to that of vitamin E. Vitamin E acts as an antioxidant that helps in chain breaking, thereby protecting the body from oxidation by free radicals. This reduces the risk of cancer and CVDs (Wu, 2015). Quinoa contains all four isoforms of tocopherol *viz*., α, β, γ, and δ, with γ-tocopherol being the dominating one that has content ranging from 47–53 mg/Kg of dry weight. Following γ-tocopherol is α-tocopherol ranging from 17–26 mg/Kg of dry weight. β-tocopherol, δ-tocopherol, and other isomers of vitamin E (i.e., α-tocotrienol and β-tocotrienol) occur in trace amounts (less than 5 mg/Kg of dry weight) in quinoa seeds (Tang and Tsao, 2017). α-tocopherol and γ-tocopherol are present in higher quantities in quinoa seeds as compared to other cereals and are well known for their antioxidant potential (Navruz-Varli and Sanlier, 2016). Tocotrienols are well known for their various properties such as hypocholestrolemic, neuro-protective, and anti-cancerous properties (Tang and Tsao, 2017).

11.5 ANTI-NUTRITIONAL FACTORS

Anti-nutritional factors, that is, saponins phytic acid (1.05–1.35%), tannins (0.1–5%), and protease inhibitors (<50 ppm) are present in Quinoa (Table 11.8). It also contains trypsin inhibitors in trace amounts and lipoxygenase activity (Wu, 2015). These anti-nutritional factors are toxic and reduce the availability of one or more nutrients by forming insoluble complexes of minerals which affect the mineral absorption, thereby adversely affecting the performance and health of an individual. Along with this, the anti-nutritional factors are somehow responsible for reducing the digestibility, bioavailability, and palatability however, the dietary saponins are not of major concern for humans and are recognized as generally safe for consumption within their limit (Tang and Tsao, 2017).

TABLE 11.8 Anti-Nutritional Factors in Quinoa Grain Fractions

Anti-Nutritional Factor	Whole	Milled	Bran	Perisperm	Embryo
Phytic acid (mg%)	163	171.3	92.5	76.2	231.7
Saponins (mg%)	263	39.5	2704.2	15.6	95.3
Trypsin inhibitor activity (units/g)	68.9	72.5	5.7	7.9	162.5
Lipoxygenase activity (units/g)	50.9	55.8	0	30.1	95.3

Source: Ando et al. (2002).

11.5.1 SAPONINS

Saponins are a wide class of steroids, glycosides, or triterpenoid glycosides which are bitter and are considered toxic if consumed in excess amounts. It comprises triterpenoidaglycone (mainly hederagenin, organic acid, oleanolic acid, and phytolaccagenic acid) with a single or more sugar moieties such as glucuronic acid, galactose, glucose, arabinose, xylose (Graf et al., 2015). Till now around 87 saponin compounds are discovered in quinoa seed coat. Sapogenin content in sweet varieties of quinoa seeds ranges from 0.02–0.04% and for bitter varieties, it ranges from 0.14–2.3%, which is higher than the sapogenin content of oats and soybean, but lower than that of yucca and green pea. The presence of saponins in Quinoa can be seen not only in the seed but throughout the plant to protect the plant from various biotic and abiotic stress conditions (James, 2009). They protect the plant from insects, birds, and microbial infection and promote the organic production of the crop (Graf et al., 2015). The saponins are majorly present in the seed coat, however, few varieties of Quinoa such as the sweet variety contain no saponins or less than 0.11%.

Saponins interfere with the palatability, digestibility, and bioavailability of Quinoa due to its astringent and bitter taste, therefore, it must be removed properly before consumption (Graf et al., 2015). As stated above, saponins are water-soluble and are removed to some extent by washing. However, it is not possible to remove them completely with normal water, but it can reduce their content to a level, which is safe for consumption. It is recommended to wash the seeds with slightly alkaline water to reduce the level of saponins and bitterness in the seeds. Other methods for reducing bitterness may include roasting, extrusion processes, or mechanical abrasion (James, 2009). Despite this, saponins also possess some biological activities that are beneficial for humankind, such as anti-viral, anti-fungal, anti-cancerous, hemolytic, hypoglycemic, analgesic, anti-microbial, anti-thrombotic, anti-inflammatory, diuretic, and has neuroprotective action (Vega-Galvez et al.,

2010; Wu, 2015). Saponins are also responsible for reducing fat absorption by enhancing the permeability of intestinal mucosa which also helps in better food intake and drug assimilation at the intestinal level (Abugoch, 2009).

11.5.2 PHYTIC ACID

Phytic acid, due to its unique structure, can strongly chelate the cations like magnesium, calcium, potassium, iron, copper, and zinc and forms insoluble salts. Thus, the high phytate content can adversely affect digestion, absorption, and availability of minerals, but as discussed above (in the mineral section), phytic acid (1.18 g/100 g) in quinoa seeds is far lower than the majority of other cereal crops and has almost double iron concentration. It has ample amount of P, Ca, Mg, Zn, and K which makes Quinoa maintain a balanced human diet irrespective of its phytic acid content (Graf et al., 2015). Unlike saponins, phytic acid is not located only on the pericarp of the seed but throughout the embryo. The embryo contains almost 60% of the total phytate content (Ando et al., 2002). In most varieties of Quinoa, the phytic acid content ranges from 1.05–1.35 g/100 g (Vega-Gálvez et al., 2010). Phytic acid can also result in decreased enzymatic activity, protein solubility, and proteolytic digestibility. Despite these negative impacts, phytic acid, if present in a low amount can act as an antioxidant, aid in the suppression of oxidative damage to the intestinal epithelium, inhibits enzyme xanthine oxidase, possess anti-carcinogenic activity by controlling the rate of cell division and proliferation, increases the activity of NK cells (natural killer cells), thereby increasing the immunity, inhibits the formation of renal stone, acts as hypoglycemic by chelation of calcium, which is a cofactor of α-amylase, lowers the cholesterol level by binding with zinc and thereby rendering it unavailable to form Zn: Cu ratio, inhibits calcification in the cardiovascular region by lowering the level of calcium and making it unavailable for such activities (Bastidas et al., 2016; Tang and Tsao, 2017).

11.5.3 OXALATES

Oxalate is a toxic constituent present in plants and involved in various metabolic processes. A diet rich in high oxalate content can pose various risks related to the formation of calcium oxalate stones. Nearly 75% of urinary stones are formed from calcium oxalate. It may also lead to acute renal failure. Plant families such as Amaranthaceae, Polygonaceae, and Chenopodiaceae are considered high in oxalate content. Normal intake of oxalate should be in the range of 50–200 mg on daily basis, exceeding which may result in problems. The soluble oxalate content of Quinoa is in the range of 123–138 mg/100 g, whereas the total oxalate content ranges from 183–185 mg/100 g (Siener et al., 2006).

11.5.4 PROTEASE AND TRYPSIN INHIBITORS

They are the proteins, which combines with a proteolytic enzyme, and forms stable complexes in the body. The quinoa seeds contain too low volume of protease inhibitors (< 50 ppm) to pose any concern. Some authors have also reported the content of trypsin inhibitors in the range of 1.36–5.04 TIU/mg of seed (TIU: trypsin inhibiting unit), which is too low compared to that of soybean (24.5 TIU/mg) and beans (12.9–14.8 TIU/mg). The content is too low to pose any risk on humans (Bastidas et al., 2016; Vega-Gálvez et al., 2010; Wu, 2015). Almost 89% of trypsin inhibitor and 62% of the total lipo-oxygenase activity is contained in the embryo (Ando et al., 2002).

11.5.5 TANNINS

Tannin is a polyphenol that possesses various pharmacological properties such as anthelmintic, anti-inflammatory, anti-microbial, anti-toxic, anti-viral, healing wounds, curing dysentery, and many more. Besides these pharmacological properties that are beneficial for human mankind, tannins have a strong anti-nutritional effect and reduced palatability due to their bitterness and astringency. As per Sharma et al. (2019), the recommended daily intake of tannin should not exceed 2.5 g and is considered safe. The tannin content of whole quinoa seeds is 0.51–0.53%, which is far lesser than the recommended value and those present in buckwheat (0.322%), *A. hybridus* (0.087%), *A. hypochondriacus* (0.060%), *A. cruentus* (0.120%) and thus poses no risk upon human consumption (Gorinstein et al., 2008). This content is further reduced during scrubbing, washing, and other processing conditions (Arneja et al., 2015).

11.6 HEALTH BENEFITS

Quinoa is a crop with various health benefits, and therapeutic properties are nutritionally rich and are an ultimate source of nutrition for people. It offers several health benefits to its consumer population even to those who are in a high-risk group including children, adults, people suffering from lactose intolerance, osteoporosis, anemia, dyslipidemia, diabetes, obesity, and so forth (Graf et al., 2015; Navruz-Varli and Sanlier, 2016). In terms of nutrition and human health, Quinoa is advantageous over other cereal grains due to the presence of numerous phytochemicals, fatty acids, minerals, vitamins, fibers, and proteins in it. Besides, it contains various bioactive compounds such as phytosterols, isoflavones, retains, a flavonol glycoside, betalains, and so on, which are responsible for various biological activities such as hypocholesterolemic, anti-inflammatory, anti-microbial, anti-fungal, diuretic, anti-obesity, antioxidant activity and many more. Various biological activities of Quinoa along with their responsible component are given in (Table 11.9).

11.6.1 ANTIOXIDANT ACTIVITY

Naturally occurring antioxidants are gaining much attention due to their inhibiting nature against free radicals and other oxidation reactions in cell membranes and tissues to reduce the impact of ROS. Antioxidants, when incorporated into food serve various functions such as delaying the formation of toxic products resulting due to oxidation, minimizing rancidity, increasing shelf-life, maintaining nutritional quality, and many more. According to Pasko et al. (2009), pseudo-cereals (sprouts and seeds) possess higher antioxidant activity. Amaranth seeds are considered an excellent and ultimate source of antioxidants; however, Quinoa's antioxidant activity was much higher than amaranth and other cereals (buckwheat and rice) when depicted using different assays (Vega-Gálvez et al., 2010). Various studies suggest that supplementation of diet with quinoa seeds decreases the oxidative stress in the heart, liver, plasma, spleen, kidney, testis, pancreas, and lungs of rats administered with fructose. It was found that quinoa seed is responsible for the maintenance of different antioxidant enzymes such as catalase, superoxide dismutase, and glutathione (GSH)-peroxidase (POD) (Tang and Tsao, 2017).

Nitric oxide is the signaling molecule acting against oxidative stress and plays an important role in repairing the damages caused by ROS. The concentration of nitrite in polysaccharide (supernatant) stimulates the macrophages, which are considered to be the reflection of the production of nitric oxide. In a study conducted by Yao and others, polysaccharide subfractions were isolated through alcohol and water (viz. QAP 1, QAP 2, QWP 1,

TABLE 11.9 Biological Activity of Quinoa

Sl. No.	Biological Activity	Component Responsible	References
1.	Antioxidant activity	Quinoa seeds contain more than 20 phenolic compounds in free or conjugated form, which is attributed to the presence of higher antioxidant activity in Quinoa. Few of these are available in the form of ferulic and vanillic acids along with their derivatives, quercetin, flavonoids, kaempferol, and glycosides.	Abderrahim et al. (2015); Tang et al. (2015, 2016)
2.	Anti-diabetic activity	Phenolic compounds are reported to possess strong inhibition of the enzyme, α-glucosidase, which plays a key role in the digestion of carbohydrates. The most prevailing phytoecdysteroid, that is, 20-HE of quinoa seeds significantly reduces the level of blood glucose in the hyperglycemic rat model.	Tang and Tsao (2017)
3.	Anti-inflammatory activity	Various carbohydrase enzymes such as pectinase, a feruloyl esterase, and xylanase liberate bound phenolics secreted by colonic bacteria, which suggest potential anti-inflammatory activity at the time of colonic fermentation in the large intestine. Saponins of quinoa seeds are associated with inhibition of TNF-α, nitric oxide, IL-6, and inflammatory mediator overproduction, which suggests that quinoa saponins are a good functional ingredient for the treatment and prevention of inflammation. Polyunsaturated fatty acids (PUFAs) and extractable phenolics of Quinoa (cooked) could efficiently downregulate mRNA expression of IL-1β, TNF-α, IL-6, IL 8, and COX-2, except IL-10, which is an anti-inflammatory cytokine). Beside down-regulation of these genes' expression, it also inhibits the release of IL-8, which is considered a major pro-inflammatory factor.	Tang et al. (2016); Tang and Tsao (2017); Yao et al. (2015)
4.	Anti-cancerous activity	Due to the presence of phytosterols in higher amounts, Quinoa imparts anti-carcinogenic activity. B-sitosterol, campesterol, and stigmasterol are the major components responsible for inducing anti-carcinogenic activity. These components are present in higher amounts in Quinoa as compared to other cereals such as rye, corn, barley, and so forth.	Ryan et al. (2007)
5.	Hypotensive activity	Membrane fractionation of protein hydrolysates into low molecular weight peptides improves the inhibition ability of angiotensin-converting enzyme (ACE), which helps the arteries and veins to be in a relaxed position, thereby lowering the flow of blood pressure.	Aluko and Monu (2003)

TABLE 11.9 *(Continued)*

Sl. No.	Biological Activity	Component Responsible	References
6.	Immunoregulatory activity	Polysaccharides extracts with water and alcohol along with their sub-fractions (QAP 1, QAP 2, QWP 1, and QWP 2) had significantly stimulated the macrophages to produce nitric oxide. As a result of which the immune system is enhanced.	Yao et al. (2014)
7.	Cytotoxic activity	Triterpenes saponins isolated from seeds and other parts of quinoa exhibit cytotoxic activity as observed by Kuljanabhagavad in HeLa cells of cervix adenocarcinoma.	Kuljanabhagavad and Wink (2009)
8.	Anti-microbial activity	Phenolic content is responsible for exhibiting anti-microbial activity against *E. coli* and *S. aureus.*	Tang and Tsao (2017)
9.	Anti-fungal activity	The total saponin content of Quinoa inhibits the growth of *Candida albicans*. Growth of mycelia *Botrytis cinerea* can also be inhibited by the same fraction by providing alkali treatment, which might be due to the formation of various derivatives of hydrophobic saponins which have a higher affinity towards sterols in cell membranes.	Vilcacundo and Hernandez (2017)
10.	Pancreatic lipase and α-glucosidase inhibiting activity	Pancreatic lipase and α-glucosidase are involved in lipid and sugar digestion, respectively. Phenolics and other bioactive compounds in quinoa seeds inhibit the activity of these enzymes, thereby helping in weight and blood sugar management (Type-II diabetes).	Tang et al. (2016)
11..	DPP-4 inhibiting activity	Protein isolates (using papain) have higher DPP-4 (dipeptidyl peptidase-4) inhibiting activity with IC_{50}, which makes them a functional promising component that has serum glucose-lowering properties.	Nongonierma et al. (2015)

and QWP 2) and possess significant immune-modulating and antioxidant activities. This study indicates the ability of these sub-fractions to stimulate the macrophages to further produce nitric oxide, in which QAP 1 was the most potent antioxidant as well as immunoregulator (Yao et al., 2014).

11.6.2 HUMANS STUDIES TO ASCERTAIN HEALTH CLAIMS

Quinoa, being an underutilized crop for centuries in many countries, is now gaining importance due to its numerous health benefits. Keeping this thing in mind, many researchers have made several products by incorporating Quinoa in one form or the other and have done clinical studies to investigate the effect of quinoa-based food products on human beings such as the reduced risk of heart-related diseases, growth, and development, role in celiac disease, and so on, as discussed in Table 11.10. A study aimed to check the role of Quinoa in the growth and development of the body in children belonging to the age group of 4–6 years using preparing quinoa-based infant food, 100 g of which was given twice a day. The study was conducted for 15 days and resulted in a significant increase in plasma insulin-like growth factors viz. IGF-1 level in children as compared to the control group. The study indicates that quinoa-based infant food provides sufficient proteins along with essential nutrients, which may play a major role in eliminating malnutrition among children (Ruales et al., 2002). Another study by De Carvalho et al. (2014) was conducted amongst 35 women who were over-weight, menopaused for 2 or more years, had 35 or more (mIU/mL) of follicle-stimulating hormone, 10–20 (pg/mL) serum estradiol, not taking any drugs since last two weeks or undergoing any hormone therapy since the past half-year. These women were given quinoa flakes (25 g/day), and a comparison was done with corn flakes. The result revealed a significant reduction in the level of serum triglyceride, vitamin E, and thiobarbituric acid reactive substances, along with a decrease in levels of LDL and total cholesterol, which depicts that quinoa flakes play an important role in reducing blood fat (De Carvalho et al., 2014).

As per the study done by Farinazzi-Machado et al. (2012), 22 students between the age group of 18–45 years for 30 days, were given quinoa-based cereal bars. At the end of the study, a significant decrease in the level of triglycerides, LDL, and total cholesterol was observed, which can be related to reducing the risk of developing coronary-related diseases. Another study by Sanchez (2012), investigates the effect of consumption of different cereal grains *viz.*, Quinoa, kaniwa, and amaranth on Type-II diabetes and weight

TABLE 11.10 Effect of Consumption of Quinoa-based Foods on Human Health

SL. No.	Therapeutic Application	Tenure of Trial	Clinical Trial	y Finding	References
1.	Development and growth	15 days	The experiment was aimed to check the role of Quinoa in the growth and development of a body of boys aged from 4–6 years by preparation of quinoa-based infant food (100 g twice a day).	• A significant increase in body weight was observed, which was due to plasma level of IGF 1 (malnutrition's marker). • Thus, quinoa-based formulated infant food plays a major role in tackling the problem of malnutrition among children.	Rules et al. (2002)
2.	Post-menopausal symptoms	4 weeks	Women having more than 80 (cm) waist circumference, menopaused for 2 or more years, having 35 or more (mIU/mL) of follicle-stimulating hormone, 10–20 (pg/mL) serum estradiol, not taking any drugs for the last 2 weeks or undergoing any hormone therapy since past semester and having excess weight. Such women were given quinoa flakes (25 g/day), and a comparison was done with corn flakes.	• Consumption of quinoa flakes results in increased fiber and protein intake, which beneficially modulates the metabolic parameters. • Other factors such as triglycerides, cholesterol, thiobarbituric acid reactive substances, and low-density lipo-protein were reduced and GSH increases.	De Carvalho et al. (2014)
3.	Risk of coronary diseases	30 days	Students in the age group of 18–45 years were given cereal bars based on Quinoa on daily basis for 30 days. The study was aimed to study the effect of Quinoa on blood pressure, anthropometric, and biochemical profile.	• Significant reduction in the levels of cholesterol, triglycerides, and LDL was observed, which is related to a reduced risk of developing coronary (heart) related diseases. • An insignificant decrease in the level of blood glucose, blood pressure, and body weight was also observed.	Farinazzi-Machado et al. (2012)

TABLE 11.10 *(Continued)*

SL. No.	Therapeutic Application	Tenure of Trial	Clinical Trial	y Finding	References
4.	Celiac disease	6 weeks	Cooked Quinoa was given to patients, 50 g per day regularly for 6 weeks.	• Intake of quinoa diet improves various gastrointestinal parameters such as cell height of surface enterocyte, intra epithelial lymphocytes, and others. • Other factors such as total cholesterol, LDL, high-density lipoprotein (HDL), and triglycerides were reduced insignificantly and at a very small rate, however, the level of serum lipids remains normal.	Zevallos et al. (2014)
5.	Increased stamina, performance, and anabolic potential	1 year	A double-blind study was conducted with 45 men (average age = 20.5 ± 3) who did exercise thrice a week for 1 year. The person was given 200 mg of 20-hydroxyecdysone per day.	• The results revealed that supplementation of 20-HE at 30 mg per day for 8 weeks neither affected the hormone status (cortisol, active, and free testosterone) nor body composition. • The study also states that the effect of 20-HE on inexperienced individual's performance is stronger as compared to experienced sportspersons.	Wilborn et al. (2006)

gain in middle-aged individuals (n = 110, diabetic = 22 and non-diabetic = 88). It was observed that a person with a normal BMI (body mass index) consumes grains more often as compared to obese or overweight persons. Also, it was concluded from the study that diabetic person consumes grains more frequently compared to non-diabetic ones.

Various other health effects have been investigated by various researchers due to the wide range of phytochemicals and bioactive compounds in Quinoa (Table 11.11). Phytoecdysteroids and their related compounds are responsible for various health effects and activities such as faster wound healing, cholesterol-lowering effects, blood glucose-lowering ability, protein synthesis and growth, enhancing physical performance, and many more. Betaine is responsible for decreased levels of plasma homocysteine and improves mitochondrial respiration (Peterson and Murphy, 2015).

11.6.3 ANIMAL STUDIES TO ASCERTAIN HEALTH CLAIMS

Several studies regarding the consumption of Quinoa on health are still limited. Few studies have been done on animals to showcase the biological activity, that is, anti-obesity, antioxidant activity, hypocholesterolemic activity, and so on, as shown in Table 11.12. Various studies have been carried out to establish the ani-obesity activity in rats fed with high-fat diets. The result showed the reduced deposits of adipose mass and lesser weight gain in animals. The major steroid responsible for this activity is 20-hydroxyecdysone which is present in abundance in the pericarp of quinoa seed and aids in decreasing the weight gain in animals (Foucault et al., 2012; Kizelsztein et al., 2009). Another study was carried out by Wang et al. (2011) on rats fed with a high-fat diet and were given the treatment of 20-HE (25 to 60 mg/Kg) for 12 weeks. As a result of which, reduced body weight accompanied by lesser fat accumulation in muscles was observed and on the other hand sensitivity of insulin was also improved.

In another study, that aimed to investigate the effect of 20-HE as an anti-diabetic agent, a hyperglycemic rat model was used and the result showed that there was a decrease in the level of blood glucose in mice, which supports Quinoa's potential to prevent or treat hyperglycemia as well as the resistance of insulin associated with metabolic syndrome (Graf et al., 2014). Similar results were observed in another study done by Pawal Pasko et al. (2010), where the rats were fed with a diet containing high fructose for 5 weeks along with quinoa seeds (310 g/Kg of fodder). The results state that the blood glucose level was reduced by 10% (0<0.001) as compared to the control

TABLE 11.11 Health Benefits of Quinoa

Sl. No.	Health Benefit	Studies/Component Responsible	References
1.	Cholesterol-lowering effects	Ecdysteroids are known to possess the cholesterol-lowering ability.	Gorellick-Feldman et al. (2008)
2.	Lowering blood glucose levels	20-hydroxyecdysone and ecdysteroids significantly reduce the level of glucose in the blood.	Tang and Tsao (2017)
3.	Protein synthesis and growth	Phytoecdysteroids increase protein synthesis by up to 20% in primary human myotubes and C2C12 murine myotubes. Ecdysteroids promote growth by enhancing protein synthesis in skeletal muscle cells across a wide range of animals, including pigs, sheep, quail, and mice. Various studies demonstrate increased protein content and growth in the kidneys and liver of ecdysteroid-treated rats.	Gorellick-Feldman et al. (2008)
4.	Improves physical performance	Ecdysteroids are known for their anabolic effects such as enhanced physical performance irrespective of training. The demonstration was done on rats using a forced swim test. An increase in the synthesis of myofibril protein was observed in digitoriumlongus and soleus.	Gorellick-Feldman et al. (2008)
5.	Anti-osteoporotic effect	Ecdysteroid (β-ecdysone) is the phytochemical responsible for the anti-osteoporotic effect in Quinoa.	Seidlova-Wuttke et al. (2008)
6.	Wound healing properties	Ecdysteroids are responsible for faster healing of wounds.	Graf et al. (2014)
7.	Gastrointestinal health	Extracted phenolics from Quinoa downregulate the expression of IL-8, IL-1β, and cytokines (TNF) when cultured in colonial epithelial Caco-2-cells. These are responsible for preventing obesity-induced inflammation and improving the gastrointestinal health in rats.	Tang and Tsao (2017)
8.	Improves mitochondrial respiration	Betaine helps in improving mitochondrial respiration as well as helps in decreasing the plasma homocysteine.	Bastidas et al. (2016)

TABLE 11.11 *(Continued)*

Sl. No.	Health Benefit	Studies/Component Responsible	References
9.	Rejuvenates skin and prevents damage	Phytoecdysteroids present in seeds of quinoa act as protective agents and prevent or delays the collagenase related skin damage. Oil along with other phytochemicals such as phytoecdysone inhibits the activity of tyrosinase enzyme, MMP 1 mRNA, and MMP 9 expression, reduces the production of intracellular ROS in dermal fibroblasts, and thereby plays important role in prevention and treatment in the aging of skin via a multiplicity of effects.	Graf et al. (2015)

group. In another 13-week study, daily administration of 20-hydroxyecdysone at a 10 mg/kg dosage was given, which decreased the expression of PEPCK (phosphoenolpyruvate) and G6Pase (glucose 6 phosphates), thus decreasing glucose synthesis. Besides this, the level of glucose tolerance and plasma insulin were lowered significantly (Kizelsztein et al., 2009). In a study done by Konishi et al. (2000) on mice, where mice were fed with a diet supplemented with 1.5 and 3% of quinoa seed coat along with the incorporation of 1% cholesterol in each case, resulting in a significant reduction in the level of liver cholesterol and serum was observed, which may be attributed to the presence of water-soluble dietary fiber content in the seed coat of Quinoa. Similar studies were done by Takao et al. (2005), where the supplementation of a diet with quinoa protein inhibits the increase in cholesterol and plasma levels. In a study performed by Pawel Pasko et al. (2010), the blood analysis at the end of the experiment indicates Quinoa's efficiency in effectively reducing the level of total serum cholesterol and triglycerides, the details of which are given in Table 11.13. Hypocholesterolemic activity in Quinoa is attributed to the presence of ecdysteroids in it, which can be explained by the enhanced conversion of cholesterol to bile acids (Dinan and Lafont, 2006). Neuroprotective effects of 20-hydroxyecdysone can be attributed to the modulation of GABA receptors (Navruz-Varli and Sanlier, 2016).

11.7 ALLERGENICITY

Very few studies have been performed on allergic effects due to the consumption of Quinoa. Among the various studies done till now, a study found that positive reaction bands were detected against antibodies of buckwheat protein, which was absent in the case of wheat protein antibodies (Asao and Watanabe, 2010). A case study was reported on 38 years old women diagnosed with eosinophilic esophagitis due to wheat and cross-reactivity was observed between Quinoa and buckwheat. Samples were prepared by extracting buckwheat flour and quinoa seeds. Skin prick test yields positive results in both cases; therefore, it is recommended to have cross-sensitization studies before recommending quinoa and buckwheat flour as a substitute to patients who are sensitive to flour (El-Qutob et al., 2014).

Another case of anaphylaxis was reported by Astier and others in France, where 52 a 52-year-old man consumed Quinoa along with bread and fish and developed a systemic reaction of urticaria, dysphagia, angioedema, and dysphonia. Upon testing ingested food samples from the patient's serum, it was observed that only the quinoa sample showed IgE (immunoglobulin)

TABLE 11.12 Effect of Consumption of Quinoa-based Foods on Animal Health

SL. No.	Biological Property	Model	Time of Study	Experiment	Key Finding	References
1.	Anti-obesity activity	Rats	3 weeks	The study aimed to highlight the role of quinoa extract supplemented with 20 HE in the prevention of obesity resulting from improper dietary habits as well as the regulation of adipocyte-specific genes.	Quinoa extract reduces the esterification in adipocytes along with a reduction in uptake of fatty acid.	Foucault et al. (2012)
		Mice	13 weeks	The study was performed to demonstrate the anti-obese effect of 20-HE in a diet-induced obesity mice model. A daily dose of 20-HE was administered at 10 mg/kg of a high-fat diet and the control group was fed an only a high-fat diet.	At the end of the study, it was observed that the group of mice fed with a 20-HE high-fat diet has gained lower body weight and deposits of adipose mass were less as compared to the control ones (fed with a high-fat diet only).	Kizelsztein et al. (2009)
	–	Rats	12 weeks	The study was done on rats fed with a high-fat diet and was given the treatment of 20-HE (25 to 50 mg/kg).	Reduced fat accumulation in muscles was observed and thus reduced body weight was noticed.	Wang et al. (2011)
2.	Hypocholesterolemic activity	Mice	5 weeks	Mice were fed with a diet supplemented with 1.5 and 3% of quinoa seed coat along with the incorporation of 1% cholesterol in both cases.	A significant reduction in the level of liver cholesterol and serum was observed, which might be due to the presence of water-soluble dietary fiber content in the seed coat of Quinoa.	Konishi et al. (2000)

TABLE 11.12 *(Continued)*

SL. No.	Biological Property	Model	Time of Study	Experiment	Key Finding	References
		Mice	4 weeks	Mice were given a diet containing quinoa protein as 0%, 2.5%, and 5% along with the incorporation of 0.5% cholesterol in all cases.	Supplementation with quinoa protein inhibits the increase in cholesterol and plasma levels.	Takao et al. (2005)
		Wistar rats	5 weeks	The experiment was aimed to study the effect of quinoa seeds on lipid profile in rats fed with a diet containing high fructose.	The blood analysis indicated that quinoa seeds were capable of effectively reducing the total serum cholesterol by 26%, triglycerides by 11% ($p<0.05$), LDL by 57% ($p<0.008$), total protein by 16% ($p<0.001$) in comparison to control group.	Pawel Pasko et al. (2010)
3.	Antioxidant property	Wistar rats	5 weeks	The experiment was aimed to study the effect of quinoa seeds enriched diet on oxidative stress in the heart, plasma, kidney, testis, liver, pancreas, and spleen of rats administered with fructose (310 g/Kg of fodder), which causes oxidative stress manifested by an increase in MDA (malondialdehyde) of plasma and other tissues.	Quinoa seeds incorporated at the rate of 310 g/Kg of fodder exhibit the normal activity of enzymes. Oxidative stress was influenced by reducing the action of various antioxidant enzymes (catalase, plasma GSH POD, erythrocyte superoxide-dismutase) and MDA in plasma. This demonstrates the action of quinoa seeds as a protective agent against various fructose-induced alterations in rats by enhancing the antioxidant potential of the heart, plasma, kidney, pancreas, lungs, testis, and thereby reducing the peroxidation.	Pawel Pasko et al. (2010)
4.	Hypoglycemic or anti-diabetic activity	Mice	15 weeks	Mice were fed with a fat-induced diet (hyperglycemic rat model).	Quinoa leachate significantly reduced the level of blood glucose in mice, which supports Quinoa's potential to prevent or treat hyperglycemia as well as the resistance of insulin associated with metabolic syndrome.	Graf et al. (2014)

TABLE 11.12 *(Continued)*

SL. No.	Biological Property	Model	Time of Study	Experiment	Key Finding	References
		Wistar rats	5 weeks	The experiment was aimed to study the effect of quinoa seeds on glucose profile in rats fed with a diet containing high fructose as a control group while the other group was given a diet enriched with quinoa seeds at 310 g/Kg of fodder.	The blood analysis indicated that quinoa seeds were capable of effectively reducing the glucose level by 10% ($p < 0.001$) in comparison to the control group.	Pawel Pasko et al. (2010)
		Mice	13 weeks	The study was performed to demonstrate the anti-diabetic effect of 20-HE in a diet-induced obesity mice model. The daily dose of 20-HE was administered at 10 mg/Kg of a high-fat diet, and the control group was fed only a high-fat diet.	The study depicts that glucose synthesis was reduced in hepatic cell culture, which was because of 20-HE in diet, as it deregulated the expression of G6Pase and PEPCK genes in a dose-dependent manner.	Kizelsztein et al. (2009)

reactivity (Astier et al., 2009). Another study on 29 a 29-year-old man revealed that the man developed rashes on his chest and arms along with itching on his palm and feet soles and angioedema in his lips after 5 min of quinoa salad consumption. A skin prick test revealed a positive result for allergy due to Quinoa in the patient (Hong et al., 2013).

11.8 TECHNOLOGICAL INNOVATIONS IN QUINOA PROCESSING

Technological innovations in food fasten the cooking time, enhance the health-beneficial phytochemicals, and generate ready-to-eat (RTE) products in today's fast-moving era. Due to the high medicinal and nutritional value of Quinoa, there is growing interest in producing various value-added products based on Quinoa. As a functional food, cosmetic, supplement, and pharmaceutical agent, quinoa-based technologies are supposed to target various therapeutic parameters of human health, such as celiac disease, fitness enhancement, sports performance, weight loss, hypertension, diabetes, obesity, cholesterol-lowering, skin care, drug absorption, and so forth.

11.8.1 FOOD PROCESSING, FORMULATION, AND PACKAGING

Various processing methods have been developed such as flattening of seeds, removal of saponins from Quinoa via superheated steam to reduce the cooking time, and so forth (Table 11.13) while other methods are being investigated to further improve the overall quality of baked goods substitution of Quinoa in other cereal grains, for instance, 10% of quinoa flour is added in wheat-based bread to enhance its nutritional quality without any change in its loaf volume (Graf et al., 2015). Roasting is another processing technique that leads to the weakening of interactions between starch and protein. As per Rothschild and others, roasting quinoa seed flour (177°C for 15 min, 30 min, and 45 min) increases the peak viscosity along with final viscosity, which is an important parameter in the manufacturing of cakes (Rothschild et al., 2014). Baking is one of the common processing techniques, baking of quinoa seeds results in an overall increase in antioxidant activity and total phenolic content (TPC), which might be due to the Maillard reaction that occurred during thermal processing (Yael et al., 2012).

Germination results in increased oleic acid content; however, germination with steeping results in decreased phytates, tannins, and other anti-nutritional contents to a greater extent. Germination, when exceeded for 72 h, results in increased antioxidant activity by 2 folds, phenolic acids by 8.57 times, and flavonoids by 4.4 times (Carciochi et al., 2016; Kanensi et al., 2011). Germination followed by subsequent oven drying improves the antioxidant and TPC of quinoa seeds and therefore, it may be employed in the formulation of various functional foods. Germinated seeds possess essential fatty acids, which are considered to be important for brain development, prostaglandin metabolism, insulin sensitivity, and lower (ω–6: ω–3) ratio, which is important for maintaining immunity, and cardiovascular health and prevent from auto-immune diseases (Vega-Gálvez et al., 2010). Malting of quinoa seeds followed by thermal treatment enriches the antioxidants in grains, which can be used later as a functional ingredient in the manufacturing of beverages and other gluten-free products (Carciochi et al., 2016). Quinoa products subjected to high pressure along with high temperature (HPHT) treatment increase the iron solubility after soaking followed by germination by 2–4 times, after fermentation (3–5 times), germination followed by fermentation (5–8 times). Cooking and baking result in increased magnesium content by 5–8 times, while processing reduced the copper content by 28% (Ruiz et al., 2016).

Tempe production has been developed by using quinoa seeds, via solid-state fermentation using *Rhizopus oligosporus* which contains a lower amount of estrogenic isoflavone with all essential amino acids in it (Graf et al., 2015). Spaghetti made from quinoa flour is not only gluten-free but also possesses reasonable physical properties along with acceptable taste as compared to spaghetti made from wheat flour. Gluten-free bread has improved texture and loaf volume when quinoa flour was substituted in place of corn or rice flour (Elgeti et al., 2014). Furthermore, maltodextrins of Quinoa possess gel-like consistency and therefore can be used as a fat or cream substitute (Caceres et al., 2017). Bread made by using quinoa flour as a starch source instead of potato starch exhibited softer crumb, improved loaf volume with increased content of protein, vitamins, minerals, antioxidants, and polyphenols (Valcárcel-Yamani et al., 2012). The gluten-free bread or other products available in the market are considerable of poor nutritional value and low quality. The fact of Quinoa is gluten-free is advantageous for patients suffering from gluten intolerance or wheat allergy (Taylor and Parker, 2002). The thermostability and compact size of quinoa starch granules make it useful in the field of thickeners (emulsion type products), malted beverages, and frozen food packaging systems (Bhargava et al., 2006). Pagno and others developed an active biofilm composed of gold-nanoparticles embedded in quinoa starch, which has an anti-bacterial effect against *Staphylococcus aureus* and *E. coli* and therefore can be used as food packaging material with anti-bacterial properties (Pagno et al., 2015).

TABLE 11.13 Domestic Processing Methods for Quinoa Grain

SL. No.	Product Derived from Quinoa	Processing Method	Outcome	Use	References
1.	Treated seeds	Superheated steam	Results in the expansion of seeds and reduction in cooking time.	Used as texturizing or flavoring agents in different food products.	Scanlin and Burnett (2010)
		Washing, mechanical abrasion	Reduction in bitterness of seeds	Used as texturizing or flavoring agents in different food products.	Scanlin and Burnett (2010)
2.	Protein concentrate	Extraction and precipitation	Results in formation of protein concentrate.	Can be used in animal food or human food as an ergogenic food.	Kruger and Inventor (2012); Scanlin and Stone (2009)
3.	Carbohydrates	Extraction of maltodextrins with enzymes or alkali.	Gel-form is produced to deliver quinoa-derived peptides.	Used as a cream substitute for better mouth feel.	Singer et al. (1990)
4.	Beverage	Soaking, malting, kilning, mashing, cooling, and fermenting with yeast.	Production of fermented and gluten-free product.	Used to make a fermented alcoholic beverage.	Kamelgard (2013)
5.	Lipids	Molecular distillation and extraction.	Production of refined oil.	Used for dermatological purposes.	Msika and Inventor (2012)
6.	Ready-to-eat products	Extrusion	Inactivation of anti-nutritional factors, aflatoxin destruction, increased fiber digestibility.	Used for making quinoa-based ready-to-eat breakfast products.	Repo-Carrasco et al. (2011)
7.	Cakes and other confectionery products	Roasting	Results in increased peak, as well as final viscosity, weakens the interactions of starch–protein.	Increased peak and final viscosity are useful in proper cake manufacturing.	Rothschild et al. (2015)

11.9 TRADITIONAL USES

Quinoa is being used for centuries. Traditionally, it was used by native people of South America, including Aymara, Chibcha, Quechua, Mapuche, and Tiahuancota (Vega-Galvez et al., 2010). Seeds of Quinoa are used in a way similar to that of rice, either by puffing to be used in breakfast or soups and can be ground to produce flour for making baked goods such as cookies, flakes, biscuits, tortilla, noodles, bread, pancakes, and so forth (Bhargava et al., 2006). The leaves of Quinoa are eaten in the same way as that of spinach whereas the germinated seeds can be used in salads. Furthermore, fermentation of quinoa seeds is done to produce beer and a traditional alcoholic beverage called chichi or chicha (traditional drink of Andes) (Graf et al., 2015), while the rest of the plant can be used to feed livestock, pigs, poultry, and cattle. Quinoa also possesses some traditional medicinal uses, such as the treatment of fractures and wounds and improving digestive health (Bhargava et al., 2006). Pungent ashes of stems, known as elliptic, are mixed with coca leaves and are chewed by farmers to retain the energy in Andean. Incan armies used a mixture of fat and Quinoa, known as war balls, to retain their energy while marching over the mountains (Graf et al., 2015). The extracts of phytoecdysteroid-containing plants were used as muscle builders, stress reducers, and adaptogens (Báthori, 2002). Saponins in Quinoa have anti-inflammatory activity and were used traditionally as medicine for its remedy (Raju and Rao, 2012).

11.10 FUTURE OF QUINOA

Quinoa, mainly produced by Peru, Bolivia, Colombia, Argentina, Chile, and Ecuador, is exported in the form of saponin-free Quinoa or dry Quinoa to the USA and Europe. Future uses of Quinoa may include textured or fermented products. In Europe, a substitute for quinoa meat has been introduced (James, 2009). Quinoa grains are being used widely in a variety of different aspects of food. A whole grain of Quinoa is used in making soups, beverages, sweets, porridges, sauces, and soufflés whereas boiled grain can be used in form of rice. Quinoa can serve as protein concentrate, and starchy material and its flour can also be produced. It is used in preparation of kispina, that is, coarse bread (Valcárcel-Yamani et al., 2012). Grains after grinding can be used as flour in making cookies, cakes, pancakes, crackers, bread, puddings, dumplings, muffins, pasta, and so forth (Alvarez-Jubete et al., 2010; Gamel et al., 2004; Mlakar et al., 2009; Schoenlechner et al., 2010). It is capable enough to be used as baby food, diet supplementation, breakfast cereals, and extrudates with high protein quality (James, 2009).

There is no doubt that only wheat protein gluten is capable of forming the cohesive dough with its ability to retain gas which gives a spongy product after cooking. Other cereals do not impart such characteristics, due to which it is suggested to mix wheat flour with flours of other cereals in bakery products. Doing the same with Quinoa not only increases its protein quality but also provides a reasonable amino-acid balance because of its high levels of methionine and lysine in it. However, it is not possible to replace wheat in the blends, as Quinoa and other pseudo-cereals do not comprise gluten, the consequence of which is a limitation for these pseudo-cereals to be used in leavened products. Therefore, these can be used in the form of composite flour with wheat flour (Taylor and Parker, 2002). Chauhan and others made quinoa bread using 10% wheat flour. TGase (transglutaminase) enzyme is promising in the field of developing protein structure, which reduces soluble-protein-fractions and low molecular weight peptides during pasta production (Kovács, 2003). TGase possesses positive reports by inducing polymerization and cross-linking of proteins in food such as soy protein, fish protein, and milk protein, which ultimately improves the physic-chemical properties of the food (James, 2009).

An optimal combination of pseudo-cereal flour, enzymes, emulsifier, and albumen is required to make high-quality pasta products with improved physical properties (Schoenlechner et al., 2010). As per Dogan and Karwe (2003), Quinoa can be used for making novel, nutrient-rich, snack-type, or extruded food products. They made the same with minimum density, maximum expansion, low WS (water-solubility) index, and higher DG (degree of gelatinization). Various novel products have been made by various researchers such as symbiotic beverages were prepared using an aqueous extract of Quinoa (Bianchi et al., 2015), quinoa milk with increased protein content and low glycemic index (GI) (Pineli et al., 2015), quinoa-based cereal bars (Farinazzi-Machado et al., 2012), gluten-free pasta using Quinoa (Kahlon and Chiu, 2015). It is important to promote and make people aware to increase its consumption so that consumers get a healthy, nutritious, and adequate balanced diet, especially the persons suffering from celiac disease. Therefore, commercial production of Quinoa should be increased and must be scaled up to an industrial level.

11.11 CONCLUSION

The intake of Quinoa and other pseudo-cereals in one or the other form in the diet increases the intake of various phytochemicals and macromolecules, which are beneficial for optimum human health. It is an alternative for patients suffering from celiac disease along with a protein of high biological value. It is useful for various medical issues such as obesity, dyslipidemia, cancer, and many more. Quinoa is a strategic crop to be used in marginal areas where the persistence of malnutrition is prevailing. Consumption of 40 g quinoa meets the daily recommended dietary allowance for phytochemicals and other essential nutrients. Saponins possess commercial as well as industrial value as these can be used in the manufacturing of detergents, soaps, shampoos, and so on. Several formulations and trials have been reported to be successful using pseudo-cereals, such as the formation of gluten-free baked or confectionary products; however, their availability in the market is still limited. More research needs to be focused on completely exploiting the functional potential of these products, which are not only healthy but also nutritionally balanced.

KEYWORDS

- **anti-nutrients**
- **health benefits**
- **loaf volume**
- **nutritional quality**
- **quinoa**
- **roasting**
- **starch-protein interactions**
- **viscosity**

REFERENCES

Abderrahim, F., Huanatico, E., Segura, R., Arribas, S., Gonzalez, M. C., & Condezo-Hoyos, L. (2015). Physical features, phenolic compounds, betalains, and total antioxidant capacity of colored quinoa seeds (*Chenopodium quinoa Willd.*) from Peruvian Altiplano. *Food Chemistry*, *183*, 83–90.

Aluko, R. E., & Monu, E. (2003). Functional and bioactive properties of quinoa seed protein hydrolysates. *Journal of Food Science*, *68*(4), 1254–1258.

Alvarez-Jubete, L., Auty, M., Arendt, E. K., & Gallagher, E. (2010). Baking properties and microstructure of pseudocereal flours in gluten-free bread formulations. *European Food Research and Technology*, *230*, 437–445.

American Diabetes Association. (2010). Ecdysterone enhances muscle insulin signaling by modulating acylcarnitine profile and mitochondrial oxidative phosphorylation complexes in mice fed a high-fat diet–Withdrawn. *Diabetes*, *DB_101605*.

Ando, H., Chen, Y. C., Tang, H., Shimizu, M., Watanabe, K., & Mitsunaga, T. (2002). Food components in fractions of quinoa seed. *Food Science and Technology Research*, *8*(1), 80–84.

Arneja, I., Tanwar, B., & Chauhan, A. (2015). Nutritional composition and health benefits of the golden grain of the 21st century, quinoa (*Chenopodium quinoa Willd.*): A review. *Pakistan Journal of Nutrition*, *14*(12), 1034.

Asao, M., & Watanabe, K. (2010). Functional and bioactive properties of quinoa and amaranth. *Food Science and Technology Research*, *16*(2), 163–168.

Astier, C., Moneret-Vautrin, D. A., Puillandre, E., & Bihain, B. E. (2009). First case report of anaphylaxis to quinoa, a novel food in France. *Allergy*, *64*(5), 819–820.

Bastidas, E. G., Roura, R., Rizzolo, D. A. D., Massanés, T., & Gomis, R. (2016). Quinoa (*Chenopodium quinoa Willd.*), from nutritional value to potential health benefits: An integrative review. *Journal of Nutrition & Food Sciences*, *6*(3), 497.

Báthori, M. (2002). Phytoecdysteroids effects on mammals: Isolation and analysis. *Mini Reviews in Medicinal Chemistry*, *2*(3), 285–293.

Bazile, D., Jacobsen, S. E., & Verniau, A. (2016). The global expansion of quinoa: Trends and limits. *Frontiers in Plant Science*, *7*, 622.

Bhargava, A., Shukla, S., & Ohri, D. (2006). *Chenopodium quinoa*: An Indian perspective. *Industrial Crops and Products*, *23*(1), 73–87.

Bianchi, F., Rossi, E. A., Gomes, R. G., & Sivieri, K. (2015). Potentially synbiotic fermented beverage with aqueous extracts of quinoa (*Chenopodium quinoa Willd.*) and soy. *Food Science and Technology International*, *21*(6), 403–415.

Bonifacio, A. (2003). *Chenopodium* sp.: Genetic resources, ethnobotany, and geographic distribution. *Food Reviews International*, *19*(1, 2), 1–7.

Caceres, J. I. E., Calderon, P. D., & Lira, F. O. (2007). *Method for the Formulation of a Gel-Format Foodstuff for use as a Nutritional Foodstuff Enriched with Peptides and Maltodextrins Obtained from Quinoa Flour* (U.S. Patent No. 9,763,463).

Carciochi, R. A., Dimitrov, K., & Galván D'Alessandro, L. (2016). Effect of malting conditions on phenolic content, Maillard reaction products formation, and antioxidant activity of quinoa seeds. *Journal of Food Science and Technology*, *53*, 3978–3985.

Chauhan, G. S., Eskin, N. A. M., & Tkachuk, R. (1992). Nutrients and antinutrients in quinoa seed. *Cereal Chemistry*, *69*(1), 85–88.

Chauhan, G. S., Zillman, R. R., & Eskin, N. M. (1992). Dough mixing and bread making properties of quinoa-wheat flour blends. *International Journal of Food Science & Technology*, *27*(6), 701–705.

Dallagnol, A. M., Pescuma, M., De Valdez, G. F., & Rollán, G. (2013). Fermentation of quinoa and wheat slurries by *Lactobacillus plantarum* CRL 778: Proteolytic activity. *Applied Microbiology and Biotechnology*, *97*, 3129–3140.

De Carvalho, F. G., Ovídio, P. P., Padovan, G. J., Jordao Junior, A. A., Marchini, J. S., & Navarro, A. M. (2014). Metabolic parameters of postmenopausal women after quinoa or corn flakes intake: A prospective and double-blind study. *International Journal of Food Sciences and Nutrition*, *65*(3), 380–385.

Elgeti, D., Nordlohne, S. D., Föste, M., Besl, M., Linden, M. H., Heinz, V., Jekle, M., & Becker, T. (2016). Volume and texture improvement of gluten-free bread using quinoa white flour. *Journal of Cereal Science*, *59*(1), 41–47.

Farinazzi-Machado, F. M. V., Barbalho, S. M., Oshiiwa, M., Goulart, R., & Pessan Junior, O. (2012). Use of cereal bars with quinoa (*Chenopodium quinoa W.*) to reduce risk factors related to cardiovascular diseases. *Food Science and Technology*, *32*, 239–244.

Foucault, A. S., Mathé, V., Lafont, R., Even, P., Dioh, W., Veillet, S., Tomé, D., Huneau, J. F., Hermier, D., & Quignard-Boulangé, A. (2012). Quinoa extract enriched in 20-hydroxyecdysone protects mice from diet-induced obesity and modulates adipokines expression. *Obesity*, *20*(2), 270–277.

Gamel, T. H., Linssen, J. P., Alink, G. M., Mosallem, A. S., & Shekib, L. A. (2004). Nutritional study of raw and popped seed proteins of *Amaranthus caudatus L.* and *Amaranthus cruentus L. Journal of the Science of Food and Agriculture*, *84*(10), 1153–1158.

González Martín, M. I., Wells Moncada, G., Fischer, S., & Escuredo, O. (2014). Chemical characteristics and mineral composition of quinoa by near-infrared spectroscopy. *Journal of the Science of Food and Agriculture*, *94*(5), 876–881.

Gorelick-Feldman, J., MacLean, D., Ilic, N., Poulev, A., Lila, M. A., Cheng, D., & Raskin, I. (2008). Phytoecdysteroids increase protein synthesis in skeletal muscle cells. *Journal of Agricultural and Food Chemistry*, *56*(10), 3532–3537.

Gorinstein, S., Lojek, A., Číž, M., Pawelzik, E., Delgado-Licon, E., Medina, O. J., Moreno, M., Salas, I. A., & Goshev, I. (2008). Comparison of composition and antioxidant capacity of some cereals and pseudocereals. *International Journal of Food Science & Technology*, *43*(4), 629–637.

Graf, B. L., Cheng, D. M., Esposito, D., Shertel, T., Poulev, A., Plundrich, N., Itenberg, D., Dayan, N., Lila, M. A., & Raskin, I. (2015). Compounds leached from quinoa seeds inhibit matrix metalloproteinase activity and intracellular reactive oxygen species. *International Journal of Cosmetic Science*, *37*(2), 212–221.

Graf, B. L., Poulev, A., Kuhn, P., Grace, M. H., Lila, M. A., & Raskin, I. (2014). Quinoa seeds leach phytoecdysteroids and other compounds with anti-diabetic properties. *Food Chemistry*, *163*, 178–185.

Graf, B. L., Rojas-Silva, P., Rojo, L. E., Delatorre-Herrera, J., Baldeón, M. E., & Raskin, I. (2015). Innovations in health value and functional food development of quinoa (*Chenopodium quinoa Willd.*). *Comprehensive Reviews in Food Science and Food Safety*, *14*(4), 431–445.

Graf, B. L., Rojo, L. E., Delatorre-Herrera, J., Poulev, A., Calfio, C., & Raskin, I. (2016). Phytoecdysteroids and flavonoid glycosides among Chilean and commercial sources of *Chenopodium quinoa*: Variation and correlation to physico-chemical characteristics. *Journal of the Science of Food and Agriculture*, *96*(2), 633–643.

Hong, J., Convers, K., Reeves, N., & Temprano, J. (2013). Anaphylaxis to quinoa. *Annals of Allergy, Asthma & Immunology*, *110*(1), 60, 61.

James, L. E. A. (2009). Quinoa (*Chenopodium quinoa Willd*): Composition, chemistry, nutritional, and functional properties. *Advances in Food and Nutrition Research*, *58*, 1–31.

Kahlon, T. S., & Chiu, M. C. M. (2015). Teff, buckwheat, quinoa and amaranth: Ancient whole grain gluten-free egg-free pasta. *Food and Nutrition Sciences*, *6*(15), 1460.

Kamelgard, J. I. (2012). Quinoa-based beverages and method of creating quinoa-based beverages. U.S. Patent Application 13/363,615.

Kanensi, O. J., Ochola, S., Gikonyo, N. K., & Makokha, A. (2011). Optimization of the period of steeping and germination for amaranth grain. *Journal of Agriculture and Food Technology*, *1*(6), 101–105.

Keservani, R. K., & Sharma, A. K. (2014). Flavonoids: Emerging trends and potential health benefits. *Journal of Chinese Pharmaceutical Sciences*, *23*(12), 815.

Keservani, R. K., Sharma, A. K., & Kesharwani, R. K. (2016a). Nutraceutical and functional foods for cardiovascular health. In *Food Process Engineering* (pp. 291–312). Apple Academic Press, CRC Press. ISBN: 9781771884020.

Keservani, R. K., Sharma, A. K., & Kesharwani, R. K. (2016b). Medicinal effect of nutraceutical fruits for cognition and brain health. *Scientifica*, *3109254*, 10. https://doi.org/10.1155/2016/3109254.

Kizelsztein, P., Govorko, D., Komarnytsky, S., Evans, A., Wang, Z., Cefalu, W. T., & Raskin, I. (2009). 20-Hydroxyecdysone decreases weight and hyperglycemia in a diet-induced obesity mice model. *American Journal of Physiology-Endocrinology and Metabolism*, *296*(3), E433–E439.

Konishi, Y., Arai, N., Umeda, J., Gunji, N., Saeki, S., Takao, T., Minoguchi, R., & Kensho, G. (2000). Cholesterol-lowering effect of the methanol-insoluble materials from the quinoa seed pericarp. *Hydrocolloids*, *417*, 417–422.

Kovács, E. (2003). Use of transglutaminase for developing pasta structure in pseudo-cereals systems. In *3rd International Symposium on Food Rheology and Structure ISFRS* (pp. 545–546).

Kozioł, M. J. (1992). Chemical composition and nutritional evaluation of quinoa (*Chenopodium quinoa Willd*). *Journal of Food Composition and Analysis*, *5*(1), 35–68.

Kruger, M. (2012). Composition, useful for treating hair, comprises quinoa protein, a quaternary ammonium compound comprising esterquats and quaternary imidazoline compound, and a nourishing fatty component comprising silicone and/or oil body. Patent Application No. DE201110085217.

Kuljanabhagavad, T., & Wink, M. (2009). Biological activities and chemistry of saponins from *Chenopodium quinoa Willd. Phytochemistry Reviews*, *8*, 473–490.

Mlakar, S. G., Turinek, M., Jakop, M., Bavec, M., & Bavec, F. (2009). Nutrition value and use of grain amaranth: Potential future application in bread making. *Agricultura*, *6*(4), 43–53.

Msika, P. (2012). Composition containing a quinoa grain extract for dermatological use. U.S. Patent Application 20100136144 (A1).

Murphy, K. S., & Matanguihan, J. (2015). *Quinoa: Improvement and Sustainable Production*. John Wiley & Sons.

Navruz-Varli, S., & Sanlier, N. (2016). Nutritional and health benefits of quinoa (*Chenopodium quinoa Willd.*). *Journal of Cereal Science*, *69*, 371–376.

Nongonierma, A. B., Le Maux, S., Dubrulle, C., Barre, C., & FitzGerald, R. J. (2015). Quinoa (*Chenopodium quinoa Willd.*) protein hydrolysates with *in vitro* dipeptidyl peptidase IV (DPP-IV) inhibitory and antioxidant properties. *Journal of Cereal Science*, *65*, 112–118.

Nowak, V., Du, J., & Charrondière, U. R. (2016). Assessment of the nutritional composition of quinoa (*Chenopodium quinoa Willd.*). *Food Chemistry*, *193*, 47–54.

Ortiz, I. (2014). Cross-reactivity between buckwheat and quinoa in a patient with eosinophilic esophagitis caused by wheat. *Journal of Investigational Allergology & Clinical Immunology*, *24*(1), 56–57.

Pagno, C. H., Costa, T. M., de Menezes, E. W., Benvenutti, E. V., Hertz, P. F., Matte, C. R., Tosati, J. V., Monteiro, A. R., Rios, A. O., & Flôres, S. H. (2015). Development of active biofilms of quinoa (*Chenopodium quinoa Willd.*) starch containing gold nanoparticles and evaluation of antimicrobial activity. *Food Chemistry*, *173*, 755–762.

Paśko, P., Bartoń, H., Zagrodzki, P., Gorinstein, S., Fołta, M., & Zachwieja, Z. (2009). Anthocyanins, total polyphenols and antioxidant activity in amaranth and quinoa seeds and sprouts during their growth. *Food Chemistry, 115*(3), 994–998.

Paśko, P., Bartoń, H., Zagrodzki, P., Izewska, A., Krosniak, M., Gawlik, M., & Gorinstein, S. (2010). Effect of diet supplemented with quinoa seeds on oxidative status in plasma and selected tissues of high fructose-fed rats. *Plant Foods for Human Nutrition, 65*, 146–151.

Paśko, P., Sajewicz, M., Gorinstein, S., & Zachwieja, Z. (2008). Analysis of selected phenolic acids and flavonoids in *Amaranthus cruentus* and *Chenopodium quinoa* seeds and sprouts by HPLC. *Acta Chromatographica, 20*(4), 661–672.

Peterson, A. J., & Murphy, K. M. (2015). Quinoa cultivation for temperate North America: Considerations and areas for investigation. In *Quinoa: Improvement and Sustainable Production* (pp. 173–192).

Pineli, L. D. L. D. O., Botelho, R. B., Zandonadi, R. P., Solorzano, J. L., de Oliveira, G. T., Reis, C. E. G., & Teixeira, D. D. S. (2015). Low glycemic index and increased protein content in a novel quinoa milk. *LWT–Food Science and Technology, 63*(2), 1261–1267.

Raju, J., & Rao, C. V. (2012). Diosgenin, a steroid saponin constituent of yams and fenugreek: Emerging evidence for applications in medicine. In *Bioactive Compounds in Phytomedicine* (pp. 125–143).

Reguera, M., Conesa, C. M., Gil-Gómez, A., Haros, C. M., Pérez-Casas, M. Á., Briones-Labarca, V., Bolaños, L., Bonilla, I., Álvarez, R., Pinto, K., & Mujica, Á. (2018). The impact of different agroecological conditions on the nutritional composition of quinoa seeds. *PeerJ, 6*, e4442.

Repo-Carrasco-Valencia, R. A. M., & Serna, L. A. (2011). Quinoa (*Chenopodium quinoa, Willd.*) as a source of dietary fiber and other functional components. *Food Science and Technology, 31*, 225–230.

Rothschild, J., Rosentrater, K. A., Onwulata, C., Singh, M., Menutti, L., Jambazian, P., & Omary, M. B. (2015). Influence of quinoa roasting on sensory and physicochemical properties of allergen-free, gluten-free cakes. *International Journal of Food Science & Technology, 50*(8), 1873–1881.

Ruales, J., Grijalva, Y. D., Lopez-Jaramillo, P., & Nair, B. M. (2002). The nutritional quality of an infant food from quinoa and its effect on the plasma level of insulin-like growth factor-1 (IGF-1) in undernourished children. *International Journal of Food Sciences and Nutrition, 53*(2), 143–154.

Ruiz, G. A. (2016). *Exploring Novel Food Proteins and Processing Technologies: A Case Study on Quinoa Protein and High Pressure–High Temperature Processing* (Doctoral dissertation, Wageningen University and Research).

Ryan, E., Galvin, K., O'Connor, T. P., Maguire, A. R., & O'Brien, N. M. (2007). Phytosterol, squalene, tocopherol content and fatty acid profile of selected seeds, grains, and legumes. *Plant Foods for Human Nutrition, 62*, 85–91.

Sanchez, K. A. (2012). *Observations Regarding Consumption of Peruvian Native Grains (Quinoa, Amaranth and Kañiwa), Weight Status, and Perceptions of Potential Risk Factors, Warning Signs and Symptoms of Type 2 Diabetes Among Peruvian Adults: A Case Study* (Master's thesis, University of Maryland, College Park).

Scanlin, L. A., & Burnett, C. (2010). Quinoa grain processing and products. U.S. Patent Application 12/756,525.

Scanlin, L. A., Stone, M. B., & Burnett, C. (2010). Quinoa protein concentrate, production and functionality. U.S. Patent Application 12/748,968.

Seidlova-Wuttke, D., Christel, D., Kapur, P., Nguyen, B. T., Jarry, H., & Wuttke, W. (2010). β-Ecdysone has bone protective but no estrogenic effects in ovariectomized rats. *Phytomedicine*, *17*(11), 884–889.

Sharma, K., Kumar, V., Kaur, J., Tanwar, B., Goyal, A., Sharma, R., Gat, Y., & Kumar, A. (2021). Health effects, sources, utilization and safety of tannins: A critical review. *Toxin Reviews*, *40*(4), 432–444.

Siener, R., Hönow, R., Seidler, A., Voss, S., & Hesse, A. (2006). Oxalate contents of species of the Polygonaceae, Amaranthaceae and Chenopodiaceae families. *Food Chemistry*, *98*(2), 220–224.

Simnadis, T. G., Tapsell, L. C., & Beck, E. J. (2015). Physiological effects associated with quinoa consumption and implications for research involving humans: A review. *Plant Foods for Human Nutrition*, *70*, 238–249.

Singer, N. S., Tang, P., Chang, H. H., & Dunn, J. M. (1990). Carbohydrate cream substitute. *European Patent No. 0 403, no. 696 (1990): A1.*

Takao, T., Watanabe, N., Yuhara, K., Itoh, S., Suda, S., Tsuruoka, Y., Nakatsugawa, K., & Konishi, Y. (2005). Hypocholesterolemic effect of protein isolated from quinoa (*Chenopodium quinoa Willd.*) seeds. *Food Science and Technology Research*, *11*(2), 161–167.

Tang, Y., & Tsao, R. (2017). Phytochemicals in quinoa and amaranth grains and their antioxidant, anti-inflammatory, and potential health beneficial effects: A review. *Molecular Nutrition & Food Research*, *61*(7), 1600767.

Tang, Y., Li, X., Chen, P. X., Zhang, B., Hernandez, M., Zhang, H., Marcone, M. F., Liu, R., & Tsao, R. (2015). Characterisation of fatty acid, carotenoid, tocopherol/tocotrienol compositions and antioxidant activities in seeds of three *Chenopodium quinoa Willd.* genotypes. *Food Chemistry*, *174*, 502–508.

Tang, Y., Zhang, B., Li, X., Chen, P. X., Zhang, H., Liu, R., & Tsao, R. (2016). Bound phenolics of quinoa seeds released by acid, alkaline, and enzymatic treatments and their antioxidant and α-glucosidase and pancreatic lipase inhibitory effects. *Journal of Agricultural and Food Chemistry*, *64*(8), 1712–1719.

U.S. Department of Agriculture (USDA). (2011). USDA national nutrient database for standard reference, Release 18. *Nutrient Data Laboratory Home Page*. Available from: http://ndb.nal.usda.gov/.

Valcárcel-Yamani, B., & Lannes, S. D. S. (2002). Applications of quinoa (*Chenopodium quinoa Willd.*) and amaranth (*Amaranthus spp.*) and their influence in the nutritional value of cereal-based foods. *Food and Public Health*, *2*(6), 265–275.

Vega-Gálvez, A., Miranda, M., Vergara, J., Uribe, E., Puente, L., & Martínez, E. A. (2010). Nutrition facts and functional potential of quinoa (*Chenopodium quinoa Willd.*), an ancient Andean grain: A review. *Journal of the Science of Food and Agriculture*, *90*(15), 2541–2547.

Vilcacundo, R., & Hernández-Ledesma, B. (2017). Nutritional and biological value of quinoa (*Chenopodium quinoa Willd.*). *Current Opinion in Food Science*, *14*, 1–6.

Wilborn, C. D., Taylor, L. W., Campbell, B. I., Kerksick, C., Rasmussen, C. J., Greenwood, M., & Kreider, R. B. (2006). Effects of methoxyisoflavone, ecdysterone, and sulfo-polysaccharide supplementation on training adaptations in resistance-trained males. *Journal of the International Society of Sports Nutrition*, *3*(2), 19.

Wright, K. H., Pike, O. A., Fairbanks, D. J., & Huber, C. S. (2002). Composition of *Atriplex hortensis*, sweet and bitter *Chenopodium quinoa* seeds. *Journal of Food Science*, *67*(4), 1383–1385.

Yael, B., Liel, G., Hana, B., Ran, H., & Shmuel, G. (2012). Total phenolic content and antioxidant activity of red and yellow quinoa (*Chenopodium quinoa Willd.*) seeds as affected by baking and cooking conditions. *Food and Nutrition Sciences*, *3*, 1150–1155.

Yao, Y., Shi, Z., & Ren, G. (2014). Antioxidant and immunoregulatory activity of polysaccharides from quinoa (*Chenopodium quinoa Willd.*). *International Journal of Molecular Sciences*, *15*(10), 19307–19318.

Yao, Y., Zhu, Y., Gao, Y., Shi, Z., Hu, Y., & Ren, G. (2015). Suppressive effects of saponin-enriched extracts from quinoa on 3T3-L1 adipocyte differentiation. *Food & Function*, *6*(10), 3282–3290.

Zevallos, V. F., Herencia, I. L., Chang, F., Donnelly, S., Ellis, J. H., & Ciclitira, P. J. (2014). Gastrointestinal effects of eating quinoa (*Chenopodium quinoa Willd.*) in celiac patients. *Official Journal of the American College of Gastroenterology | ACG*, *109*(2), 270–278.

CHAPTER 12

Edible Mushrooms as Potential Functional Food Mediators

B. L. ATTRI

Indian Council of Agriculture Research (ICAR), Directorate of Mushroom Research (DMR), Solan, Himachal Pradesh, India

ABSTRACT

Modern fresh and processed food products have paramount importance in terms of nutritional rather than food security, though consumed in smaller quantities. Mushroom, a multi-nutritional commodity, is gaining importance in India as it is a multitier crop requiring less land and water, which are dwindling faster because of the pressure of an unprecedented increase in the country's population. Apart from rich and quality protein with 9 essential amino acids along with minerals, vitamins, and low carbohydrates and fats, the mushroom is delicious for combating malnutrition and checking several chronic diseases and disorders. Although, the fresh mushrooms because of their high moisture content and short shelf-life, may be consumed through value-added products maintaining the original nutritional value in terms of prebiotics and probiotics. The functional components in the form of nutraceuticals with secondary metabolites make mushrooms an indispensable part of the food. The bioactive compounds in the form of polysaccharides (PS), alkaloids, glycosides, antioxidants, triterpenoids, and phenolic contents in different mushrooms are playing a pivotal role to counter various problems because of their miraculous properties. The therapeutic values of mushrooms have been found to check ailments like hypertension, diabetes, hypercholesterolemia, cancer, Parkinson's, atherosclerosis, cardiovascular, and neurodegeneration. The selenium-rich potential mushroom foods exhibit immune system boosting, antioxidant, and anti-cancer properties. The

Functional Compounds and Foods of Plant Origin: Krishan Dutt Sharma, Satish Kumar, Rakesh Sharma, and Raj K. Keservani (Eds.)

mushrooms rich in nutritional components with functional attributes having nutraceutical, pharmaceutical, and bioactive compounds have been regarded as superfoods. Being rich in vitamin D, mushrooms are regarded as an excellent food source for vegetarians.

12.1 INTRODUCTION

Mushrooms are the spore-bearing fruiting bodies of fungus which grow above the ground on a growth medium. Historically, they have been an integral part of diet in Greek and Rome. Mushrooms are being consumed since time immemorial because of the treasure of several nutritional and medicinal components. It has been regarded as *"Food of God"* in Rome whereas *"Elixir of Life"* by the Chinese (Bashir et al., 2014). The mushrooms, also known as macro-fungus, are found under two classes (basidiomycetes and ascomycetes), and as per the report, about 22,000 species are known, out of which about 10% have been explored worldwide (Chang and Miles, 1992). Further, it has been suggested that out of the total species, about 7000 would possibly benefit mankind on earth (Hawksworth, 2001). Apart from nutritional value and because of the various antioxidant, antitumor, antimicrobial, pharmaceutical, nutraceutical (Keservani et al., 2010a, b, 2017, 2020), and functional properties, mushrooms have been a part of the modern diet around the world. The presence of high protein, minerals, and vitamins, as well as low fat and carbohydrates, make mushrooms an important food component of the human diet (Khatun et al., 2012). Although, there is a great diversity in the mushroom species, very few of them are edible whereas others possess poisonous compounds and thus are not suitable for human consumption. A number of countries in Asia are using wild edible mushrooms for their delicious taste along with their medicinal value for curing several ailments (Saiqa et al., 2008). Because of the consciousness and awareness amongst the consumers for sound and disease-free health as well as lowering the risk of various diseases and disorders, many of them have developed an interest in their consumption, and the trend is increasing to use mushrooms as fresh as well as after value addition. As detailed information is inevitable in the present days' context related to functional components of mushrooms, various countries like the USA, EU, South Korea, and Japan have formulated and revised the regulatory guidelines which are helpful to describe their health claims (Hasler, 1996). Due to better texture coupled with typical flavor and functional attributes, mushrooms have been widely used as human foods since ancient times (Manzi et al., 2001).

In India, five mushrooms *viz.*, *Agaricus bisporus*, *Calocybe indica*, *Pleurotus* sp., *Volvariella volvalcea* and *Lentinula edodes* are being cultivated on a commercial scale with more than 72% contribution by white button mushroom (Figure 12.1). In the last couple of years there is an exponential growth of mushrooms in the country because of diverse agro-climatic conditions, sufficient availability of agri-residue, cheap labor, and awareness among the consumers for the nutritional and medicinal components of mushrooms suitable for the sound health checking several chronic diseases and disorders.

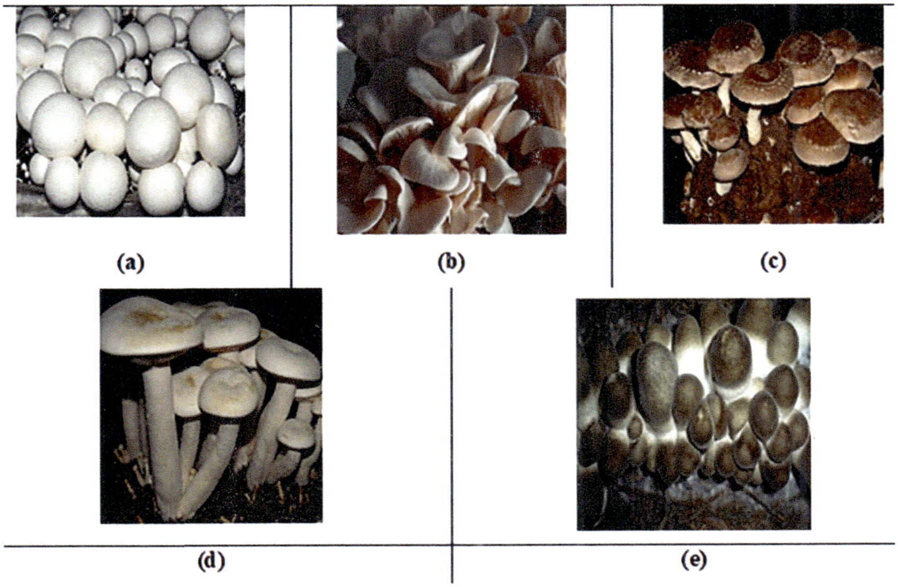

FIGURE 12.1 Commercially cultivated mushrooms in India [(a) Button; (b) oyster; (c) shiitake; (d) milky; and (e) paddy straw].

12.2 COMPONENTS OF MUSHROOMS

The moisture content of the fresh fruiting bodies of macrofungi ranges from 70% to 95%. According to Rathore et al. (2017), the dry matter primarily consists of carbohydrates (50–65%), protein (19–35%), and essential fatty acids (2–6%), along with trace amounts of vitamins and minerals. Edible mushrooms contain various nutraceuticals, such as β-glucans, lectins, unsaturated fatty acids, phenolic compounds, tocopherols, ascorbic acid, and carotenoids. Consuming edible macrofungi supports health by providing the additive and synergistic benefits of these bioactive compounds (Verghese et al., 2019).

In this context, macrofungi have the potential to be significant sources of low-calorie functional foods and nutraceuticals (Lu et al., 2020). Their fruiting bodies or cultured mycelium have been shown to contain a wide range of bioactive substances, including alkaloids, carotenoids, enzymes, lipids, folates, glycosides, lectins, minerals, organic acids, phenolics, polysaccharides, proteins, terpenoids, tocopherols, and volatile compounds in general (Valverde et al., 2015; Zeb and Lee, 2021 and Thongbai et al., 2019).

12.3 NUTRITIONAL VALUE OF MUSHROOMS

Depending upon the environmental conditions of growing and the time of crop harvest, the moisture can vary from 70–95%. The high-quality protein along with good volumes of PUFA and MUFA is available. Mushrooms have soluble fibre, β-glucan providing several health benefits (Cheung, 2013). About 50 to 60% of carbohydrates with free sugars amounting to 11% on a dry weight basis have been reported in mushroom fruit bodies (Temesgen, 2018) whereas, in oyster (*Pleurotus ostreatus*) there is about 56% (Khan and Tania, 2012). Out of the total free sugars in mushrooms, the most dominant (80%) one reported is mannitol (Wannet et al., 2000). In the functional component of mushrooms, proteins are of paramount importance which have been found higher compared to most vegetables. Because of a low GI in mushrooms, it is revealed that on blood glucose and insulin it has a very low negative effect. The essential minerals for human health in mushrooms include K, P, Na, Ca, and Mg whereas the minor are Cu, Zn, Fe, Mo, and Cd (Malinowska et al., 2004). The mineral proportion of the mushrooms depends on the type of substrates, species of the mushroom, age/maturity during harvest, and size of the mushrooms. The mineral content in mushrooms depends on the substrate used for cultivation; wild species generally contain higher mineral levels compared to cultivated mushrooms (Rudawska and Leski, 2005). Overall, the mushrooms contain good amount of proteins, phenolics, antioxidants, and vitamins (including B, C, D, nicotinic acid, niacin, riboflavin, thiamine, and folic acid). Further, they are rich in several micronutrients *viz*., K, Fe, Zn, Cu, P, and so on, and have limited sugars, sodium, cholesterol, fat, and so forth. People suffering from diseases/disorders like hypertension, diabetes as well as obesity can use mushrooms in their daily diet because of low Na and high K. It has been found that even after cooking of mushrooms there is no significant reduction in the vitamins and other vital nutritional components making them super foods.

Around 270 species of mushrooms have been identified to possess medicinal properties (Shamstyan, 2010), while some of the most commercially explored spp. includes the button mushroom (*Agaricus bisporus*), shiitake (*Lentinula edodes*), cordyceps (*Ophiocordyceps sinensis*), Maitake (*Grifola frondosa*), Reishi (*Ganoderma lucidum*), lion's mane (*Hericium erinaceus*), and turkey tail (*Trametes versicolor*) (Royse et al., 2017; Valverde et al., 2015). From ancient times, the mushroom is being used in various forms like extracts, concentrates, health tonics, tinctures, fermented beverages, soups, teas, powders, healthy food dishes as well as herbal formulas (Chang and Buswell, 1996). The mushrooms have also been described as mini pharmaceutical factories of bioactive compounds in the last decades because of their pharmaceutical potential (Patel and Goyal, 2012), and various bakery and confectionery products and other snacks have attracted consumers (Kumar, 2015; Olawuyi and Lee, 2019; Raghavendra et al., 2017). With the aim of the highest utilization and best delivery of the bioactive compounds from the mushrooms, various products have been standardized and have become popular among consumers. The increasing health awareness among the general masses with improved dietary habits have further led to increased demand of such products in the continuously reshaping food industry.

Mushrooms contain all the essential amino acids, thus entitling it as the protein reservoir with good quality protein. Besides protein, the mushrooms also have nutrients like P, Fe, and many other vitamins as discussed earlier, and so on, thus making them effective against the incidence of many diseases, that is, hypertension, hypercholesterolemia, diabetes, and cancer, and so on, are also prevented because of the therapeutic properties in mushrooms. The dietary fibre that has chitin and β-glucans contributes to the functional characteristics in mushrooms (Kumar, 2015). Further, a few mushroom species also contain antitumor, antiviral, antithrombotic, and immuno-modulating properties responsible for performing various functional activities in the human body. The common puffball (*Lycoperdon perlatum*), chanterelle (*Cantherellus cibarius*), fairy fingers (*Clavaria vermiculris*), salmon coral (*Ramaria formosa*), fairy ring (*Marasmius oreades*), and Indian oyster (*Pleurotus pulmonarius*) collected from the Western Ghats in Karnataka (Ramesh and Pattar, 2010) contained bioactive compounds such as phenols ranging from 3.20 mg/mL to 6.25 mg/ML whereas, the flavonoids were reported from 0.40 mg/mL to 2.54 mg/mL along with a very little quantity of ascorbic acid (0.06 mg/mL to 0.16 mg/mL). The water-soluble polysaccharides (PS) of mushrooms and mycelial extract have shown antitumor activity against sarcoma 180, mammary adenocarcinoma 755, leukemia L-1210 as well as other tumors in the human body (Daba and Ezeronye, 2003; Rajarathnam and Bano,

1992). The mushrooms such as willow bracket fungus (*Phellinus linteus*), oyster (*Pleurotus spp.*), button (*Agaricus blazei*), reishi (*Ganoderma lucidum*), puffball (*Calvatia utriformis*), cloud funnel (*Clitocybe nebularis*), turkey tail (*Trametes versicolor*), cordyceps (*Cordyceps militaris*), split gill (*Schizophyllum commune*), winter mushroom (*Flammulina velutips*), saffron milk (*Lactarius flavidulus*), charcoal burner (*Russula cynoxantha*) and Hoof fungus (*Fomes fomentarius*) have also been found effective against cancer due to properties like scavenging of the ROS, mitotic kinase inhibition. The mushrooms have also been reported to possess potent anti-mitotic, angiogenetic, and topoisomerase inhibitory activities, which can effectively check cancer proliferation in various tissues (Patel and Goyal, 2012).

12.4 MEDICINAL PROPERTIES OF MUSHROOM

Ajith and Janardhanan (2006) have reported significant anti-mutagenic and anti-carcinogenic properties in various mushrooms such as reishi (*Ganoderma lucidum*), willow bracket fungus (*Phellinus linteus*), oyster (*Pleurotus* spp.) which are found growing in South India. The total phenols ranging from 3.20 to 10.66 mg/mL, a flavonoid from 2.50–4.76 mg/mL, and ascorbic acid from 0.06–0.21 mg/mL were found to have anti-oxidative properties in a few mushrooms being cultivated in Bangladesh (Chowdhury et al., 2015). Further, they also demonstrated that shiitake had the best antimicrobial activity than other mushrooms. As per the study of Kosanic et al. (2013) in the extracts of wild mushrooms, the acetone extract from *Russula cynoxantha* showed the highest antioxidant activity. It was also suggested that mushrooms can be used for pharmaceutical purposes to treat several ailments. Atherosclerosis, LDL oxidation, and hypercholesterolemia are responsible for CVD and its prevention requires regulation of cholesterol levels. The edible mushrooms have been reported to check atherosclerosis due to the presence of high-fibre and low-fat content (Ishikawa et al., 1984). In shiitake mushrooms, isolation, and identification of adenosine derivative lentinacin (eritadenine) have been reported to be an effective hypocholesterolemic component (Rathee et al., 2012).

Further, high-fibre contents in *Termitomyces microcarpus* have been found to contribute to the reduction of diseases that are linked to high blood lipids (Nabubuya et al., 2010). The hepatoprotective activity in the extracts of mushrooms is exhibited to the presence of phenolics, triterpenes, PS, and peptides (Soares et al., 2013). With this awareness, there is an increasing demand for natural products that have hypoglycemic activity for treating diabetes. Button mushrooms (*Agaricus bisporus* and *A. camprestris*) having lectins showed enhancement in the release of insulin (Ahmad et al., 1984). The ethanolic extract from the fruit bodies of *Pleurotus ostreatus* showed a significant reduction in the serum glucose followed by a decrease in the creatinine and urea which resulted in the addition of the ethanolic extract of *P. ostreatus* in medicinal preparations for treating diabetes (Ravi et al., 2013). As antimicrobial agents, mushrooms contain PS, lipopolysaccharides, protein peptides, glycoproteins, nucleotides, tri-terpenoids, lectins, and lipids along with their derivatives which help provide disease control as safe and effective. Antifungal and anti-bacterial activities have been shown by various mushrooms against pathogens which are resistant in human beings (Sharma et al., 2014). Several cultural filtrates tested as antimicrobial agents revealed that filtrate of shiitake (*Lentinula edodes*) and Blewit mushroom (*Clitocybe nuda*) completely inhibited the spores of *Colletotrichum higginsianum* whereas *Ganoderma lucidum, L. edodes,* and *C. nuda* inhibited the spores of *Alternaria bassicola* (Chen and Huang, 2011). The bioactive compounds of *P. florida* were found suitable for therapeutic and antibiotic use whereas beefsteak fungus (*Fistulina hepatica*), cauliflower coral (*Russula botrytis*), and milk-white brittle gill (*Russula delica*) were found promising as antimicrobial agents (Alves et al., 2012).

12.5 SIGNIFICANCE OF SELENIUM IN MUSHROOMS

Most of the commercial spp. of mushrooms have been reported to show diverse biological properties against cholesterol, bacteria, viruses, cancer, and immune system modulation (Savić et al., 2009). Apart from the above nutritional and bioactive compounds, selenium is an essential nutrient which is present in very low concentration (40–70 mg/day). Se is essentially required for synthesis of selenoenzymes, that is, iodothyronine deiodinase, thioredoxin reductase, glutathione (GSH) POD and some important selenoprotein like W and P. These compounds are responsible for prevention of carcinogen formation, immunomodulation, detoxification, anti-inflammation, thyroid functioning as well as motility and maturation of sperms because of which selenium has been found as an important dietary supplement for human diet (Kora and Rastogi, 2016; 2017). As cultivated mushrooms have been found deficient (<1–8.5 μg/g dry weight) in Se, it may be increased substantially by cultivating mushrooms in the substrate containing inorganic selenium dioxide (SeO_2), sodium hydrogen selenite ($NaHSeO_3$), sodium selenite (Na_2SeO_3), and sodium selenate (Na_2SeO_4) or organic selenium (selenized yeast, selol) (Oliveira and Naozuka, 2019). An increase in Se content in fruit body of *Agaricus bisporus* was recorded between 3.1 and 1300 μg/g after amending Se in the substrate using Na_2SeO_3 as well as selenized yeast. In paddy straw mushroom (*Volvariella volvacea*), it was found 35 μg/g whereas, in milky mushroom (*Calocybe indica*) was 3.2, shiitake (*Lentinula edodes*) 46–170 μg/g, and reishi mushroom (*Ganoderma lucidum*) 72 μg/g after supplementing the substrate with Na_2SeO_3 (Kora, 2020).

The mushrooms are rated as potential functional foods and if the Se-fortification is done it becomes more rich exhibiting additional functions such as antioxidant and anti-cancer. Other chronic ailments like HIV infection, aging, cancer, neurodegenerative, cardiovascular, and immunological diseases are checked after adding Se-rich mushrooms as a food supplement (Cremades et al., 2012). Due to the nutraceutical nature of mushrooms, fresh as well as its products offer several physiological as well as health-related benefits against several chronic ailments. It has been reported that along with the inherent nutritional values of the mushrooms useful for human consumption, the RDA value for Se is 70 μg per day which can be easily accessed by a mushroom-based diet (Rodriguez Estrada et al., 2009) or through the purified extract of selenized mushrooms (Cremades et al., 2012).

12.6 VALIDATION OF ANTIDIABETIC PROPERTIES OF MUSHROOMS

As per the reports available globally, more than 2.8% of the population is diabetic (Wild et al., 2004), which is the major cause of several associated problems, which are secondary diseases related to diabetes (Zimmet et al., 1997). Chemical and biochemical compounds like insulin, phenformin, repaglinide, and so on are used to treat deadly diabetes; however, there are several side effects (Li et al., 2004). Both type-I and type-II diabetes are increasing at an alarming rate amongst people of all age groups, particularly in India. Dietary habits, lifestyle, and obesity are some of the major factors for diabetes mellitus (DM). Type-I diabetes is generally found because of destructive lesions of the pancreatic beta cells in the body resulting in a deficiency of insulin. On the other hand, in type-II diabetes, the mass of pancreatic β cells as well as the functions related are found to be preserved up to a certain extent, and injection of insulin is seldom required for sustaining life (Perera and Li, 2011).

The wild mushrooms offer great texture and flavor along with various nutritional compounds due to more accumulation of the secondary metabolites (Manzi et al., 1999; Sanmee et al., 2003). Apart from being rich in vitamin B, D, K, and traces of vitamin A and C, mushrooms contain good quantity of protein and minerals with less fat content (Alector, 1995; Mattila et al., 2001; Yildiz et al., 1998). Along with good source of nutrients, the mushrooms have also been found to contain therapeutic properties which are responsible for preventing several ailments including hypertension, diabetes, and cancer (Bodek and Galbavy, 1999). The mushrooms contain several functional characteristics because of dietary fibre particularly chitin and β-glucans (Manzi et al., 2001). The different mushrooms have antitumor, antiviral, anti-thrombotic as well as immune-modulatory properties, lowering the sugar level in the blood; however, advanced research is required to identify the active principles involved in reducing the blood sugar (Mau et al., 2002) and treating diabetes. Cho et al. (2007) have reported that snow fungus (*Tremella fuciformis*) exhibited hypoglycemic effect in the management of diabetes showing it as potential food. In Chinese medicine, the pachymic acid of *Poria cocos* has also shown insulin-sensitizing activity at different levels (Sato et al., 2002). Similarly, the PS of *Ganoderma lucidum* exhibited a reduction in glucose levels of the serum because of releasing insulin and facilitating the inflow into the pancreatic β-cells (Zhang and Lin, 2004). The monkey head mushroom (*Hericium erinaceus*) has shown various physiological and antioxidant activities in human beings by regulating the lipid levels and decreasing the glucose levels in the blood (Wang et al., 2005). Zhong et al. (2009) have reported a variety of bioactive metabolites in *Agrocybe aegerita* capable of dealing with free radical scavenging, anti-cancer, and antifungal activities. The caterpillar fungus (*Cordyceps sinensis*) is also used in traditional Chinese medicines and the PS extracted from the fungus exhibited a reduction in glucose level after it was administered orally (Kiho et al., 1993). Maitake or dancing mushroom (*Grifola frondosa*) which is very popular in countries like China, Japan, and Korea contain α-glucosidase inhibitor, lowering blood sugar (Matsuur et al., 2002). The different medicinal mushrooms along with their major compounds have been given in Table 12.1.

TABLE 12.1 Medicinal Mushrooms and Their Effective Compounds

SL. No.	Mushroom	Species	Compound
1.	Snow fungus	*Tremella fuciformis*	Glucuronoxylomannan
2.	Poria mushroom	*Wolfiporia extensa*	Dehydrotumulosic acid, dehydo-trametenolic acid and pachymic acid
3.	Reishi mushroom	*Ganoderma lucidum* (Curtis)	Polysaccharides (GI-PS)
4.	Reishi mushroom	*G. applanatum*	Exo-polymer (GAE)
5.	Tough shank	*Collybia confluens*	Exo-polymer (CCE)
6.	Wood ear	*Auricularia auricula-judae*	Polysaccharide (FA)
7.	Almond mushroom	*Agaricus subrufescens*	β-glucans and oligosaccharide (AO)
8.	Shaggy ink cap	*Coprinus comatus*	Vanadium
9.	Caterpillar mushroom	*Cordyceps sinensis*	Polysaccharide CSP-1
10.	Maitake mushroom	*Grifola frondosa*	α-glucan (MT-alpha-glucan)

Source: Perera and Li (2011).

12.7 NUTRACEUTICALS

The term nutraceutical or bioceutical refers to a nourishing food component with potent medicinal benefits. These food products can be categorized based on their origin, pharmacological conditions, and type of products. The nutraceuticals are often categorized into four groups including dietary supplements, functional food, medicinal food, and pharmaceuticals. Oxidative stress like allergies, Alzheimer's disease, CVDs, conditions of the eyes, obesity, Parkinson's diseases, and so on, are exhibited by nutraceuticals. On the other hand, functional foods have a potentially positive effect on human health beyond providing basic nutritional components, which may include prebiotics and probiotics. Several fruits, vegetables, nuts, seeds, and their products are included as functional foods, which are the need of the hour. The mushrooms and their products also contain several vital components and bioactive compounds which provide nutrition as well as medicinal benefits if consumed at regular intervals. Nutraceutical may refers to a complete food, a part of it or the extract obtained from the food, and it is called functional when some specific target function, and overall well-being is improved by reducing the risk of various diseases beyond the regular nutritional value of the food (Reis et al., 2017).

Because of the better and more attractive sensory quality attributes, rich nutritional and medicinal compositions as well as various functional qualities, the cultivation and consumption of the mushrooms have been reported since ancient times. The major cultivated mushrooms included are white button (*Agaricus bisporus*), oyster (*Pleurotus ostreatus*), shiitake (*Lentinula edodes*), winter mushroom (*Flammulina velutips*), jelly ear (*Auricularia auricula*) whereas, reishi mushroom (*Ganoderma lucidum*), codyceps (*Cordyceps sinensis*) and Poria (*Poria cocos*) are medicinal ones and China was the first country to cultivate the mushroom worldwide followed by USA, Italy, Netherlands, and Poland (Ma et al., 2018). The significance of different mushrooms has increased substantially over the last few decades because of their functional properties due to very high dietary fibre as well as low-fat content along with high proteins containing all the essential amino acids and several minerals and vitamins (Thatoi and Singdevsachan, 2014). The presence of various bioactive compounds in mushrooms result in wide-spectrum bio-activities against several health complications (Ma et al., 2018). Polysaccharides (PS) and proteins are bioactive nutraceuticals and are the major functional compounds that assist functional food development with very high nutritional and sensory qualities (Wang et al., 2014). The different PS which have been exploited in the mushrooms for the development of functional foods include lentinan from *Lentinula edodes*, calocyban from *Calocybe indica,* pleuran from oyster, schizophyllan from *Schizophyllum commune*, and ganoderan from *Ganoderma lucidum*. From PS of mushrooms, different monosaccharides detected were glucose, fructose, xylose, manose, galactose, fucose, rhamnase, trehalose, arabinose, and mannitol (Valverde et al., 2015). The biological activities of a few mushrooms related to PS have been shown in Table 12.2.

TABLE 12.2 Mushrooms with Their Polysaccharides (PS) and Health Benefits

SL. No.	Mushroom	Species	Characteristics	Benefits
1.	Button mushroom	*Agaricus bisporus*	Homogenous polysaccharide and heteropolysaccharide	Antitumor, anti-aging, immune-stimulatory activity
2.	Oyster mushroom	*Pleurotus ostreatus*	Heteropolysaccharide	Antidiabetic activity
3.	Shiitake mushroom	*Lentinula edodes*	Homogenous PS and heteropolysaccharide	Antitumor, lung protection, activate immunity
4.	Winter mushroom	*Flammulina velutips*	Heteropolysaccharide	Immunomodulating activity
5.	Milky mushroom	*Calocybe indica*	Heteropolysaccharide	Antioxidant and anti-aging activity
6.	King oyster mushroom	*Pleurotus eryngii*	Homogenous polysaccharide and heteropolysaccharide	Antitumor and immune modulatory activity
7.	Reishi mushroom	*Ganoderma lucidum*	Homogenous polysaccharide and heteropolysaccharide	Antitumor, antidiabetic, anti-Alzheimer's disease, anti-inflammatory

Source: Ma et al. (2018).

Studies around the world have proven that the PS of different mushrooms showed health benefits by regulating the gut microbiota along with the gastrointestinal function (Jayachandran et al., 2017) however, extensive studies are to be done in this direction in the future. The high quality of proteins and peptides in the mushrooms as nutraceuticals have been found to give multiple health benefits after binding with the lectins and performing as antitumor, antiviral, anti-bacterial, and antifungal (Singh et al., 2014). In the future, detailed studies are required for establishing a correlation between the novel mechanism related to proteins and peptides of the mushrooms which are responsible for modulating human health (Table 12.3). The terpenes are unsaturated volatile hydrocarbons and may be present as monoterpenoids, diterpenoids, sesquiterpenoids, and triterpenoids have shown various health benefits, particularly trierpenoids. From *Ganoderma lucidum* terpenes for owing the ability of anti-acetyl-cholinesterase activity has been extracted from which the various companies have developed drugs to treat the Alzheimer's disease (Lee et al., 2011). The terpenes from different mushrooms along with the health benefits have been depicted in Table 12.4.

TABLE 12.3 Bioactive Proteins and Peptides of Mushrooms with Health Benefits

SL. No.	Mushroom	Species	Bioactive Compound	Health Related Benefits
1.	Oyster	*Pleurotus ostreatus*	Lectin	Immuno-modulatory activity
2.	Reishi	*Ganoderma lucidum*	Lectin, Fungal immunomodulatory proteins, Laccases and Ribonucleases	Antitumor, inhibition of breast cancer cells, antiviral, anti-HIV activity
3.	Button	*Agaricus bisporus*	Lectin	Immuno-modulatory activity
4.	Parasol	*Macrolapiota procera*	Lectin	Antitumor activity
5.	Winter mushroom	*Flammulina velutips*	Ribosome inactivating proteins and Fungal immunomodulatory proteins	Anti-inflammatory activity, treatment of allergic airway disease, antiviral, anti-HIV-I
6.	White beech mushroom	*Hypsizigus mormoreus*	Ribosome inactivating proteins	Antifungal, antitumor, anti-HIV-I activity
7.	Coral fungus	*Ramaria formosa*	Ribonuclease	Antiviral, anti-HIV-I activity
8.	Branched oyster mushroom	*Pleurotus cornucopiae*	Laccases	Antiviral, antitumor, anti-HIV-I, anti-cancer
9.	King oyster	*Pleurotus eryngii*	Laccases	Antiviral, anti-HIV-I activity
10.	Shaggy mane	*Coprinus comatus*	Laccases	Antiviral, anti-tumor activity

Source: Ma et al. (2018).

TABLE 12.4 Isolation of Terpenes from Mushrooms and Their Health Benefits

SL. No.	Mushroom	Species	Terpenes Characters	Health Benefits
1.	Branched oyster mushroom	*Pleurotus cornucopiae*	Monoterpenoids	Inhibition of nitric oxide in activated macrophages
2.	Winter mushroom	*Flammulina velutips*	Sesquiterpenoids	Antibacterial against *E. coli, B. subtilis*
3.	Golden oyster mushroom	*Pleurotus citrinopileatus*	Sesquiterpenoids	Inhibitory effect on protein tyrosine phosphatase
4.	King oyster	*Pleurotus eryngii*	Diterpenoids	Cytotoxicity against human cancer
5.	Reishi mushroom	*Ganoderma lucidum*	Triterpenoids	Inhibition of acetylcholinesterase and butylcholinesterase

Source: Ma et al. (2018).

The phenolic compounds extracted from different mushrooms are the aromatic hydroxylated compounds have aromatic rings or hydroxyl groups which have shown significant bioactivity against free radicals as well as oxygen scavengers (Heleno et al., 2012). The phenolic compounds from different mushrooms showed a significant inhibitory effect on lipid oxidation (Palacios et al., 2011). Several other bio-activities related to phenolic compounds such as anti-microbial, anti-inflammatory, anti-cancer, and antiviral have also been reported by various workers. The ergosterol which is the major sterol in different mushrooms has been detected to have antioxidant properties as well as prevention of CVDs (Guillamon et al., 2010; Kalac, 2013). Protection against degenerative diseases, microbial as well as CVDs has also been revealed by Heleno et al. (2012).

Apart from dietary food, fresh as well as dried mushrooms have been added in other food items to increase the quality of different nutrients and the development of functional foods with acceptable sensory quality. In Turkey tail mushroom (*Trametes versicolor*), the proteoglycan (Krestin) and polysaccharide peptide has been found to show anti-cancer properties (Sabaratnum et al., 2013). *Ganoderma lucidum* has been reported to contain various functional activities such as anti-inflammation, anti-oxidation, and inhibiting prostate cancer whereas the PS increased the richness of gut microbiota and ideal prebiotic in gastrointestinal tract (Jin et al., 2017). Based on the wide range of health benefits coupled with functional components, the mushrooms will be one of the best nutraceutical foods required in the supplementation of food with the advancement in nutrigenomics and molecular nutrition (Ma et al., 2018). The nutraceutical properties in mushrooms described are the prevention or treatment of Parkinson, Alzheimer, hypertension as well a high risk of stroke (Valverde et al., 2015). Several secondary metabolites have been found in different mushrooms and shiitake (*Lentinula edodes*) has shown anti-microbial action against gram-positive and gram-negative bacteria (Alves et al., 2012). The nutritional value of different mushrooms has been depicted in Table 12.5.

TABLE 12.5 Proximate Composition of Edible Mushroom (Dry Weight Basis)

SL. No.	Mushroom	Species	Nutritional Composition				
			Protein (%)	Fat (%)	Ash (%)	Carbohydrate (%)	Energy (Kcal/Kg)
1.	Button	*Agaricus bisporus*	14.10	2.20	9.70	74.00	325.00
2.	Shiitake mushroom	*Lentinula edodes*	4.50	1.73	6.70	87.10	772.00
3.	Oyster	*Pleurotus ostreatus*	7.00	1.40	5.70	85.90	416.00
4.	King oyster	*Pleurotus eryngii*	11.00	1.50	6.20	81.40	421.00
5.	Oyster	*Pleurotus sajor-caju*	37.40	1.00	6.30	55.30	–
6.	Giant oyster	*Pleurotus ginganteus*	17.70	4.30	–	78.00	364.00
7.	Sun mushroom	*Agaricus blazei* (dry powder)	31.30	1.80	7.50	59.40	379.00
8.	Shiitake mushroom	*Lentinula edodes* (dry powder)	12.80	1.00	4.30	81.90	388.00

Source: Valverde et al. (2015).

Because of various nutraceutical compounds, mushrooms have been considered as mini-pharmaceutical factories which used to produce bioactive compounds that have miraculous properties (Patel and Goyal, 2012). It has been reported that different mushrooms and fungi possess more than 100 pharmacological benefits, that is, like antioxidant, anti-cancer, antidiabetic, anti-allergic, cardiovascular protector, antiviral, antifungal, and protection against tumor development (Chang and Wasser, 2012; Zhang et al., 2011). In the mushrooms, PS are the most important chemical compounds whereas β-glucan is the most versatile metabolic and is having a wide spectrum of various biological activities (Chen and Seviour, 2007).

High protein content, fibre, vitamins, and minerals as well as low-fat contents are found in the mushrooms because of which they are regarded as nutritionally rich. Protein is found between 200 and 250 g/Kg of dry matter that has leucine, valine, glutamine, glutamic, and aspartic acids whereas low-fat content ranging from 20–30 g/Kg of dry matter is responsible for low calorific value. The ash has been reported between 80 and 120 g/Kg of dry matter containing potassium, phosphorus, magnesium, calcium, copper, iron, and zinc. Among carbohydrates apart from chitin, glycogen, trehalose, and mannitol high fibre, β-glucans, hemicelluloses, and pectic substances are abundant whereas fructose and sucrose are low. In vitamins, high level of riboflavin, niacin, folates, and traces of vitamin C, B_1, B_{12}, and E are reported. Mushrooms are the only non-animal food source containing vitamin D rating it as an excellent food source for vegetarians. As compared to cultivated, wild mushrooms have more vitamin D as UV-B light is required for its production (Valverde et al., 2015). Due to the high contents of PS, particularly β-glucans, which are not synthesized in the human body, the mushrooms have antitumor action by activating the immune system of the host and not killing the tumor cells directly (Zhang et al., 2007). As per the report of different studies, it has been found that the sterol in the form of ergosterol in mushrooms contributes to the prevention of CVDs (Kalac, 2013). Further, the water-soluble PS found in *Ganoderma lucidum* have been found to show an effect on more than 20 types of cancers and also show inhibition in the tumor growth (Zhou et al., 2007).

Awareness among the consumers has decreased the consumption of such foods that has ingredients like high saturated fats, sugar, and salt causing health-related problems whereas the foods containing unsaturated fatty acids, vitamins, minerals, and nutraceuticals are on-demand as they promote the health of human beings (McClements et al., 2021). Due to the adoption of a vegetarian diet by a greater number of people, it is of paramount importance to use such food products which are rich in nutrients with functional components and mushroom fits here perfectly. In the present times, it has been found that edible sporophores (mushrooms) having quality proteins, dietary fibre, vitamins, and mineral and phenolic compounds have been considered healthy food ingredients to keep ourselves healthy (Lu et al., 2020). Because of the presence of umami flavor and fibrous texture in mushrooms, there is an advantage in using them in meat products as there is better compatibility which may further decrease the quantity of the animal protein during consumption (Kumar et al., 2017).

12.8 MUSHROOM AS A DELICACY FOR CULINARY PURPOSE

The mushrooms are regarded as vegetable meat around the world due to culinary delicacy. The early civilizations of various countries like Greek, Egypt, Rome, and China have recognized the beneficial effects on human health due to the presence of nutritional components (Raghavendra et al., 2017). Based on utilization, mushrooms are categorized into three major groups which are edible (54%), medicinal (38%), and wild (8%) reported by Royse et al. (2017). Although, several mushrooms are being grown around the world, however, shiitake (*Lentinula edodes*), button (*Agaricus bisporus*), oyster (*Pleurotus ostreatus*), wood ear (*Auricularia polytricha*), and winter mushroom (*Flammulina velutips*) contribute 85% (Royse et al., 2017). Apart from different nutritional components embedded in any item, the quality of the finished product is influenced by its functional properties which contribute to the health of human beings as these are derived from natural sources (Fasseas et al., 2008). The freshness, appearance as well as overall acceptability of the prepared product is improved along with the nutritional value and the additives are considered as nutraceutical as they play an important role in the prevention and treatment of various diseases besides exerting positive health benefits (Reis et al., 2017). Due to their unique nutritional and textural properties, mushrooms are considered as an alternative source of animal products like meat and fish as well as fruits and vegetables (Asgar et al., 2010). The mushrooms are also considered as bio-therapeutic agents as they contain numerous secondary metabolites or nutraceuticals that have medicinal value (Prasad et al., 2015; Rathore et al., 2017). The glutamic and aspartic acids with characteristic 5'-nucleotides sequence gives a peculiar, pleasant umami flavor in mushrooms which have been found palatable, preferable, and adaptable in different formulations in different countries (Jeng-Leun, 2005; Zhang et al., 2013). Kong et al. (2019) have isolated umami peptides from hydrolysates of shiitake mushroom which are believed for the specific and peculiar taste of *Lentinula edodes* whereas in *Agaricus bisporus* peptides like Gly-Leu-Pro-Asp and Gly-His-Gly-Asp have been reported to contribute for komumi taste (Feng et al., 2019). It has been reported that kokumi substances do not have a better taste but the basic taste like sweet, salty, and umami is enhanced by its addition (Ueda et al., 1997). The nutrient profile of important mushrooms (fresh and dried) has been depicted in Table 12.6.

TABLE 12.6 Nutrient Profile of Important Mushrooms

SL. No.	Mushroom	Species	Protein (%)	Fat (%)	Crude Fiber (%)	Ash (%)	Carbohydrate (%)	Energy (Kcal/100 g)
			Fresh (g/100 g)					
1.	Button	*Agaricus bisporus*	3.00	0.34	1.45	0.79	3.69	24.00
2.	Winter	*Flammulina velutips*	2.66	0.28	2.80	0.91	8.42	29.00
3.	Maitake	*Grifola frondosa*	1.94	0.20	2.70	0.52	2.70	29.00
4.	Oyster	*Pleurotus ostreatus*	2.00	0.99	2.10	0.24	5.35	39.00
5.	Oyster	*P. sajor caju*	23.30	3.00	35.60	3.20	65.5	–
			Dried (g/100 g)					
1.	King oyster	*Pleurotus eryngii*	28.80	3.00	–	3.50	52.20	–
2.	Winter	*Flammulina velutips*	18.42	2.94	7.81	6.33	56.3	–
3.	Button	*Agaricus bisporus*	29.29	2.22	24.56	7.12	20.57	–
4.	Oyster	*Pleurotus ostreatus*	20.04	8.65	–	7.78	60.21	421.00
5.	Maitake	*Grifola frondosa*	21.10	3.10	10.10	7.00	58.80	–
6.	Lion's mane	*Hericium erinaceus*	22.30	3.50	7.80	9.40	57.00	–
7.	Porcini	*Boletus edulis*	21.07	2.45	–	5.53	70.95	423.00
8.	Oyster	*Pleurotus florida*	34.56	2.11	11.41	7.40	31.59	–
9.	Paddy straw	*Volvariella volvacea*	29.50	5.70	–	10.40	60.00	374.00
10.	Shiitake	*Lentinula edodes*	17.50	8.00	–	8.00	67.50	387.00
11.	Wood ear	*Auricularia polytricha*	7.70	0.80	–	14.00	87.60	347.00
12.	Puffball	*Lycoperdon echinatum*	23.52	1.22	–	9.43	65.83	544.00

Source: Das et al. (2021).

For the potential health benefits PS such as chitin, β glucan, α-glucan, mannas, xylans, and galactans have been found in mushrooms (Rathore et al., 2017). It has been reported that prebiotics in the form of PS which passed through the stomach and small intestine without digestion stimulated the healthy bacteria in the colon (Chou et al., 2013).

12.9 FUTURE PERSPECTIVE

The presence of high nutrient contents like protein, fibre, vitamins, and minerals, make the mushrooms an important component of the food and are considered as a *super food*. Because of the nutraceutical as well as functional properties of mushrooms suitable for maintaining human health and awareness among the consumers, the demand for fresh as well as processed products of mushrooms including the medicinal ones is going to surge many folds in future. The mushrooms in the form of macro-fungi producing several bioactive compounds that have PS in the form of nutraceuticals have properties like anti-cancer, antidiabetic, antiviral, anti-bacterial, immunostimulatory, and immunoregulatory including hepato- and kidney-protective effects, antiaging activity, and other bio-activities. Some studies have also projected that because of the nutraceutical properties in mushrooms there will be an exponential growth in the development of mushroom products to be used as a functional food and dietary supplements. There will be a huge investment in research and development (R&D) to conduct various innovative studies including verification of the health benefits. In the future, apart from edible, medicinal mushrooms with their functional components and dietary supplementation, consumption is expected to go very high as the consumers are getting awareness and are health conscious. The medicinal mushroom market is increasing fast because of the preference of usage of natural/organic food products, herbal medicines (Rane et al., 2020a, b, c, 2023; Sharma et al., 2020) and requirements in the world sports due to very high competition among the participating countries. The only and major bottleneck in the medicinal mushroom industry would be the taste of the macro-fungi which would be required to be made palatable by adding suitable ingredients and the studies would be required to be backed up with scientific interventions as consumers mostly rely on the opinion given by the scientists around the world.

For making medicinal mushrooms more popular, there will be demand for creating awareness through innovative findings as well as effective marketing strategies covering the aspects of nutraceutical values and functional attributes. Further, as functional foods and other beverages are stacked in retail stores in developing countries, the market value of these products is expected to surge. To streamline the mushroom industry with nutraceutical and functional products, collaboration amongst various firms within the country and across will be required. As only a few medicinal mushrooms have been used for developing functional components, the remaining macro-fungi that have enormous compounds have to be exploited with proper identification of toxic elements so that a wide range of products having more benefits are available to the stakeholders. Thus, to combat and prevent various human ailments through mushrooms, a promise is hold for future innovations for the development of various drugs and food supplements. Overcoming poor awareness among the consumers regarding functional foods from medicinal mushrooms as well as very high price and poor quality of the nutraceutical products is also the need of the hour and of paramount importance which will have to be addressed in the future. The validation for the efficacy and safety of the developed products of nutraceutical and functional properties is another area that is required to be focused. Further, because of its high fibre contents along with umami flavor, the mushroom may replace cereal flours in the future, for which there is enormous potential. Compounds like ergosterol, β glucans, lentinan, and peroxidases (PODs) are also present containing bioactive properties which help in various health functionalities. As the inclusion of mushrooms in food products enhances the nutritional values, there is no surprise if nutraceutical/pharmaceutical industries are using different mushrooms as well as their bioactive compounds for the development of functional foods.

12.10 CONCLUSION

In recent times, the demand for mushrooms has surged significantly because of awareness among consumers across the globe. As these foods are used to fulfill the demand of various vital components in the body even after consuming in smaller quantities, the future of functional-rich mushrooms is very bright. As several ailments are emerging because of various factors around us posing a threat to human beings, mushrooms possess various qualities to check them substantially. The innovative modern techniques will further help to know the hidden treasure of mushrooms, thereby improving the food and living standards and increasing the average age. Apart from nutritional components required in the daily diet, the functional nutritional properties of mushrooms will play an important role in the pharmaceutical industry which is likely to expand in a big way. There is a need of the hour to invest in research and development areas taking the mushroom industry to new heights and making it one of the richest in the country that has food supplements with functional components as per requirement. In the future, the huge network among different industries in India and across the world will help to flourish the mushroom industry. The enormous potential of the medicinal mushrooms has to be exploited fully which at present is negligible. A huge market is to be developed for the functional foods of mushrooms through awareness among consumers and validation of the technologies for different stakeholders, making the sun-rising mushroom industry one of the biggest and richest.

KEYWORDS

- **anti-cancer**
- **bioactive compounds**
- **chitin**
- **functional food**
- **galactans**
- **healthy bacteria**
- **macro-fungi**
- **nutraceuticals**
- **pharmaceutical properties**
- **polysaccharides**
- **prebiotics**
- **super food**
- **vitamins**
- **xylans**

REFERENCES

Ahmad, N., Bansal, R., Rastogi, A. K., & Kidwai, J. R. (1984). Effect of PHA-B fraction of *Agaricus bisporus* lectin on insulin release and 45Ca^{2+} uptake by islets of Langerhans *in vitro*. *Acta Diabetologica Latina, 21*(1), 63–70.

Ajith, T. A., & Janardhanan, K. K. (2006). Indian medicinal mushrooms as a source of antioxidant and antitumor agents. *Journal of Clinical Biochemistry and Nutrition, 40*(3), 157–162.

Aletor, V. A. (1995). Compositional studies on edible tropical species of mushrooms. *Food Chemistry, 54*(3), 265–268.

Alves, M. J., Ferreira, I. C., Dias, J., Teixeira, V., Martins, A., & Pintado, M. (2012). A review on antimicrobial activity of mushroom (Basidiomycetes) extracts and isolated compounds. *Planta Medica, 78*(16), 1707–1718.

Alves, M. J., Ferreira, I. C., Martins, A., & Pintado, M. (2012). Antimicrobial activity of wild mushroom extracts against clinical isolates resistant to different antibiotics. *Journal of Applied Microbiology, 113*(2), 466–475.

Asgar, M., Fazilah, A., Huda, N., Bhat, R., & Karim, A. A. (2010). Nonmeat protein alternatives as meat extenders and meat analogs. *Comprehensive Reviews in Food Science and Food Safety, 9*(5), 513–529.

Bashir, A., Vaida, N., & Ahmad Dar, M. (2014). Medicinal importance of mushrooms: A review. *International Journal of Advanced Research, 2*(12), 1–4.

Bobek, P., & Galbavý, Š. (1999). Hypocholesterolemic and antiatherogenic effect of oyster mushroom (*Pleurotus ostreatus*) in rabbits. *Food/Nahrung, 43*(5), 339–342.

Chang, S. T., & Buswell, J. A. (1996). Mushroom nutriceuticals. *World Journal of Microbiology and Biotechnology, 12*, 473–476.

Chang, S. T., & Miles, P. G. (1992). Mushroom biology—A new discipline. *Mycologist, 6*(2), 64–65.

Chang, S. T., & Wasser, S. P. (2012). The role of culinary-medicinal mushrooms on human welfare with a pyramid model for human health. *International Journal of Medicinal Mushrooms, 14*(2), 95–134.

Chen, J. T., & Huang, J. W. (2011). Antimicrobial activity of edible mushroom culture filtrates on plant pathogens. *Plant Pathology Bulletin*, 19: 261–70.

Chen, J., & Seviour, R. (2012). Medicinal importance of fungal β-(1→3), (1→6)-glucans. *Mycological Research, 111*(6), 635–652.

Cheung, P. C. (2013). Mini-review on edible mushrooms as a source of dietary fiber: Preparation and health benefits. *Food Science and Human Wellness, 2*(3, 4), 162–166.

Cho, E. J., Hwang, H. J., Kim, S. W., Oh, J. Y., Baek, Y. M., Choi, J. W., Bae, S. H., & Yun, J. W. (2007). Hypoglycemic effects of exopolysaccharides produced by mycelial cultures of two different mushrooms *Tremella fuciformis* and *Phellinus baumii* in ob/ob mice. *Applied Microbiology and Biotechnology, 75*, 1257–1265.

Chou, W. T., Sheih, I. C., & Fang, T. J. (2013). The applications of polysaccharides from various mushroom wastes as prebiotics in different systems. *Journal of Food Science, 78*(7), M1041–M1048.

Chowdhury, M.M.H., Kubra, K. and Ahmed, S.R. (2015). Screening of antimicrobial, antioxidant properties and bioactive compounds of some edible mushrooms cultivated in Bangladesh. *Annals of Clinical Microbiology and Antimicrobials, 14*(8). DOI 10.1186/s12942-015-0067-3.

Cremades, O., Diaz-Herrero, M. M., Carbonero-Aguilar, P., Gutierrez-Gil, J. F., Fontiveros, E., Rodríguez-Morgado, B., Parrado, J., & Bautista, J. (2012). Preparation and characterization of selenium-enriched mushroom aqueous enzymatic extracts (MAEE) obtained from the white button mushroom (*Agaricus bisporus*). *Food Chemistry, 133*(4), 1538–1543.

Daba, A. S., & Ezeronye, O. U. (2003). Anti-cancer effect of polysaccharides isolated from higher basidiomycetes mushrooms. *African Journal of Biotechnology, 2*(12), 672–678.

Das, A. K., Nanda, P. K., Dandapat, P., Bandyopadhyay, S., Gullón, P., Sivaraman, G. K., McClements, D. J., Gullón, B., & Lorenzo, J. M. (2021). Edible mushrooms as functional ingredients for development of healthier and more sustainable muscle foods: A flexitarian approach. *Molecules*, *26*(9), 2463.

Fasseas, M. K., Mountzouris, K. C., Tarantilis, P. A., Polissiou, M., & Zervas, G. (2008). Antioxidant activity in meat treated with oregano and sage essential oils. *Food Chemistry*, *106*(3), 1188–1194.

Feng, T., Wu, Y., Zhang, Z., Song, S., Zhuang, H., Xu, Z., Yao, L., & Sun, M. (2019). Purification, identification, and sensory evaluation of kokumi peptides from *Agaricus bisporus* mushroom. *Foods*, *8*(2), 43.

Guillamón Fernández, E., García Lafuente, A. M., Lozano, M., D'Arrigo, M., Rostagno, M. A., Villares, A., & Martínez, J. A. (2010). Edible mushrooms: Role in the prevention of cardiovascular diseases. *Journal of Nutritional Biochemistry*, *81*(7), 715–723.

Hasler, C. M. (1996). Functional foods: The western perspective. *Nutrition Reviews*, *54*(11), S6.

Hawksworth, D. L. (2001). Mushrooms: The extent of the unexplored potential. *International Journal of Medicinal Mushrooms*, *3*(4).

Heleno, S.A., Barros, L., Martins, A., Queiroz, M.J.R.P., Santos-Buelga, C. and Ferreira, I.C.F.R. (2012). Phenolic, polysaccharidic and lipidic fractions of mushrooms from Northeastern Portugal: Chemical compounds with antioxidant properties. *Journal of Agriculture and Food Chemistry*, *60*(8): 4634–40.

Ishikawa, Y., Morimoto, K., & Hamasaki, T. (1984). Flavoglaucin, a metabolite of *Eurotium chevalieri*, its antioxidation and synergism with tocopherol. *Journal of the American Oil Chemists' Society*, *61*(12), 1864–1868.

Jeng-Leun, M. (2005). The Umami Taste of Edible and Medicinal Mushrooms. *International Journal of Medicinal Mushrooms*, *7*(10): 119–125.

Jayachandran, M., Xiao, J., & Xu, B. (2017). A critical review on health-promoting benefits of edible mushrooms through gut microbiota. *International Journal of Molecular Sciences*, *18*(9), 1934.

Jin, M., Zhu, Y., Shao, D., Zhao, K., Xu, C., Li, Q., Yang, H., Huang, Q., & Shi, J. (2017). Effects of polysaccharide from mycelia of *Ganoderma lucidum* on intestinal barrier functions of rats. *International Journal of Biological Macromolecules*, *94*, 1–9.

Kalač, P. (2013). A review of chemical composition and nutritional value of wild-growing and cultivated mushrooms. *Journal of the Science of Food and Agriculture*, *93*(2), 209–218.

Keservani, R. K., Kesharwani, R. K., Sharma, A. K., Vyas, N., & Chadoker, A. (2010b). Nutritional supplements: An overview. *International Journal of Current Pharmaceutical Review and Research*, *1*(1), 59–75.

Keservani, R. K., Kesharwani, R. K., Vyas, N., Jain, S., Raghuvanshi, R., & Sharma, A. K. (2010a). Nutraceutical and functional food as future food: A review. *Der Pharmacia Lettre*, *2*(1), 106–116.

Keservani, R. K., Sharma, A. K., & Kesharwani, R. K. (2017). An overview and therapeutic applications of nutraceutical and functional foods. In *Recent Advances in Drug Delivery Technology* (pp. 160–201). ISBN: 9781522507543.

Keservani, R. K., Sharma, A. K., & Kesharwani, R. K. (2020). *Nutraceuticals and Dietary Supplements: Applications in Health Improvement and Disease Management*. CRC Press. Pp. 1–342. ISBN: 9781771888738.

Khan, M. A., & Tania, M. (2012). Nutritional and medicinal importance of *Pleurotus* mushrooms: An overview. *Food Reviews International*, *28*(3), 313–329.

Khatun, S., Islam, A., Cakilcioglu, U., & Chatterje, N. C. (2012). Research on mushroom as a potential source of nutraceuticals: A review on Indian perspective. *American Journal of Experimental Agriculture*, *2*(1), 47.

Kiho, T., Hui, J. I., Yamane, A., & Ukai, S. (1993). Polysaccharides in fungi. XXXII. Hypoglycemic activity and chemical properties of a polysaccharide from the cultural mycelium of *Cordyceps sinensis*. *Biological and Pharmaceutical Bulletin*, *16*(12), 1291–1293.

Kong, Y., Zhang, L. L., Zhao, J., Zhang, Y. Y., Sun, B. G., & Chen, H. T. (2019). Isolation and identification of the umami peptides from shiitake mushroom by consecutive chromatography and LC-Q-TOF-MS. *Food Research International*, *121*, 463–470.

Kora, A. J. (2020). Nutritional and antioxidant significance of selenium-enriched mushrooms. *Bulletin of the National Research Centre*, *44*, 1–9.

Kora, A. J., & Rastogi, L. (2016). Biomimetic synthesis of selenium nanoparticles by *Pseudomonas aeruginosa* ATCC 27853: An approach for conversion of selenite. *Journal of Environmental Management*, *181*, 231–236.

Kora, A. J., & Rastogi, L. (2017). Bacteriogenic synthesis of selenium nanoparticles by *Escherichia coli* ATCC 35218 and its structural characterisation. *IET Nanobiotechnology*, *11*(2), 179–184.

Kosanic, M., Rankovic, B., & Dasic, M. (2013). Antioxidant and antimicrobial properties of mushrooms. *Bulgarian Journal of Agricultural Science*, *19*(5), 1040–1046.

Kumar, K. (2015). Role of edible mushrooms as functional foods—A review. *South Asian Journal of Food Technology & Environment*, *1*(3, 4), 211–218.

Kumar, P., Chatli, M. K., Mehta, N., Singh, P., Malav, O. P., & Verma, A. K. (2017). Meat analogues: Health promising sustainable meat substitutes. *Critical Reviews in Food Science and Nutrition*, *57*(5), 923–932.

Lee, I., Ahn, B., Choi, J., Hattori, M., Min, B., & Bae, K. (2011). Selective cholinesterase inhibition by lanostane triterpenes from fruiting bodies of *Ganoderma lucidum*. *Bioorganic & Medicinal Chemistry Letters*, *21*(21), 6603–6607.

Li, W. L., Zheng, H. C., Bukuru, J., & De Kimpe, N. (2004). Natural medicines used in the traditional Chinese medical system for therapy of diabetes mellitus. *Journal of Ethnopharmacology*, *92*(1), 1–21.

Lu, H., Lou, H., Hu, J., Liu, Z., & Chen, Q. (2020). Macrofungi: A review of cultivation strategies, bioactivity, and application of mushrooms. *Comprehensive Reviews in Food Science and Food Safety*, *19*(5), 2333–2356.

Lu, X., Brennan, M. A., Narciso, J., Guan, W., Zhang, J., Yuan, L., Serventi, L., & Brennan, C. S. (2020). Correlations between the phenolic and fiber composition of mushrooms and the glycaemic and textural characteristics of mushroom-enriched extruded products. *LWT-Food Science and Technology*, *118*, 108730.

Ma, G., Yang, W., Zhao, L., Pei, F., Fang, D., & Hu, Q. (2018). A critical review on the health promoting effects of mushrooms nutraceuticals. *Food Science and Human Wellness*, *7*(2), 125–133.

Malinowska, E., Szefer, P., & Falandysz, J. J. F. C. (2004). Metals bioaccumulation by bay bolete, *Xerocomus badius*, from selected sites in Poland. *Food Chemistry*, *84*(3), 405–416.

Manzi, P., Aguzzi, A., & Pizzoferrato, L. (2001). Nutritional value of mushrooms widely consumed in Italy. *Food Chemistry*, *73*(3), 321–325.

Manzi, P., Gambelli, L., Marconi, S., Vivanti, V., & Pizzoferrato, L. (1999). Nutrients in edible mushrooms: An inter-species comparative study. *Food Chemistry*, *65*(4), 477–482.

Matsuur, H., Asakawa, C., Kurimoto, M., & Mizutani, J. (2002). α-glucosidase inhibitor from the seeds of balsam pear (*Momordica charantia*) and the fruit bodies of *Grifola frondosa. Bioscience, Biotechnology, and Biochemistry*, *66*(7), 1576–1578.

Mattila, P., Könkö, K., Eurola, M., Pihlava, J. M., Astola, J., Vahteristo, L., Hietaniemi, V., Kumpulainen, J., Valtonen, M., & Piironen, V. (2001). Contents of vitamins, mineral elements, and some phenolic compounds in cultivated mushrooms. *Journal of Agricultural and Food Chemistry*, *49*(5), 2343–2348.

Mau, J. L., Lin, H. C., & Chen, C. C. (2002). Antioxidant properties of several medicinal mushrooms. *Journal of Agricultural and Food Chemistry*, *50*(21), 6072–6077.

McClements, D. J., Barrangou, R., Hill, C., Kokini, J. L., Lila, M. A., Meyer, A. S., & Yu, L. (2021). Building a resilient, sustainable, and healthier food supply through innovation and technology. *Annual Review of Food Science and Technology*, *12*, 1–28.

Nabubuya, A., Muyonga, J. H., & Kabasa, J. D. (2010). Nutritional and hypocholesterolemic properties of *Termitomyces microcarpus* mushrooms. *African Journal of Food, Agriculture, Nutrition and Development*, *10*(3), 2235–2257.

Olawuyi, I. F., & Lee, W. Y. (2019). Quality and antioxidant properties of functional rice muffins enriched with shiitake mushroom and carrot pomace. *International Journal of Food Science & Technology*, *54*(7), 2321–2328.

Palacios, I., Lozano, M., Moro, C., D'arrigo, M., Rostagno, M. A., Martínez, J. A., García-Lafuente, A., Guillamón, E., & Villares, A. (2011). Antioxidant properties of phenolic compounds occurring in edible mushrooms. *Food Chemistry*, *128*(3), 674–678.

Oliveira, A.P. and Naozuka, J. (2019). Preliminary results on the feasibility of producing selenium-enriched pink (*Pleurotus djamor*) and white (*Pleurotus ostratus*) oyster mushrooms: bioaccumulation, bio-accessibility and se-proteins distribution. *Microchemistry Journal*, 145: 1143–50.

Patel, S., & Goyal, A. (2012). Recent developments in mushrooms as anti-cancer therapeutics: A review. *3 Biotech*, *2*, 1–15.

Perera, P. K., & Li, Y. (2011). Mushrooms as a functional food mediator in preventing and ameliorating diabetes. *Functional Foods in Health and Disease*, *1*(4), 161–171.

Prasad, S., Rathore, H., Sharma, S., & Yadav, A. S. (2015). Medicinal mushrooms as a source of novel functional food. *International Journal of Food Science, Nutrition and Dietetics*, *4*, 221–225.

Raghavendra, V. B., Venkitasamy, C., Pan, Z., & Nayak, C. (2017). Functional foods from mushrooms. In *Microbial Functional Foods and Nutraceuticals* (pp. 65–91).

Rajarathnam, S. and Bano, Z. (1992). Utilization of edible fruiting fungi. In: Arora, D.K. and Marth, E.H. (Eds.), Handbook of applied Microbiology Vol. 3 (Food and Feeds). *Marcel Dekker Inc.*, New York. pp. 241–92.

Ramesh, C. H., & Pattar, M. G. (2010). Antimicrobial properties, antioxidant activity, and bioactive compounds from six wild edible mushrooms of Western Ghats of Karnataka, India. *Pharmacognosy Research*, *2*(2), 107.

Rane, B. R., Ahirrao, R. A., Gujarathi, N. A., Jain, A. S., & Keservani, R. K. (2023). Pathophysiology and application of herbs in the metabolic syndrome. In *The Metabolic Syndrome: Dietary Supplements and Food Ingredients* (pp. 199–230). Apple Academic Press, CRC Press, Taylor & Francis Group. ISBN: 978-1-77491-111-2.

Rane, B. R., Bharath, M. S., Patil, R. R., Keservani, R. K., & Jain, A. S. (2020). Novel approaches in nutraceuticals. In *Enhancing the Therapeutic Efficacy of Herbal Formulations through Novel Drug Delivery Systems* (pp. 241–266). IGI Global. ISBN13: 9781799844532.

Rane, B. R., Patil, A. S., Keservani, R. K., & Jain, A. S. (2020). Novel approaches in herbal formulation. In *Enhancing the Therapeutic Efficacy of Herbal Formulations Through Novel Drug Delivery Systems* (pp. 43–68). IGI Global. ISBN13: 9781799844532.

Rane, B. R., Tadavi, S. A., & Keservani, R. K. (2020). Naturopathy. In *Herbal Product Development* (pp. 321–347). Apple Academic Press, CRC Press, Taylor & Francis Group. ISBN: 9781771888776.

Rathee, S., Rathee, D., Rathee, D., Kumar, V., & Rathee, P. (2012). Mushrooms as therapeutic agents. *Revista Brasileira de Farmacognosia*, *22*, 459–474.

Rathore, H., Prasad, S., & Sharma, S. (2017). Mushroom nutraceuticals for improved nutrition and better human health: A review. *Pharma Nutrition*, *5*(2), 35–46.

Ravi, B., Renitta, R. E., Prabha, M. L., Issac, R., & Naidu, S. (2013). Evaluation of antidiabetic potential of oyster mushroom (*Pleurotus ostreatus*) in alloxan-induced diabetic mice. *Immunopharmacology and Immunotoxicology*, *35*(1), 101–109.

Reis, F. S., Martins, A., Vasconcelos, M. H., Morales, P., & Ferreira, I. C. (2017). Functional foods based on extracts or compounds derived from mushrooms. *Trends in Food Science & Technology*, *66*, 48–62.

Rodriguez Estrada, A. E., Lee, H. J., Beelman, R. B., Jimenez-Gasco, M. D. M., & Royse, D. J. (2009). Enhancement of the antioxidants ergothioneine and selenium in *Pleurotus eryngii* var. *eryngii* basidiomata through cultural practices. *World Journal of Microbiology and Biotechnology*, *25*, 1597–1607.

Royse, D. J., Baars, J., & Tan, Q. (2017). Current overview of mushroom production in the world. In *Edible and Medicinal Mushrooms: Technology and Applications* (pp. 5–13).

Rudawska, M., & Leski, T. (2005). Macro-and microelement contents in fruiting bodies of wild mushrooms from the Notecka forest in west-central Poland. *Food Chemistry*, *92*(3), 499–506.

Sabaratnam, V., Kah-Hui, W., Naidu, M., & David, P. R. (2013). Neuronal health–can culinary and medicinal mushrooms help? *Journal of Traditional and Complementary Medicine*, *3*(1), 62–68.

Saiqa, S., Haq, N. B., Muhammad, A. H., Muhammad, A. A., & ATA, U. R. (2008). Studies on chemical composition and nutritive evaluation of wild edible mushrooms. *South Asian Journal of Food Technology and Environment*, *1*(3, 4), 211–218.

Sanmee, R., Dell, B., Lumyong, P., Izumori, K., & Lumyong, S. (2003). Nutritive value of popular wild edible mushrooms from northern Thailand. *Food Chemistry*, *82*(4), 527–532.

Sato, M., Tai, T., Nunoura, Y., Yajima, Y., Kawashima, S., & Tanaka, K. (2002). Dehydrotrametenolic acid induces preadipocyte differentiation and sensitizes animal models of noninsulin-dependent diabetes mellitus to insulin. *Biological and Pharmaceutical Bulletin*, *25*(1), 81–86.

Savić, M. D., Petrović, J., Klaus, A., Nikšić, M., Rajković, M., Filipović, N., & Antić-Mladenović, S. (2009). Growth and fruit body formation of *Pleurotus ostreatus* on media supplemented with inorganic selenium. *Zbornik Matice Srpske za Prirodne Nauke*, *116*, 209–215.

Shamtsyan, M. (2010). Bioactive compounds in mushrooms. In D. R. Heldman & D. G. Hoover (Eds.), *Encyclopedia of Biotechnology in Agriculture and Food* (pp. 76–81). CRC Press.

Sharma, A. K., Jana, A. M., Srivastav, A., Gupta, M., Sharma, S., & Gill, S. S. (2014). Antimicrobial properties of some edible mushrooms: A review. *World Journal of Pharmaceutical and Pharmaceutical Sciences*, *3*(5), 1009–1023.

Sharma, A. K., Keservani, R. K., & Gautam, S. P. (2020). *Herbal Product Development*. Apple Academic Press, CRC Press, Taylor & Francis Group.

Singh, S. S., Wang, H., Chan, Y. S., Pan, W., Dan, X., Yin, C. M., Akkouh, O., & Ng, T. B. (2014). Lectins from edible mushrooms. *Molecules*, *20*(1), 446–469.

Soares, A.A., Sa-Nakanishi, A.B., Bracht, A., Costa, S.M.G., Koehnlein, E.A., De Souza, C.G.M. and Peralta, R.M. (2013). Hepatoprotective effects of mushrooms. *Molecules*, *18*(7):7609–30.

Thatoi, H., & Singdevsachan, S. K. (2014). Diversity, nutritional composition and medicinal potential of Indian mushrooms: A review. *African Journal of Biotechnology*, *13*(4).

Thongbai, B., Rapior, S., Hyde, K. D., Wittstein, K., & Stadler, M. (2015). *Hericium erinaceus*, an amazing medicinal mushroom. *Mycological Progress*, *14*, 1–23.

Ueda, Y., Yonemitsu, M., Tsubuku, T., Sakaguchi, M., & Miyajima, R. (1997). Flavor characteristics of glutathione in raw and cooked foodstuffs. *Bioscience, Biotechnology, and Biochemistry*, *61*(12), 1977–1980.

Valverde, M. E., Hernández-Pérez, T., & Paredes-López, O. (2015). Edible mushrooms: Improving human health and promoting quality life. *International Journal of Microbiology*, 376387.

Varghese, R., Dalvi, Y. B., Lamrood, P. Y., Shinde, B. P., & Nair, C. K. K. (2019). Historical and current perspectives on therapeutic potential of higher basidiomycetes: An overview. *3 Biotech*, *9*, 1–15.

Wang, J. C., Hu, S. H., Wang, J. T., Chen, K. S., & Chia, Y. C. (2005). Hypoglycemic effect of extract of *Hericium erinaceus*. *Journal of the Science of Food and Agriculture*, *85*(4), 641–646.

Wang, X. M., Zhang, J., Wu, L. H., Zhao, Y. L., Li, T., Li, J. Q., Wang, Y. Z., & Liu, H. G. (2014). A mini-review of chemical composition and nutritional value of edible wild-grown mushroom from China. *Food Chemistry*, *151*, 279–285.

Wannet, W. J., Hermans, J. H., van der Drift, C., & Op den Camp, H. J. (2000). HPLC detection of soluble carbohydrates involved in mannitol and trehalose metabolism in the edible mushroom *Agaricus bisporus*. *Journal of Agricultural and Food Chemistry*, *48*(2), 287–291.

Wild, S., Roglic, G., Green, A., Sicree, R., & King, H. (2004). Global prevalence of diabetes: Estimates for the year 2000 and projections for 2030. *Diabetes Care*, *27*(5), 1047–1053.

Yildiz, A., Karakaplan, M., & Aydin, F. (1998). Studies on *Pleurotus ostreatus* (Jacq. ex Fr.) Kum. var. *salignus* (Pers. ex Fr.) Konr. et Maubl.: Cultivation, proximate composition, organic and mineral composition of carpophores. *Food Chemistry*, *61*(1, 2), 127–130.

Zeb, M., & Lee, C. H. (2021). Medicinal properties and bioactive compounds from wild mushrooms native to North America. *Molecules*, *26*(2), 251.

Zhang, H. N., & Lin, Z. B. (2004). Hypoglycemic effect of *Ganoderma lucidum* polysaccharides. *Acta Pharmacologica Sinica*, *25*(2), 191–195.

Zhang, L., Fan, C., Liu, S., Zang, Z., Jiao, L., & Zhang, L. (2011). Chemical composition and antitumor activity of polysaccharide from *Inonotus obliquus*. *Journal of Medicinal Plants Research*, *5*(7), 1251–1260.

Zhang, M., Cui, S. W., Cheung, P. C. K., & Wang, Q. (2007). Antitumor polysaccharides from mushrooms: A review on their isolation process, structural characteristics, and antitumor activity. *Trends in Food Science & Technology*, *18*(1), 4–19.

Zhang, Y., Venkitasamy, C., Pan, Z., & Wang, W. (2013). Recent developments on umami ingredients of edible mushrooms–A review. *Trends in Food Science & Technology*, *33*(2), 78–92.

Zhong, J. J., Feng-Wu, B., & Wei, Z. (2009). *Biotechnology in China I: From Bioreaction to Bioseparation and Bioremediation*. Springer.

Zhou, X.W., Lin, J., Li, Q.Z., Yin, Y.Z., Sun, X.F. and Tang, K.X. (2007). Study progress on bioactive proteins from *Ganoderma spp. Natural Products Research Development*, 19:916–24.

Zimmet, P. Z., McCarty, D. J., & De Courten, M. P. (1997). The global epidemiology of non-insulin-dependent diabetes mellitus and the metabolic syndrome. *Journal of Diabetes and its Complications*, *11*(2), 60–68.

CHAPTER 13

Functional Food Products of Dragon Fruit: An Insight

ABINAYA VENKATESH,[1] S. ANANDAKUMAR,[1*] and SOUMITRA BANERJEE[2]

[1]*Department of Food Packaging and System Development, NIFTEM, Thanjavur, Tamil Nadu, India*

[2]*Center for Incubation, Innovation, Research, and Consultancy, Jyothy Institute of Technology, Thathaguni, Karnataka, India*

*Corresponding author.

ABSTRACT

Dragon fruit is an exotic high value tropical fruit that has market throughout the globe and highly prevailed in Asian countries. Dragon fruit is also known as pitahaya fruit or strawberry pear. The fruit attained popularity among consumers owing to attractive skin color with mellow mouth-watering taste. In addition, it is an abundant source of vital nutrients and an essential bioactive compound. Major compounds in dragon fruit are glucose, dietary fiber, polyphenols, betacyanin, essential fatty acids, carotene, vitamin C, calcium, and phosphorous. The presence of these compounds not only provides nutrition that requires for the body but also promotes good health by preventing various chronic diseases thus is also a functional food. In dragon fruit, the whole part is utilized in the development of products like flavored milk, yogurt, ice cream, cookies, pastries, and so on, as they are enriched with nutritional and bioactive compounds that act as antioxidants. The cultivation, nutritional, and biochemical constituents of dragon fruit along with their properties, therapeutic value, and minimally processed (MP) products are discussed in detail in this chapter. Finally, their potential applications in food industries, as well as other industries, are further discussed.

Functional Compounds and Foods of Plant Origin:Krishan Dutt Sharma, Satish Kumar, Rakesh Sharma, and Raj K. Keservani (Eds.)

13.1 INTRODUCTION

The daily intake of fruits in adequate amounts is highly recommended to maintain good health. It is low in calories and is rich in various nutritive compounds such as carbohydrates, vitamins, minerals, dietary fiber, and non-nutritive phytochemicals such as polyphenols, flavonoids (Keservani and Sharma, 2014; Keservani et al., 2016a, b), and carotenoids. These bioactive compounds have been reported to possess several therapeutic properties and can decrease oxidative stress and metabolic disorders by neutralizing the free radicals that cause damage to health. Fruits in the regular diet have a positive impact that prevents several chronic diseases such as diabetes, obesity, cancer, neurodegenerative, and CVD (Oz and Kafkas, 2017; Septembre-Malaterre et al., 2018). Dragon fruit has a wide spectrum of bioactive ingredients that perform specific functions in the body beyond the provision of essential nutrients, in specific it improves health and reduces the risk of serious illness and thus is also referred to as a potent functional food (Quiroz-Iturra et al., 2017). In fruits, mostly the flesh is available for consumption, and the remaining parts, such as pulp, seeds, and peels, are considered waste, as well as the by-products obtained from fruit processing industries are further utilized in food, pharma, or cosmetics industries. The major studied fruits are apples, pomegranates, oranges, lemons, pineapples, papayas, tomatoes, bananas, grapes, cranberries, and kiwi (Fierascu et al., 2020; Kodagoda and Marapana, 2017).

In recent times, increased health awareness of the consumer has resulted in a significant increase in the consumption of fruits and utilization of their by-products has become popular due to the knowledge regarding their nutritional (Keservani et al., 2010a, b, 2017, 2020) as well as their medicinal properties (Li et al., 2016). The presence of nutrients and various bioactive compounds present in fruits perform various bioactivities such as neuroprotective effects, anti-inflammatory, antimicrobial, antioxidant, and anti-cancerous activity as well as reduce CVDs. It also prevents malnutrition, and various fatal illnesses and promotes good health (Cornara et al., 2020). Although, the consumption of exotic fruits is on a growing trend still some of the fruits are less known or underutilized and their availability is limited to the respective production areas (Bhat and Paliyath, 2016; Cornara et al., 2020).

In tropical exotic fruits, the dragon fruit has reached much global attention because of its unique taste and exotic appearance. It also exhibits high nutritional and therapeutic values (Perween et al., 2018). Although, the dragon fruit is native to Mexico, Southern, and Central America because of its attractive features and economic value production is increased worldwide and has gained popularity in various countries such as Indonesia, Thailand, Philippines, Vietnam, Taiwan, Malaysia, and Sri Lanka (Mercado-Silva, 2018; Perween et al., 2018).

The dragon fruit is produced by the night-blooming cactus plant species named *Hylocereus cacti*, a semi-epiphytic plant. The fruit is identified by its oval or oblong shape, and it has a white or red-purple flesh with its outer bright red or yellow skin. Dragon fruit is referred to as pitahaya or pitaya, which means scaly fruit. The triangular shape of green scales or bracts on the fruit's surface resembles a dragon head-like appearance, and hence, it is named dragon fruit (Hossain et al., 2021; Wall and Khan, 2008). Other names for dragon fruit include strawberry pear, night-blooming cereus, *paniniō kapunahou* or *pāpipi pua* (Hawaiian), *la pitahaya rouge* (French), *buah naga* (Malaysian), *thang loy* (Vietnamese), and belle of the night (Gunasena et al., 2007; Jalgaonkar et al., 2020).

It is rich in glucose, vitamins, minerals, unsaturated fatty acid, and antioxidant constituents especially flavonoids, carotenoids, tannins, anthocyanidins, and phenolic compounds that helps to prevent diabetes, cancer, cardiovascular, gastrointestinal, respiratory, and urinary-related problems. It is also recommended as a diet food because it has low calories and is rich in fibers (Hossain et al., 2021; Kumar et al., 2018). The whole fruit is fleshy and edible with a bland taste and mild sweetness. It can be eaten directly as fresh fruit or juice. In terms of processing, several products—such as jam, jelly, wine, candy, and pectin—have been standardized from dragon fruit and are reported to have high nutritional value (Jalgaonkar et al., 2020).

13.2 CULTIVATION, MORPHOLOGY, AND CLASSIFICATION OF DRAGON FRUIT

Dragon fruit is a non-climacteric fruit, and it is the most extensively distributed variety in the Cactaceae family which is native to Mexico. It is a semi-epiphytic plant, which climbs, crawls, or attaches to any natural or artificial support. The fruits emerge from the climbing cacti which can grow up to a height of 6 meters (stem). The cultivation of the dragon fruit is mostly in areas that have less annual rainfall (Jalgaonkar et al., 2020; Mercado-Silva,

2018). During the cultivation of dragon fruit, it requires an average rainfall in the range of 500–2000 mm/year and requires an optimum temperature of about 20°C–30°C but can tolerate up to 38–40°C and the low temperature of 0°C for short periods. A flower bud of a plant starts to grow in mid-May until June and fruits begin to grow in August–December. Depending on the cultivar it takes around 10–15 days to 25–30 days for the attaining the harvest maturity (Arivalagan et al., 2019; Gunasena et al., 2007). As it is non-climacteric fruit it should, therefore, be harvested once it attains physiological maturity. In the mature state, the color of the fruit's skin changes from green to red or pinkish-red, and the taste of the fruit become sweeter (Liaotrakoon, 2013). Jalgaonkar et al. (2020) have reviewed the post-harvest profile of dragon fruit, its processing, method of grading, and the utilization of their by-products. The dragon fruit is widely cultivated and distributed worldwide, and the major countries are Thailand, Australia, Philippines, Bermuda, Indonesia, Vietnam, southern China, Hawaii, Colombia, Philippines, Israel, Okinawa (Japan), Taiwan, and West Indies. Vietnam is the world's largest exporter of dragon fruit (Le Bellec et al., 2006; Perween et al., 2018).

13.2.1 FRUIT MORPHOLOGY

Dragon fruit consists of three main parts, including skin or peel, flesh or pulp, and seeds. The whole fruit comprises 36.70–37.60% of skin, 47.40–73.76% of flesh, and 2.70–14.67% of seeds. The weight of the dragon fruit ranges from 100–800 g (Jalgaonkar et al., 2020; Le Bellec and Vaillant, 2011). The fruit has a thin rind of skin with attractive color and is usually covered with scales or bracts on the fruit surface. It has mellow mouth-melting flesh with many edible seeds embedded in it while the texture mimics the texture of kiwi fruit. The flesh color varies from white to varying hues of red and purple and color of the skin is deeper for red flesh fruit than for white flesh fruit and it usually varies depending on the cultivar. The tiny edible seeds are black-colored with soft (Lichtenzveig et al., 2000; Panchal et al., 2018; Pawde et al., 2020).

13.2.2 VARIETIES OF DRAGON FRUIT

It is widespread and classified under two different genera namely *Hylocereus* and *Selenicereus*. The genus *Hylocereus* is the most commercially cultivated variety comprising 16 different species. Among them, the four species from the genus *Hylocereus* and one species from the *Selenicereus* varieties are major known edible varieties (Gunasena et al., 2007; Le Bellec and Vaillant, 2011). The fruit morphology is mentioned in Figure 13.1. Bauer (2003) revised the genus of species *Selenicereus megalanthus* and placed it in the species *Hylocereus* and renamed it *Hylocereus megalanthus*, *Hylocereus polyrhizus*, and *Hylocereus monocanthus* (Mercado-Silva, 2018).

FIGURE 13.1 Morphology of dragon fruit.

13.3 NUTRITIONAL IMPORTANCE OF DRAGON FRUIT

Dragon fruit is nutrient-rich and is primarily composed of glucose, vitamins, minerals, dietary fiber, and unsaturated fatty acids, all of which are essential for regulating body metabolism. The nutritional value of the fruit is influenced by growth conditions, preharvest factors, environmental conditions, species, and harvesting stage (Hossain et al., 2021; Liaotrakoon, 2013). The comparative nutritional profile of three major species of dragon fruits—*Hylocereus undatus*, *Hylocereus polyrhizus*, and *Selenicereus megalanthus*—is shown in Table 13.1 (Jalgaonkar et al., 2020; Mercado-Silva, 2018). The flesh of the fruit is rich in ascorbic acid and minerals such as calcium and phosphorous. The vitamin C content is higher in *H. undatus* followed by *H. polyrhizus* and *S. megalanthus*, respectively. The pulp of *H. polyrhizus* contains higher phosphorus content than the other two varieties. Nurul and Asmah (2014) reported the diverse nutritional and physicochemical properties of red dragon fruit which include mineral content, total flavonoid content (TPC), and betacyanin content (Nurul and Asmah, 2014).

TABLE 13.1 Nutritional Composition of Three Major Varieties of Dragon Fruit

Composition	Variety		
	Hylocereus undatus	***Hylocereus polyrhizus***	***Selenicereus megalanthus***
Water (%)	89	82.5–83	85
Protein (g)	0.5	0.16–0.23	0.4
Fat (g)	0.1	0.21–0.61	0.1
Crude fiber (g)	0.3	0.7–0.9	0.5
Ash (g)	0.5	0.28	0.4
Carotene (mg)	NR	0.005–0.012	NR
Vitamin B_1 (mg)	0	0.028–0.043	0
Vitamin B_2 (mg)	0	0.043–0.045	0
Vitamin B_3 (mg)	0.2	0.297–0.430	0.2
Vitamin C (mg)	25	8–9	4
Iron (mg)	0.4	0.55–0.65	0.3
Calcium (mg)	6	6.3–8.8	10
Phosphorous (mg)	19	30.2–36.1	16

Note: Values are expressed per 100 g of fruit pulp; NR represents not reported.

Source: Gunasena et al. (2007); Mercado-Silva (2018).

13.4 BIOCHEMICAL CONSTITUENTS

The biochemical constituents and the functional compounds present in the fruits are responsible for the functional properties of the fruit and have an important role in promoting good health. In dragon fruit, the whole fruit (flesh, skin, seeds) has immense biochemical components in terms of minerals, vitamins, dietary fiber, phytochemicals such as polyphenols, flavonoids, and betacyanin content (Mercado-Silva, 2018; Nurliyana et al., 2010; Senadheera and Abeysinghe, 2015). The important bioactive compounds and nutrients available in various parts of the dragon fruit are mentioned in Table 13.2 (Jalgaonkar et al., 2020). The reported values are concerned with the main varieties such as *H. undatus*, *H. polyrhizus,* and *S. megalanthus,* respectively.

The fruits have high quality carbohydrates such as hemicellulose, oligosaccharides, and pectin. Hemicellulose is composed of glucose, galactose, and xylose, while pectin consists of arabinose and galactose (Ramírez et al., 2011; Wichienchot et al., 2010). Phytochemicals are chemical compounds that involve protecting human health against various diseases by acting as

TABLE 13.2 Biochemical Composition of Dragon Fruit

Component	Skin	Flesh	Seeds
Carbohydrate (%)	NR	6.11–6.29	20.2–50.2
Glucose (g)	4.12–4.18	491.4–1039.5	NA
Fructose (g)	0.84–0.88	192.0–289.7	NA
Pectin (mg/g)	NA	0.64–1.36	NA
Dietary fiber (g/100 g)	69.3	1.1–3.20	11.2–49.2
Protein (g)	0.64–0.66	12–12.5	14.6–28.6
Fat (%)	0.02–0.07	0.10–0.61	23.6–35.6
Minerals (%)	0.17–0.22	NA	NA
Ash (%)	14.29	0.28–0 50	1.1–3.1
Ascorbic acid (mg/100 g)	NR	13.0–55.80	NA
Citric acid (g)	0.08	9.5–21.1	NA
Malic acid (g)	0.64	60.8–82.0	NA
Betacyanin content (mg)	41.55	NA	NA
Total phenolic content (mg/100 g)	28.16–36.12	3.75–19.72	1353.96–1358.0

Note: Values expressed in grams (g) are per 100 mL juice; NA represents not available.

antibacterial, antiviral, antifungal, and anti-inflammatory effects. Some of the phytochemicals that are naturally present in dragon fruit are oligosaccharides, flavonoids, phytosterols, phenolic acids, tocopherols, and carotenoids (Elmarzugi et al., 2016; Oz and Kafkas, 2017).

The polyphenols are the secondary metabolites of phytochemicals which consists of phenolic acids and flavonoids. According to Kugler et al. (2007), the TPC is higher in *H. polyrhizus* (61 mg GAE/100 mL) followed by *S. megalanthus* (22.7 mg GAE/100 mL) and *H. undatus* (18.2 mg GAE/100 mL). The seven phenolic acids that are identified and present in dragon fruit seeds are gallic, syringic, vanillic, p-hydroxybenzoic, p-coumaric, protocatechuic, and caffeic. The available flavonoid compounds are quercetin, kaempferol, catechin, isorhamnetin, and also their glycosylated derivatives (Akram and Mushtaq, 2019; Cornara et al., 2020).

Carotenoids are the plant pigments that results in various color shades in fruits and vegetables. In dragon fruit, they are available as β-carotene, betalains, and lycopene (Cornara et al., 2020). Betalains are water-soluble coloring matter, available in flower-bearing plants and their fruits. Betalains are classified into betacyanins and betaxanthins which impart a red-purple and yellow color to the fruits and flowers (Polturak and Aharoni, 2018). In dragon fruit, the skin and flesh consist of specific compounds such as lycopene, betacyanin, betanin, and betaxanthins resulting in red-purple and yellow pigment of fruit. It is also a source of antioxidant that helps in regulating body metabolism (Hawa, 2010; Phebe et al., 2009; Wybraniec and Mizrahi, 2002).

The tiny edible black seeds in dragon fruit are rich in oil content that provides polyunsaturated fatty acid (PUFA) as a lipid source. Among the three varieties of dragon fruit, the higher seed oil content was found in *H. undatus* (27.5%) followed by *H. polyrhizus* (22.8%) and *S. megalanthus* (18.8%), respectively (Chemah et al., 2010). The major fatty acids that are available in seed oil is linoleic acid followed by oleic acid and palmitic acid which are considered to be essential fatty acid that helps in promoting good health and preventing various diseases (Ariffin et al., 2009; Lim et al., 2010). The major tocopherols content present in the seeds is α tocopherol, β tocopherol, γ tocopherol, and δ tocopherol (Lim et al., 2010). The phytosterols in dragon fruit are cholesterol, stigmasterol, campesterol, and β-sitosterol (Akram and Mushtaq, 2019; Cornara et al., 2020; Lim et al., 2010).

13.5 HEALTH BENEFITS OF DRAGON FRUIT

13.5.1 ANTIOXIDANT PROPERTY

Antioxidants help prevent various oxidative reactions that are naturally present in fruits and vegetables. It prevents oxidation reaction and thus inhibits the formation of free radicals and their chain reactions that may harm the cells and causes serious illness which results in diabetes, cancer, and brain and heart-related diseases. The antioxidants have extensive food applications in the processing industry as an additive that prevents the spoilage of food products from bacterial or fungal attacks (Joshi and Prabhakar, 2020; Singh and Immanuel, 2014; Wilson et al., 2017). In dragon fruit, various bioactive components are present which are not a part of nutrition but have various functional roles in human health. The major compounds that are present in the fruit that exhibit antioxidant activities are polyphenols, flavonoids, carotenoids, betalains tocopherol, and ascorbic acid. The highest antioxidant capacity was found in *H. polyrhizus* followed by *S. megalanthus* and *H. undatus*. These antioxidant compounds in dragon fruit help to prevent various chronic diseases and are also involved in promoting good health (Choo and Yong, 2011; Mercado-Silva, 2018; Nurliyana et al., 2010).

13.5.2 PRE-BIOTIC PROPERTY

Prebiotics are the non-digestible food components that enter the lower gastrointestinal tract without getting digested or absorbed. They act as a source of nutrition for the gut microbiota as it not only promotes the growth and/or activities of beneficial microorganisms but also reduce the availability of undesirable bacteria. The most known prebiotics are galactooligosaccharides (GOS), trans-galactic-oligosaccharides (TOS), fructooligosaccharides (FOS), and inulin, and the fruits are found to be the rich source of prebiotics (Davani-Davari et al., 2019; Patel and Goyal, 2012; Septembre-Malaterre et al., 2018). In dragon fruit, the major carbohydrates that are present are fructose, glucose, and oligosaccharides. The skin and flesh also act as a source of prebiotics (Pansai et al., 2020; Rohin et al., 2014; Wichienchot et al., 2010). The maximum yield of oligosaccharides has been reported in *H. polyrhizus* followed by *H. undatus* at 89.6 g/Kg and 86.2 g/Kg (Wichienchot et al., 2010). The consumption of dragon fruit-based oligosaccharides improves the growth and proliferation of gut microbiota, that is, *Lactobacillus* and *Bifidobacterium* and also decreases the harmful bacteria including *Clostridia* and *Bacteroides*. It also enhances the concentration of plasma immunoglobulin A and G and thus it exhibits immune-boosting properties (Pansai et al., 2020).

13.5.3 ANTIMICROBIAL PROPERTY

The major polyphenol compounds that are present in fruits such as flavonoids, tannins, phenolic acids, and carotenoids have shown antimicrobial effects against various bacterial strains (Adamczak et al., 2020). In dragon fruit, both skin and flesh have proved as a rich source of phytochemicals such as polyphenols, flavonoids, carotenoids, and betalains that showed antioxidant effects and also can have potential as antimicrobial activity (Patel and Ishnava, 2019; Sudha et al., 2017). The skin extracts of *H. undatus and H. polyrhizus* have potent antibacterial activity and has been reported to inhibit the activity of *Staphylococcus aureus, Escherichia coli, Listeria monocytogenes, Bacillus cereus, Klebsiella pneumoniae* and *Campylobacter jejuni* which are important food spoilage causing microorganisms. The red flesh dragon fruit (*H. polyrhizus*) showed higher antibacterial effects than the white flesh dragon fruit (*H. undatus*) against both gram-positive and gram-negative bacteria (Nurmahani et al., 2012). The extract of the dragon fruit can be used as a natural antimicrobial agent in inhibiting the growth of food-borne pathogens including food spoilage bacteria, yeasts, and molds (Hendra et al., 2019; Tenore et al., 2012).

13.6 THERAPEUTIC PROPERTIES OF DRAGON FRUIT

13.6.1 ANTI-DIABETIC PROPERTIES

Diabetes mellitus (DM) is commonly referred to as diabetes, a metabolic disorder that increases the blood glucose level due to the insufficient production of insulin. Research in the areas of herbal (Rane et al., 2020a, b, c, 2023; Sharma et al., 2020) or natural remedies from the source of plant material is increasing to treat diabetes. The consumption of dragon fruit and its extracts has shown a positive impact on carbohydrate metabolism, and thus significantly helps in treating diabetes. In specific, the phytoconstituents that are present in dragon fruit are effectively reducing the blood glucose level (Joshi and Prabhakar, 2020; Patel and Goyal, 2012; Poolsup et al., 2017). The fruits of *H. undatus* contain flavonoid compounds which can mediate the hypoglycemic effect upon regular consumption. Three main mechanisms involved in this activity include reducing the oxidative stress by the effect of antioxidants, including the inhibition of intestinal mucosal GLUT 2 and cAMP phosphodiesterase, and thus results in better retention of insulin for a longer period (Ajie, 2015; Joshi and Prabhakar, 2020). The red flesh dragon fruit (*H. polyrhizus*) also showed a positive effect on blood glucose levels and maintained a good lipid profile specific for people of Type-2 diabetics. The extracts of dragon fruit in the modern world can not only help to maintain healthy insulin levels in the individual in general but can be an effective tool to simplify diabetic complications (Abd Hadi et al., 2016).

13.6.2 ANTI-CANCEROUS PROPERTIES

Cancer is one of the leading diseases that involve the transformation of normal cells into uncontrolled abnormal cells that continue to grow and spread anywhere in a body. It causes serious health illnesses worldwide that result in death. The major components such as polyphenols, flavonoids, and hydroxycinnamate betalains from natural food sources have shown antiproliferative effects in cancer cell growth (Cornara et al., 2020; Joshi and Prabhakar, 2020). The extract from the skin and flesh of *Hylocereus polyrhizus* exhibits antiproliferative activity against melanoma cancer cells

(B16F10), with the skin showing a stronger effect than the flesh (Wu et al., 2006). The skin extract of *H. undatus* and *H. polyrhizus* showed cytotoxic effect against various cancer cell lines such as human prostate cancer cell line (PC3), gastric cancer cell line (MGC-803), and breast cancer cell line (Bcap-37) and found that *H. polyrhizus* showed better effect than *H. undatus* on MGC-803 cells. Among many phytoconstituents, specific ones such as β-sitosterol, β-amyrin, and stigmast-4-en-3-one have shown good anticancer activity against these three cell lines (Luo et al., 2014). The flesh extract of *H. undatus* has shown promising anti-apoptotic and anticancer activity against human liver cancer cell HepG-2 (Padmavathy et al., 2021). In dragon fruit, the skin extract proved a better antiproliferative effect than flesh as it contains 3 to 5 times higher content of flavonoids and polyphenols as compared to the flesh (Joshi and Prabhakar, 2020; Kim et al., 2011).

13.6.3 AORTIC STIFFNESS AND OXIDATIVE STRESS

Cardiovascular disease (CVD) is a condition that majorly affects the heart and blood vessels and is usually associated with the deposition of fats inside the layer of arteries and also results in a risk of blood clots. Aortic stiffness is a common CVD related to the rigidity of the arterial blood vessel wall that causes stiffer and thicker blood vessels, which results in hypertension, high blood pressure, or vascular complications. Oxidative stress is an important factor in vascular complications that are also associated with diabetes (Baynes, 1991; Joshi and Prabhakar, 2020). In fruits, various phytochemicals such as polyphenols, flavonoids, carotenoids, and betalains have shown to increase microvascular activity, decrease vascular stiffness, and thus help in preventing CVD (Oz and Kafkas, 2017). The aqueous extract of *H. undatus* flesh can result in significant reduction of the pulse wave velocity (PWV), pulse pressure, and systolic pressure of streptozotocin (STZ)-induced diabetic rats as it proves its effectiveness in treating aortic stiffness along with diabetic complications (Swarup et al., 2010).

13.6.4 CHOLESTEROL CONTROL

The oral consumption of red flesh dragon fruit (*H. polyrhizus*) extract regularly showed a significant reduction in the total cholesterol level and triglycerides level in plasma of hypercholesterolemia rats. It also increases the HDL-C level in plasma for the intake of 1.17% of fruit extract. The decrease in the cholesterol level is due to the excretion of bile acid at a higher level. This hypocholesterolemic effect of dragon fruit is due to the presence of unsaturated fatty acids, soluble fiber, and minerals such as sodium, potassium, magnesium, phosphorous, and zinc (Rohin et al., 2009).

13.6.5 ROLE IN WOUND HEALING

The healing of the wound is one of the major complications associated with diabetes patients. The aqueous extract of flesh, leaves, rind, and flowers dragon fruit treated the wounds in STZ-induced diabetic rats and also notable increase in the hydroxyproline content, DNA collagen content and total protein content, tensile strength, and also improved epithelialization (Cornara et al., 2020; Perez et al., 2005). In brief, the functions of antioxidant compounds in dragon fruit that perform various biological activities in protecting human health are mentioned in Table 13.3.

TABLE 13.3 Functions of Antioxidant Compounds in Dragon Fruit

Compound		Function(s)	References
Dietary fiber		Helps decreases the blood cholesterol level, improves bowel movements, prevents irritable bowel syndrome (IBS), and improves insulin sensitivity and thus, helps to prevent diabetes.	Hernawati et al. (2018)
Oligosaccharides		Acts as a pre-biotic source that helps to improve gut microbiota and exhibits immune-boosting properties.	Pansai et al. (2020)
Vitamins	Vitamin B_1 (Thiamine)	Helps to generate energy from carbohydrates, helps in the proper functioning of nerves, and also plays an important role in DNA and RNA production.	Kumar et al. (2018)
	Vitamin B_2 (Riboflavin)	Involves in fatty acids catabolism process and also in tricarboxylic acid cycle.	
	Vitamin B_3 (Niacin)	Important in the energy transfer reactions.	
	Ascorbic acid (Vitamin C)	Prevents respiratory problems like asthma and cough, strengthens the immune system, reduces the joint pain for arthritis, and also exhibits an anti-inflammatory effect.	Elmarzugi et al. (2016)
	γ-tocopherol (Vitamin E)	Reduces the risk of heart diseases and exhibits anti-inflammatory properties.	Liaotrakoon (2013)
Calcium and phosphorous		Improve bone strength and also helps in tissue formation and also maintain dental health.	Verma et al. (2017)

TABLE 13.3 *(Continued)*

Compound	Function(s)	References
Iron	Increases hemoglobin and erythrocytes levels and thus improves blood count.	Widyaningsih et al. (2017)
Poly-unsaturated fatty acids	Reduces the formation of triglycerides and thus reduces the CVDs.	Sonawane (2017)
Carotenoids	Reduces the risk of cancer and cardio-related diseases.	Verma et al. (2017)
Lycopene	Helps in inhibiting the growth of various human-related cancer cell lines.	Zainoldin and Baba (2009)
Hydroxycinnamtes and phytoalbumins	Exhibit anticancer properties that involve in the prevention of cancer.	Hossain et al. (2021)
Flavonoids	Acts on brain cells and blood vessels that help to maintain brain health and blood pressure. It also reduces the risk of heat-related illness.	Harahap et al. (2020)
Betalains	Reduces LDL cholesterol level in the blood, prevents cancer cell formation and reduces oxidative stress, and exhibits anti-inflammatory effects.	Hossain et al. (2021); Le et al. (2021)

13.7 CONSUMPTION OF DRAGON FRUIT

The attractive color and pigmentation of the fruits coupled with its nutritional profile have resulted in increased popularity and consumption. It has attractive color, pleasant aroma with a juicy and sweet taste, and also has various nutritional benefits. In dragon fruit, each part (skin, flesh, seeds, dried flower, buds) is edible and has enormous nutrition in terms of vitamin C, fiber, antioxidants, essential fatty acids, minerals in specific phosphorous, and calcium. The fruit can be best eaten in a raw fresh or dried form of fruit. It is also available as ready-to-eat (RTE) fresh-cut fruit after the removal of peel and sliced that can be stored at below 8°C and available for consumption. The dried form of fruit flesh is also used as coloring agent in various beverages (Jalgaonkar et al., 2020; Le Bellec and Vaillant, 2011). Apart from the raw matured fruit, the fresh flower buds and young stem are also edible and can be consumed as a vegetable (Hossain et al., 2021; Perween et al., 2018).

Various processing technologies have been used for shelf-life extension, quality retention and value addition of the dragon fruits with the highest level of sensory attributes. Various processed products from dragon fruit are shown in Figure 13.2. In fruit, the pulp is used to make jellies, jam, juice, and wine and then, the peel is rich in betacyanin content which can be extracted and utilized as a natural food color and also a source of pectin. The seeds can be utilized for the extraction of oil, which contains 50% of essential fatty acids and has got diverse food applications in various functional food products such as ice cream, sherbet, syrup, yogurt, candy, and pastries (Jalgaonkar et al., 2020; Liaotrakoon, 2013).

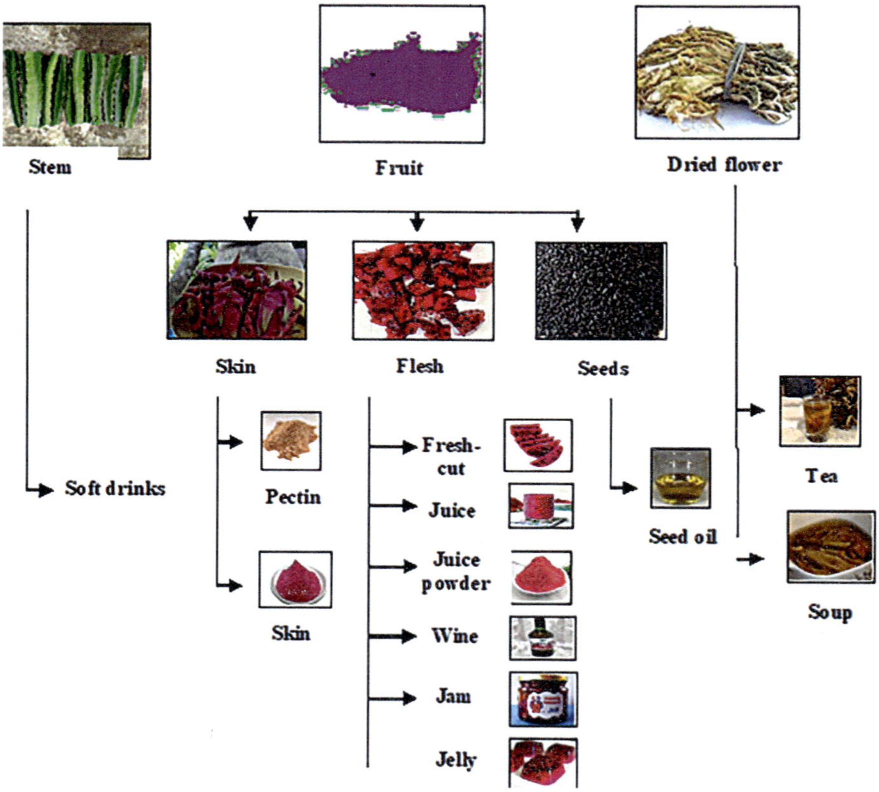

FIGURE 13.2 Processed products from dragon fruit.

13.7.1 FERMENTED BEVERAGE

Fermentation of fresh fruit juice can enhance the shelf-life and storability of the product and also the sensory attributes and its antioxidant content. The fermented drink must be transparent, clear, with improved sensory attributes, and has functional properties. The flesh of dragon fruit is utilized for the preparation of wine, a fermented beverage drink (Huan et al., 2020; Jalgaonkar et al., 2020). The processing and development of wine from dragon fruit pulp was shown in Figure 13.3. According to the consumer acceptability, for a semi-sweet wine, about 250 g of sugar and 2% yeast is required per Kg of dragon fruit however, for sweet wine, about 500 g with 2% yeast is required. About 5 mL of 10% of sodium metabisulfite is added to the juice which acts as a preserver and stabilizer of wine (Dimero and Tepora, 2018).

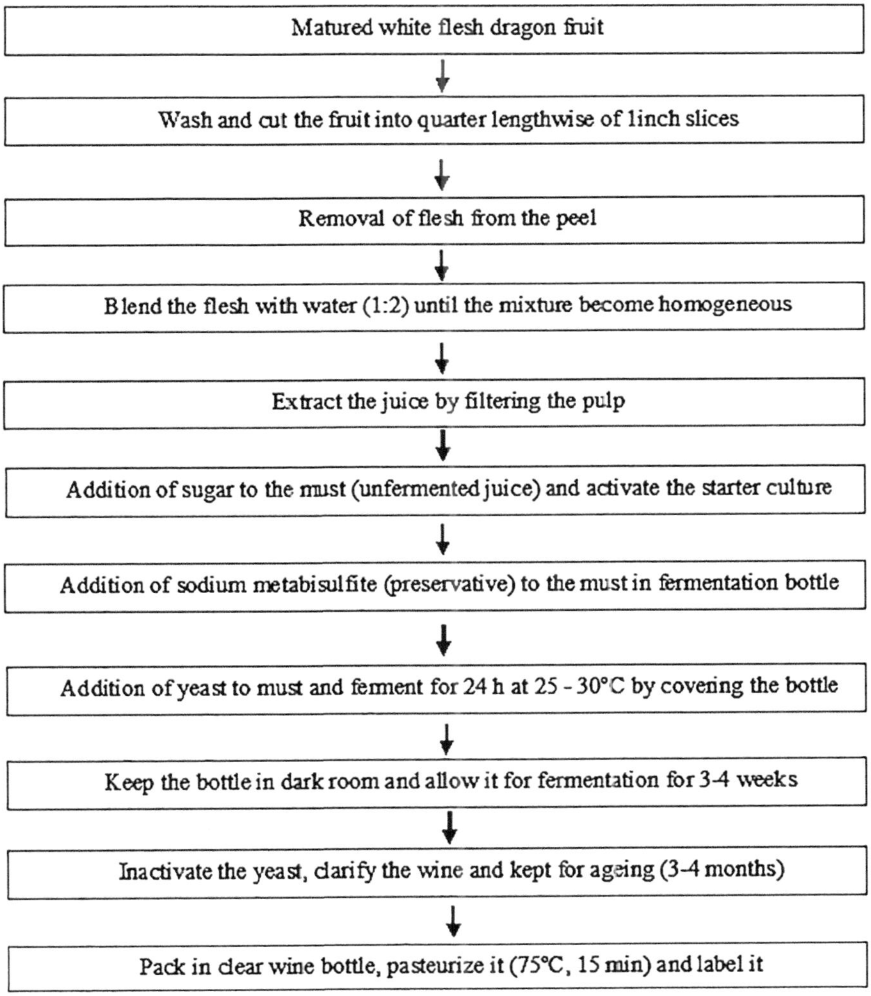

FIGURE 13.3 Flow chart of winemaking from dragon fruit pulp.

The fermented liquid drink from red flesh dragon fruit is prepared along with the pasteurization process shown in Figure 13.4. The fermentation of juice followed by the pasteurization step significantly improves the bioactive compounds in juice such as acetic acid, phytosterols (stigmasterol, campesterol, β-sitosterol), betacyanins (isobetanin, betanin), and TFC and also the juice exhibits higher free radical scavenging activity as compared to unpasteurized fermented juice (Foong et al., 2012).

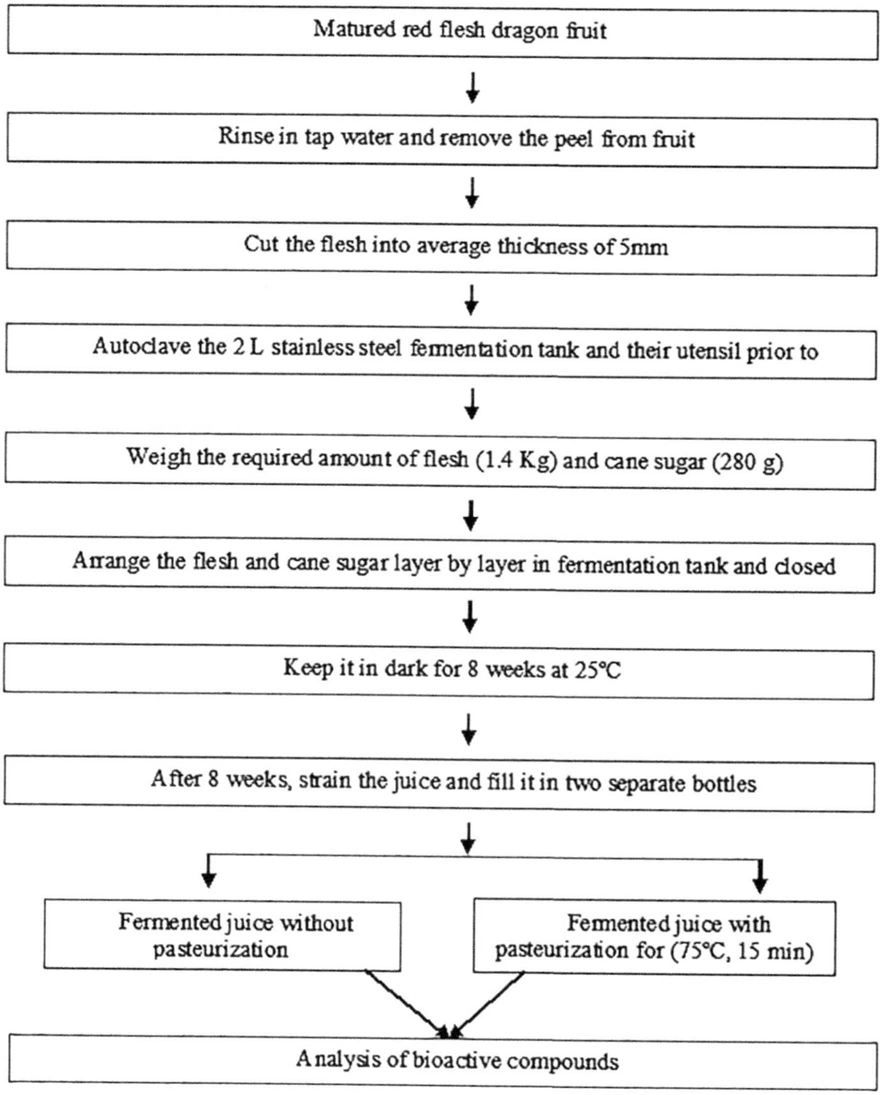

FIGURE 13.4 Flow chart of preparation of fermented drink from red flesh dragon fruit.

The white flesh dragon fruit-based fermented drink is prepared by differing total soluble solids (16, 18, and 20°B), yeast (*S. cerevisiae* 135) concentration, v/v (1, 2, and 3%), and fermentation time (20, 22, and 48 h) and studied their quality and sensory characteristics. The drink prepared with an initial TSS of 18°Brix, 2% of yeast concentration, and fermented the extract for 44 h showed a better result of alcohol content of 3.54% (v/v) and remaining sugar content (14.6°Brix) with better sensory results as compared to other treatment (Huan et al., 2020).

The fermented drink from the dragon fruit (*H. costaricensis*) peel powder (1, 1.5, and 2%) along with the addition of skim milk powder (5 and 10%) and 6% (v/v) yeast culture (*S. thermophilus, L. bulgaricus, and L. plantarum*) in different ratios (1:1:1, 1:2:1, 1:1:2, 2:1:1) were fermented under 42°C for 14 h. The drink with 1% of peel powder with 5% of skim milk powder and 6% of 1:1:1 ratio of yeast concentration showed increased antioxidant activity by 241% with acceptable sensory attributes and also proved as safe for consumption (Eveline and Audina, 2019). The fresh dragon fruit juice along with the addition of 10% of fermented dragon fruit juice enhanced the bioactive compounds present in the juice and also maintained the sensory characteristics such as aroma, taste, and color till the storage period of 3 months at 8°C. The high antioxidant and antibacterial activity of developed juice can be applied to potential functional beverages (Muhialdin et al., 2020).

13.7.2 TEA

The preparation of semi-fermented tea from peel of *H. undatus* is shown in Figure 13.5. The peels have good amount of vitamin C 93.87 mg/g as compared to the flesh of dragon fruit. The phenolics and antioxidants of dragon fruit peel tea after different withering times, such as 30 min, 60 min, and 120 min, and different rolling times of 10, 20, and 30 sec were examined. The longer withering time of 120 min with a shorter rolling time of 10 sec showed higher retention of the TPC of 223.70 mg/L and antioxidant activity of 41.19% compared to other treatments. The phenolic content was higher as compared to black tea (223.70 mg/L) and lower than the green tea of 649.33 mg/L, respectively (Sari and Hardiyanti, 2013).

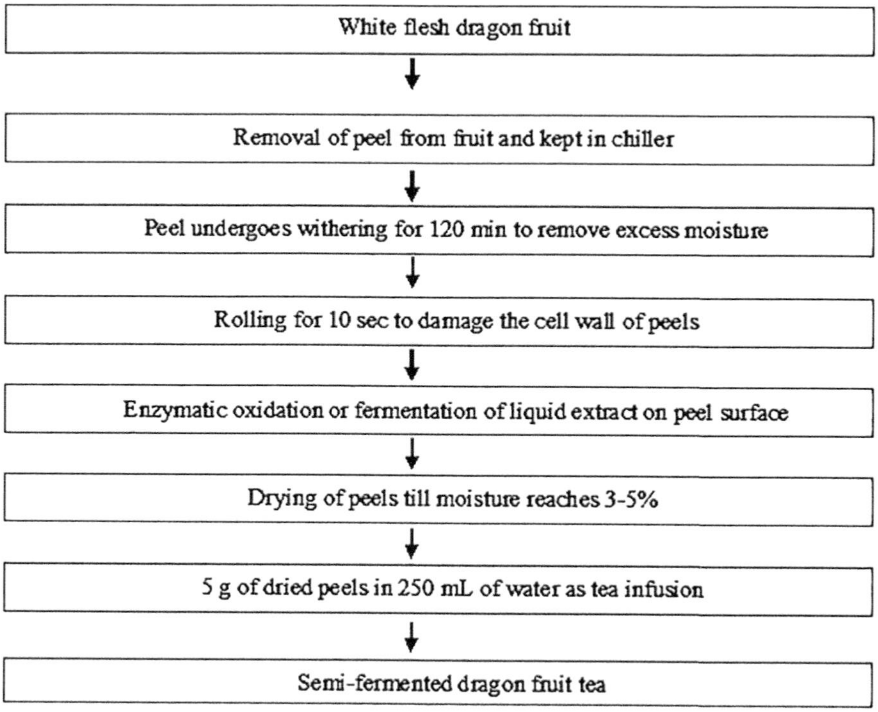

IGURE 13.5 Flowchart of preparation of tea from the peel of dragon fruit.

The tea beverage prepared from the powder of the matured flower of the dragon fruit plant was prepared by drying the flower at 100°C for 40 min. The dried flower of 1 g with water and allowed to boil for 7 min to extract the aqueous solution. The extract of dried flower along with 9% (w/v) of sugar content, 0.2% of citric acid with 0.03% of litchi flavor, and 0.2% of liquorice juice showed a better color of tea as brown-red color with a slightly sweet taste and good aroma as compared with the other market available products such as instant green tea (Dam, 2012).

13.7.3 JAM

The red dragon fruit jam is prepared as shown in Figure 13.6. The shelf-life of jam stored at different temperatures such as 30, 40, 50, and 60C were evaluated using Arrhenius model based on ASLT (accelerated shelf-life test) for 28 days and found that the jam stored at low temperatures (30°C) retained better antioxidant activity as compared to other storage condition. From the models, the shelf-life of the prepared jam predicted that the jam stored at 28°C was estimated to be 99 days (Zaidiyah et al., 2019).

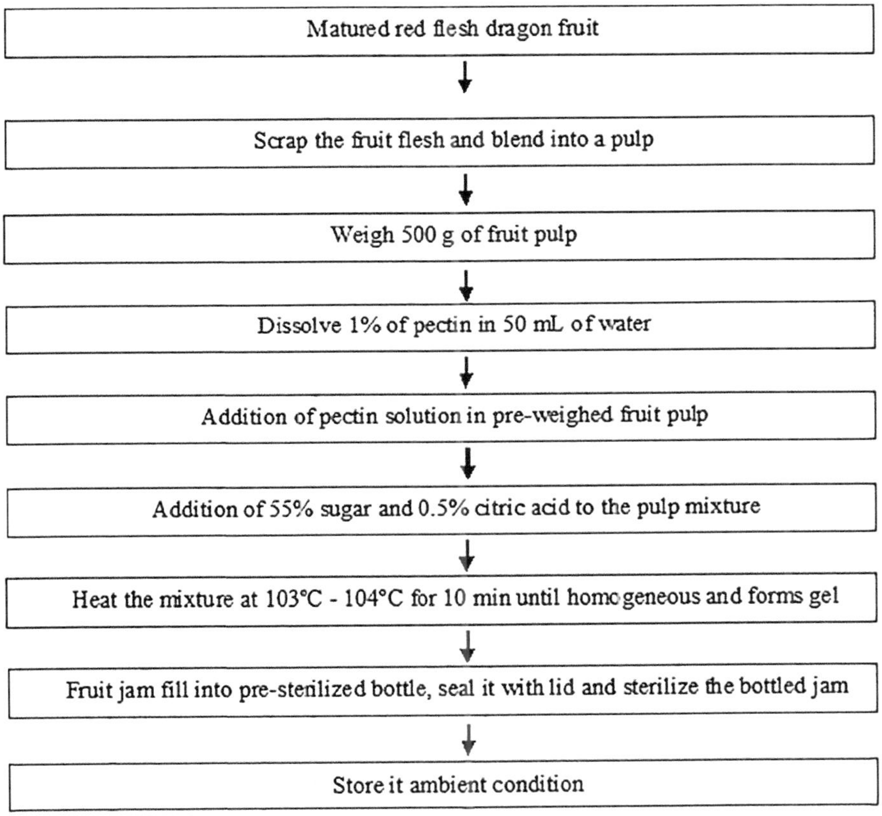

FIGURE 13.6 Flowchart for the preparation of jam from dragon fruit.

The fruit spread from the peel and pulp of red dragon fruit were prepared and evaluated their chemical composition, microbial, and consumer acceptability. The prepared spread consists of moisture (46.43%), pH (4.90), soluble solids (38.65%), fat (0.44%), and water activity (0.82). The mean general score of dragon fruit spread was 8.0 and is highly accepted by the consumers (Barcelon et al., 2015).

13.7.4 JELLY

The preparation of jelly from white flesh dragon fruit were shown in Figure 13.7. The chemical composition of jelly consists of 28.9% moisture, 67.26%

TSS, 0.51% TA, 27.77% reducing sugar, and ascorbic acid of 2.96 mg/100 g, respectively. According to Panchal et al. (2018), the prepared jelly retains better chemical composition and sensory properties with good storage stability up to a storage period of 90 days in both ambient (30 ± 2°C) and refrigerated condition (5 ± 2°C).

Matured red flesh dragon fruit

↓

Scrap the fruit flesh and blend into a pulp

↓

Weigh 500 g of fruit pulp

↓

Dissolve 1% of pectin in 50 mL of water

↓

Addition of pectin solution in pre-weighed fruit pulp

↓

Addition of 55% sugar and 0.5% citric acid to the pulp mixture

↓

Heat the mixture at 103°C-104°C for 10 min until homogeneous and forms gel

↓

Fruit jam fill into pre-sterilized bottle, seal it with lid and sterilize the bottled jam

↓

Store it ambient condition

FIGURE 13.7 Flowchart for the preparation of jelly from dragon fruit.

13.7.5 SEED OIL

The dragon seeds contain good amount of essential fatty acids. Seeds are embedded in the pulp of the fruit and there are various methods for the separation of seeds and extraction of oil (Figure 13.8). The seed oil extracted from the *H. polyrhizus* and *H. undatus* showed a seed yield of 14.27% and 14.67% and higher seed oil of 28.37% from *H. undatus* followed by 18.33% from *H. polyrhizus*, respectively (Lim et al., 2010). The seed oil extracted from the dragon fruit is yellow due to the presence of yellow-colored pigment. The oil contains essential fatty acids such as palmitic acid, linoleic acid, stearic acid, and oleic acid and it also exhibits high antioxidant activity (Adnan et al., 2011; Liaotrakoon et al., 2013; Murugesu et al., 2013).

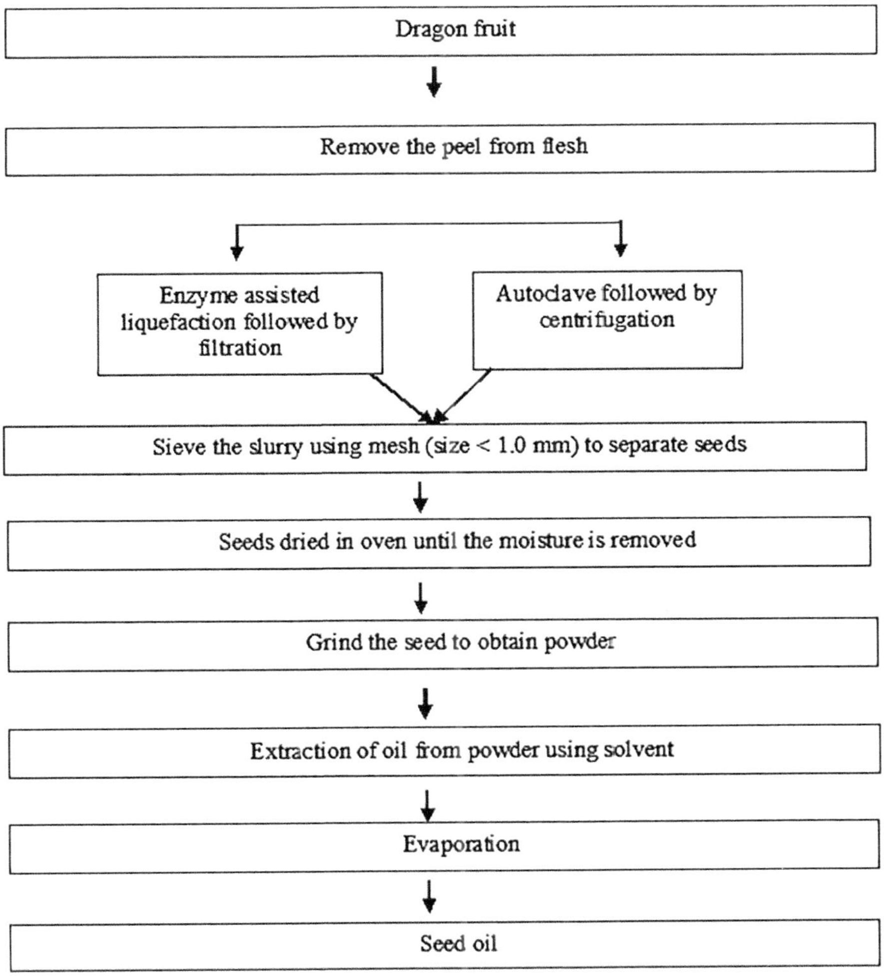

FIGURE 13.8 Flowchart on the extraction of oil from dragon fruit seeds.

13.8 POTENTIAL INDUSTRIAL APPLICATION AND WASTE UTILIZATION

13.8.1 FOOD COLORANT

The skin and flesh of dragon fruit consist of betacyanin, an anthocyanin pigment that gives the fruit its characteristic red color. The flesh of dragon fruit contains higher betacyanin content than skin and can be used as a natural red dye. It is also a rich source of vitamins, flavonoids, and antioxidant compounds (Harivaindaran et al., 2008; Phebe et al., 2009; Vaillant et al., 2005). As a waste utilization, the powder of skin can be utilized as a natural food colorant in various food products such as pastille (Yee et al., 2017), ice cream (Hafids et al., 2019), flavored milk (Gengatharan et al., 2016), yogurt (Gengatharan et al., 2017), and pastries (Wahyuningtias, 2015).

13.8.2 THICKENER

Pectin is the complex carbohydrate that is present in the cell wall of fruits that are used as a natural gelling agent, stabilizer, and thickener. The peel of the dragon fruit is a rich source of pectin and is extracted by various techniques and can be used as a natural pectin source (Abang Zaidel et al., 2017; Tang et al., 2011). In comparison with commercially available apple and citrus pectin, the dragon fruit peel pectin exhibits a light red color and textural properties similar characteristics. Dragon fruit pectin used in the preparation of pineapple jelly showed a good sensory characteristic and can be used as a thickening agent in low-viscous food and beverages (Muhammad et al., 2016).

13.8.3 DIETARY FIBER

The skin of dragon fruit is usually considered as waste that exhibits a rich source of dietary fiber and various other bioactive compounds (Jamilah et al., 2011; Madane et al., 2020). The fiber-rich biscuits have been prepared by incorporating the peel powder of white flesh dragon fruit. The prepared biscuit from 50% of dragon fruit peel powder in replacement of 50% of wheat flour showed a fivefold increase in fiber content of the biscuit with acceptable sensory attributes. The consumption of 100 g of biscuits provides 20% of the recommended daily intake of dietary fiber (Pawde et al., 2020).

13.8.4 SOURCE OF ANTIOXIDANT

Dragon fruit is a source of phytochemical constituents that exhibit antioxidant activity. It can be incorporated as an antioxidant source in various food products (Choo and Yong, 2011; Madane et al., 2020). The antioxidant-rich fruit-based yogurt has been prepared by incorporating the flesh of red and white dragon fruit. The white flesh fruit yogurt showed a higher TPC of 64.43% as compared to red flesh fruit yogurt (49.61%). The addition of dragon fruit flesh also enhanced lactic acid content, antioxidant activity, and sensory properties (Zainoldin and Baba, 2009). The incorporation of white flesh dragon fruit peel power (DFP) in the preparation of chicken nuggets improves the nutritional composition, dietary fiber, and antioxidant activity compared to the nuggets without DFP. It also enhances palatability with health-promoting benefits (Madane et al., 2020).

The whole dragon fruit is reported to be a source of bioactive compounds such as vitamin C, polyphenols, flavonoids, and antioxidant pigments that show numerous health benefits (Senadheera and Abeysinghe, 2015). People perform heavy physical exercise with high intensity causing an increase in the production of free radicals that results in various cardiovascular diseases (CVDs). The regular intake of red flesh dragon fruit juice as an antioxidant source showed a decrease in cholesterol level followed by an increase in the maximum aerobic capacity of oxygen (VO_2), a measure of cardio-respiratory state (Harahap et al., 2020). Iron deficiency is the main cause of anemia and is mostly affected by pregnant women. Dragon fruit is a rich source of iron content, and the consumption of red and purple flesh dragon fruit juice regularly improved the hemoglobin content and erythrocyte levels in pregnant women with anemia (Kusmiyati et al., 2019; Widyaningsih et al., 2017).

13.8.5 OTHER APPLICATIONS

The extract of skin and flesh of dragon fruit contains a high level of betacyanins pigment that is utilized in various cosmetics, textile, and paint industries as a natural colorant. The flesh extract of *H. polyrhizus* and *H. costsricencis* used to formulate a natural colorant lipstick that exhibits antioxidant and also antibacterial activity (Azwanida et al., 2014). The lipstick prepared from betacyanin pigment of dragon fruit showed a low level of lead content as compared to synthetic dyes-based lipstick (Lwin et al., 2020). The peel extract of dragon fruit is also used in textile industries as a natural dye, absorbent, sensitizer dye, and a natural coloring agent alternative to chemical paints (Ali et al., 2010; Azizan, 2020; Jawad et al., 2018).

13.9 CONCLUSION

This chapter aimed to bring forth detailed insight into dragon fruit as a source of functional food. Dragon fruit is a tropical fruit known for its attractive skin color, bracts, and seed-speckled pulp with sweet. Although it is a seasonal and localized fruit, the demand in domestic and international markets is very high due to its nutritional value and bioactive compounds that help contribute to various health benefits. The fruit is the best source of vitamins, minerals, dietary fiber, oligosaccharides, polyphenols, betacyanin pigments, and flavonoids. The presence of bioactive compounds as functional ingredients improves human health. It helps in maintaining gut health, preventing CVDs, decreasing cholesterol levels, and better bone health and thus all together strengthening the immune system of the human body. In dragon fruit, not only the flesh but also skin, skin is proved to be a source of bioactive compounds. The peel of dragon fruit is a waste source utilized in food industries as a thickener, colorant, antioxidant, and dietary fiber in various food products and is also used in cosmetics and drug industries. Thus, dragon fruit as a natural functional food ingredient not only provides a nutritional source but also contains bioactive compounds that help in promoting good health. In addition, research on dragon fruit should be extended in understanding the role of various bioactive compounds and the utilization of by-products in the food, drug, and cosmetic industries.

KEYWORDS

- **antioxidant activity**
- **betacyanin**
- **colorant**
- **dragon fruit**
- **essential fatty acids**
- **food colorant**
- **functional food**
- **linoleic acid**
- **oil extraction**
- **oleic acid**
- **seed extraction**
- **seed yield**
- **stearic acid**
- **waste utilization**

REFERENCES

Abang Zaidel, D. N., Md Rashid, J., Hamidon, N. H., Md Salleh, L., & Mohd Kassim, A. S. (2017). Extraction and characterization of pectin from dragon fruit (*Hylocereus polyrhizus*) peels. *Chemical Engineering Transactions*, *56*, 805–810.

Abd Hadi, N., Mohamad, M., Rohin, M. A. K., & Yusof, R. M. (2016). Effects of red pitaya fruit (*Hylocereus polyrhizus*) consumption on blood glucose level and lipid profile in type 2 diabetic subjects. *Borneo Science*, *31*, 143–151.

Adamczak, A., Ożarowski, M., & Karpiński, T. M. (2019). Antibacterial activity of some flavonoids and organic acids widely distributed in plants. *Journal of Clinical Medicine*, *9*(1), 109.

Adnan, L., Osman, A., & Abdul Hamid, A. (2011). Antioxidant activity of different extracts of red pitaya (*Hylocereus polyrhizus*) seed. *International Journal of Food Properties*, *14*(6), 1171–1181.

Ajie, R. B. (2015). White dragon fruit (*Hylocereus undatus*) potential as diabetes mellitus treatment. *Jurnal Majority*, *4*(1), 69–72.

Akram, S., & Mushtaq, M. (2019). Dragon (*Hylocereus megalanthus*) seed oil. In *Fruit Oils: Chemistry and Functionality* (pp. 675–689).

Ali, R. A. M., & Nayan, N. (2010). Fabrication and analysis of dye-sensitized solar cell using natural dye extracted from dragon fruit. *International Journal of Integrated Engineering*, *2*(3).

Ariffin, A. A., Bakar, J., Tan, C. P., Rahman, R. A., Karim, R., & Loi, C. C. (2009). Essential fatty acids of pitaya (dragon fruit) seed oil. *Food Chemistry*, *114*(2), 561–564.

Arivalagan, G., Karunakaran, M., S. Sriram, & G. Karunakaran. (2019). Dragon fruit country report from India. *FFTC Agricultural Policy Platform*, FFTC-AP, Food and Fertilizer Technology for the Asian and Pacific Region.

Azizan, M. Z. (2020). The extraction of dragon fruit pericarp as coloring agent for painting. *Voice of Academia (VOA)*, *16*(2), 41–52.

Azwanida, N. N., Normasarah, N., & Afandi, A. (2014). Utilization and evaluation of betalain pigment from red dragon fruit (*Hylocereus Polyrhizus*) as a natural colorant for lipstick. *Jurnal Teknologi*, *69*(6), 139–142.

Barcelon, E., Carreon, L., Guillermo, J., Jacob, E., Jocson, S., Panopio, J. G., & Rosalinas, S. (2015). Consumer acceptability and physico-chemical content of red flesh dragon fruit spread. *Australian Journal of Basic and Applied Sciences*, *9*(2), 18–21.

Bauer, R. (2003). A synopsis of the tribe *Hylocereus* F. Buxb. *Cactaceae Syst. Initiat.*, *17*, 3–63.

Baynes, J. W. (1991). Role of oxidative stress in development of complications in diabetes. *Diabetes*, *40*(4), 405–412.

Bhat, R., & Paliyath, G. (2016). Fruits of tropical climates: Dietary importance and health benefits. In *Encyclopedia of Food and Health* (Vol. 1, pp. 144–149).

Choo, W. S., & Yong, W. K. (2011). Antioxidant properties of two species of *Hylocereus* fruits. *Advances in Applied Science Research*, *2*(3), 418–425.

Cornara, L., Xiao, J., Smeriglio, A., Trombetta, D., & Burlando, B. (2020). Emerging exotic fruits: New functional foods in the European market. *EFood*, *1*(2), 126–139.

Dam, S. M. (2012). Development of different processed products from the edible and inedible parts of the dragon fruit (*Hylocereus undatus*). In *Southeast Asia Symposium on Quality Management in Postharvest Systems* and *Asia Pacific Symposium on Postharvest Quality* (Vol. 989, pp. 271–278).

Davani-Davari, D., Negahdaripour, M., Karimzadeh, I., Seifan, M., Mohkam, M., Masoumi, S. J., Berenjian, A., & Ghasemi, Y. (2019). Prebiotics: Definition, types, sources, mechanisms, and clinical applications. *Foods*, *8*(3), 92.

Dimero, F. N., & Tepora, T. F. (2018). Processing and development of dragon fruit wine. *International Journal of Environment, Agriculture and Biotechnology*, *3*(5), 1943–1947.

Elmarzugi, N. A., Cheah, L., Eid, A., Aziz, A., & Ariffin, F. (2016). Phytochemical properties and health benefits of *Hylocereus undatus*. *Nanomed Nanotechnol*, *1*(1), 1–10.

Eveline, & Audina, M. (2019). Utilization of Super Red Dragon Fruit Peel (*Hylocereus Costaricensis* (FAC Weber) Britton & Rose) in the making of fermented beverage. In *IOP Conference Series: Earth and Environmental Science* (Vol. 292, No. 1, 012037).

Fierascu, R. C., Sieniawska, E., Ortan, A., Fierascu, I., & Xiao, J. (2020). Fruits by-products–A source of valuable active principles: A short review. *Frontiers in Bioengineering and Biotechnology*, *8*, 319.

Foong, J. H., Hon, W. M., & Ho, C. W. (2012). Bioactive compounds determination in fermented liquid dragon fruit (*Hylocereus polyrhizus*). *Borneo Sci.*, *31*, 37–56.

Gengatharan, A., Dykes, G. A., & Choo, W. S. (2016). Stability of betacyanin from red pitahaya (*Hylocereus polyrhizus*) and its potential application as a natural colourant in milk. *International Journal of Food Science & Technology*, *51*(2), 427–434.

Gengatharan, A., Dykes, G. A., & Choo, W. S. (2017). The effect of pH treatment and refrigerated storage on natural colourant preparations (betacyanins) from red pitahaya and their potential application in yoghurt. *LWT*, *80*, 437–445.

Gunasena, H. P. M., Pushpakumara, D. K. N. G., & Kariyawasam, M. (2007). Dragon fruit *Hylocereus undatus* (Haw.) Britton and Rose. Underutilized fruit trees in Sri Lanka, *1*, 110–141.

Hafids, S., Rahmi, S. L., & Chairunisah, A. R. (2019). Study of low-fat ice cream with the substitution of super red dragon (*Hylocereus costaricensis*) fruit peel. *Indonesian Food Science & Technology Journal*, *3*(1), 23–28.

Harahap, N. S., Sunarno, A., & Simatupang, N. (2020). The effect of red dragon fruit juice towards cholesterol level and maximum aerobic capacity (VO_2max) on sport science students treated with heavy physical exercise. *Journal of Physics: Conference Series*, *1462*(1), 012030.

Harivaindaran, K. V., Rebecca, O. P., & Chandran, S. (2008). Study of optimal temperature, pH, and stability of dragon fruit (*Hylocereus polyrhizus*) peel for use as potential natural colorant. *Pakistan Journal of Biological Sciences*, *11*(18), 2259–2263.

Hawa, M. M., Salleh, B., & Latiffah, Z. (2010). Characterization and intraspecific variation of *Fusarium semitectum* (Berkeley and Ravenel) associated with red-fleshed dragon fruit (*Hylocereus polyrhizus* [Weber] Britton and Rose) in Malaysia. *African Journal of Biotechnology*, *9*(3), 273–284.

Hendra, R., Masdeatresa, L., Abdulah, R., & Haryani, Y. (2019). Antibacterial activity of red dragon peel (*Hylocereus polyrhizus*) pigment. *Journal of Physics: Conference Series*, *1351*(1), 012042.

Hernawati, Setiawan, N. A., Shintawati, R., & Priyandoko, D. (2018). The role of red dragon fruit peel (*Hylocereus polyrhizus*) to improvement blood lipid levels of hyperlipidaemia male mice. *Journal of Physics: Conference Series*, *1013*(1), 012167.

Hossain, F. M., Numan, S. M. N., & Akhtar, S. (2021). Cultivation, nutritional value, and health benefits of dragon fruit (*Hylocereus spp.*): A review. *International Journal of Horticultural Science and Technology*, *8*(3), 259–269.

Huan, P. T., Hien, N. M., & Anh, N. H. T. (2020). Optimization of alcoholic fermentation of dragon fruit juice using response surface methodology. *Food Research*, *4*(5), 1529–1536.

Izalin, M. N., Kharidah, M., Jamilah, B., & Noranizan, M. A. (2016). Functional properties of pectin from dragon fruit (*Hylocereus polyrhizus*) peel and its sensory attributes. *J. Trop. Agric. and Fd. Sc.*, *44*(1), 95–101.

Jalgaonkar, K., Mahawar, M. K., Bibwe, B., & Kannaujia, P. (2022). Postharvest profile, processing, and waste utilization of dragon fruit (*Hylocereus spp.*): A review. *Food Reviews International*, *38*(4), 733–759.

Jamilah, B., Shu, C. E., Kharidah, M., Dzulkily, M. A., & Noranizan, A. (2011). Physico-chemical characteristics of red pitaya (*Hylocereus polyrhizus*) peel. *International Food Research Journal*, *18*(1), 279–285.

Jawad, A. H., Kadhum, A. M., & Ngoh, Y. S. (2018). Applicability of dragon fruit (*Hylocereus polyrhizus*) peels as low-cost biosorbent for adsorption of methylene blue from aqueous solution: Kinetics, equilibrium, and thermodynamics studies. *Desalination and Water Treatment*, *109*, 231–240.

Joshi, M., & Prabhakar, B. (2020). Phytoconstituents and pharmaco-therapeutic benefits of pitaya: A wonder fruit. *Journal of Food Biochemistry*, *44*(7), e13260.

Keservani, R. K., & Sharma, A. K. (2014). Flavonoids: Emerging trends and potential health benefits. *Journal of Chinese Pharmaceutical Sciences*, *23*(12), 815.

Keservani, R. K., Kesharwani, R. K., Sharma, A. K., Vyas, N., & Chadoker, A. (2010b). Nutritional supplements: An overview. *International Journal of Current Pharmaceutical Review and Research*, *1*(1), 59–75.

Keservani, R. K., Kesharwani, R. K., Vyas, N., Jain, S., Raghuvanshi, R., & Sharma, A. K. (2010a). Nutraceutical and functional food as future food: A review. *Der Pharmacia Lettre*, *2*(1), 106–116.

Keservani, R. K., Sharma, A. K., & Kesharwani, R. K. (2016a). Nutraceutical and functional foods for cardiovascular health. In *Food Process Engineering* (pp. 291–312). Apple Academic Press, CRC Press.

Keservani, R. K., Sharma, A. K., & Kesharwani, R. K. (2016b). Medicinal effect of nutraceutical fruits for cognition and brain health. *Scientifica, 3109254*, 10. https://doi.org/10.1155/2016/3109254.

Keservani, R. K., Sharma, A. K., & Kesharwani, R. K. (2017). An overview and therapeutic applications of nutraceutical and functional foods. In *Recent Advances in Drug Delivery Technology* (pp. 160–201). ISBN: 9781522507543.

Keservani, R. K., Sharma, A. K., & Kesharwani, R. K. (2020). *Nutraceuticals and Dietary Supplements: Applications in Health Improvement and Disease Management*. CRC Press. ISBN: 9781771888738.

Kim, H., Choi, H. K., Moon, J. Y., Kim, Y. S., Mosaddik, A., & Cho, S. K. (2011). Comparative antioxidant and antiproliferative activities of red and white pitayas and their correlation with flavonoid and polyphenol content. *Journal of Food Science*, *76*(1), C38–C45.

Kodagoda, K. H. G. K., & Marapana, R. A. U. J. (2017). Utilization of fruit processing by-products for industrial applications: A review. *International Journal of Food Science and Nutrition*, *2*(6), 24–30.

Kugler, F., Stintzing, F. C., & Carle, R. (2012). Evaluation of the antioxidant capacity of betalainic fruits and vegetables. *Journal of Applied Botany and Food Quality*, *81*(1), 69–76.

Kumar, S. B., Issac, R., & Prabha, M. L. (2018). Functional and health-promoting bioactivities of dragon fruit. *Drug Invention Today*, *10*, 3307–3310.

Kusmiyati, Tirtawati, G. A., & Alow, G. (2019). Dragon fruit (*Hylocereus polyrhizus*) increases hemoglobin levels in pregnant mothers. In *Proceedings of the 2nd Manado Health Polytechnic International Conference* (pp. 189–192).

Le Bellec, F., & Vaillant, F. (2011). Pitahaya (pitaya) (*Hylocereus spp.*). In *Postharvest Biology and Technology of Tropical and Subtropical Fruits* (pp. 247–273). Woodhead Publishing.

Le Bellec, F., Vaillant, F., & Imbert, E. (2006). Pitahaya (*Hylocereus spp.*): A new fruit crop, a market with a future. *Fruits*, *61*(4), 237–250.

Le, T. L., Huynh, N., & Quintela-Alonso, P. (2021). Dragon fruit: A review of health benefits and nutrients and its sustainable development under climate changes in Vietnam. *Czech Journal of Food Sciences*, *39*(2), 71–94.

Li, Y., Zhang, J. J., Xu, D. P., Zhou, T., Zhou, Y., Li, S., & Li, H. B. (2016). Bioactivities and health benefits of wild fruits. *International Journal of Molecular Sciences*, *17*(8), 1258.

Liaotrakoon, W. (2013). Characterization of dragon fruit (*Hylocereus spp.*) components with valorization potential (Doctoral dissertation, Ghent University).

Liaotrakoon, W., De Clercq, N., Van Hoed, V., & Dewettinck, K. (2013). Dragon fruit (*Hylocereus spp.*) seed oils: Their characterization and stability under storage conditions. *Journal of the American Oil Chemists' Society*, *90*(2), 207–215.

Lichtenzveig, J., Abbo, S., Nerd, A., Tel-Zur, N., & Mizrahi, Y. (2000). Cytology and mating systems in the climbing cacti *Hylocereus* and *Selenicereus*. *American Journal of Botany*, *87*(7), 1058–1065.

Lim, H. K., Tan, C. P., Karim, R., Ariffin, A. A., & Bakar, J. (2010). Chemical composition and DSC thermal properties of two species of *Hylocereus* cacti seed oil: *Hylocereus undatus* and *Hylocereus polyrhizus*. *Food Chemistry*, *119*(4), 1326–1331.

Luo, H., Cai, Y., Peng, Z., Liu, T., & Yang, S. (2014). Chemical composition and *in vitro* evaluation of the cytotoxic and antioxidant activities of supercritical carbon dioxide extracts of pitaya (dragon fruit) peel. *Chemistry Central Journal*, *8*(1), 1–7.

Lwin, T., Myint, C. Y., Win, H. H., Oo, W. W., & Chit, K. (2020). Formulation and evaluation of lipstick with betacyanin pigment of *Hylocereus polyrhizus* (Red Dragon Fruit). *Journal of Cosmetics, Dermatological Sciences and Applications*, *10*(4), 212–224.

Madane, P., Das, A. K., Nanda, P. K., Bandyopadhyay, S., Jagtap, P., Shewalkar, A., & Maity, B. (2020). Dragon fruit (*Hylocereus undatus*) peel as antioxidant dietary fiber on quality and lipid oxidation of chicken nuggets. *Journal of Food Science and Technology*, *57*(4), 1449–1461.

Mercado-Silva, E. M. (2018). Pitaya—*Hylocereus undatus* (Haw). In *Exotic Fruits* (pp. 339–349).

Muhialdin, B. J., Kadum, H., Zarei, M., & Hussin, A. S. M. (2020). Effects of metabolite changes during lacto-fermentation on the biological activity and consumer acceptability for dragon fruit juice. *LWT*, *121*, 108992.

Murugesu, S., Ariffin, A. A., Ping, T. C., & Chern, B. H. (2013). Physicochemical properties of oil extracted from the hot and cold extracted red pitaya (*Hylocereus polyrhizus*) seeds. *Journal of Food Chemistry and Nutrition*, *1*(2), 78–83.

Nurliyana, R. D., Syed Zahir, I., Mustapha Suleiman, K., Aisyah, M. R., & Kamarul Rahim, K. (2010). Antioxidant study of pulps and peels of dragon fruits: A comparative study. *International Food Research Journal*, *17*(2), 367–375.

Nurmahani, M. M., Osman, A., Hamid, A. A., Ghazali, F. M., & Dek, M. S. (2012). Antibacterial property of *Hylocereus polyrhizus* and *Hylocereus undatus* peel extracts. *International Food Research Journal*, *19*(1), 77–84.

Nurul, S. R., & Asmah, R. (2014). Variability in nutritional composition and phytochemical properties of red pitaya (*Hylocereus polyrhizus*) from Malaysia and Australia. *International Food Research Journal, 21*(4), 1689–1697.

Oz, A. T., & Kafkas, E. (2017). Phytochemicals in fruits and vegetables. In V. Waisundara (Ed.), *Superfood and Functional Food* (pp. 75–184). London: IntechOpen.

Padmavathy, K., Sivakumari, K., Karthika, S., Rajesh, S., & Ashok, K. (2021). Phytochemical profiling and anticancer activity of dragon fruit *Hylocereus undatus* extracts against human hepatocellular carcinoma cancer (HepG2) cells. *International Journal of Pharmaceutical Science & Research, 12*(5), 2770–2778.

Panchal, J. B., Gaikwad, R. S., Dhemre, J. K., & Chavan, U. D. (2018). Studies on preparation and storage of jelly from dragon fruit (*Hylocereus undatus*). *Journal of Pharmacognosy and Phytochemistry, 7*(4), 2648–2655.

Pansai, N., Chakree, K., Takahashi Yupanqui, C., Raungrut, P., Yanyiam, N., & Wichienchot, S. (2020). Gut microbiota modulation and immune boosting properties of prebiotic dragon fruit oligosaccharides. *International Journal of Food Science & Technology, 55*(1), 55–64.

Patel, S. K., & Ishnava, K. B. (2019). *In-vitro* antioxidant and antimicrobial activity of fruit pulp and peel of *Hylocereus undatus* (Haworth) Britton and Rose. *Asian Journal of Ethnopharmacology and Medicinal Foods, 5*(2), 30–34.

Patel, S., & Goyal, A. (2012). The current trends and future perspectives of prebiotics research: A review. *3 Biotech, 2*, 115–125.

Pawde, S., Talib, M. I., & Parate, V. R. (2020). Development of fiber-rich biscuit by incorporating dragon fruit powder. *International Journal of Fruit Science, 20*(sup3), S1620–S1628.

Perez G, R. M., Vargas S, R., & Ortiz H, Y. D. (2005). Wound healing properties of *Hylocereus undatus* on diabetic rats. *Phytotherapy Research: An International Journal Devoted to Pharmacological and Toxicological Evaluation of Natural Product Derivatives, 19*(8), 665–668.

Perween, T., Mandal, K. K., & Hasan, M. A. (2018). Dragon fruit: An exotic super fruit of India. *Journal of Pharmacognosy and Phytochemistry, 7*(2), 1022–1026.

Phebe, D., Chew, M. K., Suraini, A. A., Lai, O. M., & Janna, O. A. (2009). Red-fleshed pitaya (*Hylocereus polyrhizus*) fruit color and betacyanin content depend on maturity. *International Food Research Journal, 16*(2), 233–242.

Polturak, G., & Aharoni, A. (2018). "La Vie en Rose": Biosynthesis, sources, and applications of betalain pigments. *Molecular Plant, 11*(1), 7–22.

Poolsup, N., Suksomboon, N., & Paw, N. J. (2017). Effect of dragon fruit on glycemic control in prediabetes and type 2 diabetes: A systematic review and meta-analysis. *PloS One, 12*(9), e0184577.

Quiroz-Iturra, L., Rosas-Saavedra, C., & Klein, C. S. (2017). Functional fruits through metabolic engineering. In N. Shiomi & V. Waisundara (Eds.), *Superfood and Functional Food: The Development of Superfoods and their Roles as Medicine* (pp. 47–74).

Ramírez-Truque, C., Esquivel, P., & Carle, R. (2011). Neutral sugar profile of cell wall polysaccharides of pitaya (*Hylocereus* sp.) fruits. *Carbohydrate Polymers, 83*(3), 1134–1138.

Rane, B. R., Ahirrao, R. A., Gujarathi, N. A., Jain, A. S., & Keservani, R. K. (2023). Pathophysiology and application of herbs in the metabolic syndrome. In R. K. Keservani, D. Yadav, R. K. Keservani, S. Singh, & S. Kumar (Eds.), *The Metabolic Syndrome: Dietary*

Supplements and Food Ingredients (pp. 199–230). Apple Academic Press, CRC Press, Taylor & Francis Group. ISBN: 978-1-77491-111-2.

Rane, B. R., Bharath, M. S., Patil, R. R., Keservani, R. K., & Jain, A. S. (2020b). Novel approaches in nutraceuticals. In R. K. Keservani, R. K. Keservani, A. K. Sharma (Eds.), *Enhancing the Therapeutic Efficacy of Herbal Formulations Through Novel Drug Delivery Systems* (pp. 241–266). IGI Global International Publisher. ISBN13: 9781799844532.

Rane, B. R., Patil, A. S., Keservani, R. K., & Jain, A. S. (2020a). Novel approaches in herbal formulation. In R. K. Keservani, R. K. Keservani, A. K. Sharma (Eds.), *Enhancing the Therapeutic Efficacy of Herbal Formulations Through Novel Drug Delivery Systems* (pp. 43–68). IGI Global International Publisher. ISBN13: 9781799844532.

Rane, B. R., Tadavi, S. A., & Keservani, R. K. (2020c). Naturopathy. In A. K. Sharma, R. K. Keservani, & S. P. Gautam (Eds.), *Herbal Product Development* (pp. 321–347). Apple Academic Press, CRC Press, Taylor & Francis Group. ISBN: 9781771888776.

Rohin, M. A. K., Abu Bakar, C. A., & Ali, A. M. (2014). Isolation and characterization of oligosaccharides composition in organically grown red pitaya, white pitaya, and papaya. *International Journal of Pharmacy and Pharmaceutical Sciences*, *6*(S2), 131–136.

Sari, A. R., & Hardiyanti, R. (2013). Antioxidant level and sensory of dragon fruit (*Hylocereus undatus*) peel tea infusion made by partially fermented process. *Agroindustrial Journal*, *2*(1), 63.

Senadheera, S. P. N. M. K., & Abeysinghe, D. C. (2015). Bioactive compounds and total antioxidant capacity of different tissues of two pitaya (dragon fruit) species grown in Sri Lanka. *Journal of Food and Agriculture*, *8*(1, 2), 33–40.

Septembre-Malaterre, A., Remize, F., & Poucheret, P. (2018). Fruits and vegetables as a source of nutritional compounds and phytochemicals: Changes in bioactive compounds during lactic fermentation. *Food Research International*, *104*, 86–99.

Sharma, A. K., Keservani, R. K., & Gautam, S. P. (2020). *Herbal Product Development*. Apple Academic Press, CRC Press, Taylor & Francis Group. ISBN: 9781771888776.

Singh, S., & Immanuel, G. (2014). Extraction of antioxidants from fruit peels and its utilization in paneer. *Journal of Food Processing & Technology*, *5*(7), 1000349.

Sonawane, M. S. (2017). Nutritive and medicinal value of dragon fruit. *Asian Journal of Horticulture*, *12*(2), 267–271.

Sudha, K., Baskaran, D., Ramasamy, D., & Siddharth, M. (2017). Evaluation of functional properties of *Hylocereus undatus* (White dragon fruit). *International Journal of Agricultural Science and Research*, *7*(5), 451–456.

Swarup, K. R. A., Sattar, M. A., Abdullah, N. A., Abdulla, M. H., Salman, I. M., Rathore, H. A., & Johns, E. J. (2010). Effect of dragon fruit extract on oxidative stress and aortic stiffness in streptozotocin-induced diabetes in rats. *Pharmacognosy Research*, *2*(1), 31.

Tang, P. Y., Wong, C. J., & Woo, K. K. (2011). Optimization of pectin extraction from peel of dragon fruit (*Hylocereus polyrhizus*). *Asian Journal of Biological Sciences*, *4*(2), 189–195.

Tenore, G. C., Novellino, E., & Basile, A. (2012). Nutraceutical potential and antioxidant benefits of red pitaya (*Hylocereus polyrhizus*) extracts. *Journal of Functional Foods*, *4*(1), 129–136.

Vaillant, F., Perez, A., Davila, I., Dornier, M., & Reynes, M. (2005). Colorant and antioxidant properties of red-purple pitahaya (*Hylocereus* sp.). *Fruits*, *60*(1), 3–12.

Verma, D., Yadav, R. K., Rani, M. Y. B., Punar, S., Sharma, A., & Maheshwari, R. K. (2017). Miraculous health benefits of exotic dragon fruit. *Research Journal of Chemical and Environmental Sciences*, *5*(5), 94–96.

Wall, M. M., & Khan, S. A. (2008). Postharvest quality of dragon fruit (*Hylocereus* spp.) after X-ray irradiation quarantine treatment. *HortScience*, *43*(7), 2115–2119.

Wichienchot, S., Jatupornpipat, M., & Rastall, R. A. (2010). Oligosaccharides of pitaya (dragon fruit) flesh and their prebiotic properties. *Food Chemistry*, *120*(3), 850–857.

Widyaningsih, A., Setiyani, O., Umaroh, U., Sofro, M. A. U., & Amri, F. (2017). Effect of consuming red dragon fruit (*Hylocereus costaricensis*) juice on the levels of hemoglobin and erythrocyte among pregnant women. *Belitung Nursing Journal*, *3*(3), 255–264.

Wilson, D. W., Nash, P., Buttar, H. S., Griffiths, K., Singh, R., De Meester, F., Horiuchi, R., & Takahashi, T. (2017). The role of food antioxidants, benefits of functional foods, and influence of feeding habits on the health of the older person: An overview. *Antioxidants*, *6*(4), 81.

Wu, L. C., Hsu, H. W., Chen, Y. C., Chiu, C. C., Lin, Y. I., & Ho, J. A. A. (2006). Antioxidant and antiproliferative activities of red pitaya. *Food Chemistry*, *95*(2), 319–327.

Wybraniec, S., & Mizrahi, Y. (2002). Fruit flesh betacyanin pigments in *Hylocereus* cacti. *Journal of Agricultural and Food Chemistry*, *50*(21), 6086–6089.

Yee, L. P., Ping, T. C., Kui, L. P., & Wah, C. S. (2017). Application of red pitaya powder as a natural food colourant in fruit pastille. *Jurnal Gizi Klinik Indonesia*, *13*(3), 111–120.

Zaidiyah, A. S., Ayun, S. Q., & Nasution, I. S. (2019). Shelf life estimation of red dragon fruit jam using accelerated shelf life testing (ASLT) method. *IOP Conference Series: Earth and Environmental Science*, *365*, 012029.

Zainoldin, K. H., & Baba, A. S. (2009). The effect of *Hylocereus polyrhizus* and *Hylocereus undatus* on physicochemical, proteolysis, and antioxidant activity in yogurt. *World Academy of Science, Engineering and Technology*, *60*(3), 361–366.

PART III

Nutraceuticals for Disease Management

CHAPTER 14

Emergence of Curcumin as a Disease Management Food Ingredient

RACHNA GUPTA,[1] RAKESH KUMAR AMETA,[2] and PARTH MALIK[3*]

[1]*Department of Biotechnology, Visva-Bharti Shantiniketan, Bolpur, West Bengal, India*

[2]*Sri M. M. Patel Institute of Sciences and Research, Kadi Sarva Vishwavidyalaya, Gandinhgar, Gujarat, India*

[3]*School of Chemical Sciences, Central University of Gujarat, Gandhinagar, Gujarat, India*

*Corresponding author.

ABSTRACT

Curcumin, the major polyphenolic constituent of turmeric, is a prominent constituent of the ginger family Zingiberaceae and is widely recognized for its wide spectrum of bioactivities. For more than 5 decades, the antioxidant, anti-inflammatory, anti-diabetic, and anti-cancer attributes of curcumin have been receiving overwhelming attention from nutritionists and biochemists. Rigorous studies in animal models have illustrated the vicious immunomodulatory attributes of curcumin, typically highlighted by distinct controls on proteins and transcription factors at the same instant. A unanimous concern pointed out herein pertains to the poor aqueous solubility of this polyphenol, impairing its physiological absorption to a great extent. However, proven non-toxic intake till 15 g/Kg level in mice models has manifolded the interest in curcumin bioactivities. There is no dearth herewith of the bioavailability augmenting approaches for curcumin aimed at its delivery through altered carriers, such as oil-in-water nanoemulsions, nanoparticle encapsulation, nanofiber adsorption, and several others. Stress-driven complications in

Functional Compounds and Foods of Plant Origin:Krishan Dutt Sharma, Satish Kumar, Rakesh Sharma, and Raj K. Keservani (Eds.)

our present routine, have been a major cause of accumulating physical and psychological deviations. To cope well with the changing food intake on advancing age, it is a pre-requisite to include curcumin as a dietary ingredient, which could mitigate the free radical damage and delay aging-induced wearing of body tissues and organs. With such insights in mind, this chapter sheds light on the development of curcumin as a potential food ingredient via distinct chemical environments.

14.1 INTRODUCTION

Curcumin usage as a dietary spice and curing agent for non-communicable disorders like cardiovascular disease (CVD), diabetes, metabolic syndrome, arthritis, and others has been in practice for more than centuries in Asian countries. However, the biological attributes of this natural polyphenol were not in the public domain till the late 20th century. The first documentation of curcumin bioactivities was reported by Schraufstatter and associates in their 1949 Nature publication wherein significant antibacterial activities were reported against the *Staphylococcus aureus*, *Salmonella paratyphi,* and *Mycobacterium tuberculosis* (Schraufstatter and Bernt, 1949). The journey progressed at a moderate rate till the next two decades wherein only five major works on curcumin bioactivities got published. It was in the 1970s that the pioneering efforts of three independent research groups made curcumin the talk of scientific investigation, with the discovery of diversified characteristics comprising, cholesterol-lowering, antioxidant, anti-inflammatory, and anti-diabetic responses (Patil and Srinivasan, 1971; Sharma, 1976; Srinivasan, 1972). Subsequently, Kuttan et al. (1987) reported curcumin anti-cancer potential *via* distinct *in vitro* and *in vivo* conditions. The anti-inflammatory traits of curcumin *via* the suppressed activity of nuclear factor-κB along with a demonstration of the mechanism have been reported by Singh and Aggarwal (1995).

With gradual progress, interest in curcumin chemistry grew by leaps and bounds, as inferred by the staggering number of articles (> 4000) being listed in the National Institute of Health PubMed database. Significant studies have unanimously elucidated curcumin biological potency in manifold animal models of human diseases. The physiological non-toxicity of this plant polyphenol has paved way for US Food and Drug Administration approval as a "generally regarded as safe" (GRAS) compound. A persisting concern about the manifold curcumin biological activities has been its relatively low systemic bioavailability caused by its structural hydrophobicity. Flanked by terminal methoxyl groups, the centrally positioned symmetric $-CH_2$ group together contribute to its hydrophobic behavior (Figure 14.1), that is, terminal methoxyl groups and the centrally positioned symmetric $-CH_2$ group together contribute to its hydrophobic behavior.

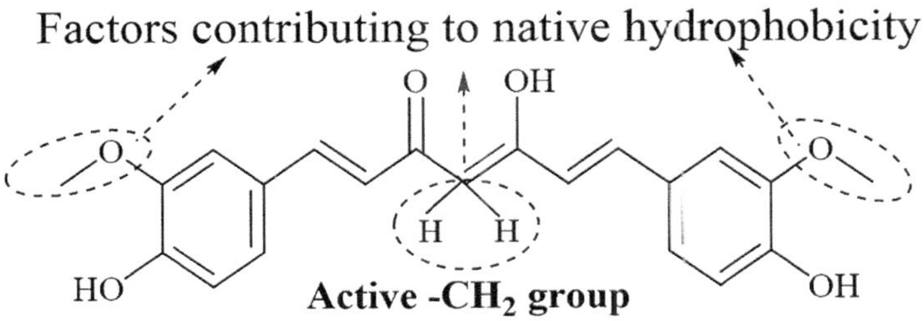

FIGURE 14.1 Hydrophobic receptivity of curcumin with the terminal $-OCH_3$ and the central $-CH_2$ group.

Though, at the electronic scale, the compound is a continuous string of electrons, characterized by the pi-conjugation of phenyl rings and extended alternate double bond placement of its 7C-linkage. Studies have convincingly described the implicit role of this 7C-linkage and the terminal phenyl rings for the constitutive curcumin bioactivities (belonging to the general class of diarylheptanoids that has manifold useful pharmacological properties), but even then, a clear-cut structure-function understanding of curcumin does not have any consensus (Sharifi et al., 2020; Shen et al., 2016). Analysis reveals a substantial extent of consumed curcumin as remaining unabsorbed in the stomach from where it is progressively discarded through feces. Studies aimed toward understanding curcumin therapeutic attributes reveal a beneficial deliverance at micro-molar extents although, the metabolism analysis elucidates its nano-molar plasma ranges on being orally consumed. These observations infer a profuse curcumin metabolism within the intestine and liver, owing to which a very less amount of delivered dosage is available to the diseased cells and tissues (Dei Cas and Ghidoni, 2019; Stohs et al., 2020). So, targeted and stealth curcumin delivery is the real challenge that minimizes its early metabolism and utilization at undesired physiological activities.

Indeed, modifications for better curcumin intake are swiftly modulating its bioactivities, wherein concomitant intake of piperine revealed an astounding more than 2000% increment in localized curcumin bioavailability (Shoba et al., 1998). Of more interest is the higher piperine normalization (for improved curcumin activities) in humans than in rats (Anand et al., 2007). Though traditionally curcumin has been used in almost every household (as a spice ingredient) to a significant extent its use as a sheer therapeutic agent has never surfaced. The traditional beliefs of applying curcumin-mustard oil combination over a vicious wound or cut injury to consuming turmeric-laced milk for throat and cough infections also deserve a mention in shaping up the present status of curcumin. To understand and address the quick degradation of curcumin from physiological boundaries, it is pertinent to know about the metabolism of curcumin and the various intermediates formed in its absorption. Erstwhile this, the pH labile nature of this polyphenol is well-known with a spontaneous hydrolytic degradation at > 7 pH (Barclay et al., 2000). Therefore, it must screen the *in vitro* pH sensitivity of carrier-bound curcumin before it is *in vivo* administered. Apart from this, due caution must be exercised that curcumin binding with the carrier is stabilized via moderate binding forces (BF) so that the release of curcumin happens spontaneously with the changes in pH and redox environment. A too strong binding with the carrier molecules could induce an irreversible structural loss owing to which a proportionate functionality loss is inevitable. Another noteworthy aspect of curcumin pertains to its concentration-dependent anti- and pro-oxidant responses, and such uncertainties become more pronounced at similar concentrations, that is, 5–50 μM of protective and toxic responses. Despite these concerns, the incentive for using curcumin stems from its remarkable ability to select the tumor cells whereby normal cells receive much less concentrations that could induce toxicity. It is pertinent to mention here that the anti-cancer activity of curcumin manifests from its intracellular reactive oxygen species (ROS) generation that leads to apoptosis *via* autophagic induction. So, the prooxidant activities of curcumin are the sources of its anti-cancer activities therefore, delivery attempts must be targeted for a maximum internalization of delivered extent so that the ROS enhancement exclusively affects only the tumor cells (Malik and Mukherjee, 2014). We begin the discussion of curcumin food ingredients (in further sections) by discussing its distribution in turmeric, extraction methods available (optimum recovery solvents), and the spectrum of curcumin bioactivities for its beneficial *in vivo* activities. Thereafter, the various routes of curcumin delivery for maximizing its sustainability and longevity are included wherein intent similar to food component is maximized.

14.2 DISTRIBUTION, EXTRACTION, AND METABOLISM OF CURCUMIN

Turmeric owes its antioxidant characteristics to its polyphenolic constituents, together referred to as curcuminoids, of which curcumin is the major component with ~77% occurrence probability. The other two sister fractions are demethoxycurcumin (DMC) and bisdemethoxycurcumin (BDMC) with 17% and 3% prevalence, respectively. The yellowish-orange appearance of turmeric substantially emanates from its constitutive curcuminoids, manifesting their activities as pigments (Sharma, 1976; Sharifi et al., 2020). In line with the quantification approaches aimed at isolation and characterization of phenolic compounds, curcumin separation from a blend of turmeric polyphenols is accomplished via solvent extraction succeeded by column chromatography. Briefly, 20 g ground commercial turmeric is magnetically stirred with 50 mL dichloromethane along with heating at reflux for an hour. The mixture is thereafter suction filtered followed by concentration of filtrate at 50°C. The oily reddish-yellow remainder is triturated with 20 mL hexane, retrieving the resultant solid (~ 0.57 g) through suction filtration. Thereafter, the crude material obtained was dissolved in a minimum quantity of 99% dichloromethane and 1% methanol on a volume-by-volume basis and was loaded over a 30 g silica gel packed column. The column was eluted with the same solvent after which the Thin Layer Chromatography analysis separated the three constituents corresponding to 0.49, 0.22, and 0.085 resolution frequencies, respectively. The fractions comprising the least polar components were mixed before water bath aided solvent elimination to retrieve 0.16 g yellow solid (Anderson and Ng, 2000). To date, ethanol is perhaps the most suitable solvent recommended for curcumin extraction, using Soxhlet extraction, ultrasonic extraction, and microwave treatment (Paulucci et al., 2013; Li et al., 2014). Up to 1 ng/mL, detection is made *via* 400–450 nm fluorescent excitation whereas, higher extents are ascertained using 350–450 nm absorption.

Similar to many conventional carotenoids, curcumin is primarily quantified using high-pressure liquid chromatography (HPLC) equipped with reverse-phase C18 columns as the stationary phase and acetonitrile-water or chloroform-methanol gradients as mobile phases (Ali et al., 2014; Lee et al., 2013; Paramasivam et al., 2009). Low curcumin extents in bio-fluids can be screened using HPLC and liquid chromatography (LC) coupled mass spectrometry (LC/MS), aiding in monitoring curcumin pharmacokinetics, bio-distribution, and metabolism (Pentea et al., 2015).

The physiological activity of curcumin is affected by multiple factors, ranging from changes in pH to differential metabolizing extents. The functional expression (Figure 14.2) is an implicit function of the keto-enol tautomerism ability of this plant polyphenol, wherein < 7 pH is the optimum condition for keto form whereas > 7 pH favors the enol form (Lestari and Indrayanto, 2014). The mild acidic character of this polyphenol is described by three pKa (dissociation constants), the more prominent being 7.7–8.5 and 8.5–10.4 ranged phenolic –OH groups whereas that of enolic proton ranges within 9.5–10.7 (Priyadarsini, 2014). Overall, the investigations have shown that enol is majorly prevalent under alkaline conditions, contrary to keto in acidic conditions. Higher solution stability of enol form in the solution phase contributes to its H^+ donating nature.

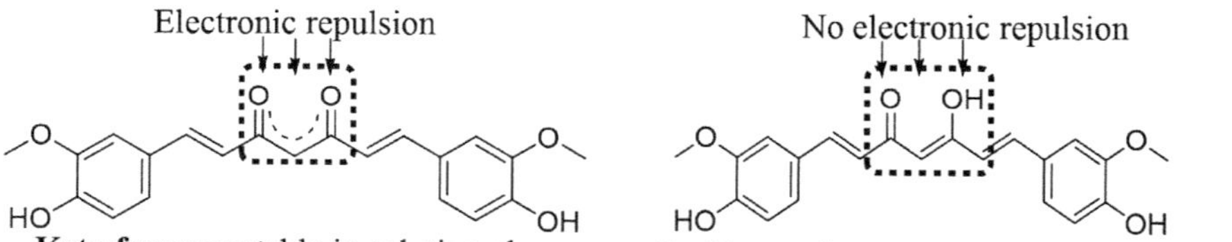

Keto form: unstable in solution phase
(a)

Enol form: H^+ rearrangement conferred stability
(b)

FIGURE 14.2 The distinctions between keto and enol isomers of curcumin arise from the adjacently positioned oxygen atoms, which contribute to their instability in the solution phase.

Ironically, despite being chemically stable within (1–6) pH range, the water solubility of curcumin is practically nil under these conditions. This paradox hinders an efficient absorption of native curcumin owing to which knowledge about the biological activities of its different metabolism products assumes significance. Physiological pH causes curcumin autoxidation to form a series of bicyclopentadione products having oxygenated and double cyclized 7-C. It is pertinent to note here that early curcumin autoxidation products are responsible for curcumin activity on DNA topoisomerase, a characteristic mechanism of several chemotherapeutic agents. A minor pathway of this autoxidation is the cleavage in the alkaline conditions, causing degradation to trans-6-(4'-hydroxy-3'-methoxyphenyl)-2,4-dioxo-5-hexanal and subsequently into vanillin, ferulic acid and feruloyl-methane (Figure 14.3). A careful look at the structures of these metabolites infers their augmenting antioxidant attributes with an invariable possession of –OH groups accompanied with diffused electronic movements and delocalized pi-conjugation, that is, the presence of gapped –OH groups in all these compounds along with alternately positioned oxygen contributes to a curcumin similar antioxidant nature of these compounds.

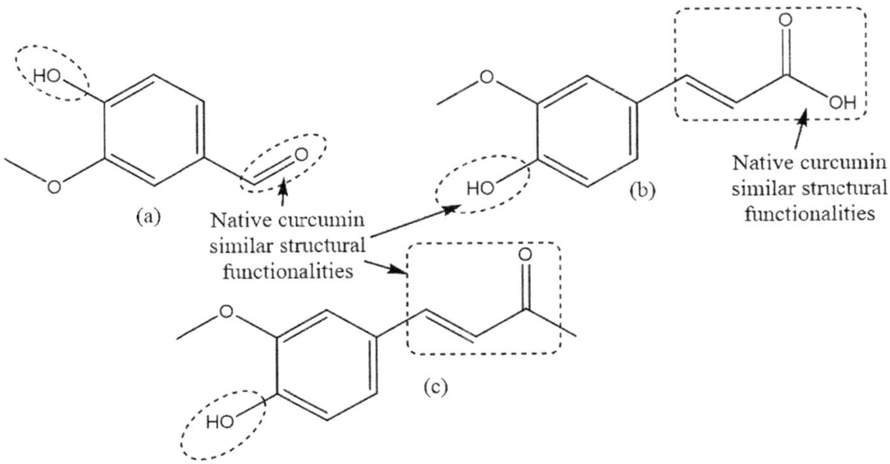

FIGURE 14.3 Chemical structures of alkaline autoxidation of curcumin [(a) vanillin; (b) ferulic acid; and (c) feruloyl-methane].

Accordingly, these compounds exhibit free radical scavenging and cholesterol dilution by their absorption and chemical degradation. Unaltered –OH groups are highly needed for the native curcumin functional activities and stability is indeed affected by their dissociation. Oxidative degradation and modification are likely to be induced *via* light exposure, lending a photosensitive functioning to curcumin. The degradation reactions are widely responsible for changing the native curcumin structure and properties, inducing sharp deterioration of the native pharmacokinetics and pharmacodynamics.

Phase I and II biotransformation are highly susceptible for a curcumin functioning with liver and intestine along with gut micro-biota as the exclusive metabolism sites (Luca et al., 2020). The double bonds in curcumin structure are afterward reduced in enterocytes and hepatocytes *via* reductase activity to yield dihydrocurcumin, tetrahydrocurcumin, hexahydrocurcumin, and octahydrocurcumin (Figure 14.4(a)). A glance at the structures of these metabolites to reveals an intact β-diketonic backbone with the modifications in terminal phenyl rings. So, it is inferred from these structures that metabolites of curcumin are likely to express the majority of curcumin bioactivities. However, an urgent need to conduct more authentic and formidable studies to study the pH effects on the stability of these intermediates. Such an analysis would be beneficial towards understanding the direct assimilation of these compounds wherein suitable modifications could be made in the native curcumin structure. Although this is not a 100% untouched aspect, with several of the analogs being known, it is needed for an improved structure-function understanding of curcumin and its metabolites.

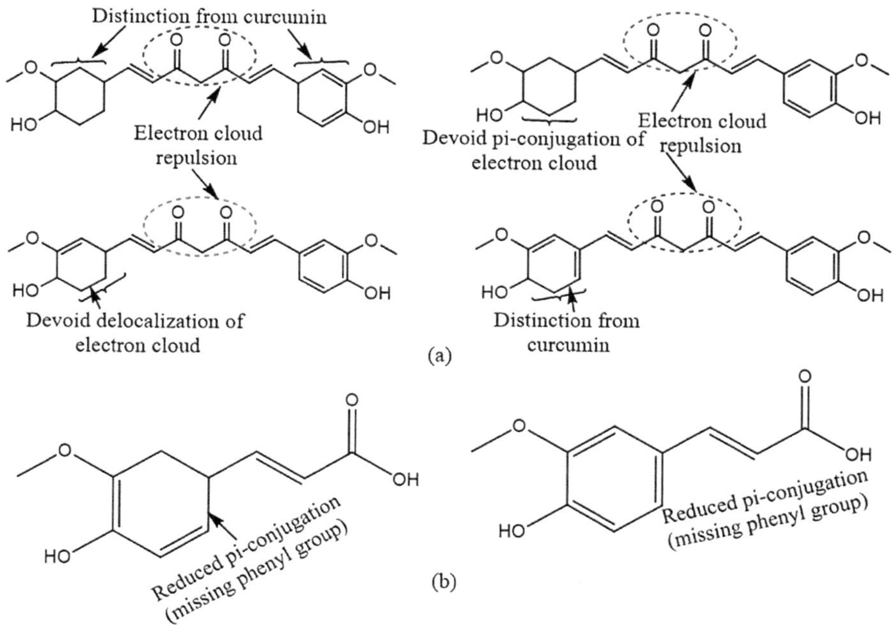

FIGURE 14.4 Chemical structures of (a) enterocytes and hepatocyte reduced curcumin metabolites; and (b) secondary biliary metabolism of curcumin.

The secondary biliary metabolism of curcumin yields dihydro-ferulic acid and ferulic acid (Figure 14.4(b)), yet again exhibiting polyphenolic functionality with a stronger antioxidant nature (Esatbeyoglu et al., 2012; Sharma et al., 2001). Phase II metabolism of curcumin as well as its phase I metabolites is usually very active and operates *via* intestinal and hepatic cytosol, involving phenolic site conjugation with glucuronic aid and sulfate. Curcumin is sulfated in the cytosol by the activities of SULTs, majorly SULT1A1 and SULT1A3 while the UGTs catalyze its glucuronidation within the intestinal and hepatic microsomes. An alternative course to curcumin metabolism is *via* intestinal microbiota such as *Escherichia coli* and *Blautia* sp., with the former being activated by NADPH dependent reductase in a two-step reduction cycle, generating dihydrocurcumin and tetrahydrocurcumin (Hussaninasab et al., 2011). The *Blautia* sp. demethylates curcumin to demethylcurcumin and bis-demthylcurcumin, which are also polyphenolic like the earlier metabolites (Burapan et al., 2017). Thereby, it is not ambiguous to conclude that the bioactivities of curcumin are due to the diverse functional attributes of its metabolites, generated amidst its metabolism.

14.3 A SPECTRUM OF CURCUMIN BIOACTIVITIES

The bioactivities of curcumin substantiate from its pleiotropic functioning through which it simultaneously exercises control on the functioning of proteins, transcription factors, and genes. The essence of curcumin bioactivities manifests from the antioxidant and anti-inflammatory functions whereby the functioning of organs and tissues are rejuvenated and the ROS degradation of physiological events, is prevented. The spectrum of such activities would be better understood through its description at the molecular, cellular, and *in vivo* scales. At the molecular platform, curcumin is known to regulate multiple signaling mediators and the reported targets either via direct binding or by exerting an indirect modulation on their activity. Amongst the indirect targets, the prominent biomolecules include transcription factors, enzymes, inflammatory mediators, protein kinases, drug resistance proteins, adhesion molecules, growth factors, receptors, cell-cycle regulatory proteins, cell-survival proteins, chemokines, and chemokine receptors. Direct targets at the molecular level comprise inflammatory regulators, cell-survival proteins, protein kinases, protein reductases, histone acetyltransferases, histone decetylase, glyoxylase I, xanthine oxidase, proteasomes, HIV1 integrase, HIV1 protease endoplasmic reticulum Ca^{2+} ATPase, DNA methyltransferase 1, carrier proteins and metal ions. The most important curcumin targets at the molecular level include the pro-inflammatory transcription factors, such as NF-κB, activator protein 1, signal transducer, and activator of transcription (STAT) proteins (Shishodia et al., 2007). The importance of these transcription factors is well-versed through the regulation of gene expression that controls tumorigenesis, cell survival, proliferation, invasion, and angiogenesis. Curcumin extensively exercises a negative control on the transcription factors, with the major targets being protein kinases (Shishodia et al., 2007). Such activities are expressed in course of anti-cancer expression of curcumin, with downregulation of epidermal growth factor receptor

(EGFR) and extracellular signal-regulated kinase ½ (ERK1/2) in pancreatic and lung cancer cells (Lev-Ari et al., 2006). Similarly, malignant glioma cells are retarded in their growth and metastasis by curcumin inhibited Phosphatidylinositol 3 kinase/AKT pathway (Aoki et al., 2007). Other prominent protein kinases inhibited by curcumin include phosphorylase kinase, protein kinase C, protamine kinase, autophosphorylation activated protein kinase, and tyrosine kinase (Aggarwal et al., 2006; Shoshodia et al., 2007).

The chemical actions of curcumin on several inflammation-promoting molecules majorly, including NF-κB and STAT3 transcription factors, signaling molecules, cell-cycle regulatory proteins, and ROS enhancing precursors are well-known. Concerning tackling inflammation, studies have revealed curcumin's ability to moderate an inflammatory response as well as completely act as an anti-inflammatory agent. For instance, in mouse fibroblast cells, curcumin inhibits the phorbol 12-myristate 13-acetate (PMA)-induced inflammation (Gupta et al., 2010). Similarly, curcumin inhibition of pro-inflammatory cytokines has been observed in lipopolysaccharide-stimulated peripheral blood monocytes and alveolar macrophages (Abe et al., 1999). From a structural viewpoint, investigations have observed a characteristic role of hydroxyl group substituted terminal phenyl rings in setting up the anti-inflammatory response. This was counter-screened through a deliberate small-sized alkyl or methoxyl group substitutions on the 3' and 5' phenyl ring positions, through an enhancement in inflammatory response was observed (Nurfina et al., 1997). Similarly, the antioxidant activity of curcumin has been linked with the –OH groups of terminal phenyl groups as well as the central $–CH_2$ group of the β-diketonic backbone. There is a pulsating debate regarding this structure-function correlation and computational studies have comprehensively elucidated that the bond dissociation energy of terminal –OH groups is low as compared to the central $–CH_2$ group. So, despite being more vulnerable, the –OH group of terminal phenyl rings were once touted as chief antioxidant contributors. Nevertheless, H^+ release does happen from $–CH_2$ group even though it may be later in expression. Amongst the major physiological domains of curcumin antioxidant expression, blood plasma and platelets in manifold cell lines have been noticed. Apart from this, curcumin has been shown to entirely inhibit the superoxide anions, hydrogen peroxide, and nitrite radical formation in rat macrophages (Joe and Lokesh, 1994). More recently, the oxidative activation of G proteins in human brain membranes by metallic pro-oxidants like homocysteine and hydrogen peroxide has been observed as being considerably suppressed by curcumin (Jefremov et al., 2007). The actions of curcumin on tumor cells have been noted as the functions of specific cell types, administered dosage, and the typical treatment duration. The anti-cancer actions of curcumin have been observed through its conventional delivery and combined delivery with chemotherapeutic drugs, illustrating the chemo-sensitizing ability of curcumin delivery through nanocarriers of only curcumin as well as its varying stoichiometric combinations with chemotherapeutic drugs. The noted anti-cancer mechanisms of curcumin include inhibition of cell-cycle regulating proteins and activation of pro-apoptotic factors. An important observation noted in the studies of combined curcumin delivery with chemotherapeutic drugs is the ability to specifically select and recognize the tumor cells and enhanced tumor cell internalization. These attributes ensure a targeted expression of chemotherapeutic drugs within the tumor cells only, thereby preventing the normal cells from the possible risks and side effects. This is accomplished through the incentive of non-toxic curcumin nature, whereby there is no restriction in its delivered dosage, so masking of low concentration of chemotherapeutic drugs with a high concentration of curcumin is the practiced mechanism. Our recent compilation of curcumin activities for non-small cell lung cancer treatment has comprehensively demonstrated a curcumin-improved activity of cisplatin, carboplatin, doxorubicin, and paclitaxel (all chemotherapeutic drugs) through moderating the delivered dosages along with enhanced cell death (Malik et al., 2021). Lately, many studies have demonstrated the antimicrobial potential of curcumin whereby the responses towards 14 distinct Candida strains have been screened, and surprisingly, all strains exhibited suppressed growth (Neelofar et al., 2011). A separate study has revealed curcumin improved the response of azole and polyene anti-fungal compounds whereas in cell-culture systems, potent anti-viral responses have been observed (Sharma et al., 2010). Potential curcumin activities demonstrated by *in vitro* cell culture studies comprise wound healing in skin fibroblasts, anti-nociceptive functioning in ganglion neurons, anti-parasitic effect against African trypanosomes, schistosomicidal potential against *Schistosoma mansoni* adult worms, antimalarial and nematocidal activities (Ji and Shen, 2009; Kiuchi et al., 1993; Koide et al., 2002; Magalhães et al., 2009; Yeon et al., 2010).

The investigation of curcumin activities in animals and humans has also been quite significant, illustrating its protective influence against several diseases in particular over the last five decades. The conditions where most of the curcumin benefits have been noted to be actively expressed include hyper-inflammation, diabetes, obesity, cancer prevention, and a combination of multiple organs and general stressful circumstances *viz.*, asthma, fatigue, neuropathic pain, cognitive deficit, memory weakening, morphine addiction, and muscle wasting. The majority of these curcumin immuno-protective activities have been reported or noticed in mice and rats. There is a direct correlation between these activities with adequate food assimilation as most of these problems are associated with aggravated stress and are quenched by the antioxidant ability of curcumin. As a result, there is a much lower formation of free radicals, and the bioavailability of nutrients taken via food is enhanced. One characteristic attribute amongst these is the moderation of intestinal inflammation in multidrug-resistant gene-devoid mice, eventually resulting in the formation of colitis (Muriithi et al., 2020). Similarly, curcumin administration suppressed colitis aggravations in dinitrobenzene sulfonic acid-administered mice. Likewise, curcumin administration to trinitrobenzene sulfonic acid-fed mice relieved diarrhea and colonic hierarchy disruption (Salh et al., 2003; Ukil et al., 2003). Multiple studies have also screened the anti-diabetic activities of curcumin, wherein Sprague Dawley rats were induced with insulin resistance (IR) and subsequent manifestation of Type-II diabetes mellitus (TDM), through administering a high-fat diet for two to two and a half months. Intake of curcumin during the last 15 days of this trial revealed an antihyperglycemic response with improved insulin feeling, exhibiting anti-inflammatory and anti-lipolytic functions. An added incentive for these curcumin activities is shown by the lack of modulation of similar activities in normal cells (El-Moselhy et al., 2011; Seo et al., 2008). For instance, no glucose-lowering response of curcumin is noticed in normal cells contrary to the TDM-infected mice models. Studies screening the curcumin-controlled obesity in mice models observed that combining 500 mg/Kg curcumin with the high-fat diet of mice caused no change in food intake but surprisingly reduced the gain in body weight, adipose tissue, and micro-vessel density. These varied patterns were accompanied by reduced vascular endothelial growth factor and its receptor-2, peroxisome proliferator-activated receptor-γ (PPARγ), and CCAT/enhancer-binding protein α, activities. Hence, it is quite evident that curcumin intake with the food regulates the needed fat absorption by the body tissue and mediated a swift elimination of excess fat by the metabolic pathways of the physiological system.

Apart from the above, curcumin has also been reported to induce a protective influence on neurological conditions, as well as chemopreventive and chemotherapeutic responses in cancers and psychiatry disorders. The various cancers towards which curcumin has been found effective include the colon, esophageal, lung, kidney, stomach, liver, mouth, breast, bladder, leukemia, skin, small intestine, breast, pancreatic, and prostate (Azuine et al., 1992;

Chuang et al., 2000; Hecht et al., 1999; Huang et al., 1994, 1998; Ikezaki et al., 2001; Limtrakul et al., 1997; Narayanan et al., 2009; Okazaki et al., 2005; Purkayastha et al., 2009; Sindhwani et al., 2001; Swamy et al., 2008; Ushida et al., 2000). The salient findings of curcumin-treated abnormal neurological conditions are listed in Table 14.1, characteristics amongst which are Alzheimer's disease, Parkinson's disease, and epilepsy (Garcia-Alloza et al., 2007; Gupta et al., 2009; Mehla et al., 2010; Yang et al., 2005; Yu et al., 2010; Zbarsky et al., 2005). It is pertinent to note here that animal studies comprise a significant component of therapeutic trials wherein the optimum dosage regimen in humans is decided by a series of optimizations in animal models.

TABLE 14.1 Spectrum of Curcumin Bioactivities Towards Neurological Activities in Various Animal Models

SL. No.	Disease Cured, Dosage, and Animal Models/Alternative Subjects	Key Observation	References
1.	Alzheimer's disease in a mouse model subjected to curcumin treatment for 7 days.	Reduced plaque formation and amyloid beta accumulation.	Garcia-Alloza et al. (2007)
2.	Pretreatment of rats with curcumin protected against pentazole-induced seizure.	Reduced severity of seizures along with less cognitive impairment.	Mehla et al. (2010)
3.	Kainic acid treatment induced seizures in rats.	Dose-dependent anti-epileptic activity.	Gupta et al. (2009)
4.	Pretreatment with curcumin in rat models.	Improved dopamine levels in striata of rat brain, neuroprotection towards 6-hydroxydopamine via antioxidant activity, and brain penetrating ability.	Zbarsky et al. (2005)
5.	Protection against dopaminergic neurotoxicity induced by 1-methyl-4-phenylpyridinium ion in C57BL/6N mice.	Improved behavioral deficits, enhanced neuron survival in substantia nigra of mouse affected with Parkinson's disease.	Yu et al. (2010)
6.	The mouse model administered curcumin.	Curcumin crossed the blood-brain barrier, reduced amyloid levels, and plaque burden, and protected against Alzheimer's disease.	Yang et al. (2005)

It is important to note that the bioactivities of any phenolic compound (many like curcumin) are an implicit function of native curcumin structure. Of late, interest in curcumin biochemistry correlation with its immunomodulatory attributes has grown by leaps and bounds, with a screening of manifold structurally modified analogs. Such an alternative has been majorly screened via modifications in terminal phenyl group substituents and β-diketonic backbone. To mention a few of these attempts, a study focused on analyzing comparative bis-1,7-(2-hydroxyphenyl)-hepta-1,6-diene-3,5-dione (a BDMC analog) and curcumin towards 1,2-dimethylhydrazine induced colon cancer suppression in male Wistar rats (Devasena et al., 2003). Analysis revealed a similar tumor-suppressing potency of curcumin and BDMC analog, whereby it was concluded that the potential activities are the consequences of the terminal –OH groups and the alternate double bonds in central β-diketonic linkage. The realization and reasons for such inferences could be due to the intact terminal –OH groups along with the central β-diketonic backbone in the BDMC analog and curcumin structures and drew confidence from the similar treatment duration. The study generates significant interest regarding a gradual modulation in β-diketonic backbone by keeping the terminal –OH groups intact and vice versa. It is because the reported low bioavailability of curcumin is due to hydrophobicity aggravating terminal $-OCH_3$ substitution of phenyl rings. Another significant effort whereby curcumin and DMC diminished a Pb-induced memory deficiency in male Wistar rats has been reported by Flora et al. (2012). The investigation works to minimize the Pb impaired neurobehavioral response caused by the Pb interference with N-methyl-D-aspartate (NMDA) receptors in the hippocampal formation and cerebral cortex. Two possible explanations for a curcuminoid inhibited Pb activity, either by scavenging the generated free radicals which reduced the neuron damage or there could be a possibility of curcuminoid engaging Pb which forbade Pb interaction with NMDA receptors. Such specific mechanisms need an urgent compilation to further improve the structure-function understanding of curcumin.

Studies analyzing curcumin potency in humans are well-versed in a screening of preclinical data from several years in the past. To date as many as 55 clinical trials in human subjects have been made, with the majority of conclusions terming curcumin as safe and effective in multiple diseases. The distinctive curcumin activities have been noted in inflammation moderation, cancer treatment, skin, ophthalmic disorders, neurological disorders, diabetic nephropathy, and pain. The first major attempt herewith reported a symptomatic relief following curcumin administration in 62 patients that have external cancerous lesions. Monitoring revealed a persisting protective effect for several months with an exhibition of adverse reaction only in one sufferer (Kuttan et al., 1987). A subsequent major investigation is a 2008 study wherein oral curcumin efficacy was screened in 25 pancreatic cancer sufferers as a part of a phase II clinical trial. In the course of treatment, the patients were administered 8 g of curcumin daily before the disease progression ceased. While the disease staging was done every 60 days, flowing curcumin was detected in glucuronide and sulfate conjugate forms with minor level steady-state extents, inferring an inadequate bioavailability. The study revealed two sufferers with a clinical biological response, one exhibiting a 73% tumor regression while another developed persisting disease stability over 18 months. To match with presumed beliefs, none of the patients exhibited toxicity, inferring nutraceutical (Keservani et al., 2010a, b, 2017, 2020) similar traits of curcumin. In terms of treating cancers, the impact of curcumin is well suited, wherein enhanced activities of several chemotherapeutic drugs have been reported. In a characteristic study among these attempts, the efficacy of curcumin and gemcitabine combined was screened in 17 advanced pancreatic cancer patients. To all patients, 8 g of curcumin was administered through the mouth daily. The gemcitabine was simultaneously administered through an intravenous route at 1000 mg per square meter thrice a week for 4 weeks. Receiving the treatment, five candidates stopped taking curcumin within one to two weeks, due to a stubborn abdominal entirety. In two sufferers, curcumin dosage was reduced to 4 g per day owing to abdominal complications. While one of the 11 remaining fits to analyze sufferers exhibited a partial response, four developed a stable disease while 6 continued to show tumor progression. On the whole, it was observed that 8 g curcumin intake daily in combination with systemic extents of gemcitabine could moderate a high dose of oral curcumin required to attain a systemic effect (Epelbaum et al., 2010). Similar observations characterized a study focused on monitoring the safety and efficacy of curcumin-gemcitabine combined in pancreatic using phase I/II experimental settings. Curcumin administration moderated the gemcitabine toxicity once again, with 8 g daily curcumin extent being deemed safe for pancreatic cancer sufferers. This study analyzed 21 gemcitabine-resistant pancreatic cancer patients. The primary endpoint was safety for phase I trials, while for phase II, it was to analyze the feasibility of oral curcumin intake. While phase I revealed no ill-effects to 8 g extents, the phase II analysis screened 5 patients with all exhibiting an extent between 29 and 412 ng/mL. Similar attributes of curcumin were also observed in colorectal cancer sufferers where an intake of 360 mg on daily basis for 10–30 days, resulted in improved body weight along with a considerable decrement in serum tumor necrosis factor (TNF)-α extents (He et al., 2011). Besides aggravated inflammation and cancer, the major physiological complications studied to have been normalized through curcumin use include moderation of lipid content in sufferers of acute coronary syndrome, abnormal eye conditions *viz*., uveitis, uvea inflammation, infected conjunctiva pain and decreased vision, and pain resulting from oxidative stress. The salient prospects of these studies are described in Table 14.2, and readers are recommended to consult additional explicit literature sources to know about the specific mechanisms involved (Allegri et al., 2010; Alwi et al., 2008; Durgaprasad et al., 2005; Usharani et al., 2008).

TABLE 14.2 Curcumin Bioactivities Towards Ophthalmological and Oxidative Stress Elevated Disorders

SL. No.	Administered Curcumin Dosages with Frequency in Specific Kinds of Disorders	Key Observation	References
1.	75 patients with acute coronary syndrome were administered curcumin at low (15 mg per day), moderate (30 mg per day), and high (60 mg per day), each thrice a day.	Low doses of curcumin were observed as more effective in reducing total cholesterol and low-density lipoprotein (LDL) extents in all the sufferers.	Alwi et al. (2008)
2.	72 sufferers of T2D were administered two capsules containing 150 mg curcumin twice a day.	Standardized curcuminoid administration was observed as effective towards oxidative stress and inflammatory stress markers in T2D patients.	Usharani et al. (2008)
3.	106 uveitis patients from Italy were administered 600 mg curcumin twice a day through oral means.	Well-tolerated curcumin dosages with reduced eye discomfort in more than 80% of patients on a few weeks of treatment.	Allegri et al. (2010)
4.	20 patients were administered 500 mg curcumin with 5 mg piperine, thrice a day for 6 weeks.	Curcumin-piperine combined therapy was well tolerated with a considerable reduction in erythrocyte MDA content and increased GSH secretion in the tropical pancreatitis sufferers.	Durgaprasad et al. (2005)

The above discussion pinpoints and reassures curcumin's potency to mitigate oxidative stress and inflammatory aggravations by relieving aggravated oxidative stress. The edible nature, along with no toxicity at the extent proportionate to needful pharmacological responses, deciphers the nutraceutical equivalent potential of curcumin. The subsequent paragraphs specifically focus on the latest progress made in developing functional foods with curcumin.

14.4 CURCUMIN-BASED FUNCTIONAL FOODS

The usage of curcumin as a food ingredient is aimed at its inclusion to enhance the food quality in terms of nutrition and better absorption *via* minimized free radical generation and conferring an easier absorptive ability to the food texture. The very first aspect which strikes me after knowing this is the water insolubility of curcumin due to which it is not possible to accomplish its additive or synergistic activities. So, before incorporating curcumin as a food ingredient, we must have a thorough knowledge of the final storage stability and the processing which has to be undergone for making that functional combination. Though implicit knowledge about curcumin is known the ultimate fate of its native structure on its inclusion in the modified delivery vehicles or food systems of varying textures and compositions, is a big uncertainty. This is because the metabolic fractions of curcumin under varying chemical environments may retain the curcumin functions but the same is not true about its processing before its consumption. For instance, the effect of heating, storage conditions, and mixing compositions are likely to change the pH of the storage environment owing to which there is a likelihood of structural degradation. Furthermore, even if a sustainable and shelf life enhancing combination is optimized, it mandates elaborate testing on animal models and terminally ill volunteers before being successfully launched in the market on a commercial scale. Thereby, the progress toward commercializing curcumin-supplemented or fortified foods is much less as compared to *in vitro* and *in vivo* analysis of modified formulations on a laboratory scale. Functional foods are distinct from normal food sources in providing multiple health benefits erstwhile of their nutritive value. Eminent examples are vitamins, minerals, probiotics, or fiber-fortified foods. Entities that can be consumed as such also comprise these materials, vis-à-vis fruits, vegetables, nuts, seeds, and grains. In general, the essence other than nutritive significance unlike the common foods relates to a therapeutic significance, such as the oats typically contain a fiber, β-glucan known to moderate inflammation, improve immune functioning and heart performance. Likewise, fruits and vegetables are often enriched with antioxidants in their commercial cadre availability which not only maintains their shelf-life but also guards against excessive free radical generation. The two distinct classes of functional foods include conventional and modified, with the first and major being the vitamins, minerals, antioxidants, and essential fats rich in natural, whole-food ingredients. To date, the most significant progress in developing curcumin-based functional foods has been through encapsulation in emulsified textures, as several edible plant-grade metabolites can be included as emulsifiers (surfactant alternatives). Some studies have no doubt, shown immense potential for adsorption-driven nanoscale distribution of curcumin in the plant extract capped or stabilized, metal or metal oxide nanoparticles (NPs). Further sections describe the salient findings of such attempts.

14.5 EMULSIONS

Emulsions are the chemical environments intended to distribute a molecule, usually hydrophobic across the philic-phobic force gradients. Food-grade emulsions have rapidly progressed as the carriers of curcumin, wherein curcumin is initially dissolved into the oil phase, which is thereafter dispersed in water as the dispersion medium. The process intends to distribute the dispersed phase across the philic-phobic force gradients which improve and sustain a structural expression and hence the bioavailability. It is an obvious concern here that without the application of external energy, how one can homogenize the hydrophilic and hydrophobic phases. This requirement distinguishes the preparation methods of emulsions, characteristically within low and high-energy methods. Though low energy methods need more time to homogenize the dispersed phase their long-term stability is high. Curcumin is a pH-sensitive and thermolabile compound, and therefore, dispersing curcumin across the public-phobic force gradients needs a proper initial optimization as heating the oil and curcumin mixture may induce irreversible damage to the curcumin structure. Furthermore, after being prepared, there is an urgent requirement to choose the optimum storage conditions, as many emulsions undergo phase separation if they are kept at room temperature. Just like medicinal syrups, these systems mandate storage at low temperatures. Our studies used cottonseed, peanut, mustard, and linseed oils as curcumin dispersion media with sodium dodecyl sulfate, SDS, dodecyl trimethylammonium bromide, DTAB, Tween-20, and Tween-40 as surfactants enabled improved curcumin antioxidant activities (Malik et al., 2014; Malik and Singh, 2017). These studies used ethanol as a co-solvent and glycerol as a co-surfactant. With distinct fatty acid hydrophobicities, there were noticeable variations in improved antioxidant activities (analyzed using DPPH free radical), with peanut oil-Tween-20 enabling the highest (92%) activity (Malik and Singh, 2017). In general, non-ionic surfactants or emulsifiers are more suited for food-grade emulsions as these bind weakly with the dispersed phase (here, curcumin) and do not hinder their release under *in vivo* conditions. Contrary to this, much energy is needed to weaken the binding of ionic surfactants over the dispersed phase, as well as the risk of structural tampering due to stronger BF. In a separate study cetyl trimethylammonium bromide (CTAB, a cationic surfactant) alone did not produce much improvement in curcumin antioxidant activity. However, when Tween-20 and Tween-20 dispersed Ag nanoparticles (NPs) which were also capped with *Syzygium cumini* leaf extract in 1:1 proportion with 0.004 m aq CTAB, the antioxidant activities improved to much higher extents with > 100% improvements in the majority of cases (Malik et al., 2019). Likewise, the improved antioxidant expression on combining with ultrasonication-driven graphene oxide and 5-aminoindazole functionalized graphene oxide (Maktedar et al., 2017). However, no bioavailability tests were conducted in any of these studies although the graphene oxide and functionalized graphene oxide were separately screened as non-toxic. Thereby, these systems having millimolar curcumin extents require extensive validation on animal models and volunteers along with a concentration optimization before being recommended to the general public as medicine. One crucial attempt toward commercialization was made by Zheng and colleagues who assessed the comparative efficacy of oil-loading, heating, and pH variation methods for ascertaining the bioaccessibility of commercially available turmeric extracts. The pH modification is best for mono dispersion attainment with the highest bioaccessibility extent for a commercial product like CurcuWin (Jäger et al., 2014).

14.6 CURCUMIN NANOPARTICLES (NPS)

Growing interest in metal and metal oxide NPs usage for biopharmaceutical and therapeutic purposes is swiftly gaining momentum as the high surface areas of NPs confer them multifunctional interaction attributes. Apart from this, the shape and size-dependent physicochemical activities of NPs make them robust candidates for achieving desired applications in distinct biochemical environments. For example, ZnO NPs are now familiar ingredients of toothpaste, sunscreen lotions, soaps, and many processed food items. The ZnO and Ag NPs are rapidly emerging antiseptic agents wherein combinations with curcumin for additive anti-inflammatory and wound healing activities are being made with huge expectations. In one recent study, Perera and colleagues loaded the spherical and rod morphologies of ZnO NPs on a curcumin surface and the grafted combinations were characterized using SEM, TGA, FT-IR, XRD, and UV-vis spectrophotometry. Screening the antimicrobial potential of curcumin grafted Zn O NPs, the investigators found them as active toward gram-positive and gram-negative bacterial stains. Larger inhibition zones compared to ZnO, and curcumin alone inferred a synergistic potential of the optimized combination. It is pertinent to note for the readers here that the used ZnO NPs were prepared through chemical reduction, using Zn $(NO_3)_2 6H_2O$ as a precursor and NaOH as a reducing agent (Perera et al., 2020). Contrary to the chemically prepared ZnO NPs, the plant extract-driven ZnO NPs can be readily incorporated as food quality supplementing agents. Furthermore, the polyphenols in plant extract exhibit a higher likelihood of having additive activities with curcumin. Our studies revealed significant antioxidant activities of Psidium Guajava capped ZnO NPs, screened *via* DPPH free radical assay which improved when analyzed in 1:1 combination with curcumin encapsulated o/w nanoemulsions (Gupta et al., 2019). Perhaps, the findings of a 2021 study project emergent hope toward using curcumin as a reducing agent for the formation of metal and metal oxide NPs from their respective precursors (Patra and Kurdi, 2021). Thus, curcumin functionalized NPs can be more readily included as food ingredients in light of their non-toxic and edible nature. Furthermore, having curcumin-capped NPs (deemed fit to be included in foods) argues well to guard against likely free radical or rancidity-induced damage. Shome et al. (2016) observed the benefits of incorporating curcumin in the solid-lipid and polymeric NPs. In general, the solid lipid texture resembles the lipidic gesture and facilitates easier and enhanced intracellular access. Similarly, polymeric NPs like poly-lactic-co-glycolytic acid, polycarbonate, and chitosan-loaded curcumin can be tailored for their inclusion as food constituents. This is because these polymeric textures are biocompatible and are therefore non-toxic although no studies of including polymeric NPs encapsulated curcumin directly as food ingredients are yet reported. Similar anticipations could be explored with hydrogels and nanofibers, wherein the choice of the base material can be tailored for their edible sensitivity. The combinations of gelatine-curcumin, and guar gum-curcumin along with several other carbohydrates and protein-based NPs can be readily explored as food quality enhancing agents. Although, such incentives are not much explored in developing and under-developed countries and are at risk of societal backlash, a systematic evaluation, and screening of animal models can be a significant step toward using curcumin as a potential food ingredient on a commercial scale.

14.7 ADDITIVE ACTIVITIES

The edible nature along with an ancient history of usage has been the major incentive for the development of commercial-grade curcumin-based functional foods. Lately, United States (US) based Sabinsa's branded curcumin C3 complex has been accorded a GRAS certification by US FDA. This complex has an extract of turmeric roots standardized as 95% curcuminoids, thereby contributing to splendid antioxidant and anti-inflammatory activities, together manifesting as a healthy immune system. The recommended extent of the curcumin C3 complex have been 0.1255 on a weight-by-weight basis, with an investigated safe intake extent being 1.755 g per day, estimated at the 90th percentile. The conferred GRAS status makes the Sabinsa's ingredient safe for use as a dietary supplement, making it eligible for food and drinking purposes. The fortified textures hereby comprise baked foods, milk derivatives, fats, oils, jams, gelatins, puddings, potages, nibbles, frozen and imitated dairy products, along with non-alcoholic beverages, seasonings, and flavors. In 2018, an attempt reported by scientists at the National Dairy Research Institute (NDRI) supplemented the buttermilk (traditionally referred to as Lassi) made using approved cow milk curd *via* 1% commercial yoghurt culture comprising *Streptococcus thermophilus* and *Lactobacillus bulgaricus* (the noted probiotics). The investigators made an herbal variant of this buttermilk by adding a maximum concentration of herbal (Rane et al., 2020a, b, c, 2023; Sharma et al., 2020) extract amidst the processing and optimizing it with 1% turmeric (having curcumin as a major phenolic constituent) through sensory evaluation. Subsequently, the total phenolic content (TPC) of turmeric-supplemented buttermilk was ascertained as gallic acid equivalent (GAE) using a double-beam UV-visible spectrophotometer at 765 nm. Quite interestingly, the TPC of turmeric-laced buttermilk was observed as 0.226 ± 0.001 mg GAE for each gram and differed significantly from the native sample. The reference sample that has no added turmeric developed a TPC of 0.124 ± 0.001 mg GAE for every 1 g. The high TPC of turmeric-laced buttermilk was thereby, inferred as a good sensory trait and remained intact for 9 days upon being stored at 7 ± 2°C in a glass bottle (Maji et al., 2018). Thus, this study elucidates the significant improvement in antioxidants via increased polyphenol content of turmeric-laced buttermilk. Similarly, the supplementation of whey (dairy by-product) with ethanolic extract of curcumin has also been reported. Whey-based beverages are in increasing demand because of their nutraceutical traits, and supplementing such material with curcumin presents a significant additive logic. The fortified beverage was developed with optimized contents of curcumin, sweetener, and flavor extents and was tested for antioxidant activity using the standard 2,2-diphenyl-1-picrylhydrazyl (DPPH) radical assay. Even though the results of the combination were not significant (< 50% DPPH˙ scavenging), the 60.65% DPPH˙ scavenging for ethanol-extracted curcumin delivered a DPPH scavenging of 43.51% on being fortified with whey. The ethanolic extract of curcumin was more effective than standard curcumin towards an antioxidant activity improvement of whey. The 16-day shelf-life of fortified beverage without a preservative addition does indicate a concern for the commercial use of this beverage but the combination could be bettered by varying the mutual stoichiometries of curcumin and whey protein. Since white coagulation was noticed after 18 days along with an unpleasant odor on 21 days, the inclusion of edible plant-based NPs or even curcumin NPs could further enhance the stability as well as the antioxidant activity. The findings of both these studies argue well for a sustainable curcumin/turmeric supplementation to other consumable daily-grade food items, such as curd, ice creams, milk, and its derived soft drinks.

Another significant attempt in 2018 reported the inclusion of curcumin-containing lipophilic turmeric extract in the diet of animals and birds. The rationale for this idea was the emergence of pathogenic strains due to indiscriminate antibiotic usage in domestic animals and poultry farming. The study evaluated the efficacy of standardized lipophilic turmeric extract that has curcumin along with numerous (TF-136) when used as a natural growth promoter in the diet of poultry animals. The sample size comprised 180, one-day-old chicks which were distributed in three groups. The control group was administered a basal diet while the supplementary groups were given a 0.5 and 1% TF-36 fortified basal diet for 42 days. Each dietary composition is comprised of 10 birds with six replicates. The effect of administered food was analyzed via monitoring body weight, food intake, food conversion ratio (FCR), skin color, a biochemical inspection of blood, and serum antioxidant efficacy. Analysis revealed a considerable improvement in body weight in 1% TF-36 animals with a 10% FCR decrement than the reference standard. It was noticed that TF-36 enhancement improved the antioxidant activity with a decreased lipid peroxidation. Though mild yellow skin pigmentation was observed no variation in meat texture was noticed, thereby conveying enhanced curcumin bioavailability. Unaltered extents of serum creatinine, net protein content, and liver enzyme activities were noticed which inferred null toxicity induction. So, this study demonstrated that the antioxidant and antimicrobial essence of curcuminoid and turmerones supplemented poultry and bird diets, which resulted in improved growth. The healthcare benefits of curcumin and turmerones could be attributed to their gastroprotective and anti-inflammatory as reported earlier (Nm et al., 2018).

A noteworthy attempt of 2021 itself reported the pH-optimized curcumin in plant-based milk. The rationale for this fortification was attributed to the presence of a triglyceride-rich core in the oil bodies of the studied milk, which are coated by a phospholipidic (proteinaceous) shell. The study was conducted on gaining encouragement through earlier loading of curcumin in the soymilk oil bodies using a pH-driven method only. In the present attempt, the loading was made in other potential plant-derived milk substitutes including coconut, cashew, almond, and goat milk. To fortify these milk oil bodies with curcumin, initially, the effect of pH-optimized curcumin loading was screened on their physicochemical properties. Thereafter, curcumin bioavailability was screened through a simulated gastrointestinal tract administration. Inspection revealed a structurally compatible curcumin loading over the oil bodies which maintained their stability. The synergism concerning curcumin inclusion was also inferred by >86% encapsulation efficacy (conveying a curcumin localization within the oil bodies). All the milk supplemented with curcumin exhibited >60% bioaccessibility, which was significantly higher than the 5% extent for the crystalline form of curcumin (Zeng et al., 2021). Thus, this investigation also conveyed a food-grade utilization of curcumin wherein the textures of plant-derived milk could be improved along with minimized free radical generation possibilities. The results are extensions of curcumin-loaded o/w emulsions wherein oil-mixed curcumin is included as a dispersed phase. This environment not only serves as the modified carrier for curcumin delivery but also preserves oil deterioration by preventing triglyceride generation. The triglycerides are contributory elements of bad cholesterol which affects the healthy growth of cells and tissues. So, curcumin inclusion within the edible oils for cooking purposes does not merely safeguard the triglyceride generation but also enhances the shelf-life of cooked vegetables.

Idowu-Adebayo et al. (2021) studied the enrichment of the street-vendor zobo drink with turmeric to improve its healthcare-promoting attributes. Turmeric inclusion was attempted *via.* (i) boiled turmeric root in zobo; and (ii) adding fresh turmeric extract to zobo to varying extents. The vitamin C content of turmeric-fortified zobo was within 496–725 µg/100 mL, while delphinidin-3-sambubioside and cyanin-3-sambubioside remained within 52–69 mg/100 mL and 21–27 mg/100 mL, respectively. The Fe and Zn contents ranged from 10.9–14 mg/L and 2.19–2.67 mg/L, respectively. Folic acid, vitamin C, anthocyanins, and Fe exhibited maximum extents in 2% boiled turmeric zobo samples. The improvements in ferulic acid of 0.16–2.03 mg/100 mL and chlorogenic acid (CGA) of 20–24 mg/100 mL, were statistically different for 2% boiled turmeric enriched zobo. The samples with turmeric extract exhibited lesser vitamins, polyphenols, and minerals than the boiled turmeric fortified zobo samples. Thereby, turmeric-enriched zobo could be a significant booster of health-promoting benefits, with a 500 mL bottle providing 63–88% Fe and 18–23% Zn.

14.8 FUTURE OF CURCUMIN

Despite high hopes and overwhelming expectations, much of the curcumin activities remain within the laboratory domain as the protocols to transit it to the ambit of a common man are lengthy, elaborate, and need to be reproduced concerning dosage optimization. There is an urgent need to precede each repetitive study with curcumin to the pilot scale with an enthusiasm to be conducted in animal models as the minimum necessary condition. An incentive behind curcumin usage is the aesthetic belief of the society at large about which newer and better combinations must be pursued without any funding crunch. Recently, in a compilation focused on anti-cancer platinum complexes, the authors proposed curcumin combined delivery as a toxicity moderator with a higher cell internalization ability (Ameta and Malik, 2020). The electronically vibrant force contributions from curcumin make it compatible to function in an integrated manner with the electronically deficient molecules. Not being restricted to the food arena, the applications of curcumin for improved healthcare are immense and multiple, with dye degradation, waste pollutant minimization, and proven biological activities at length. The findings for improved nutrient content, texture, shelf-life, and better absorption of curcumin supplemented foods are of immense significance since more than 30 years for unanimity, curcumin is being used as a nutraceutical in many parts of the world. It is a divine gift that the natural source of this polyphenol does not make it patentable whereby it is comparatively easy and within reach of a common man. For instance, in Japan, curcumin is famously included in tea while in Thailand, it is used as a cosmetic. Similarly, in China, it is used as a colorant, dished up in drinks in Korea, and used as an antiseptic in Malaysia.

Encouragements for strengthening industry-academia partnerships must be provided for traditional plant products like those of curcumin so that the research findings could be replicated on a larger scale and in diverse conditions. Indeed, pharmaceutical industries do have a reservation to pursue curcumin trials as it is a natural product and the findings can't be patented, unlike conventional medicines and vaccines. Therefore, ministries and regulatory commissions must increase the funding and provide adequate justifications for pursuing curcumin research as the timid needs of emergencies like we had with the COVID-19 pandemic. A sustainable solution can arrive only when the developed economies of the world partner with their developing counterparts for uniform and integrated objectives and aims. This is important because we collectively need to reach a step higher than our common belief which inevitably prevails in society at large. Furthermore, research proposals for laboratory-scale working should be sanctioned only for novel aims as we frequently notice multiple studies of curcumin-treated different cancers, CVD, and other troubles. This would streamline the research mandate as well as direct stringent efforts to pursue further the direction of curcumin chemistry.

14.9 CONCLUSION

The development of curcumin-based functional foods is growing steadily but needs swift attempts in light of the proven spectrum of bioactivities. Natural origin and edible nature are the assets of this plant-derived polyphenol which collectively inspire for pursuing the healthcare research. The multiple bioactivities are the most crucial evidence of the food-grade therapeutic relevance of curcumin, but regulatory norms are stiff and mandate the completion on a pilot scale. So, there is a need to sharpen laboratory research and aim at sustainable combinations to minimize the side effects of chemotherapeutic drugs. Though research attempts are swiftly underway for incorporating curcumin as a functional food ingredient commercialization requires elaborate optimization and prior validation in animal models and human volunteers. Integrated efforts, therefore, are the only visible solutions that can accomplish these requirements in the minimum possible time. On a global scale, it is equally essential that developing nations are in active partnership with developed counterparts so that the trials could be channelized specifically, and mutual complementation of efforts can help in achieving the very next goal.

KEYWORDS

- **absorption**
- **anti-cancer**
- **bioactivities**
- **chemical environments**
- **curcumin**
- **food quality**
- **free radical generation**
- **functional foods**
- **metabolic fractions**
- **water insolubility**

REFERENCES

Abe, Y., Hashimoto, S. H. U., & Horie, T. (1999). Curcumin inhibition of inflammatory cytokine production by human peripheral blood monocytes and alveolar macrophages. *Pharmacological Research*, *39*(1), 41–47.

Aggarwal, S., Ichikawa, H., Takada, Y., Sandur, S. K., Shishodia, S., & Aggarwal, B. B. (2006). Curcumin (diferuloylmethane) down-regulates expression of cell proliferation and antiapoptotic and metastatic gene products through suppression of IκBα kinase and Akt activation. *Molecular Pharmacology*, *69*(1), 195–206.

Ali, I., Haque, A., & Saleem, K. (2014). Separation and identification of curcuminoids in turmeric powder by HPLC using phenyl column. *Analytical Methods*, *6*(8), 2526–2536.

Allegri, P., Mastromarino, A., & Neri, P. (2010). Management of chronic anterior uveitis relapses: Efficacy of oral phospholipidic curcumin treatment. Long-term follow-up. *Clinical Ophthalmology*, 1201–1206.

Alwi, I., Santoso, T., Suyono, S., Sutrisna, B., Suyatna, F. D., Kresno, S. B., & Ernie, S. (2008). The effect of curcumin on lipid level in patients with acute coronary syndrome. *Acta Med Indones*, *40*(4), 201–210.

Ameta, R. K., & Malik, P. (2020). Platinum complexes in medicine and in the treatment of cancer. In *Advances in Metallodrugs: Preparation and Applications in Medicinal Chemistry* (pp. 203–245).

Anand, P., Kunnumakkara, A. B., Newman, R. A., & Aggarwal, B. B. (2007). Bioavailability of curcumin: Problems and promises. *Molecular Pharmaceutics*, *4*(6), 807–818.

Anderson, A. K., & Ng, P. K. W. (2000). Changes in disulfide and sulfhydryl contents and electrophoretic patterns of extruded wheat flour proteins. *Cereal Chemistry*, *77*(3), 354–359.

Aoki, H., Takada, Y., Kondo, S., Sawaya, R., Aggarwal, B. B., & Kondo, Y. (2007). Evidence that curcumin suppresses the growth of malignant gliomas *in vitro* and *in vivo* through induction of autophagy: Role of Akt and extracellular signal-regulated kinase signaling pathways. *Molecular Pharmacology, 72*(1), 29–39.

Azuine, M. A., & Bhide, S. V. (1992). Protective single/combined treatment with betel leaf and turmeric against methyl (acetoxymethyl) nitrosamine-induced hamster oral carcinogenesis. *International Journal of Cancer, 51*(3), 412–415.

Barclay, L. R. C., Vinqvist, M. R., Mukai, K., Goto, H., Hashimoto, Y., Tokunaga, A., & Uno, H. (2000). On the antioxidant mechanism of curcumin: Classical methods are needed to determine antioxidant mechanism and activity. *Organic Letters, 2*(18), 2841–2843.

Burapan, S., Kim, M., & Han, J. (2017). Curcuminoid demethylation as an alternative metabolism by human intestinal microbiota. *Journal of Agricultural and Food Chemistry, 65*(16), 3305–3310.

Chuang, S. E., Cheng, A. L., Lin, J. K., & Kuo, M. L. (2000). Inhibition by curcumin of diethylnitrosamine-induced hepatic hyperplasia, inflammation, cellular gene products and cell-cycle-related proteins in rats. *Food and Chemical Toxicology, 38*(11), 991–995.

Dei Cas, M., & Ghidoni, R. (2019). Dietary curcumin: Correlation between bioavailability and health potential. *Nutrients, 11*(9), 2147.

Devasena, T., Rajasekaran, K. N., Gunasekaran, G., Viswanathan, P., & Menon, V. P. (2003). Anticarcinogenic effect of bis-1, 7-(2-hydroxyphenyl)-hepta-1, 6-diene-3, 5-dione, a curcumin analog on DMH-induced colon cancer model. *Pharmacological Research, 47*(2), 133–140.

Durgaprasad, S., Pai, C. G., & Alvres, J. F. (2005). A pilot study of the antioxidant effect of curcumin in tropical pancreatitis. *Indian Journal of Medical Research, 122*(4), 315.

El-Moselhy, M. A., Taye, A., Sharkawi, S. S., El-Sisi, S. F., & Ahmed, A. F. (2011). The antihyperglycemic effect of curcumin in high fat diet fed rats: Role of TNF-α and free fatty acids. *Food and Chemical Toxicology, 49*(5), 1129–1140.

Epelbaum, R., Schaffer, M., Vizel, B., Badmaev, V., & Bar-Sela, G. (2010). Curcumin and gemcitabine in patients with advanced pancreatic cancer. *Nutrition and Cancer, 62*(8), 1137–1141.

Esatbeyoglu, T., Huebbe, P., Ernst, I. M., Chin, D., Wagner, A. E., & Rimbach, G. (2012). Curcumin—from molecule to biological function. *Angewandte Chemie International Edition, 51*(22), 5308–5332.

Flora, G., Gupta, D., & Tiwari, A. (2012). Toxicity of lead: A review with recent updates. *Interdisciplinary Toxicology, 5*(2), 47–58.

Garcia-Alloza, M., Borrelli, L. A., Rozkalne, A., Hyman, B. T., & Bacskai, B. J. (2007). Curcumin labels amyloid pathology *in vivo*, disrupts existing plaques, and partially restores distorted neurites in an Alzheimer mouse model. *Journal of Neurochemistry, 102*(4), 1095–1104.

Gupta, R., Malik, P., Das, N., & Singh, M. (2019). Antioxidant and physicochemical study of *Psidium guajava* prepared zinc oxide nanoparticles. *Journal of Molecular Liquids, 275*, 749–767.

Gupta, S. C., Kim, J. H., Prasad, S., & Aggarwal, B. B. (2010). Regulation of survival, proliferation, invasion, angiogenesis, and metastasis of tumor cells through modulation of inflammatory pathways by nutraceuticals. *Cancer and Metastasis Reviews, 29*, 405–434.

Gupta, Y. K., Briyal, S., & Sharma, M. (2009). Protective effect of curcumin against kainic acid induced seizures and oxidative stress in rats. *Indian Journal of Physiology and Pharmacology, 53*(1), 39–46.

Hassaninasab, A., Hashimoto, Y., Tomita-Yokotani, K., & Kobayashi, M. (2011). Discovery of the curcumin metabolic pathway involving a unique enzyme in an intestinal microorganism. *Proceedings of the National Academy of Sciences, 108*(16), 6615–6620.

He, Z. Y., Shi, C. B., Wen, H., Li, F. L., Wang, B. L., & Wang, J. (2011). Upregulation of p53 expression in patients with colorectal cancer by administration of curcumin. *Cancer Investigation, 29*(3), 208–213.

Hecht, S. S., Kenney, P. M., Wang, M., Trushin, N., Agarwal, S., Rao, A. V., & Upadhyaya, P. (1999). Evaluation of butylated hydroxyanisole, myo-inositol, curcumin, esculetin, resveratrol, and lycopene as inhibitors of benzo[a]pyrene plus 4-(methylnitrosamino)-1-(3-pyridyl)-1-butanone-induced lung tumorigenesis in A/J mice. *Cancer Letters, 137*(2), 123–130.

Huang, M. T., Lou, Y. R., Ma, W., Newmark, H. L., Reuhl, K. R., & Conney, A. H. (1994). Inhibitory effects of dietary curcumin on forestomach, duodenal, and colon carcinogenesis in mice. *Cancer Research, 54*(22), 5841–5847.

Huang, M. T., Lou, Y. R., Xie, J. G., Ma, W., Lu, Y. P., Yen, P., Zhu, B. T., Newmark, H., & Ho, C. T. (1998). Effect of dietary curcumin and dibenzoylmethane on formation of 7, 12-dimethylbenz[a]anthracene-induced mammary tumors and lymphomas/leukemias in Sencar mice. *Carcinogenesis, 19*(9), 1697–1700.

Idowu-Adebayo, F., Toohey, M. J., Fogliano, V., & Linnemann, A. R. (2021). Enriching street-vended zobo (*Hibiscus sabdariffa*) drink with turmeric (*Curcuma longa*) to increase its health-supporting properties. *Food & Function, 12*(2), 761–770.

Ikezaki, S., Nishikawa, A., Furukawa, F., Kudo, K., Nakamura, H., Tamura, K., & Mori, H. (2001). Chemopreventive effects of curcumin on glandular stomach carcinogenesis induced by N-methyl-N'-nitro-N-nitrosoguanidine and sodium chloride in rats. *Anti-cancer Research, 21*(5), 3407–3411.

Jäger, R., Lowery, R. P., Calvanese, A. V., Joy, J. M., Purpura, M., & Wilson, J. M. (2014). Comparative absorption of curcumin formulations. *Nutrition Journal, 13*(1), 1–8.

Jefremov, V., Zilmer, M., Zilmer, K., Bogdanovic, N., & Karelson, E. (2007). Antioxidative effects of plant polyphenols: From protection of G protein signaling to prevention of age-related pathologies. *Annals of the New York Academy of Sciences, 1095*(1), 449–457.

Ji, H. F., & Shen, L. (2009). Interactions of curcumin with the PfATP6 model and the implications for its antimalarial mechanism. *Bioorganic & Medicinal Chemistry Letters, 19*(9), 2453–2455.

Joe, B., & Lokesh, B. R. (1994). Role of capsaicin, curcumin, and dietary n—3 fatty acids in lowering the generation of reactive oxygen species in rat peritoneal macrophages. *Biochimica et Biophysica Acta (BBA)-Molecular Cell Research, 1224*(2), 255–263.

Keservani, R. K., Kesharwani, R. K., Sharma, A. K., Vyas, N., & Chadoker, A. (2010). Nutritional supplements: An overview. *International Journal of Current Pharmaceutical Review and Research, 1*(1), 59–75.

Keservani, R. K., Kesharwani, R. K., Vyas, N., Jain, S., Raghuvanshi, R., & Sharma, A. K. (2010). Nutraceutical and functional food as future food: A review. *Der Pharmacia Lettre, 2*(1), 106–116.

Keservani, R. K., Sharma, A. K., & Kesharwani, R. K. (2017). An overview and therapeutic applications of nutraceutical and functional foods. In *Recent Advances in Drug Delivery Technology* (pp. 160–201). ISBN: 9781522507543.

Keservani, R. K., Sharma, A. K., & Kesharwani, R. K. (2020). *Nutraceuticals and Dietary Supplements: Applications in Health Improvement and Disease Management*. CRC Press. ISBN: 9781771888738.

Kiuchi, F., Goto, Y., Sugimoto, N., Akao, N., Kondo, K., & Tsuda, Y. (1993). Nematocidal activity of turmeric: Synergistic action of curcuminoids. *Chemical and Pharmaceutical Bulletin, 41*(9), 1640–1643.

Koide, T., Nose, M., Ogihara, Y., Yabu, Y., & Ohta, N. (2002). Leishmanicidal effect of curcumin *in vitro*. *Biological and Pharmaceutical Bulletin, 25*(1), 131–133.

Kuttan, R., Sudheeran, P. C., & Josph, C. D. (1987). Turmeric and curcumin as topical agents in cancer therapy. *Tumori Journal, 73*(1), 29–31.

Lee, K. J., Kim, Y. S., Jung, P. M., & Ma, J. Y. (2013). Optimization of the conditions for the analysis of curcumin and a related compound in *Curcuma longa* with mobile-phase composition and column temperature via RP-HPLC. *Asian Journal of Chemistry, 25*(11).

Lestari, M. L., & Indrayanto, G. (2014). Curcumin. In *Profiles of Drug Substances, Excipients and Related Methodology* (Vol. 39, pp. 113–204).

Lev-Ari, S., Starr, A., Vexler, A., Karaush, V., Loew, V., Greif, J., Fenig, E., Aderka, D., & Ben-Yosef, R. (2006). Inhibition of pancreatic and lung adenocarcinoma cell survival by curcumin is associated with increased apoptosis, down-regulation of COX-2 and EGFR, and inhibition of Erk1/2 activity. *Anti-cancer Research, 26*(6B), 4423–4430.

Li, M., Ngadi, M. O., & Ma, Y. (2014). Optimisation of pulsed ultrasonic and microwave-assisted extraction for curcuminoids by response surface methodology and kinetic study. *Food Chemistry, 165*, 29–34.

Limtrakul, P., Lipigorngoson, S., Namwong, O., Apisariyakul, A., & Dunn, F. W. (1997). Inhibitory effect of dietary curcumin on skin carcinogenesis in mice. *Cancer Letters, 116*(2), 197–203.

Luca, S. V., Macovei, I., Bujor, A., Miron, A., Skalicka-Woźniak, K., Aprotosoaie, A. C., & Trifan, A. (2020). Bioactivity of dietary polyphenols: The role of metabolites. *Critical Reviews in Food Science and Nutrition, 60*(4), 626–659.

Magalhães, L. G., Machado, C. B., Morais, E. R., Bueno de Carvalho Moreira, É., Soares, C. S., da Silva, S. H., Da Silva Filho, A. A., & Rodrigues, V. (2009). *In vitro* schistosomicidal activity of curcumin against *Schistosoma mansoni* adult worms. *Parasitology Research, 104*, 1197–1201.

Maji, S., Ray, P. R., Ghatak, P. K., & Chakraborty, C. (2018). Total phenolic content (TPC) and quality of herbal lassi fortified with turmeric (*Curcuma longa*) extract. *Asian Journal of Dairy and Food Research, 37*(4), 273–277.

Maktedar, S. S., Malik, P., Avashthi, G., & Singh, M. (2017). Dispersion enhancing effect of sonochemically functionalized graphene oxide for catalysing antioxidant efficacy of curcumin. *Ultrasonics Sonochemistry, 39*, 208–217.

Malik, P., & Mukherjee, T. K. (2014). Structure-function elucidation of antioxidative and prooxidative activities of the polyphenolic compound curcumin. *Chinese Journal of Biology, 2014*, 1–8.

Malik, P., & Singh, M. (2017). Study of curcumin antioxidant activities in robust oil–water nanoemulsions. *New Journal of Chemistry, 41*(21), 12506–12519.

Malik, P., Ameta, R. K., & Singh, M. (2014). Preparation and characterization of bionanoemulsions for improving and modulating the antioxidant efficacy of natural phenolic antioxidant curcumin. *Chemico-Biological Interactions, 222*, 77–86.

Malik, P., Hoidal, J. R., & Mukherjee, T. K. (2021). Recent advances in curcumin-treated non-small cell lung cancers: An impetus of pleiotropic traits and nanocarrier-aided delivery. *Current Medicinal Chemistry*, *28*(16), 3061–3106.

Malik, P., Inwati, G. K., Mukherjee, T. K., Singh, S., & Singh, M. (2019). Green silver nanoparticle and Tween-20 modulated pro-oxidant to antioxidant curcumin transformation in aqueous CTAB stabilized peanut oil emulsions. *Journal of Molecular Liquids*, *291*, 111252.

Mehla, J., Reeta, K. H., Gupta, P., & Gupta, Y. K. (2010). Protective effect of curcumin against seizures and cognitive impairment in a pentylenetetrazole-kindled epileptic rat model. *Life Sciences*, *87*(19–22), 596–603.

Muriithi, W., Macharia, L. W., Heming, C. P., Echevarria, J. L., Nyachieo, A., Niemeyer Filho, P., & Neto, V. M. (2020). ABC transporters and the hallmarks of cancer: Roles in cancer aggressiveness beyond multidrug resistance. *Cancer Biology & Medicine*, *17*(2), 253.

Narayanan, N. K., Nargi, D., Randolph, C., & Narayanan, B. A. (2009). Liposome encapsulation of curcumin and resveratrol in combination reduces prostate cancer incidence in PTEN knockout mice. *International Journal of Cancer*, *125*(1), 1–8.

Neelofar, K., Shreaz, S., Rimple, B., Muralidhar, S., Nikhat, M., & Khan, L. A. (2011). Curcumin as a promising anticandidal of clinical interest. *Canadian Journal of Microbiology*, *57*(3), 204–210.

Nm, J., Joseph, A., Maliakel, B., & Im, K. (2018). Dietary addition of a standardized extract of turmeric (TurmaFEED™) improves growth performance and carcass quality of broilers. *Journal of Animal Science and Technology*, *60*, 1–9.

Nurfina, A. N., Reksohadiprodjo, M. S., Timmerman, H., Jenie, U. A., Sugiyanto, D., & Van der Goot, H. (1997). Synthesis of some symmetrical curcumin derivatives and their anti-inflammatory activity. *European Journal of Medicinal Chemistry*, *32*(4), 321–328.

Okazaki, Y., Iqbal, M., & Okada, S. (2005). Suppressive effects of dietary curcumin on the increased activity of renal ornithine decarboxylase in mice treated with a renal carcinogen, ferric nitrilotriacetate. *Biochimica et Biophysica Acta (BBA)–Molecular Basis of Disease*, *1740*(3), 357–366.

Paramasivam, M., Poi, R., Banerjee, H., & Bandyopadhyay, A. (2009). High-performance thin layer chromatographic method for quantitative determination of curcuminoids in *Curcuma longa* germplasm. *Food Chemistry*, *113*(2), 640–644.

Patil, T. N., & Srinivasan, M. (1971). Hypocholesteremic effect of curcumin in induced hypercholesteremic rats. *Indian Journal of Experimental Biology*, *9*(2), 167–169.

Patra, D., & El Kurdi, R. (2021). Curcumin as a novel reducing and stabilizing agent for the green synthesis of metallic nanoparticles. *Green Chemistry Letters and Reviews*, *14*(3), 474–487.

Paulucci, V. P., Couto, R. O., Teixeira, C. C., & Freitas, L. A. P. (2013). Optimization of the extraction of curcumin from *Curcuma longa* rhizomes. *Revista Brasileira de Farmacognosia*, *23*(1), 94–100.

Pentea, M., Butu, M., Samfira, I., Cristina, R. T., & Butnariu, M. (2015). Extraction and analytical study of salvinorin from leaves of *Salvia divinorum*. *Digest Journal of Nanomaterials and Biostructures*, *10*(1), 291–297.

Perera, W. P. T. D., Dissanayake, R. K., Ranatunga, U. I., Hettiarachchi, N. M., Perera, K. D. C., Unagolla, J. M., De Silva, R. T., & Pahalagedara, L. R. (2020). Curcumin loaded zinc oxide nanoparticles for activity-enhanced antibacterial and anti-cancer applications. *RSC Advances*, *10*(51), 30785–30795.

Priyadarsini, K. I. (2014). The chemistry of curcumin: From extraction to therapeutic agent. *Molecules*, *19*(12), 20091–20112.

Purkayastha, S., Berliner, A., Fernando, S. S., Ranasinghe, B., Ray, I., Tariq, H., & Banerjee, P. (2009). Curcumin blocks brain tumor formation. *Brain Research*, *1266*, 130–138.

Rane, B. R., Ahirrao, R. A., Gujarathi, N. A., Jain, A. S., & Keservani, R. K. (2023). Pathophysiology and application of herbs in the metabolic syndrome. In R. K. Keservani, D. Yadav, R. K. Keservani, S. Singh, & S. Kumar (Eds.), *The Metabolic Syndrome: Dietary Supplements and Food Ingredients* (pp. 199–230). Apple Academic Press, CRC Press, Taylor & Francis Group. ISBN: 978-1-77491-111-2.

Rane, B. R., Bharath, M. S., Patil, R. R., Keservani, R. K., & Jain, A. S. (2020b). Novel approaches in nutraceuticals. In R. K. Keservani, R. K. Keservani, A. K. Sharma (Eds.), *Enhancing the Therapeutic Efficacy of Herbal Formulations through Novel Drug Delivery Systems* (pp. 241–266). IGI Global International Publisher. ISBN: 9781799844532.

Rane, B. R., Patil, A. S., Keservani, R. K., & Jain, A. S. (2020a). Novel approaches in herbal formulation. In R. K. Keservani, R. K. Keservani, A. K. Sharma (Eds.), *Enhancing the Therapeutic Efficacy of Herbal Formulations through Novel Drug Delivery Systems* (pp. 43–68). IGI Global International Publisher. ISBN: 9781799844532.

Rane, B. R., Tadavi, S. A., & Keservani, R. K. (2020c). Naturopathy. In A. K. Sharma, R. K. Keservani, & S. P. Gautam (Eds.), *Herbal Product Development* (pp. 321–347). Apple Academic Press, CRC Press, Taylor & Francis Group. ISBN: 9781771888776.

Sahl, B., Assi, V., & Templeman, V. (2003). Curcumin attenuates DBN-induced colitis. *American Journal of Physiology-Gastrointestinal and Liver Physiology*, *285*, G235–G243.

Schraufstätter, E., & Bernt, H. (1949). Antibacterial action of curcumin and related compounds. *Nature*, *164*(4167), 456–457.

Seo, K. I., Choi, M. S., Jung, U. J., Kim, H. J., Yeo, J., Jeon, S. M., & Lee, M. K. (2008). Effect of curcumin supplementation on blood glucose, plasma insulin, and glucose homeostasis-related enzyme activities in diabetic db/db mice. *Molecular Nutrition & Food Research*, *52*(9), 995–1004.

Sharifi-Rad, J., Rayess, Y. E., Rizk, A. A., Sadaka, C., Zgheib, R., Zam, W., Sestito, S., Rapposelli, S., Neffe-Skocińska, K., Zielińska, D., & Salehi, B. (2020). Turmeric and its major compound curcumin on health: Bioactive effects and safety profiles for food, pharmaceutical, biotechnological and medicinal applications. *Frontiers in Pharmacology*, *11*, 1021.

Sharma, A. K., Keservani, R. K., & Gautam, S. P. (2020). *Herbal Product Development* (pp. 1–376). Apple Academic Press, CRC Press, Taylor & Francis Group. ISBN: 9781771888776.

Sharma, M., Manoharlal, R., Negi, A. S., & Prasad, R. (2010). Synergistic anticandidal activity of pure polyphenol curcumin I in combination with azoles and polyenes generates reactive oxygen species leading to apoptosis. *FEMS Yeast Research*, *10*(5), 570–578.

Sharma, O. P. (1976). Antioxidant activity of curcumin and related compounds. *Biochemical Pharmacology*, *25*(15), 1811–1812.

Sharma, R. A., McLelland, H. R., Hill, K. A., Ireson, C. R., Euden, S. A., Manson, M. M., Pirmohamed, M., Marnett, L. J., Gescher, A. J., & Steward, W. P. (2001). Pharmacodynamic and pharmacokinetic study of oral Curcuma extract in patients with colorectal cancer. *Clinical Cancer Research*, *7*(7), 1894–1900.

Shen, L., Liu, C. C., An, C. Y., & Ji, H. F. (2016). How does curcumin work with poor bioavailability? Clues from experimental and theoretical studies. *Scientific Reports*, *6*(1), 1–10.

Shishodia, S., Singh, T., & Chaturvedi, M. M. (2007). Modulation of transcription factors by curcumin. In *The Molecular Targets and Therapeutic Uses of Curcumin in Health and Disease* (pp. 127–148).

Shoba, G., Joy, D., Joseph, T., Majeed, M., Rajendran, R., & Srinivas, P. S. S. R. (1998). Influence of piperine on the pharmacokinetics of curcumin in animals and human volunteers. *Planta Medica*, *64*(4), 353–356.

Shome, S., Talukdar, A. D., Choudhury, M. D., Bhattacharya, M. K., & Upadhyaya, H. (2016). Curcumin as potential therapeutic natural product: A nanobiotechnological perspective. *Journal of Pharmacy and Pharmacology*, *68*(12), 1481–1500.

Sindhwani, P., Hampton, J. A., Baig, M. M., Keck, R., & Selman, S. H. (2001). Curcumin prevents intravesical tumor implantation of the MBT-2 tumor cell line in C3H mice. *The Journal of Urology*, *166*(4), 1498–1501.

Singh, S., & Aggarwal, B. B. (1995). Activation of transcription factor NF-κB is suppressed by curcumin (diferuloylmethane). *Journal of Biological Chemistry*, *270*(42), 24995–25000.

Srinivasan, M. (1972). Effect of curcumin on blood sugar as seen in a diabetic subject. *Indian Journal of Medical Sciences*, *26*(4), 269–270.

Stohs, S. J., Chen, O., Ray, S. D., Ji, J., Bucci, L. R., & Preuss, H. G. (2020). Highly bioavailable forms of curcumin and promising avenues for curcumin-based research and application: A review. *Molecules*, *25*(6), 1397.

Swamy, M. V., Citineni, B., Patlolla, J. M., Mohammed, A., Zhang, Y., & Rao, C. V. (2008). Prevention and treatment of pancreatic cancer by curcumin in combination with omega-3 fatty acids. *Nutrition and Cancer*, *60*(S1), 81–89.

Ukil, A., Maity, S., Karmakar, S., Datta, N., Vedasiromoni, J. R., & Das, P. K. (2003). Curcumin, the major component of food flavor turmeric, reduces mucosal injury in trinitrobenzene sulphonic acid-induced colitis. *British Journal of Pharmacology*, *139*(2), 209–218.

Usharani, P., Mateen, A. A., Naidu, M. U. R., Raju, Y. S. N., & Chandra, N. (2008). Effect of NCB-02, atorvastatin and placebo on endothelial function, oxidative stress and inflammatory markers in patients with type 2 diabetes mellitus: A randomized, parallel-group, placebo-controlled, 8-week study. *Drugs in R & D*, *9*, 243–250.

Ushida, J., Sugie, S., Kawabata, K., Pham, Q. V., Tanaka, T., Fujii, K., Takeuchi, H., Ito, Y., & Mori, H. (2000). Chemopreventive effect of curcumin on N-nitrosomethylbenzylamine-induced esophageal carcinogenesis in rats. *Japanese Journal of Cancer Research*, *91*(9), 893–898.

Yang, F., Lim, G. P., Begum, A. N., Ubeda, O. J., Simmons, M. R., Ambegaokar, S. S., Chen, P. P., Kayed, R., Glabe, C. G., Frautschy, S. A., & Cole, G. M. (2005). Curcumin inhibits formation of amyloid β oligomers and fibrils, binds plaques, and reduces amyloid *in vivo*. *Journal of Biological Chemistry*, *280*(7), 5892–5901.

Yeon, K. Y., Kim, S. A., Kim, Y. H., Lee, M. K., Ahn, D. K., Kim, H. J., Kim, J. S., Jung, S. J., & Oh, S. B. (2010). Curcumin produces an antihyperalgesic effect via antagonism of TRPV1. *Journal of Dental Research*, *89*(2), 170–174.

Yu, S., Zheng, W., Xin, N., Chi, Z. H., Wang, N. Q., Nie, Y. X., Feng, W. Y., & Wang, Z. Y. (2010). Curcumin prevents dopaminergic neuronal death through inhibition of the c-Jun N-terminal kinase pathway. *Rejuvenation Research*, *13*(1), 55–64.

Zbarsky, V., Datla, K. P., Parkar, S., Rai, D. K., Aruoma, O. I., & Dexter, D. T. (2005). Neuroprotective properties of the natural phenolic antioxidants curcumin and naringenin but not quercetin and fisetin in a 6-OHDA model of Parkinson's disease. *Free Radical Research*, *39*(10), 1119–1125.

Zheng, B., Zhou, H., & McClements, D. J. (2021). Nutraceutical-fortified plant-based milk analogs: Bioaccessibility of curcumin-loaded almond, cashew, coconut, and oat milks. *Lwt*, *147*, 111517.

CHAPTER 15

Food Polyphenolics for the Management of Diabetes

NEHA SHARDA,[1] RITIKA VERMA,[2] POOJA RANI,[1] DEEPAK KUMAR,[1] SHAILJA SANKHYAN,[3] AJAY SHARMA,[4#] and ANIL KUMAR[1,5*]

[1]Department of Microbiology, DAV University, Jalandhar, Punjab, India

[2]ICAR–Indian Agricultural Research Institute, Regional Station, Shimla, Himachal Pradesh, India

[3]Department of Biotechnology, Chandigarh University, Mohali, Punjab, India

[4]Department of Chemistry, Career Point University, Hamirpur, Himachal Pradesh, India

[5]School of Chemical and Metallurgical Engineering, University of the Witwatersrand, Johannesburg, South Africa

*Corresponding author.
#Co-corresponding author.

ABSTRACT

Diabetes is characterized by increased levels of blood glucose and inadequacy in the production and action of insulin (also referred to as β-cell dysfunction). Globally, diabetes mellitus (DM) is a prime cause of disability and discomfort. Robust evidence links diabetes with the increased formation of "reactive oxygen species (ROS)" and decreased levels of antioxidants. ROS dysregulates the expression of genes responsible for insulin signaling and secretion. Polyphenols comprise the largest group among phytochemicals, and several of them are present in plant-based foods. Plant-associated food-linked polyphenols are being used for the treatment of diabetes due to their biological properties of scavenging free radicals and maintaining antioxidant potential. Food-linked polyphenols help in preventing and managing type 2 diabetes

Functional Compounds and Foods of Plant Origin: Krishan Dutt Sharma, Satish Kumar, Rakesh Sharma, and Raj K. Keservani (Eds.)

(T2D) via stimulation of insulin secretion, insulin signaling activation, shielding of islet β-cell (pancreatic), promotion of β-cell proliferation, reduction of β-cell apoptosis and oxidative stress, inhibition of digestive enzymes and glucose absorption, and regulation of intestinal microbiota. Additionally, dietary polyphenols also provide protection from other diabetic-related complications such as coronary diseases, cardiac, kidney, and liver-related diseases. Further, the antidiabetic potential of food-linked polyphenols has been investigated in cell, animal, human, and clinical studies.

15.1 INTRODUCTION

Diabetes has become one of the most rapidly growing health emergencies of today. According to the estimates by the International Diabetes Federation (IDF), 542 million people are reported to have diabetes in the year 2021, and it is projected to reach 783 million by the year 2045 (IDF, 2021). T2D is the most prevalent form of diabetes, accounting for over 90% of all diabetes globally. The cause of T2D is not completely understood but can be linked to several unhealthy lifestyle factors, such as overweight/obesity, unhealthy diets, sedentary lifestyle, smoking, and excessive alcohol intake (Farias et al., 2021; Ferre et al., 2017).

Preventing T2D is more effective rather than treating its manifestation and complications. A healthy diet abundant in dietary functional foods and their respective bioactive compounds is regarded as a key lifestyle factor in the management and prevention of T2D. Polyphenols, the vital components of vegetables, plant-derived food products, fruits, and beverages, are widely reported as naturally occurring antioxidants, which may explain their apparent medicinal promise in T2D prevention (Rehman et al., 2017). Phenolics are naturally occurring complex bioactive molecules produced by acetate and shikimate pathways. These compounds are present in almost all edible and medicinal plants.

Numerous epidemiological studies show that T2D can be prevented using a diet abundant in polyphenolic compounds (Cao et al., 2019; Porto et al., 2021). These compounds may offer protection against oxidative stress and T2D as evidenced by their apparent low toxicity and extensive presence in the diet of humans (Xiao and Hogger, 2015). The present chapter reviews the role of food-linked polyphenolics in the management of diabetes. This includes food sources of polyphenolic compounds, metabolism and bioavailability, and antidiabetic mechanisms. Further, the clinical studies performed to demonstrate the efficacy of food-linked polyphenols for diabetes management are discussed in the chapter.

15.2 FOOD SOURCES OF ANTIDIABETIC POLYPHENOLICS

Foods of plant origin include fruits, vegetables, nuts, seeds, oils, whole grains, legumes, beans, and beverages are not only rich sources of nutrients, but also contain significant levels of non-nutrient bioactive compounds known as phytochemicals. Among the different classes of phytochemicals, phenolics like flavonoids (Keservani and Sharma, 2014; Keservani et al., 2016a, b) and non-flavonoids are non-essential bioactive dietary compounds and are extensively present in plant-based diets (Figure 15.1). This class of compounds comprises an aromatic ring having one or more hydroxyl groups bonded to the aromatic hydrocarbon group. The structural diversity is associated with phenolics and polyphenolics class (phytophenols) of plant secondary metabolites as well as accountable for the aliments and health beneficial action of the functional foods in various diseases including diabetes, neurodegenerative, and cardiovascular diseases (CVDs), cancer, and so forth. These plant secondary metabolites are generally classified into flavonoids and a non-flavonoid group of compounds. Their chemical structures are characterized by one/more aromatic rings and one/more hydroxyl [–OH] substituents commonly called phenol ring units. The classification of these compounds and further their subdivision into subclass depend on (i) the number of phenol ring units and/or linkage type between phenol ring units; and (ii) the position and pattern of substituents within their molecular structures. The presence of the phenol ring unit in these compounds is accountable for their biological action and therapeutic role. They can scavenge free radicals, chelate metals, and other pro-oxidative compounds and neutralize the toxin by protecting the biological tissues against their damaging action. There are various types of phytophenols, namely phenolic acids, flavonoids, stilbenes, lignin, tannins, coumarins, curcuminoids, xanthones, and so forth. Among them, flavonoids are the most abundant phytophenols, having various subclasses, viz. flavanols, anthocyanidins, flavones, flavan-3-ols (catechins), flavanones, isoflavones, and so forth (Sharma et al., 2019, 2021). The commonly studied classes of phenolics and polyphenolic compounds with their plant-based foods and common structural features are summarized in Table 15.1.

15.3 METABOLISM AND BIOAVAILABILITY OF POLYPHENOLS

Despite being a reservoir of pharmacotherapeutic properties, polyphenols have not been exploited to the verge of their actual worth. In this context, understanding the bioavailability of food-linked polyphenolic compounds and their metabolism will trigger the expansion of their health benefits and role in various diseases, including diabetes, neurodegenerative, CVDs, and cancer (Rasouli et

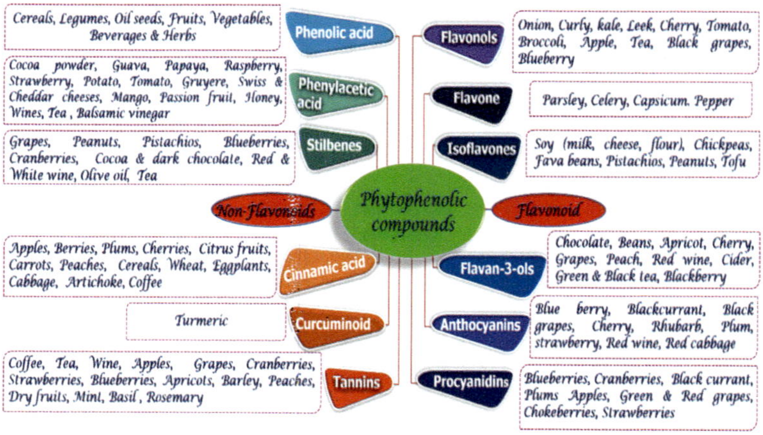

FIGURE 15.1 Common classes of phenolics of plants/plant-based foods.

al., 2017). The complexity of polyphenols in terms of structural diversity and biological activities is yet another challenge in understanding the absorption, metabolism, and bioavailability of these significant compounds. Also, the interactions of polyphenols with other food ingredients, such as fatty acids, dietary fibers, and proteins, affect the metabolic activity and bioactivity of polyphenols (Renard et al., 2017). In order to understand the diverse biotransformation processes that polyphenols undergo inside the human body, it is necessary to understand the synchronization between the metabolic pathway of polyphenols and other biochemical pathways (proteins, carbohydrates, and lipids) that are being regulated within the human system (Cao et al., 2019).

The major fraction of food-linked polyphenols lingers inside the colon, where the residing gut microbiota transform the polyphenolic compounds into metabolites that are further metabolized after entering the systemic circulation. However, the structural diversity of polyphenols has also complicated the clear vision of their metabolism because the bioavailability varies among different categories of dietary polyphenols. The most abundant category of food-linked polyphenols comprises flavonoids. The physicochemical properties (size, lipophilicity, configuration, solubility, and pKa) of flavonoids determine their absorption. Flavonoids either have to enter the colon before absorption or they can be directly absorbed from the small intestine. Based on structural diversity, flavonoids exist as glycosides or aglycones. Their glycoside form has to be converted to aglycon form before absorption while aglycons are directly absorbed from the small intestine (Hollman et al., 1999).

Another reported theory says that inside the small intestine, the metabolism of flavonoids includes cleavage of sugar moieties from the phenolic

TABLE 15.1 Type of Phytophenols, Structural Features, and Their Actions

Class of Phytophenol		Structural Feature	Action	References
Flavonoids	Flavone Flavanol Anthocyanins Flavanones Isoflavones	[Phenyl substituted chromones; (C_6–C_3–C_6) diphenylpropane carbon skeleton] 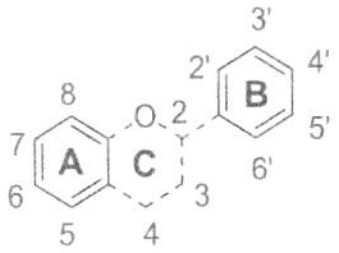 • B ring either connected at 2nd or 3rd-position of C ring • B ring is monohydroxylated (4′-OH), pyrocatechol (3′-, 4′-OH), resorcinol (3′-, 5′-OH), pyrogallol (3′-, 4′-, 5′-OH). • A ring is either *meta*-dihydroxylated (5- and 7-OH groups) or monohydroxylated • Presence or absence of C_2=C_3 double bond, 3-OH and 4-oxo group in ring C	Glucose absorption regulation, intestinal micro-biota regulation, and glucose metabolism regulation, by reducing and increasing the expression of IL-6 and MCP-1 and anti-inflammatory adipokine and adiponectin, respectively, inhibit digestive enzymes and inhibiting end product formation of glycation.	Sun et al. (2020)
	Isoflavones	[3-phenylchromen-4-one; C_6–C_3–C_6 carbon skeleton] • Ring B is either *mono*-hydroxylated (4′-OH group), methoxy substituted (4′-OCH_3 group). • Ring A is *mono*-hydoxlated (7-OH groups), *meta*-dihydroxylated (5- and 7-OH groups). • C_2=C_3 double bond and 4-oxo group on ring C. • Ring B linked at the 3-position of ring C.	Antidiabetic and pleiotropic effects, reduces cholesterol and triglycerides and improves glycemic control.	Duru et al. (2019)
	Catechins (flavan-3-ols)	[2,3-dihydro-2-phenylchroman-3-ol; C_6–C_3–C_6 carbon skeleton] • Ring B is monohydroxylated (4′-OH), pyrocatechol (3′-, 4′-OH), resorcinol (3′-, 5′-OH), pyrogallol (3′-, 4′-, 5′-OH). • Ring A is either monohydroxylated, *meta*-dihydroxylated (5- and 7-OH groups) C_2–C_3 single bond, 3-OH or 3-gallate ester, chiral center at C_2 and C_3, 4-oxo group on ring C. • Ring B linked at 2-position of ring C.	Decrease inflammation by modulating the expression of MCP-1 and IL-6 and increased production of adipokine and adiponectin.	Duru et al. (2019)
	Anthocyanins/ anthocyanidins	[3-hydroxy-2-phenylchromenylium; C_6–C_3–C_6 carbon skeleton] • Ring B is monohydroxlated (4′-OH), pyrocatechol (3′-, 4′-OH), resorcinol (3′-, 5′-OH), pyrogallol (3′-, 4′-, 5′-OH). • Ring A is either momohydroxlated or *meta*-dihydroxylated (5- and 7-OH groups) Ring C a pyrylium ion with a 3-OH group. • Ring B linked at 2-position of ring C.	Reducing CVD, cancer, hyperlipidemia, and other insulin-resistance-related diseases, reduces glucose absorption, adipose triglyceride lipase expressions, lipid synthesis, and FoxO1, and improves insulin secretion, sensitivity, glucose tolerance, uptake, consumption, and antioxidant activity.	Sivamaruthi (2018)

TABLE 15.1 *(Continued)*

Class of Phytophenol		Structural Feature	Action	References
	Procyanidins	[Condensed flavan-3-ols (catechin or epicatechin)] • Polymeric compounds of catechin or epicatechin monomers, composed of at least 4 or more monomers. • Under oxidative conditions, procyanidin is depolymerized into anthocyanidin.	• AMPK signaling and insulin pathways, cellular NAD^+ and SIRT1 concentrations, expression of pro-inflammatory cytokine and target digestive enzymes.	Sun et al. (2020)
Stilbenes Resveratrol		[1,2-diphenylethylene; C_6–C_2–C_6 carbon skeleton] • Two aromatic rings connected through ethylene moiety. • The different patterns of substituents, such as hydroxyl, methoxy, isoamylene, *O*-glycosides, and so on, on both aromatic rings. • Exist as *trans*- and *cis*-stilbenes.	• Antioxidants and anti-inflammatory, decreasing expression of mRNA and secreting various adipokines such as plasminogen activator inhibitor, and increases function of pancreatic β-cell by inhibition of phosphodiesterase activity.	Sun et al. (2020)
Cinnamic acid	Hydroxycinnamic acids	[Phenylpropenoic acid; C_6–C_3 carbon skeleton] • Organic aromatic compounds containing an aromatic ring and a carboxylic acid group forming 3-phenylprop-2-enoic acid. • The aromatic ring is substituted either by hydroxyl or methoxy. • Ester of these acids with quinic acid is also included in this class.	Prevent β-cell destruction, increase production of insulin, inhibit carbohydrate digestive enzymes, increase glycogensynthesis, and enhance uptake of glucose in adipocytes by AMPK phosphorylation and increased GLUT4 content.	Alam et al. (2016)
	Caffeoylquinic acids		Neuroprotective, antioxidant, stimulates the secretion of pancreatic enzymes, slows down aging, and regulates lipid metabolism.	Woźniak et al. (2020)
Tannins		[Hydroxylated and polymerized high molecular weight polyphenols] • Basic units are Gallic acid, Phloroglucinol, and Flavan-3-ol's. • Compounds consist of at least 12 hydroxyl groups and at least five phenyl groups to function as protein binders.	Inhibits adipogenesis and enhances glucose uptake, improves the pathological oxidative state of a diabetic situation, AMPK phosphorylation, and glucose uptake in adipose tissue via IRS-1 phosphorylation.	Kumari and Jain (2012); Sun et al. (2020)
Curcuminoids Curcumin		[1,7-Bis (substituted phenyl)-1,6-heptadiene-3,5-dione; C_6–C_7–C_6 carbon skeleton] • A two aromatic rings system linked by seven carbon linkers containing an *α*, *β*-unsaturated *β*-diketones motif. • Para-hydroxyl groups and meta-methoxy groups either on one or both aromatic rings.	Anti-inflammatory, antioxidant, and anti-carcinogenic effects increase adiponectin secretion, protect pancreatic β-cell, prevent translocation of the "insulin-regulated glucose transporter (GLUT4)" and glucose transport.	Naheed et al. (2017)
Coumarin Cinnamon		[2H-chromen-2-one; C_6–C_3 carbon skeleton] • 1-benzopyran-2-one chemical class and pondered as a lactone. • Various substituents including alkyl, alkoxy, hydroxyl, and phenyl, and so on, are present, and different types such as phenyl, furano, simple, pyrano, and biscoumarins.	Natural blood thinner, patients with T2D had their blood sugar levels lowered by half a teaspoon and decreased fasting blood glucose levels.	Naheed et al. (2017)

backbone, which are then absorbed here. Intestinal enzymes, that is, "lactase phlorizin hydrolase or β-glucosidase" hydrolyze glycosylated flavonoids into aglycones which invade intestinal epithelial cells through passive diffusion (Marin et al., 2015). Those flavonoid glycosides are not the substrate for the hydrolyzing intestinal enzymes; they enter the colon where the gut microbes hydrolyze them, but simultaneously these microbes degrade the liberated flavonoid aglycones. Thus, a very small amount of polyphenol absorption occurs in the colon as compared to that in the small intestine. Similarly, the other category of polyphenols namely hydroxycinnamic acids are not hydrolyzed by the intestinal enzymes rather their conjugate moieties are cleaved by the colon microbiota. Further, the microbiota metabolizes the resultant aglycones; leading to the production of hydroxyphenyl acetic acids (Marin et al., 2015). After absorption, most of the dietary polyphenols enter systematic circulation upon conjugating with sulfate, glucuronide, and methyl groups in the gastrointestinal mucosa (Han et al., 2007) as shown in Figure 15.2.

The products released after absorption cross the bloodstream through the portal vein leading to entry into the liver. In the liver, various enzymatic transformations occur and further from there, the absorbed products are directed towards other tissues and organs, either via the blood or digestive tract towards urine excretion. In the transit of absorbed polyphenolic products along various tissues and organs, strong antibacterial, antiviral, and antiparasitic activities are shown by these products (Marin et al., 2015).

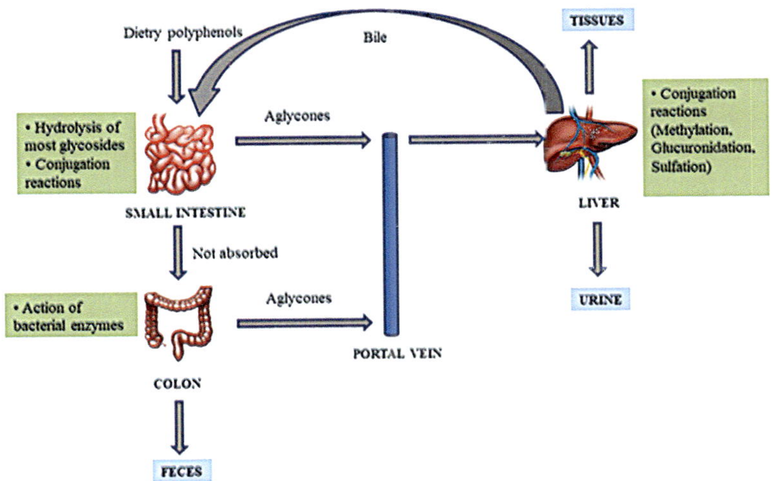

FIGURE 15.2 Routes for absorption and metabolism of food-linked polyphenols in humans.
Source: Adapted from Marin et al. (2015).

Several studies both *in-vitro* and *in-vivo* are being conducted and required to further elucidate the absorption/metabolism/and bioavailability of food-linked polyphenols. Even the deployment of metabolomics and microbiomics along with the implementation of humanized mouse models, *in vitro* gut models, and human intervention trials can unravel the bioconversion of polyphenols by colonic microbiota (Dyunhoven et al., 2011).

15.4 ANTIDIABETIC MECHANISM OF POLYPHENOLS

Diabetes is a chronic, metabolic disease that occurs due to an interaction between genetic factors and an unhealthy lifestyle. There are a variety of drugs available to combat this chronic disorder; however, they impose side effects on the patients. Thus, there is a pressing demand for a new therapeutic agent that has negligible side effects that can treat this metabolic disorder. In this context, plant-derived polyphenols are beneficial to control and prevent diabetes through different mechanisms, that is, (i) glucagon-like peptide-1 (GLP-1); (ii) antiglycation potential; (iii) improvement of insulin sensitivity; (iv) inhibition of the digestive enzyme; and (v) pre-biotic potential. Food-linked polyphenols help in preventing and managing type 2 diabetes (T2D) via stimulation of insulin secretion, insulin signaling activation, shielding of islet β-cell (pancreatic), promotion of β-cell proliferation, reduction of oxidative stress and β-cell apoptosis, inhibition of digestive enzymes and glucose absorption, and regulation of intestinal microbiota (Table 15.2). More details on these mechanisms are given in the subsections.

15.4.1 GLUCAGON-LIKE PEPTIDE-1 (GLP-1)

Glucagon-like peptide-1 (GLP-1) is an incretin hormone secreted by enteroendocrine L-cells in response to food ingestion. GLP-1 stimulates pancreatic beta-cells to secrete insulin in a glucose-dependent manner (Lee et al., 2014). GLP-1 is an agonist in the treatment of T2D. The effects of GLP-1 on the treatment of diabetes are increasing glucose transporter-2 expression, inducing beta cells to secrete insulin, restricting alpha cells to secrete glucagon, and reducing pro-inflammatory cytokine secretion (Lee et al., 2017). Therefore, GLP-1 is an important target for the treatment of diabetes. The short half-life of 1.5 to 5 minutes in blood serum is a major disadvantage of GLP-1 (Hui et al., 2002). The short half-life of GLP-1 is mainly due to DPP-4. Dipeptidyl peptidase-4 (DPP-4) regulates glucose metabolism and degrades GLP-1. The absence of DPP4 in mice improves glycemic control. DPP4 is approved as an antidiabetic drug target (Zilleßen et al., 2016). DPP-4 inhibitors are molecules that inhibit the degradation of the incretins, and glucose-dependent insulinotropic peptides (Thornberry et al., 2009).

Pomegranate polyphenols ellagic acid, punicalagin, and urolithin are involved in carbohydrate metabolism and alpha-GLU and DPP-4. A study using the 3T3-L1 cell line revealed that pomegranate polyphenol inhibits DPP-4 in a dose-dependent manner and modulates gene expression of genes such as GLUT-4 and PABA-Y (Les et al., 2018). Sprouted quinoa yogurt modulates glucagon-like peptide levels. *In-vitro* study in NCI-H716 cells shows that polyphenol-polysaccharide conjugate isolated from sprout upregulates expression of GLP-1. The expression of proglucagon mRNA is also upregulated in cells (Obaroakpo et al., 2020).

15.4.2 ANTIGLYCATION AND ANTIOXIDANT PROPERTY

Glycation can be defined as a non-enzymatic and spontaneous nucleophilic addition reaction between reducing sugar and proteins to form a reversible Schiff base. The Schiff base endures changes and forms a stable product named an Amadori product, and further irreversible changes such as oxidation, polymerization, and cross-linking result in "advanced glycation end products (AGEs)" formation (Kim et al., 2017). Non-enzymatic glycation plays a crucial physiological role in aging, diabetes, cardiometabolic disorder, and neurodegenerative disease. The receptor for AGEs (RAGE) is present in many cells, and interaction between them causes the release of free radicals, pro-inflammatory molecules, alteration in gene expression, and intracellular signaling, which leads to diabetic complications (Ahmed, 2005).

Polyphenol extract has anti-glycation properties and possible therapeutic potential for diabetic complications. The generation of intracellular ROS by albumin-derived AGEs also inhibits glucose uptake and oxidative alteration of intracellular protein. The phenolic and flavonoid content of four medicinal plants, for example, *Fumaria officinalis* have antiglycation and antioxidant properties. The extract inhibits albumin glycation at three stages, that is, fructosamine generation, beta-amyloid generation, and carbonyl formation (Safari et al., 2018). Polyphenolic extracts of nutmeg, ginger, and alligator pepper also show potential antiglycation and antioxidant activity (Kazeem et al., 2012). *In-vivo* study revealed that polyphenol extract of *Myrica faya* inhibits enzymes alpha-glucosidase, alpha-amylase, aldose reductase, lipase, and glycation of bovine serum. The enzymes are associated with T2D, obesity, and the formation of AGEs (Spinola et al., 2019).

15.4.3 METHYLGLYOXAL AND GLYOXAL

By various metabolic pathways, AGEs produce reactive intermediates, that is, methylglyoxal, 3-deoxyglucose, and glyoxal. The intermediate can bind DNA, protein, and other molecules. These can disrupt the function and structure of the molecule and lead to diabetes and other disease complications (Desai et al., 2007). Methylglyoxal is an extremely reactive dicarbonyl compound. It is produced as a by-product of the non-enzymatic degradation of glucose-3 phosphate and dihydroxy acetone phosphate (Schalkwijk et al., 2019). Glycation is also an important therapeutic target to prevent diabetes complications.

15.4.4 INSULIN SENSITIVITY

Insulin resistance (IR) refers to the decreased ability of insulin to regulate glucose metabolism. The primary tissues are the liver, muscle, and adipose tissue. IR leads to hyperinsulinemia, hyperglycemia, hypertension, and an increase in inflammatory markers (Reynolds et al., 2020). During normal insulin signaling insulin binds to insulin receptors and leads to the activation of glucose transporter-4 (GLUT-4). GLUT-4 imports glucose into the cell. Insulin-senstive tissues in healthy indivdual, adipose tissue, liver, and skeletal muscle are highly specific to insulin whereas in diabetic patients the tissues are insulin-resistant.

Apple polyphenol extract improves insulin sensitivity in an insulin-resistant animal model. The *in-vivo* and *in-vitro* studies revealed that apple polyphenol extract increases insulin sensitivity and increases the uptake of glucose in muscle cells through GLUT-4 translocation. Phosphatidyl inositol 3-kinase (PI3K) and PPARγ signaling pathways are involved in translocation (Manzano et al., 2016). A study revealed that the combined effect of grape polyphenol and exercise prevents IR in obese rats with a high-fat diet. Exercise and polyphenols increase oxidative pathways in skeletal muscle. After 8 weeks of treatment AMPK activates, phosphorylation level induced, mitochondrial enzymes citrate synthase and cytochrome oxidase activated. The synergistic effect of grape polyphenol and exercise increases lipid oxidation in muscle and glycogen utilization (Lambert et al., 2018).

15.4.5 INHIBITION OF DIGESTIVE ENZYME

Dietary polyphenol extract can inhibit various digestive enzymes involved in starch digestion, glucose transport, amylases, and glucosidases. The inhibition depends on doses, structure, and food. The inhibition by polyphenols leads to reduced formation of carbohydrates such as maltose, dextrins, and maltotriose. Thereby, polyphenols reduce the transport and formation of glucose in the membrane and reduce glycemia and insulin secretion (Ayua et al., 2021). The polyphenolic compound and digestive enzyme interact through non-covalent binding involving Van der Waals forces, hydrophobic binding, hydrogen bond, and electrostatic interaction. The interaction inhibits enzyme activity by non-competitive type inhibition (Martinez-Gonzalez et al., 2017).

15.4.6 OTHERS

Berry polyphenol controls diabetes by insulin-dependent and insulin-independent mechanisms. The insulin-dependent mechanism involves enhancing pancreatic beta cell and peripheral tissue insulin and the insulin-independent mechanism involves inhibition of glucose absorption and digestive enzymes, and modification of energy metabolism (Edirisinghe et al., 2016). *In-vivo* and human studies revealed the role of anthocyanin in controlling diabetes. Anthocyanin controls diabetes by different molecular mechanisms, that is, inhibits digestive enzymes (alpha-amylase), inhibits DPP-IV, and reduces IR, inflammation, and oxidative stress (Les et al., 2021). Curcumin polyphenols undergo degradation by intestinal microbiota. The degradative products of curcumin have a direct effect on tissues as they alter microbial flora, improve intestinal wall permeability, and have an anti-inflammatory and antidiabetic effect (Jin et al., 2018).

TABLE 15.2 Examples of Polyphenolic Foods, Their Target, and Mechanism of Action

Polyphenol Food Example	Target	Mechanism of Action	References
Pomegranate	GLP-1	Inhibits DPP-4	Les et al. (2018)
Sprouted quinoa yoghurt	GLP-1	Up-regulate expression of GLP-1.	Obaroakpo et al. (2020)
Fumaria officinalis	Antiglycation	Inhibits albumin glycation at three stages, that is, fructosamine generation, beta-amyloid generation, and carbonyl formation.	Safari et al. (2018)
Myrica faya	Advanced glycation end product	Inhibits enzymes alpha-glucosidase, alpha-amylase, aldose reductase, lipase, and glycation of bovine serum.	Spinola et al. (2019)
Apple	Increase insulin sensitivity (GLUT-4)	"PI3K and PPARγ" signaling pathways are involved in translocation.	Manzano et al. (2016)
Grape polyphenol	Prevents IR	AMPK activates, phosphorylation level induced, mitochondrial enzymes citrate synthase and cytochrome oxidase activated.	Lambert et al. (2018)
Turmeric	Intestinal micro-biota	Anti-inflammatory	Jin et al. (2018)

15.5 CLINICAL APPLICATIONS

Phytophenols are naturally occurring micronutrients and are well known for their antioxidant and anti-inflammatory properties. They are compounds having one or more phenolic rings commonly found in green tea, alcoholic beverages like wine, various fruits, and veggies.

A trial on healthy adults (20 in number) showed a decrease in "C-reactive protein (CRP)" in plasma and "tumor necrosis factor (TNF)" after 6 weeks of supplementation of resveratrol-containing plant extract (Semba et al., 2014). Resveratrol inhibits the expression of tyrosine phosphatase 1B in the muscles and liver of diabetic rats by improving IRS-1-mediated insulin signaling (González et al., 2015). It was reported that tetrahydrocurcumin administrated at a body weight of 80 mg/kg decreased blood glucose level in streptozotocin (STZ) diabetic rats. It might be owing to pancreatic insulin secretion from existing β-cell or to increased blood glucose transport to peripheral tissue (Pari and Murugan, 2007). Studies showed catechins and epigallocatechin 3-gallate affect glucose tolerance, insulin sensitivity, and insulin secretion in STZ diabetic rats, and it has also shown improvement in those diabetic patients who drink green tea four times a day (Mousavi et al., 2013). The bioavailability of these phytophenols depends upon their consumption, preparation of food, absorption, and metabolism in the GI tract. During metabolism, dietary polyphenols act differently and improve the health of diabetic patients. A few clinical studies performed on the effect of various phytophenols are summarized in Table 15.3.

TABLE 15.3 Clinical Effect of Various Phytophenols with Respect to Diabetes

Sl. No.	Phytophenols	Daily Dose	Effect	References
1.	Curcumin	22.1 mg	Curcumin for 2 months shows reduction of proteinuria, IL-8 levels and TGF-β.	Khajehdehi et al. (2011)
		100 mg/Kg/12 weeks	Reduce diabetic nephropathy	Sun et al. (2014)
2.	Resveratrol	Supplementation of 1 g/day of RSV for 45 days	Decrease in metabolic rate, reduction in IR, reduction in levels of inflammatory markers.	Oyenihi et al. (2016)
3.	Catechin	400–500 mg EGCG/day	Drop in body weight, fat mass, body mass index, waist circumference, abdominal fat, and blood pressure, decrease in postprandial blood glucose level.	Chen et al. (2016); Liu et al. (2014)
4.	Genistein	30–60 mg/day	Boost in serum adiponectin levels, decrease in blood glucose levels.	Yamagata and Yamori (2021)

TABLE 15.3 *(Continued)*

Sl. No.	Phytophenols	Daily Dose	Effect	References
5.	Quercetin	150–1000 mg/day	Reduction in HDL, total cholesterol level, effectively, and safely alleviates early diabetic renal injuries possibly via improving the lipid metabolism through the SCAP-SREBP2-LDLr signaling pathway.	Conquer et al. (1998)
6.	Ginger (Gingerol).	1–2 gm/day	Decrease in triglyceride, cholesterol, LDL, VLDL, Reduction in HbA1c, HOMA, and CRP.	Navael Reza et al. (2008)

15.6 CONCLUSION

Food-linked polyphenolics have tremendous therapeutic promise in the management and prevention of diabetes. The action mechanism includes several mechanisms like the stimulation of insulin secretion, insulin signaling activation, islet β-cell (pancreatic) protection, promotion of β-cell proliferation, reduction of β-cell apoptosis and oxidative stress, inhibition of digestive enzymes and glucose absorption, and regulation of intestinal microbiota. The antidiabetic potential of food-linked polyphenolics depends on factors like concentration/dosage in food, absorption, bio-accessibility, bioavailability, and metabolism. It is evident from clinical and epidemiological studies that food-linked polyphenolics improve glucose homeostasis by a variety of mechanisms. However, greater clinical shreds of evidence are required to prove the efficacy of these bioactive components for diabetes control and to elucidate the precise action mechanism underlying their antidiabetic effect.

KEYWORDS

- **albumin glycation**
- **aldose reductase**
- **antiglycation**
- **antioxidant properties**
- **beta-amyloid generation**
- **blood glucose**
- **citrate synthase**
- **glucose absorption**
- **insulin resistance**
- **insulin signaling**
- **lipase**
- **micronutrients**
- **mitochondrial enzymes**
- **phosphorylation**
- **polyphenols**
- **reactive oxygen species**

REFERENCES

Ahmed, N. (2005). Advanced glycation end-products—Role in pathology of diabetic complications. *Diabetes Research and Clinical Practice*, *67*(1), 3–21.

Alam, M. A., Subhan, N., Hossain, H., Hossain, M., Reza, H. M., Rahman, M. M., & Ullah, M. O. (2016). Hydroxycinnamic acid derivatives: A potential class of natural compounds for the management of lipid metabolism and obesity. *Nutrition & Metabolism*, *13*, 1–13.

Alizadeh-Navaei, R., Roozbeh, F., Saravi, M., Pouramir, M., Jalali, F., & Moghadamnia, A. A. (2008). Investigation of the effect of ginger on the lipid levels. *Saudi Med J*, *29*(9), 1280–1284.

Aryaeian, N., Sedehi, S. K., & Arablou, T. (2017). Polyphenols and their effects on diabetes management: A review. *Medical Journal of the Islamic Republic of Iran*, *31*, 134.

Ayua, E. O., Nkhata, S. G., Namaumbo, S. J., Kamau, E. H., Ngoma, T. N., & Aduol, K. O. (2021). Polyphenolic inhibition of enterocytic starch digestion enzymes and glucose transporters for managing type 2 diabetes may be reduced in food systems. *Heliyon*, *7*(2), e06245.

Bacanli, M., Dilsiz, S. A., Başaran, N., & Başaran, A. A. (2019). Effects of phytochemicals against diabetes. *Advances in Food and Nutrition Research*, *89*, 209–238.

Bahadoran, Z., Mirmiran, P., & Azizi, F. (2013). Dietary polyphenols as potential nutraceuticals in management of diabetes: A review. *Journal of Diabetes & Metabolic Disorders*, *12*, 1–9.

Cao, H., Ou, J., Chen, L., Zhang, Y., Szkudelski, T., Delmas, D., Daglia, M., & Xiao, J. (2019). Dietary polyphenols and type 2 diabetes: Human study and clinical trial. *Critical Reviews in Food Science and Nutrition*, *59*(20), 3371–3379.

Chen, I. J., Liu, C. Y., Chiu, J. P., & Hsu, C. H. (2016). Therapeutic effect of high-dose green tea extract on weight reduction: A randomized, double-blind, placebo-controlled clinical trial. *Clinical Nutrition*, *35*(3), 592–599.

Conquer, J. A., Maiani, G., Azzini, E., Raguzzini, A., & Holub, B. J. (1998). Supplementation with quercetin markedly increases plasma quercetin concentration without effect on selected risk factors for heart disease in healthy subjects. *The Journal of Nutrition*, *128*(3), 593–597.

Da Porto, A., Cavarape, A., Colussi, G., Casarsa, V., Catena, C., & Sechi, L. A. (2021). Polyphenols rich diets and risk of type 2 diabetes. *Nutrients*, *13*(5), 1445.

de Paulo Farias, D., de Araujo, F. F., Neri-Numa, I. A., & Pastore, G. M. (2021). Antidiabetic potential of dietary polyphenols: A mechanistic review. *Food Research International*, *145*, 110383.

Desai, K., & Wu, L. (2007). Methylglyoxal and advanced glycation endproducts: New therapeutic horizons? *Recent Patents on Cardiovascular Drug Discovery (Discontinued)*, *2*(2), 89–99.

Duru, K. C., Kovaleva, E. G., Danilova, I. G., Van der Bijl, P., & Belousova, A. V. (2018). The potential beneficial role of isoflavones in type 2 diabetes mellitus. *Nutrition Research*, *59*, 1–15.

Edirisinghe, I., & Burton-Freeman, B. (2016). Antidiabetic actions of berry polyphenols—Review on proposed mechanisms of action. *Journal of Berry Research*, *6*(2), 237–250.

Fan, J., Johnson, M. H., Lila, M. A., Yousef, G., & De Mejia, E. G. (2013). Berry and citrus phenolic compounds inhibit dipeptidyl peptidase IV: Implications in diabetes management. *Evidence-Based Complementary and Alternative Medicine*, *2013*, *479505*.

González-Rodríguez, Á., Santamaría, B., Mas-Gutierrez, J. A., Rada, P., Fernández-Millán, E., Pardo, V., Álvarez, C., Cuadrado, A., Ros, M., Serrano, M., & Valverde, Á. M. (2015).

Resveratrol treatment restores peripheral insulin sensitivity in diabetic mice in a Sirt1-independent manner. *Molecular Nutrition & Food Research, 59*(8), 1431–1442.

Guasch-Ferré, M., Merino, J., Sun, Q., Fitó, M., & Salas-Salvadó, J. (2017). Dietary polyphenols, Mediterranean diet, prediabetes, and type 2 diabetes: A narrative review of the evidence. *Oxidative Medicine and Cellular Longevity, 2017, 6723931.*

Haldar, S., Chia, S. C., Lee, S. H., Lim, J., Leow, M. K. S., Chan, E. C. Y., & Henry, C. J. (2019). Polyphenol-rich curry made with mixed spices and vegetables benefits glucose homeostasis in Chinese males (Polyspice Study): A dose-response randomized controlled crossover trial. *European Journal of Nutrition, 58*, 301–313.

Han, X., Shen, T., & Lou, H. (2007). Dietary polyphenols and their biological significance. *International Journal of Molecular Sciences, 8*(9), 950–988.

Hui, H., Farilla, L., Merkel, P., & Perfetti, R. (2002). The short half-life of glucagon-like peptide-1 in plasma does not reflect its long-lasting beneficial effects. *European Journal of Endocrinology, 146*(6), 863–869.

International Diabetes Federation. (2021). *IDF Diabetes Atlas 10th Edition*. Retrieved from: https://diabetesatlas.org (accessed on 5 November 2024).

Jin, T., Song, Z., Weng, J., & Fantus, I. G. (2018). Curcumin and other dietary polyphenols: Potential mechanisms of metabolic actions and therapy for diabetes and obesity.

Kawakami, Y., Watanabe, Y., Mazuka, M., Yagi, N., Sawazaki, A., Koganei, M., Natsume, M., Kuriki, K., Morimoto, T., Asai, T., & Arai, H. (2021). Effect of cacao polyphenol-rich chocolate on postprandial glycemia, insulin, and incretin secretion in healthy participants. *Nutrition, 85*, 111128.

Kazeem, M. I., Akanji, M. A., Hafizur, R. M., & Choudhary, M. I. (2021). Antiglycation, antioxidant, and toxicological potential of polyphenol extracts of alligator pepper, ginger, and nutmeg from Nigeria. *Asian Pacific Journal of Tropical Biomedicine, 2*(9), 727–732.

Keservani, R. K., & Sharma, A. K. (2014). Flavonoids: Emerging trends and potential health benefits. *Journal of Chinese Pharmaceutical Sciences, 23*(12), 815.

Keservani, R. K., Sharma, A. K., & Kesharwani, R. K. (2016a). Nutraceutical and functional foods for cardiovascular health. In *Food Process Engineering* (pp. 291–312). Apple Academic Press, CRC Press. (pp. 257–278). ISBN: 9781771884020.

Keservani, R. K., Sharma, A. K., & Kesharwani, R. K. (2016b). Medicinal effect of nutraceutical fruits for cognition and brain health. *Scientifica, 3109254*, 10. https://doi.org/10.1155/2016/3109254.

Khajehdehi, P., Pakfetrat, M., Javidnia, K., Azad, F., Malekmakan, L., Nasab, M. H., & Dehghanzadeh, G. (2011). Oral supplementation of turmeric attenuates proteinuria, transforming growth factor-β and interleukin-8 levels in patients with overt type 2 diabetic nephropathy: A randomized, double-blind and placebo-controlled study. *Scandinavian Journal of Urology and Nephrology, 45*(5), 365–370.

Kim, C. S., Park, S., & Kim, J. (2017). The role of glycation in the pathogenesis of aging and its prevention through herbal products and physical exercise. *Journal of Exercise Nutrition & Biochemistry, 21*(3), 55.

Kumari, M., & Jain, S. (2012). Tannins: An antinutrient with a positive effect to manage diabetes. *Research Journal of Recent Sciences, ISSN, 2277*, 2502.

Lambert, K., Hokayem, M., Thomas, C., Fabre, O., Cassan, C., Bourret, A., Bernex, F., Feuillet-Coudray, C., Notarnicola, C., Mercier, J., & Avignon, A. (2018). Combination of nutritional polyphenols supplementation with exercise training counteracts insulin resistance and improves endurance in high-fat diet-induced obese rats. *Scientific Reports*, *8*(1), 2885.

Lee, S., & Lee, D. Y. (2017). Glucagon-like peptide-1 and glucagon-like peptide-1 receptor agonists in the treatment of type 2 diabetes. *Annals of Pediatric Endocrinology & Metabolism*, *22*(1), 15–26.

Lee, Y. S., & Jun, H. S. (2014). Antidiabetic actions of glucagon-like peptide-1 on pancreatic beta-cells. *Metabolism*, *63*(1), 9–19.

Les, F., Arbonés-Mainar, J. M., Valero, M. S., & López, V. (2018). Pomegranate polyphenols and urolithin A inhibit α-glucosidase, dipeptidyl peptidase-4, lipase, triglyceride accumulation, and adipogenesis related genes in 3T3-L1 adipocyte-like cells. *Journal of Ethnopharmacology*, *220*, 67–74.

Les, F., Cásedas, G., Gómez, C., Moliner, C., Valero, M. S., & López, V. (2021). The role of anthocyanins as antidiabetic agents: From molecular mechanisms to *in vivo* and human studies. *Journal of Physiology and Biochemistry*, *77*(1), 109–131.

Li, X. W., Chen, H. P., He, Y. Y., Chen, W. L., Chen, J. W., Gao, L., Hu, H. Y., & Wang, J. (2018). Effects of rich-polyphenols extract of *Dendrobium loddigesii* on antidiabetic, anti-inflammatory, antioxidant, and gut microbiota modulation in db/db mice. *Molecules*, *23*(12), 3245.

Liu, C. Y., Huang, C. J., Huang, L. H., Chen, I. J., Chiu, J. P., & Hsu, C. H. (2014). Effects of green tea extract on insulin resistance and glucagon-like peptide 1 in patients with type 2 diabetes and lipid abnormalities: A randomized, double-blinded, and placebo-controlled trial. *PLOS One*, *9*(3), 91163.

Manzano, M., Giron, M. D., Vilchez, J. D., Sevillano, N., El-Azem, N., Rueda, R., Salto, R., & Lopez-Pedrosa, J. M. (2016). Apple polyphenol extract improves insulin sensitivity *in vitro* and *in vivo* in animal models of insulin resistance. *Nutrition & Metabolism*, *13*(1), 1–10.

Marín, L., Miguélez, E. M., Villar, C. J., & Lombó, F. (2015). Bioavailability of dietary polyphenols and gut microbiota metabolism: Antimicrobial properties. *BioMed Research International*, *2015*, 905215.

Martínez-González, A. I., Díaz-Sánchez, Á. G., de la Rosa, L. A., Vargas-Requena, C. L., Bustos-Jaimes, I., & Alvarez-Parrilla, E. (2017). Polyphenolic compounds and digestive enzymes: *In vitro* non-covalent interactions. *Molecules*, *22*(4), 669.

Mousavi, A., Vafa, M., Neyestani, T., Khamseh, M., & Hoseini, F. (2013). The effects of green tea consumption on metabolic and anthropometric indices in patients with Type 2 diabetes. *Journal of Research in Medical Sciences: The Official Journal of Isfahan University of Medical Sciences*, *18*(12), 1080.

Obaroakpo, J. U., Liu, L., Zhang, S., Lu, J., Liu, L., Pang, X., & Lv, J. (2020). *In vitro* modulation of glucagon-like peptide release by DPP-IV inhibitory polyphenol-polysaccharide conjugates of sprouted quinoa yoghurt. *Food Chemistry*, *324*, 126857.

Oyenihi, O. R., Oyenihi, A. B., Adeyanju, A. A., & Oguntibeju, O. O. (2016). Antidiabetic effects of resveratrol: The way forward in its clinical utility. *Journal of Diabetes Research*, 9737483. https://doi.org/10.1155/2016/9737483.

Rehman, K., Al-Gubory, K. H., Laher, I., & Akash, M. S. H. (2017). Dietary polyphenols in the prevention and treatment of diabetes mellitus. In K. Al-Gubory & I. Laher (Eds.), *Nutritional Antioxidant Therapies: Treatments and Perspectives* (pp. 345–369). Springer. https://doi.org/10.1007/978-3-319-67625-8_15.

Renard, C. M., Watrelot, A. A., & Le Bourvellec, C. (2017). Interactions between polyphenols and polysaccharides: Mechanisms and consequences in food processing and digestion. *Trends in Food Science & Technology*, *60*, 43–51.

Reynolds, A. N., Akerman, A. P., & Mann, J. (2020). Dietary fibre and whole grains in diabetes management: Systematic review and meta-analyses. *PLOS Medicine*, *17*(3), 1003053.

Safari, M. R., Azizi, O., Heidary, S. S., Kheiripour, N., & Ravan, A. P. (2018). Antiglycation and antioxidant activity of four Iranian medical plant extracts. *Journal of Pharmacopuncture*, *21*(2), 82.

Schalkwijk, C. G., & Stehouwer, C. D. A. (2020). Methylglyoxal, a highly reactive dicarbonyl compound, in diabetes, its vascular complications, and other age-related diseases. *Physiological Reviews*, *100*(1), 407–461.

Semba, R. D., Ferrucci, L., Bartali, B., Urpí-Sarda, M., Zamora-Ros, R., Sun, K., Cherubini, A., Bandinelli, S., & Andres-Lacueva, C. (2014). Resveratrol levels and all-cause mortality in older community-dwelling adults. *JAMA Internal Medicine*, *174*(7), 1077–1084.

Sharma, A., Sharma, S., Kumar, A., Kumar, V., & Sharma, A. K. (2022). Plant secondary metabolites: An introduction to their chemistry and biological significance with physicochemical aspects. In *Plant Secondary Metabolites: Physico-Chemical Properties and Therapeutic Applications* (pp. 1–45). Springer Nature Singapore.

Sharma, A., Tuli, H. S., & Sharma, A. K. (2019). Chemistry and synthetic overview of flavonoids. In *Current Aspects of Flavonoids: Their Role in Cancer Treatment* (pp. 23–38).

Shi, Y., & Hu, F. B. (2014). The global implications of diabetes and cancer. *Lancet (London, England)*, *383*(9933), 1947–1948.

Sivamaruthi, B. S., Kesika, P., Subasankari, K., & Chaiyasut, C. (2018). Beneficial effects of anthocyanins against diabetes mellitus-associated consequences: A mini review. *Asian Pacific Journal of Tropical Biomedicine*, *8*(10), 471.

Spínola, V., Llorent-Martínez, E. J., & Castilho, P. C. (2019). Polyphenols of *Myrica faya* inhibit key enzymes linked to type II diabetes and obesity and formation of advanced glycation end-products (*in vitro*): Potential role in the prevention of diabetic complications. *Food Research International*, *116*, 1229–1238.

Sun, C., Zhao, C., Guven, E. C., Paoli, P., Simal-Gandara, J., Ramkumar, K. M., Wang, S., Buleu, F., Pah, A., Turi, V., & Damian, G. (2020). Dietary polyphenols as antidiabetic agents: Advances and opportunities. *Food Frontiers*, *1*(1), 18–44.

Sun, L. N., Yang, Z. Y., Lv, S. S., Liu, X. C., Guan, G. J., & Liu, G. (2014). Curcumin prevents diabetic nephropathy against inflammatory response via reversing caveolin-1 Tyr14 phosphorylation influenced TLR4 activation. *International Immunopharmacology*, *23*(1), 236–246.

Thomas, T., & Pfeiffer, A. F. (2012). Foods for the prevention of diabetes: How do they work? *Diabetes/Metabolism Research and Reviews*, *28*(1), 25–49.

Thornberry, N. A., & Gallwitz, B. (2009). Mechanism of action of inhibitors of dipeptidyl-peptidase-4 (DPP-4). *Best Practice & Research Clinical Endocrinology & Metabolism*, *23*(4), 479–486.

van Duynhoven, J., Vaughan, E. E., Jacobs, D. M., Kemperman, R. A., van Velzen, E. J., Gross, G., Roger, L. C., Possemiers, S., Smilde, A. K., Doré, J., & Westerhuis, J. A. (2011). Metabolic fate of polyphenols in the human superorganism. *Proceedings of the National Academy of Sciences*, *108*(supplement_1), 4531–4538.

Williamson, G., & Sheedy, K. (2020). Effects of polyphenols on insulin resistance. *Nutrients*, *12*(10), 3135.

Woźniak, D., Nawrot-Hadzik, I., Kozłowska, W., Ślusarczyk, S., & Matkowski, A. (2020). Caffeoylquinic acids. In J. Xiao, S. Sarker, & Y. Asakawa (Eds.), *Handbook of Dietary Phytochemicals* (pp. 23–1). Springer Singapore. https://doi.org/10.1007/978-981-13-1745-3_23-1.

Xiao, J. B., & Hogger, P. (2015). Dietary polyphenols and type 2 diabetes: Current insights and future perspectives. *Current Medicinal Chemistry*, *22*(1), 23–38.

Yamagata, K., & Yamori, Y. (2021). Potential effects of soy isoflavones on the prevention of metabolic syndrome. *Molecules*, *26*(19), 5863.

Zilleßen, P., Celner, J., Kretschmann, A., Pfeifer, A., Racké, K., & Mayer, P. (2016). Metabolic role of dipeptidyl peptidase 4 (DPP4) in primary human (pre) adipocytes. *Scientific Reports*, *6*(1), 1–12.

CHAPTER 16

Plant-Based Foods for the Management of Cancer

PRATAP G. KENCHAPPA,[1] POORNIMA D. VIJENDRA,[2*] HARSHINI V. KUMAR,[3] and PRATHAP H. MANJUNATHA SWAMY[1]

[1]*Department of Studies in Biochemistry, Davangere University, Shivagangothri, Davangere, Karnataka, India*

[2]*Department of Studies in Food Technology and Biochemistry, Davangere University, Shivagangothri, Davangere, Karnataka, India*

[3]*Jagadguru Jayadeva Murugharajendra Medical College, Davangere, Karnataka, India*

*Corresponding author.

ABSTRACT

The beneficial food ingredients for the prevention of diseases are present in a broad range of functional foods. Nowadays, researchers are more focused to find out the relation between foods with human diseases and their impact at the gene level. Because of the anticipated function they would play in the treatment of several pathologies, including breast cancer, colorectal cancer, and so on, they have gained remarkably significant importance in the modern era. In many industrialized and developing nations, cancer has emerged as one of the leading reasons for mortality. In the world, there are about 21 million patients suffering from various types of cancers. When ingested at effective amounts as a regular part of the diet, functional foods are regarded as enhanced meals that offer health advantages beyond the requirement of vital elements (e.g., bioactive, protein, vitamins, and minerals). They are the elements of food that offer additional health advantages beyond only providing basic nutrition and lowering the risk of illness. Since there is a

Functional Compounds and Foods of Plant Origin:Krishan Dutt Sharma, Satish Kumar, Rakesh Sharma, and Raj K. Keservani (Eds.)

high likelihood that certain items will be consumed as part of a healthy diet, this chapter explores foods such berries, seasonal fruits, vegetables, cereals, dried fruits, wild edible fruits, fruit wine, and beverages.

16.1 INTRODUCTION

Cancer leads the chronic diseases causing more deaths and disabilities globally compared to other diseases. An estimated 21 million people worldwide are affected by various cancers (Iqbal et al., 2017). Cancer can be classified into several forms, including bladder cancer, lung cancer, melanoma, leukemia, non-small cell lymphoma, kidney cancer, breast cancer, pancreatic cancer, colorectal cancer, thyroid cancer, prostate cancer, and liver cancer. A broad range of illnesses in any body part, which is characterized by uncontrolled cell division and growth, is collectively referred to as cancer. Rapidly forming aberrant cells develop outside of their normal bounds, infect nearby bodily components, and spread rapidly while the process is known as metastasis. Cancer develops when healthy cells transform into tumor cells over the course of a multi-stage process that frequently progresses from a precancerous lesion to a malignant tumor. The interaction of a person's genetic factors, ultraviolet radiation, ionizing radiation, alcohol, and tobacco smoke results in these alterations (Ng et al., 2020), aflatoxin, viruses, bacteria, an unhealthy diet, inactivity, and air pollution (Peters et al., 2019). As a result, these factors are risk factors for cancer. By 2030, there will be 23.9 million cancer-related deaths annually; a rise of 69.5% can be seen in the number of cases (Custódio et al., 2016). By altering one's lifestyle, increasing the consumption of functional foods, reducing exposure to carcinogens, and using medications or vaccines, among other things, one can reduce their risk of developing cancer. Functional foods are receiving increasing attention today since they can prevent or delay the development of cancer (Islam and Siddiqua, 2020).

Foods which beyond basic nutrition also provides additional health advantages and protection against diseases are referred to as functional foods. They have been explicitly found to have a positive impact on health and lower the chance of illness through studies. Due to their anticipated significance in the treatment of numerous illnesses, including cancer, they have grown significantly in relevance in recent years (Gupta, 2016). The nutraceutical market has expanded over the past 10 years, which may be attributed to consumers being more aware of these compounds and their use in the treatment and prevention of various ailments (Daliu et al., 2018). Numerous clinical research is being done to determine how omega-3 fatty acids affect the body's ability to deal with chronic illnesses like psoriasis, Crohn's disease, cancer, cognitive dysfunction, and autoimmune disease (Gogus and Smith, 2010; Gupta et al., 2013).

Currently, the available cancer treatments are less effective, and their efficacy is restricted to the side effects and/or drug resistance development. Some common side effects of cancer chemotherapy like diarrhea, fatigue, nausea, vomiting, swallowing problems, taste and smell changes, hair loss, constipation, and easy bruising and bleeding. Because of the potential of functional food products to maintain general well-being and exert the highest bioactivity against specific health conditions like different types of cancer, their consumption has seen an upsurge in the recent past. One of the most common chronic illnesses in the world that has no cure is cancer. However, various herbal food ingredients contain certain bioactive compounds which have shown potential to treat or manage various types of cancers. Keeping in view different types of functional foods which have the potential to manage certain types of cancer, the present chapter has been conceptualized to provide an overview of the developments in the plant-based food industry for the treatment of cancer in the last decade.

16.2 NUTRIENTS, NON-NUTRIENTS, AND NUTRITIONAL THERAPY

Phytonutrients are the chemicals that can be found in foods made from plants that humans don't require to survive but that may be beneficial to their health. Carbohydrates, proteins, minerals, vitamins, fats, oils, and organic acids are the primary nutrients in plants. They also contain anti-nutrients, including tannin, tannic acid, and trypsin inhibitors (Das et al., 2020). These offer nutritious value as well as taste and scent. However, there is currently a lot of research being done on meals to find any extra physiological advantages that can lower the risk of chronic diseases (Hasler, 2002b). The biological effects of fresh and processed food products are often linked to their ability to bind with proteins and inhibit or regulate specific enzymes and their antioxidant activity (Panche et al., 2016). Furthermore, bioactive chemicals from plant diets are extensively studied for their antioxidant capacities due to their association with oxidative stress-mediated diseases (Chandran and Abrahamse, 2020). Plants contain over 8000 different bioactive substances, such as polyphenols, carotenoids, omega-3 fatty acids, alkaloids, terpenoids, and polyunsaturated fatty acids (PUFAs) (Pratap and Shantaram, 2020). Many bioactive elements become accessible after digestion and have extra-nutritional benefit qualities (Adams et al., 2007). Antioxidant, anti-inflammatory, antibacterial, anti-hypertensive, and anti-cancer characteristics are among the properties of such chemicals, which also serve as regulators of intracellular and extracellular signaling pathways (Andrade et al., 2018). The nutraceutical industry has risen in the previous decade, owing to increased consumer awareness of these chemicals and their utility in the prevention and treatment of a variety of ailments (Farías et al., 2014). Consumption of high-protein foods, fruits, vegetables, and dry fruits can prevent the side effects of chemical treatments. Eating nutrient foods will provide nutrients to our body to fight cancer.

Water and fluids are vital to health because our body is made up of more than 70% of water. All body cells need water to function and drinking the required amount of water daily is essential for good health and prevents cancer. During chemotherapy treatment for cancer, drinking more quality water helps in recovery. Carbohydrates are the main constituent of most of the food items that we consume and the body's primary energy source, which are required for routine physical activity, regulatory processes, and functioning of the body's organs. The proper kind and quantity of carbohydrates should be consumed to reduce the risk of developing cancer, although high glycemic index carbohydrates have a direct and positive association with lung cancer. However, an adequate intake of carbohydrates is crucial for good health. Fruits, vegetables, and whole grains are the finest sources of carbohydrates since they can also provide the body's cells with essential minerals, fiber, vitamins, and phytonutrients. Bread, peas, spaghetti, cereals, potatoes, rice, corn, pasta, and beans are other sources of carbohydrates.

Plant polysaccharides (PS) and lignin, which are not hydrolyzed by digestive enzymes in humans, are commonly referred to as dietary fibers. Dietary fibers are made up mainly of plant cell wall elements such as cellulose, hemicellulose, pectic compounds, and lignin. Gums and mucilage are also considered dietary fibers (Sarker and Rahman, 2017). Water-soluble fibers are another name for dietary fibers. Guar, psyllium, pectin, and oat bran are examples of soluble fibers with hypocholesterolemic characteristics. Wheat bran and cellulose are commonly thought to give protection against cancer (McRorie, 2015). In the current climate of rising cancer diagnoses, nutritional supplements and side-effect-free therapies are a top priority. Natural anti-cancer PS have demonstrated enormous promise. Polysaccharides (PS), a type of bio-macromolecule produced by a wide range of organisms including plants, mammals, fungus, bacteria, and algae, have piqued researchers' interest in recent years due to their anti-cancer properties and fewer effects (Li et al., 2021).

Food proteins, which are essential for growth, tissue repair, and immune system maintenance, are another crucial component of food. Our body may break down muscle for fuel if we don't receive enough protein. More protein is frequently required when a person has cancer. Extra protein is frequently needed to repair tissues and combat infection during chemotherapy, radiation therapy, or surgery. Fish, eggs, poultry, lean red meat, nuts and nut butter, low-fat dairy products, peas, lentils, dried beans, and soy products are all excellent sources of protein. Nutraceutical peptides found in food proteins enhance health and prevent disease. Peptides (protein fragments) with a variety of biological functions have been produced from legume seed proteins (González-Montoya et al., 2017). The peptide can be used directly as a cytotoxic agent in a variety of ways, or it can act as a carrier of cytotoxic chemicals and radioisotopes by targeting cancer cells preferentially. Peptide-based hormone therapy for the treatment of breast and prostate cancers has received a lot of attention and use (Thundimadathil, 2012). The three amino acids that are most frequently present in peptides with anti-cancer effects are glycine, lysine, and leucine (Chiangjong et al., 2020).

The body uses fats as a rich source of energy, and they have a significant place in nutrition. Fats are broken down by the body and employed for a variety of functions, including energy storage, tissue insulation, and blood-borne delivery of several vitamins. Saturated fats can be found in cheese, butter, full or low-fat milk, meat and poultry, and some vegetable oils such as palm kernel, coconut, and palm oil. Instead of saturated or trans fats, it is better to choose monounsaturated fats like canola, olive, and peanut oil and polyunsaturated fats like sunflower, corn, safflower, and flaxseed oil. Since they are regarded as immunological nutrients and have a wide range of biological effects, omega-3 PUFAs are used for cancer patients as nutritional treatment. Omega-3 fatty acids are crucial for maintaining membrane fluidity, cell shape, and cell signaling.

Vitamins, that is, A, C, E, minerals, that is, Zn, Se, and some of the enzymes has the ability to absorb and bind to free radicals (destructive chemicals) can all act as antioxidants and save our cells from oxidative harm. For example, vitamin C can exploit the differences between cancer and healthy cells and kills only cancer cells and helps healthy cells. Fresh horticultural produce has richness of antioxidants. Phytonutrients or phytochemicals includes a diverse range bioactive compounds which have health-protecting and promoting qualities. Several such compounds are proven to be useful in cancer treatment (Pratap et al., 2021).

Vitamins and minerals are the most important food constituents with a significant role in the management of the health and prevention of various diseases. Balanced food with the intake of sufficient calories and protein and an ample supply of essential growth factors like vitamins and minerals protects our body from several diseases, including cancer. Vitamins are compounds that our body requires for appropriate growth and development. Vitamin A is a fat-soluble vitamin. B vitamins are essential for good health. Although vitamin B_{12} has not been demonstrated to be a cancer-fighting agent, there is some indication that it may be beneficial. SC-3 tumors put into mice were suppressed by methylcobalamin (Shetty et al., 2020). Folic acid, vitamin D, vitamin C, vitamin B_{12}, selenium, chlorophyll, and antioxidants including carotenoids (lycopene, lutein, carotene, and cryptoxanthin) are all beneficial components in a cancer prevention diet. Ascorbic acid provides little benefit when taken orally, however it may be quite helpful when given intravenously (Shetty et al., 2020). Vitamin C is involved in a number of activities, including antioxidant defense, transcription, and epigenetic control of gene expression, as a co-factor for enzymes involved in processes and effects crucial for the transformation of cancer. Vitamin C is also said to have anti-inflammatory and immune-boosting properties, which are important in the battle against precancerous and cancerous cells (Pawlowska et al., 2019). There are a number of minerals that have been linked to both our health and the emergence of cancer in our bodies. The body's natural defense against cancer is the immune system, which also includes complement, phagocytosis, immunoglobulins, interferon, cell-mediated immunity, lysosomes, metabolic function, hormones, and metabolic and respiratory alkalosis. This natural system is supported by minerals such as magnesium, vanadium, zinc, selenium, and germanium (Ciulla et al., 2019).

Plant secondary metabolites are produced through metabolic pathways that appear to not seem to be crucial for plant development but are important biologically and pharmacologically (Crozier et al., 2006). Plant phenols are classified into several groups based on phenolic polymers, which are commonly referred to as tannins. These high-mass compounds are separated into two categories: hydrolysable and condensed tannins (Oksana et al., 2012). Because of their antioxidant properties, several phenolic chemicals found in natural foods may lessen the risk of serious health disorders. Plant phenolic compounds are utilized to treat oxidative stress as well as chronic disorders including cancer, diabetes, and colon cancer (Tewari et al., 2018). With varying degrees of success, herbs have been used to cure illnesses for hundreds of years. In numerous items today, including liquid extracts, tablets, ointments, and teas, herbs can be found. Many herbs now hinder the healing process after surgery and block effective cancer therapies.

16.3 SPECIFIC FOODS FOR CANCER

No food protects people against cancer completely, but prevents the risk of developing cancer, by adding them to the daily diet. Plant-based foods play a significant role in cancer prevention and management (Aghajanpour et al., 2017). Some of the plant-based foods are mentioned below:

- Vegetables that are dark green and yellow, such as squash, spinach, kale, and yellow capsicum, contain lutein, which can obstruct the advancement of the cancer cell cycle and limit the formation of cancer cells.
- Lycopene, which is found in tomatoes, guavas, apricots, water-melons, and peaches, also slows the growth of many cancer cells.
- Cruciferous vegetables like cauliflower, kale, and broccoli, aid in the process of detoxification and reduce the incidence of certain malignancies, including as bowel, liver, and breast cancers.
- Flavonoids, which are abundant in fruits and vegetables including onions, grapes, apples, tea, and berries, are helpful in preventing and treating certain malignancies.
- Squalene, a substance found in legumes like lentils, inhibits the formation of polyphenols, hence protecting against the development of colorectal cancer.
- Blackberries and blueberries in particular have anthocyanins that inhibit the activation of NF-κB, a potent pro-inflammatory factor that promotes the formation of cancer.

16.3.1 FRUITS AND VEGETABLES

It is commonly advised for its health-promoting features that fruits and vegetables rich foods can reduce the risk of various types of cancers (He et al., 2008). Since ancient times, fruits and vegetables have been recommended as part of a healthy diet because they contain good amount of vitamins, particularly vitamin A and C, minerals, electrolytes, and phytochemicals. As a source of dietary fiber, fruits, and vegetables are also advised (Reiland and Slavin, 2015; Slavin and Lloyd, 2012). All types of fresh, frozen, canned, and frozen dark green leafy vegetables, including spinach, broccoli, lettuce, sweet potatoes, winter squash, collard greens, romaine, tomatoes, mustard greens, red peppers, potatoes, carrots, pumpkin, peas, kidney beans, lentils, chickpeas, pinto beans, and corn, have anti-cancer properties (Slavin and Lloyd, 2012; Warren, 2001). Fruits include all fresh, canned, frozen, and fruit juices (bananas, melons, grapes, oranges, apples, berries, and raisins). Most of the fruits contains good volume of vitamin C, however the sources of vitamin C are diverse in nature and the consumers may need to relay on fruits of different origin for reaching the optimum levels (Guruvayoorappan et al., 2015; Slavin, 2013). Stomach, breast, liver, skin, colon, and prostate cancers are all prevented by the anti-cancer capabilities of fruits, vegetables, and the substances that make them up. They are a rich source of triterpenoids and phenolic antioxidants such as epicatechin, quercetin, ursolic acid, procyanidin B2, phloretin, and maslinic acid. According to Lin et al. (2008), the veggies have anti-cancer characteristics that protect against skin, liver, and breast cancers (Lin et al., 2008). Fruit is high in cyanogenic glycosides, vitamins A and vitamin C dietary fiber, carotenes, cynaropicrin, volatile oils, flavone glycosides, phenylpropanoids, sesquiterpene lactones, and taraxasterol (Beidokhti, 2013; Kolniak-Ostek et al., 2020). Berries are a great source of these bioactive substances, which are thought to have a wide range of health benefits, such as antioxidant and anti-cancer characteristics (Nile and Park, 2014). Strawberries contain a wide range of phytochemicals, mostly phenolic acids like hydroxybenzoic acid and hydroxycinnamic acid, as well as flavonoids, primarily anthocyanins. In terms of quantity, anthocyanins are the most significant polyphenolic component found in berries (Nile and Park, 2014; Skrovankova et al., 2015). Table 16.1 contains the information regarding foods that fight cancer, including cereals, fruits, and vegetables.

16.3.2 WILD EDIBLE FRUITS

Wild edible fruits are employed in the ancient medical practices of India to cure different illnesses. Microbial, nutritional, and metabolic illnesses are among the conditions that are addressed. Fruits are thought to have phenolic chemicals and vitamins that have anti-inflammatory, anti-microbial, antioxidant, anti-cancer, anti-mutagenic, and neuroprotective activities. The

TABLE 16.1 Cereal, Fruit, and Vegetable-based Foods Effective against Cancer

Food Source	Bioactive Component	Type of Study	Type of Cancer Prevented	Beneficial Effect	References
Cereal, Fruit, and Vegetable					
Blueberries, blackberries, and strawberries	Phenolic compounds like flavonoids, mainly anthocyanins like gallic acid, Pterostilbene, and resveratrol derivative	*In vitro* with breast cancer cell lines and *in vivo* with the mouse model	Breast and colon cancer	Antioxidant, anti-cancer, and inhibit NF-κB, Wnt/β-catenin, PI3K/Akt/mTOR, and ERK/MAPK signaling pathways	Baby et al. (2018); Chikwere (2017); Tao et al. (2018)
Cranberry	Proanthocyanide	Rat model	Breast, lungs, and Lymph nodes cancer	Anti-carcinogenic activity	Montbriand (2004)
Apple	The flavonoid quercetin, other polyphenols, and fiber	Mouse model	Lung and breast cancer	Antioxidant, anti-cancer, and anti-inflammatory	Dhalaria et al. (2020); Montbriand (2004)
Grapes	Polyphenols like resveratrol	*In vitro* with breast cancer cell lines	Breast cancer and colon cancer	Pro-apoptotic, anti-proliferative, anti-invasive, anti-angiogenic, and anti-metastatic action	Dhalaria et al. (2020); Tiwary and Hussain (2021); Zhou and Raffoul (2012)
Cherry	Anthocyanins, quercetin, and carotene	*In vitro* colon and breast cancer studied in mice	Breast and colon cancer	Anti-inflammatory and antioxidant	Dhalaria et al. (2020)
Custard apple	Alkaloids, protein, amino acids carbohydrates	Rat histiocytic tumor cell line AK-5 and human cell lines of various cancer	Lung, breast, liver, and cervical cancer	Down-regulation of anti-apoptotic genes Bcl-2 and Bclxi	Mónica (2020); Sasikumar and Eagappan (2014)

TABLE 16.1 *(Continued)*

Food Source	Bioactive Component	Type of Study	Type of Cancer Prevented	Beneficial Effect	References
Pomegranate, plum, date palm, and gooseberries	Dietary fibers, polyphenolics, hydroxycinnamic acid and ellagic acid	Mouse model	Colon cancer	Antioxidant and anti-cancer	Baby et al. (2018); Tao et al. (2018)
Orange, peach, papaya, mango, watermelon, plum, grapefruit, black olive	β-cryptoxanthin, carotene	Mouse model	Lung, breast, and cervical cancer	Scavenging of free radicals, ↑ of gap junctions, immune modulation	Aghajanpour et al. (2017); Dhalaria et al. (2020); Kapinova et al. (2018)
Cruciferous vegetables	Isothiocyanates and phenethyl isothiocyanates sulforaphane, genistein, and indole-3-carbinol	Various animal models, in mouse xenograft models	Lungs, breast, liver, esophagus, stomach, small intestine, prostate, and colon cancer	Nuclear factor erythroid 2-related factor 2 (NRFF2)/antioxidant response element (ARE) pathway, increase apoptosis, anti-angiogenic, and anti-metastasis potential, and increase autophagy	Aghajanpour et al. (2017); Guruvayoorappan et al. (2015); Rose and Strombom (2019); Tao et al. (2018)
Cucumber	Cucurbitacins, vitexin, cucumber in A and B, divinyl reductase proteins, cucumegastigmanes I and II, orientin, and glucosides	Animal studies	Skin cancer	Anti-cancer, anti-diabetic, hemostatic, lipid-lowering, tonic, and antioxidant activities	Guruvayoorappan et al. (2015)
Carrot, pumpkin, and winter squash	α-carotene and β-carotene	Various animal models	Skin, lungs, liver, colorectal, prostate, breast, and stomach cancer	Inhibiting tumor genesis, antioxidant, and anti-cancer properties	Aghajanpour et al. (2017); Guruvayoorappan et al. (2015)

TABLE 16.1 *(Continued)*

Food Source	Bioactive Component	Type of Study	Type of Cancer Prevented	Beneficial Effect	References
Beetroot	Betanin and the betacyanin	Various animal models	Prostate cancer and anemia	Anti-cancer, antioxidant, and free radical scavenger	Tiwary and Hussain (2021)
Green leafy vegetables	β-carotene	*In vitro* study in bronchial epithelial cells, various animal models	Lungs, and breast cancer	Invert the downregulation of RARβ by smoke-borne carcinogens, decrease peroxy radicals, show antioxidant and anti-cancer activity	Aghajanpour et al. (2017); Kapinova et al. (2018); Madigan and Karhu (2018)
Turmeric	Curcumin	Various animal models	Colorectal cancer leukemia and Glioblastoma	Anti-septic and anti-cancer properties	Hussain et al. (2022); Montbriand (2004); Ooko et al. (2017)
Garlic	Sulfur-containing active compounds like diallyl trisulfide	Transgenic mouse model and cohort study	Prostate, lung, liver, and blood cancer and carcinoma of the mammary gland	Inhibit the expression of the androgen receptor	Bayan, Koulivand, and Gorji (2014); Hasler (2002b)
Chili	Luteolin	*In vitro* and *in vivo* study	Colon cancer	Antioxidant and anti-cancer	Lin et al. (2008)
Ginger	Gingerol	*In vitro* and *in vivo* study	Ovary, cervix, colon, liver, and urinary caner	Antioxidant, anti-cancer, and anti-inflammatory	Kumara et al. (2017)
Artichoke	Tannins	Cancer cell line studies	Breast, liver, and skin cancers	Anti-cancer	Guruvayoorappan et al. (2015)
Bean	Fiber and folate	Several animal studies	Colon, breast, and pancreatic cancer	Anti-cancer	Guruvayoorappan et al. (2015)

TABLE 16.1 *(Continued)*

Food Source	Bioactive Component	Type of Study	Type of Cancer Prevented	Beneficial Effect	References
Maize, leafy green vegetables like spinach, sweet peppers, sweet corn, pea, and kiwi fruit	Lutein, zeaxanthin, and phytic acid	Various animal models	Breast and colon cancer	Anti-carcinogenic and chemoprotective activity	Aghajanpour et al. (2017); Rose and Strombom (2019)
Tomato, watermelon, grapefruit, apricot, pink guava, and pawpaw	Lycopene	*In vitro* studies in various human cancer cell lines and *in vivo* studies in mouse models	Breast, liver, gastric, skin, and lungs cancer	Antioxidant and free radical quencher, decrease Bcl-2/Bax and increase gap-junctional intercellular communication	Sasikumar and Eagappan (2014); Aghajanpour et al. (2017); Kapinova et al. (2018)
Soy	Isoflavones	*In-vitro* with cultured human cancer cells	Breast, colorectal, and prostate cancer	Increase mammary gland differentiation, decrease activation of procarcinogens to carcinogens	Badger et al. (2005); Chikwere (2017)
Whole grain cereals like wheat, rice, maize oat, rye, and barley	Phenolics, flavonoids, glucans, and pigments	Various animal models	Colorectal and liver cancer	Anti-cancer activity	Gupta (2016); Campbell (2015)

fractionation of fruit may encourage the creation of novel medicinal biological ingredients which can prevent and treat a variety of illnesses and disorders (Rouamba et al., 2018). Wild fruits are frequently uncommon, underappreciated, or less well-known. Though, many of them are safe to eat, and some have even been used in the development of medications. Wild fruits have a high phytochemical content, including anthocyanins and flavonoids, due to varied genotypes and environmental issues. Consequently, it is common knowledge that wild fruits are nutritious diets. The bioactive benefits of wild fruits, such as anti-microbial, antioxidant, anti-inflammatory, and anti-cancer activities, have drawn the attention of researchers in recent years. Numerous studies have indicated that wild fruits are effective against several chronic illnesses, and the phytochemicals present in the wild fruits are responsible for the positive effects on health (Li et al., 2016).

16.3.3 DRY FRUITS

Fruit dehydration procedures are used to create a wide range of products in both traditional and contemporary food technology (Ahmed et al., 2013). Traditional and commercial dehydrated fruits, that is, prunes, dates, raisins, dried figs, and so on, and sweetened dried fruits (dried cranberries and candied pineapple) are two kinds of such items that are widely known. In contemporary societies, dried fruits make up just a tiny share of daily meals, while this percentage increases in hot, dry areas (Mossine et al., 2020; Sijtsema et al., 2012). In addition to the usual basic nutrients, dried fruit is crucial for determining its water-soluble fiber, fructosamine, and reactive carbonyl species contents (Mossine et al., 2020). Amygdalin is abundant in almonds, which are also cyanogenic glycosides. Amygdalin has anti-cancer properties against breast, cervical, colon, and prostate cancers (Kolniak-Ostek et al., 2020).

16.3.4 FRUIT WINE

The consumption of wine has long been widespread across the world. Red wine has abundance of various bioactive compounds with health promoting properties and is relished by people all over the world and is advantageous. The art of winemaking and drinking have been customs in many countries while the health benefits of wine were discovered and established by the Romans, who led the spread of wine across the globe popular (Snopek et al., 2018). Wine is the primary product of grapes. According to Champ and Kundu-Champ, the primary components of wine include water, carbohydrates, organic acids, minerals, alcohol, polyphenols, and aromatics (Champ and Kundu-Champ, 2019). A moderate amount of wine drinking each day may help to avoid a number of chronic illnesses. This is because red wine contains significant antioxidants, which is why study has concentrated on them (Snopek et al., 2018). The most potent wine antioxidants are wine polyphenols, notably resveratrol, anthocyanin, and catechin. Resveratrol prevents cancer by scavenging reactive nitrogenous and free oxygen radicals (Xiang et al., 2014). The most recent research on the protective effects of wine drinking against various malignancies is summarized in this chapter. Their use in food and drink lowers the risk of several malignancies (Lukas et al., 2018).

Polyphenols are most noteworthy biologically active ingredients in the wine. The grape varietal, for example, has an impact on the polyphenol composition and precise content. Red wine polyphenols, as per certain studies (Lukas Snopek et al., 2018; Visioli et al., 2020), appear to benefit human organs. Resveratrol, catechin, anthocyanins, and tannins (proanthocyanidins and ellagitannins) are the four most significant polyphenols found in red wine. Red wine and wine grapes are two foods that include resveratrol, which has biological activity, a significant impact on cancer, and is found in a small number of other foods. The favorable effects of wine are positively influenced by additional wine polyphenols as well (Champ and Kundu-Champ, 2019; Lukas et al., 2018).

16.3.5 NON-ALCOHOLIC BEVERAGES

Worldwide use of tea and coffee has been linked to positive health effects, including anti-cancer properties, as evidenced by epidemiological and clinical research. The primary constituents of green tea include epigallocatechin gallate (EGCG) and chlorogenic acid (CGA), are thought to be primarily responsible for the majority of these benefits (Hayakawa et al., 2020; Saibabu et al., 2015). Green tea is made by processing *Camellia sinensis* (Theaceae) plant leaves and is used widely all over the world. The anticancer, anti-infectious, anti-cardiovascular, anti-diabetic, anti-obesity, and anti-neurodegenerative properties of green tea have been demonstrated to benefit human health (Suzuki et al., 2012). The majority of these biological benefits are thought to be primarily attributed to EGCG, the most prevalent catechin in green tea (Tanaka et al., 2013). Green tea generally comprises 2.5 g of tea leaves, which yields 240–320 mg of catechins, of which 60–65% are EGCG (Shrivastava et al., 2018).

The fermentation of *Camellia sinensis* using intrinsic enzymes and microorganisms yields black tea. During this process, catechins are often polymerized into various high molecular weight catechin derivatives such theaflavins and theasinensins (Li et al., 2013). Due to its lower EGCG level than green tea, black tea has been found to produce physiological effects that are similar to but less pronounced than those of green tea (Takemoto and Takemoto, 2018). Coffee is used all over the world and has a number of health benefits. About 2000 distinct compounds may be found in it, but CGA and its derivatives, which make up about 3% of the dry weight of roasted coffee powder, are the main polyphenols (Cavalli and Tavani, 2016; Hayakawa et al., 2020). Table 16.2 describes where wild edible fruits, dry fruits, fruit wines, and drinks come from, what they do, how they do it, and how it affects anti-cancer activity.

16.4 PRESENT STATUS AND FUTURE OF PLANT FOODS AGAINST CANCER

Nature has provided mankind with plant food as a gift to aid in their quest for improved health. Since ancient times, medical techniques have used plants and their bioactive components. Numerous types of medicinal plants and the phytochemicals they contain prevent the spread and growth of cancer. Over 2,50,000 different plant species are known in the in the plant kingdom, however, only 10% of them have been examined for their potential to cure various ailments (Greenwell and Rahman, 2015). A variety of pharmacological actions are carried out by phytochemicals and their derived analogs, which are found in many plant components, including the pericarp, sprouts, flower, flower stigmas, roots, rhizomes, stem, leaf, fruits, seeds, embryo, bark (Pratap et al., 2017). Many plant substances, including vitamins, minerals, gums, oils, flavonoids, lignans, saponins, alkaloids, terpenes, taxanes, glycosides, and biomolecules (Pratap and Shantaram, 2020), play important roles in inhibiting cancer cell-activating enzymes, proteins, and signaling pathways (Sebastian et al., 2020), the topoisomerase enzyme, COX-2, Bcl-2, the mechanistic target of rapamycin by activating DNA repair mechanism and stimulating the formation of protective enzymes (Caspase-3, 7, 8, 9, 10, 12), promoting antioxidant activity, exhibiting high anti-cancer effects

TABLE 16.2 Bioactive Components of Wild Edible Fruits, Dry Fruits, Wine, and Beverages and Their Effect on Cancer Prevention

Plant Food	Bioactive Present	Cancer Prevention/Beneficial Effects
	Wild Edible Fruits	
Canthium parviflorumlam	β-carotene, α-tocopherol, ascorbic acid, and phytochemicals *viz*., total polyphenols, GSH, and flavonoids.	Anti-cancer (Reddy Palvai et al., 2014) activity against the Caco-2 human epithelial colorectal adenocarcinoma cell lines and liver cancer (Khasim et al., 2020).
Carissa carandas L.	Alkaloids, flavonoids, glycosides, reducing sugars, steroids, terpenoids, and tannins.	Anti-malarial activity, antioxidant, and anti-cancer activity (Tesfave and Ravichandran, 2018). Good anti-cancer activity against Caov-3 (ovarian carcinoma) and remarkable activity against lung cancer cells (Sharma et al., 2019).
Campomanesia cambessedeana, *Byrsonoma verbascifolia* and *Pouteria guardneriana*	Phenolics	Inhibiting the growth of hepG-2 human liver cancer cells (Li et al., 2016).
Piper capense	Essential oil	Inhibiting the growth of human breast adenocarcinoma, malignant melanoma, and colon carcinoma cells (Li et al., 2016).
Rubus rosifolius	Hypotatic acid B and 4-epi-nigaichigoside F1.	Inhibiting the growth of colon tumor cells (Li et al., 2016).
Vaccinium angustifolium	Oligomeric proanthocyanidins fraction.	Inhibiting the growth of the human prostate and mouse liver cancer cell lines (Li et al., 2016).
Syzgium cumini	Malic acid, oxalic acid, gallic acid, tannins, cyanidin glycoside oleanolic acid, flavonoids, and essential oils.	Anti-cancer and anti-clastogenic (Ramya, 2012). Breast adenocarcinoma and ovarian adenocarcinoma (Santos et al., 2020; Yadav et al., 2011).
Ximenia americana L.	Yellow flavonoids and polyphenol.	Anti-ulcer, anti-inflammation of uterus and ovaries, anemia, and cancer (Rouamba et al., 2018).
Opuntia ficus-indica	Ascorbic acid, vitamin E, carotenoids, and so on.	Anti-cancer (El-Mostafa et al., 2014).

TABLE 16.2 *(Continued)*

Plant Food	Bioactive Present	Cancer Prevention/Beneficial Effects
Vitex doniana (sweet)	Alkaloid, saponin, terpenoid, and flavonoid.	Rheumatism, cancer (Rouamba et al., 2018).
	Dry Fruits	
Prunus dulcis (almonds), *Anacardium occidentale* (cashews), *Phoenix dactylifera* L., (dates), and *Juglans* (walnuts)	Carotenoids, urolithins, polyphenols, and minerals.	Antioxidant and anti-cancer. Effective against breast cancer (Bolling et al., 2014).
	Fruit Wine	
Vitis vinifera "Red Globe" (Red wine)	Polyphenols, especially resveratrol and quercetin, anthocyanins, and catechins.	Antioxidant and anti-cancer properties (Lukas et al., 2018). Resveratrol suppresses colon cancer cell and quercetin induces apoptosis of cancer cells (Oi et al., 2010).
	Beverages	
Camellia sinensis (black tea)	Catechin derivatives such as theaflavins and theasinensins.	Anticancer, antioxidant (Hayakawa et al., 2020). These are effective against colon cancer, lung cancer, gastric, breast cancer, and leukemia (Vijay and Anu, 2017) and transgenic adenocarcinoma of the mouse prostate (TRAMP) model and in malignant human bladder cells (Chikwere, 2017; Gumbs et al., 2021).
Camellia sinensis (green tea)	(–)-Epigallocatechin gallate (EGCG) and polyphenols.	
Coffee arabica (coffee)	Polyphenols are chlorogenic acid (CGA).	

in terms of their effectiveness on the aforementioned proteins, enzymes, and signaling pathways (Iqbal et al., 2017; Sebastian et al., 2020).

The current chapter makes clear that phytochemical research is a viable and fruitful field with a promising future. The high cost and high toxicity of the current anti-cancer drugs, along with other limitations in conventional therapy brought about by the rising incidence of cancer, have made it extremely difficult for researchers to come up with new, environmentally friendly, biocompatible, and cost-effective treatments that are also more environmentally friendly. According to this hypothesis, phytomolecules will change cancer treatment over the next 10 years (Choudhari et al., 2020). These phytomolecules are now more effective in treating cancer because of their high biodegradability and biocompatibility. This extensive chapter contains information on medicinal plants and their bioactive substances that may be able to treat various cancers. Curcumin, garcinol, isothiocyanates, epigallocatechin, gossypol, sulforaphane, and other potential anti-cancer phytochemicals should be further investigated in clinical trials for their efficacy and toxicity documentation on various models. These phytochemicals are covered in this thorough review paper. Furthermore, in-depth investigation into these phytochemicals is needed to assess their potential uses, toxicological profiles, and unique genotoxic profiles against a variety of cancers in both *in vitro* and *in vivo* settings (Iqbal et al., 2017).

16.5 CONCLUSION

Many developed and developing nations both suffer greatly from cancer, which is a leading cause of mortality and disability. There are a number of negative effects from regularly used cancer treatments. Nowadays, researchers focus on the association between functional foods and cancer. They have gained remarkable significance in current times for their foreseen visualized role in the treatment of different pathologies such as breast cancer, colorectal cancer, and so forth. When taken at effective amounts as a regular element of the diet, plant foods are thought to offer health advantages beyond the requirement of essential nutrie*nts.* They are the food components that provide health benefits beyond basic nutrition and reduce the risk of cancer. There is a high potential for the consumption of plant-based food products for a healthy diet, such as berries, seasonal fruits, vegetables, dry fruits, wild edible fruits, fruit wine, and beverages in the population to prevent different types of cancer.

KEYWORDS

- **alkaloids**
- **anti-cancer**
- **bioactive components**
- **biomolecules**
- **cancer**
- **flavonoids**
- **glycosides**
- **lignans**
- **medicinal plants**
- **non-alcoholic beverage**
- **phytochemicals**
- **rhizomes**
- **terpenes**

REFERENCES

Adams, M., Gmünder, F., & Hamburger, M. (2007). Plants traditionally used in age-related brain disorders—A survey of ethnobotanical literature. *Journal of Ethnopharmacology, 113*(3), 363–381.

Aghajanpour, M., Nazer, M. R., Obeidavi, Z., Akbari, M., Ezati, P., & Kor, N. M. (2017). Functional foods and their role in cancer prevention and health promotion: A comprehensive review. *American Journal of Cancer Research, 7*(4), 740.

Ahmed, N., Singh, J., Chauhan, H., Anjum, P. G. A., & Kour, H. (2013). Different drying methods: Their applications and recent advances. *International Journal of Food Nutrition and Safety, 4*(1), 34–42.

Andrade, V., Guzman-Martinez, L., Cortes, N., Gonzalez, A., & Farias, G. (2018). The emergency of nutraceutical compounds in the preventive medicine scenario: Potential for treatment of Alzheimer's disease and other chronic disorders. *Journal of Alzheimer's Disease & Parkinsonism, 8*(456), 2161–0460.

Baby, B., Antony, P., & Vijayan, R. (2018). Antioxidant and anti-cancer properties of berries. *Critical Reviews in Food Science and Nutrition, 58*(15), 2491–2507.

Badger, T. M., Ronis, M. J., Simmen, R. C., & Simmen, F. A. (2005). Soy protein isolate and protection against cancer. *Journal of the American College of Nutrition, 24*(2), 146S–149S.

Bayan, L., Koulivand, P. H., & Gorji, A. (2014). Garlic: A review of potential therapeutic effects. *Avicenna Journal of Phytomedicine, 4*(1), 1.

Beidokhti, M. N. (2013). 55 anti-cancer fruits, vegetables, beverages, oils, and spices. *International Journal of Phytomedicine*, *5*(4), 415.

Bolling, B. W., Chen, C. Y. O., McKay, D. L., & Blumberg, J. B. (2011). Tree nut phytochemicals: Composition, antioxidant capacity, bioactivity, and impact factors. A systematic review of almonds, Brazils, cashews, hazelnuts, macadamias, pecans, pine nuts, pistachios, and walnuts. *Nutrition Research Reviews*, *24*(2), 244–275.

Campbell, T. (2015). *The Campbell Plan: The Simple Way to Lose Weight and Reverse Illness, Using the China Study's Whole-Food, Plant-Based Diet*. Rodale.

Cavalli, L., & Tavani, A. (2016). Coffee consumption and its impact on health. In *Beverage Impacts on Health and Nutrition: Second Edition* (pp. 29–47).

Champ, C. E., & Kundu-Champ, A. (2019). Maximizing polyphenol content to uncork the relationship between wine and cancer. *Frontiers in Nutrition*, *6*, 44.

Chandran, R., & Abrahamse, H. (2020). Identifying plant-based natural medicine against oxidative stress and neurodegenerative disorders. *Oxidative Medicine and Cellular Longevity*, *2020*, 8648742.

Chiangjong, W., Chutipongtanate, S., & Hongeng, S. (2020). Anti-cancer peptide: Physicochemical property, functional aspect, and trend in clinical application. *International Journal of Oncology*, *57*(3), 678–696.

Chikwere, P. (2017). Functional foods and nutraceuticals, wonders in cancer risks—a review. *World Scientific News*, *64*, 17–33.

Choudhari, A. S., Mandave, P. C., Deshpande, M., Ranjekar, P., & Prakash, O. (2020). Phytochemicals in cancer treatment: From preclinical studies to clinical practice. *Frontiers in Pharmacology*, *10*, 1614.

Ciulla, M., Marinelli, L., Cacciatore, I., & Stefano, A. D. (2019). Role of dietary supplements in the management of Parkinson's disease. *Biomolecules*, *9*(7), 271.

Crozier, A., Jaganath, I. B., & Clifford, M. N. (2007). Phenols, polyphenols, and tannins: An overview. In *Plant Secondary Metabolites: Occurrence, Structure, and Role in The Human Diet* (pp. 1–25).

Custódio, I. D. D., Marinho, E. D. C., Gontijo, C. A., Pereira, T. S. S., Paiva, C. E., & Maia, Y. C. D. P. (2016). Impact of chemotherapy on diet and nutritional status of women with breast cancer: A prospective study. *PLOS ONE*, *11*(6), e0157113.

Daliu, P., Santini, A., & Novellino, E. (2018). A decade of nutraceutical patents: Where are we now in 2018? *Expert Opinion on Therapeutic Patents*, *28*(12), 875–882.

Das, P., Adak, S., & Lahiri Majumder, A. (2020). Genetic manipulation for improved nutritional quality in rice. *Frontiers in Genetics*, *11*, 776.

Dhalaria, R., Verma, R., Kumar, D., Puri, S., Tapwal, A., Kumar, V., Nepovimova, E., & Kuca, K. (2020). Bioactive compounds of edible fruits with their anti-aging properties: A comprehensive review to prolong human life. *Antioxidants*, *9*(11), 1123.

El-Mostafa, K., El Kharrassi, Y., Badreddine, A., Andreoletti, P., Vamecq, J., El Kebbaj, M. H. S., Latruffe, N., Lizard, G., Nasser, B., & Cherkaoui-Malki, M. (2014). Nopal cactus (Opuntia ficus-indica) as a source of bioactive compounds for nutrition, health, and disease. *Molecules*, *19*(9), 14879–14901.

Farias, G. A., Guzman-Martinez, L., Delgado, C., & Maccioni, R. B. (2014). Nutraceuticals: A novel concept in prevention and treatment of Alzheimer's disease and related disorders. *Journal of Alzheimer's Disease*, *42*(2), 357–367.

Gogus, U., & Smith, C. (2010). n-3 Omega fatty acids: A review of current knowledge. *International Journal of Food Science & Technology*, *45*(3), 417–436.

González-Montoya, M., Cano-Sampedro, E., & Mora-Escobedo, R. (2017). Bioactive peptides from legumes as anti-cancer therapeutic agents. *International Journal of Cancer Clinical Research, 4*, 81.

Greenwell, M., & Rahman, P. K. S. M. (2015). Medicinal plants: Their use in anti-cancer treatment. *International Journal of Pharmaceutical Sciences and Research, 6*(10), 4103.

Gumbs, A. A., Gogol, M., Spolverato, G., Taher, H., & Chouillard, E. K. (2021). Systematic review of the integrative medicine recommendations for patients with pancreatic cancer. *Surgeries, 2*(2), 216–230.

Gupta, P. (2016). Functional foods for cancer therapeutics. *Natural Products Chemistry & Research, 4*(2), 6836.

Gupta, S. C., Patchva, S., & Aggarwal, B. B. (2013). Therapeutic roles of curcumin: Lessons learned from clinical trials. *The AAPS Journal, 15*, 195–218.

Guruvayoorappan, C., Sakthivel, K. M., Padmavathi, G., Bakliwal, V., Monisha, J., & Kunnumakkara, A. B. (2015). Cancer preventive and therapeutic properties of fruits and vegetables: An overview. In *Anti-cancer Properties of Fruits and Vegetables: A Scientific Review* (pp. 1–52).

Hasler, C. M. (2002). Functional foods: Benefits, concerns and challenges—A position paper from the American Council on Science and Health. *The Journal of Nutrition, 132*(12), 3772–3781.

Hayakawa, S., Ohishi, T., Miyoshi, N., Oishi, Y., Nakamura, Y., & Isemura, M. (2020). Anti-cancer effects of green tea epigallocatchin-3-gallate and coffee chlorogenic acid. *Molecules, 25*(19), 4553.

He, S., Sun, C., & Pan, Y. (2008). Red wine polyphenols for cancer prevention. *International Journal of Molecular Sciences, 9*(5), 842–853.

Iqbal, J., Abbasi, B. A., Mahmood, T., Kanwal, S., Ali, B., Shah, S. A., & Khalil, A. T. (2017). Plant-derived anti-cancer agents: A green anti-cancer approach. *Asian Pacific Journal of Tropical Biomedicine, 7*(12), 1129–1150.

Islam, S. R., & Siddiqua, T. J. (2020). Functional foods in cancer prevention and therapy: Recent epidemiological findings. In *Functional Foods in Cancer Prevention and Therapy* (pp. 405–433).

Kapinova, A., Kubatka, P., Golubnitschaja, O., Kello, M., Zubor, P., Solar, P., & Pec, M. (2018). Dietary phytochemicals in breast cancer research: Anti-cancer effects and potential utility for effective chemoprevention. *Environmental Health and Preventive Medicine, 23*, 1–18.

Khasim, S. M., Long, C., Thammasiri, K., & Lutken, H. (2020). *Medicinal Plants: Biodiversity, Sustainable Utilization and Conservation.* Springer.

Kolniak-Ostek, J., Kłopotowska, D., Rutkowski, K. P., Skorupińska, A., & Kruczyńska, D. E. (2020). Bioactive compounds and health-promoting properties of pear (*Pyrus communis* L.) fruits. *Molecules, 25*(19), 4444.

Kumara, M., Shylaja, M. R., Nazeem, P. A., & Babu, T. (2017). 6-Gingerol is the most potent anticancerous compound in ginger (*Zingiber officinale* Rosc.). *Journal of Developing Drugs, 6*(1), 1–6.

Li, N., Wang, C., Georgiev, M. I., Bajpai, V. K., Tundis, R., Simal-Gandara, J., Lu, X., Xiao, J., Tang, X., & Qiao, X. (2021). Advances in dietary polysaccharides as anti-cancer agents: Structure-activity relationship. *Trends in Food Science & Technology, 111*, 360–377.

Li, S., Lo, C. Y., Pan, M. H., Lai, C. S., & Ho, C. T. (2013). Black tea: Chemical analysis and stability. *Food & Function, 4*(1), 10–18.

Li, Y., Zhang, J. J., Xu, D. P., Zhou, T., Zhou, Y., Li, S., & Li, H. B. (2016). Bioactivities and health benefits of wild fruits. *International Journal of Molecular Sciences*, *17*(8), 1258.

Lin, Y., Shi, R., Wang, X., & Shen, H. M. (2008). Luteolin, a flavonoid with potential for cancer prevention and therapy. *Current Cancer Drug Targets*, *8*(7), 634–646.

Madigan, M., & Karhu, E. (2018). The role of plant-based nutrition in cancer prevention. *J. Unexplored Med. Data*, *3*(9), 1–16.

McRorie Jr, J. W. (2015). Evidence-based approach to fiber supplements and clinically meaningful health benefits, part 1: What to look for and how to recommend an effective fiber therapy. *Nutrition Today*, *50*(2), 82.

Mónica, C. G. (2020). Antiproliferative activity of total extracts from *Annona squamosa*, *Petiveria alliacea*, and *Punica granatum* on cancer cells. *Pharmacology OnLine*, *3*, 7–18.

Montbriand, M. J. (2004). Herbs or natural products that protect against cancer growth: Part three of a four-part series. *Number 6/2004*, *31*(6), E127–E146.

Mossine, V. V., Mawhinney, T. P., & Giovannucci, E. L. (2020). Dried fruit intake and cancer: A systematic review of observational studies. *Advances in Nutrition*, *11*(2), 237–250.

Ng, R., Sutradhar, R., Yao, Z., Wodchis, W. P., & Rosella, L. C. (2020). Smoking, drinking, diet, and physical activity—Modifiable lifestyle risk factors and their associations with age to first chronic disease. *International Journal of Epidemiology*, *49*(1), 113–130.

Nile, S. H., & Park, S. W. (2014). Edible berries: Bioactive components and their effect on human health. *Nutrition*, *30*(2), 134–144.

Oi, N., Jeong, C. H., Nadas, J., Cho, Y. Y., Pugliese, A., Bode, A. M., & Dong, Z. (2010). Resveratrol, a red wine polyphenol, suppresses pancreatic cancer by inhibiting leukotriene A4 hydrolase. *Cancer Research*, *70*(23), 9755–9764.

Oksana, S., Marian, B., Mahendra, R., & Bo, S. H. (2012). Plant phenolic compounds for food, pharmaceutical, and cosmetic production. *Journal of Medicinal Plants Research*, *6*(13), 2526–2539.

Ooko, E., Kadioglu, O., Greten, H. J., & Efferth, T. (2017). Pharmacogenomic characterization and isobologram analysis of the combination of ascorbic acid and curcumin—two main metabolites of *Curcuma longa*—in cancer cells. *Frontiers in Pharmacology*, *8*, 38.

Panche, A. N., Diwan, A. D., & Chandra, S. R. (2016). Flavonoids: An overview. *Journal of Nutritional Science*, *5*, e47.

Pawlowska, E., Szczepanska, J., & Blasiak, J. (2019). Pro-and antioxidant effects of vitamin C in cancer in correspondence to its dietary and pharmacological concentrations. *Oxidative Medicine and Cellular Longevity, 2019, 7286797.*

Peters, R., Ee, N., Peters, J., Beckett, N., Booth, A., Rockwood, K., & Anstey, K. J. (2019). Common risk factors for major noncommunicable diseases: A systematic overview of reviews and commentary: The implied potential for targeted risk reduction. *Therapeutic Advances in Chronic Disease*, *10*, 2040622319880392.

Pratap, G. K., & Shantaram, M. (2020). An alternative approach for anti-Alzheimer's compounds from plant extracts. In *Phytomedicine and Alzheimer's Disease* (pp. 5–40). CRC Press.

Pratap, G. K., Ashwini, S., & Manjula, S. (2017). Alzheimer's disease: A challenge in managing with certain medicinal plants—A review. *International Journal of Pharmaceutical Sciences and Research*, *12*(8), 4960–4972.

Pratap, G. K., Jayakar, V., Lokapur, V., & Shantaram, M. (2021). Role of functional foods in the amelioration of chronic and neurodegenerative diseases. In *Bentham Publishers* (pp. 1–20).

Ramya, S., Neethirajan, K., & Jayakumararaj, R. (2012). Profile of bioactive compounds in *Syzygium cumini*—A review. *Journal of Pharmaceutical Research, 5*(8), 4548–4553.

Reddy Palvai, V., Mahalingu, S., & Urooj, A. (2014). *Canthium parviflorum* leaves: Antioxidant activity in food and biological systems, pH, and temperature stability. *Chinese Journal of Biology*.

Reiland, H., & Slavin, J. (2015). Systematic review of pears and health. *Nutrition Today, 50*(6), 301.

Rose, S., & Strombom, A. (2019). Colorectal cancer prevention with a plant-based diet. *Cancer Therapy & Oncology International Journal, 15*(2), 50–67.

Rouamba, A., Compaoré, M., & Kiendrebeogo, M. (2018). Ethnomedicinal use, phytochemical constituents and bioactivity of wild edible fruits commonly consumed in Burkina Faso. *Asian Journal of Research in Biochemistry, 2*, 1–15.

Saibabu, V., Fatima, Z., Khan, L. A., & Hameed, S. (2015). Therapeutic potential of dietary phenolic acids. *Advances in Pharmacological Sciences, 2015, 823539*.

Santos, C. A., Almeida, F. A., Quecán, B. X., Pereira, P. A., Gandra, K. M., Cunha, L. R., & Pinto, U. M. (2010). Bioactive properties of *Syzygium cumini* (L.) Skeels pulp and seed phenolic extracts. *Frontiers in Microbiology, 11*, 990.

Sarker, M., & Rahman, M. (2017). Dietary fiber and obesity management–A review. *Advances in Obesity, Weight Management & Control, 7*(3), 00199.

Sasikumar, S., & Eagappan, K. (2014). Nutri-cognosy in cancer. *International Journal of Pharmacy and Pharmaceutical Sciences, 6*(3), 23–29.

Sebastian, R., Jaykar, B., & Gomathi, V. (2020). Current status of anti-cancer research in the *Fabaceae* family. *Pathways, 6*(7).

Sharma, V., Heer, A., Kour, N., Sharma, A., & Singh, S. K. (2019). Karonda and Jamun seeds' *in vitro* anti-cancer efficacy. *Indian Journal of Traditional Knowledge, 18*(3), 573–578.

Shetty, R., Ghosh, A., Honavar, S. G., Khamar, P., & Sethu, S. (2020). Therapeutic opportunities to manage COVID-19/SARS-CoV-2 infection: Present and future. *Indian Journal of Ophthalmology, 68*(5), 693.

Shrivastava, R. R. S. R. R., Pateriya, P. P. P., & Singh, M. S. M. (2018). Green tea—A short review. *International Journal of Indigenous Herbs and Drugs*, 12–21.

Sijtsema, S. J., Jesionkowska, K., Symoneaux, R., Konopacka, D., & Snoek, H. (2012). Perceptions of the health and convenience characteristics of fresh and dried fruits. *LWT-Food Science and Technology, 49*(2), 275–281.

Skrovankova, S., Sumczynski, D., Mlcek, J., Jurikova, T., & Sochor, J. (2015). Bioactive compounds and antioxidant activity in different types of berries. *International Journal of Molecular Sciences, 16*(10), 24673–24706.

Slavin, J. L. (2013). Carbohydrates, dietary fiber, and resistant starch in white vegetables: Links to health outcomes. *Advances in Nutrition, 4*(3), 351S–355S.

Slavin, J. L., & Lloyd, B. (2012). Health benefits of fruits and vegetables. *Advances in Nutrition, 3*(4), 506–516.

Snopek, L., Mlček, J., Fic, V., Hlaváčová, I., Škrovánková, S., Fišera, M., Velichová, H., & Ondrášová, M. (2018). Interaction of polyphenols and wine antioxidants with its sulfur dioxide preservative. *Potravinarstvo Slovak Journal of Food Sciences, 12*(1), 180–185.

Snopek, L., Mlcek, J., Sochorova, L., Baron, M., Hlavacova, I., Jurikova, T., Kizek, R., Sedlackova, E., & Sochor, J. (2018). Contribution of red wine consumption to human health protection. *Molecules, 23*(7), 1684.

Suzuki, Y., Miyoshi, N., & Isemura, M. (2012). Health-promoting effects of green tea. *Proceedings of the Japan Academy, Series B*, *88*(3), 88–101.

Takemoto, M., & Takemoto, H. (2018). Synthesis of theaflavins and their functions. *Molecules*, *23*(4), 918.

Tanaka, T., Matsuo, Y., & Kouno, I. (2013). Biochemical and physicochemical characteristics of green tea polyphenols. In *Green Tea Polyphenols* (pp. 19–38). CRC Press.

Tao, J., Li, Y., Li, S., & Li, H. B. (2018). Plant foods for the prevention and management of colon cancer. *Journal of Functional Foods*, *42*, 95–110.

Tesfaye, T., & Ravichadran, Y. D. (2018). Traditional uses, pharmacological action, and phytochemical analysis of *Carissa carandas* Linn. A review. *Natural Products Chemistry and Research*, *6*, 334.

Tewari, D., Stankiewicz, A. M., Mocan, A., Sah, A. N., Tzvetkov, N. T., Huminiecki, L., Horbańczuk, J. O., & Atanasov, A. G. (2018). Ethnopharmacological approaches for dementia therapy and significance of natural products and herbal drugs. *Frontiers in Aging Neuroscience*, *10*, 3.

Thundimadathil, J. (2012). Cancer treatment using peptides: Current therapies and future prospects. *Journal of Amino Acids*, *967347*.

Tiwary, S., & Hussain, M. S. (2021). Functional foods for prevention and treatment of cancer. *Asian Journal of Pharmaceutical Clinical Research*, *14*(3), 4–10.

Vijay, M., & Anu, Y. (2017). Anti-cancer activity of *Camellia sinensis*-mediated copper nanoparticles against HT-29, MCF-7, and MOLT-4 human cancer cell lines. *Asian Journal of Pharmaceutical Clinical Research*, *10*(2), 71–77.

Visioli, F., Panaite, S. A., & Tomé-Carneiro, J. (2020). Wine's phenolic compounds and health: A Pythagorean view. *Molecules*, *25*(18), 4105.

Warren, G. L. (2001). Vegetables: Dark-green leafy, deep yellow, dry beans and peas (legumes), starchy vegetables and other vegetables. *University of Florida Cooperative Extension Service, Institute of Food and Agricultural Sciences, EDIS*.

Xiang, L., Xiao, L., Wang, Y., Li, H., Huang, Z., & He, X. (2014). Health benefits of wine: Don't expect resveratrol too much. *Food Chemistry*, *156*, 258–263.

Yadav, S. S., Meshram, G. A., Shinde, D., Patil, R. C., Manohar, S. M., & Upadhye, M. V. (2011). Antibacterial and anti-cancer activity of bioactive fraction of *Syzygium cumini* L. seeds. *HAYATI Journal of Biosciences*, *18*(3), 118–122.

Zhou, K., & Raffoul, J. J. (2012). Potential anti-cancer properties of grape antioxidants. *Journal of Oncology*, *2012*, *803294*.

Index

C

D

E

F

H

J

K

L

O

Q

R

S

W